Africa

For many, Africa is regarded as a place of mystery and negative images, where reports of natural disasters and civil strife dominate media attention, with relatively little publicity given to any of the continent's more positive attributes. Africa has at last begun to receive the depth of interest it has long deserved, in the shape of debates about trade, aid and debt, the 'Make Poverty History' campaign, and the UK's 'Commission on Africa'. But, behind the superficial media façade, Africa is a diverse, complex and dynamic place, with a rich history and a colonial engagement that, although short-lived, was fundamental in determining the long-term future of the continent.

At the start of the second decade of the twenty-first century, when the world is engulfed in a major financial crisis, Africa has the dubious distinction of being the world's poorest continent. This book introduces and de-mystifies Africa's diversity and dynamism, and considers how its peoples and environments have interacted through time and space. The background and diversity of Africa's social, cultural, economic, political and environmental systems is examined as well as key development issues which have affected Africa in the past, and are likely to be significant in shaping the future of the continent. These include: the impact of HIV/AIDS; sources of conflict and post-conflict reconstruction; the state and governance; the nature of African economies in a global context; and future development trajectories.

Africa: Diversity and Development is a refreshing interdisciplinary text which enhances understanding of the background to Africa's current position and clarifies possible future scenarios. It is richly illustrated throughout with diagrams and plates, and contains a wealth of detailed case studies and current data.

Tony Binns is Ron Lister Professor of Geography at the University of Otago, Dunedin, New Zealand, and Visiting Professorial Fellow at the University of Sussex, Brighton, UK.

Alan Dixon is Senior Lecturer in Geography at the University of Worcester, UK.

Etienne Nel is Associate Professor in the Department of Geography at the University of Otago, Dunedin, New Zealand, and Research Associate of the School of Tourism, University of Johannesburg, South Africa.

Africa

Diversity and Development

Tony Binns, Alan Dixon and Etienne Nel

LONDON AND NEW YORK

First published 2012
by Routledge
2 Park Square, Milton Park, Abingdon, Oxon OX14 4RN

Simultaneously published in the USA and Canada
by Routledge
711 Third Avenue, New York, NY 10017

Routledge is an imprint of the Taylor & Francis Group, an informa business

British Library Cataloguing in Publication Data
A catalogue record for this book is available from the British Library

Library of Congress Cataloging in Publication Data
Binns, Tony.
Africa / Tony Binns, Alan Dixon, and Etienne Nel.
p. cm.
"Simultaneously published in the USA and Canada"—T.p. verso.
Includes bibliographical references and index.
1. Africa—Geography. 2. Africa—Social conditions. 3. Social change—Africa. 4. Social conflict—Africa. 5. Cultural pluralism—Africa. 6. Africa—Economic conditions. 7. Economic development—Africa. 8. Africa—Environmental conditions. I. Dixon, Alan
II. Nel, E. L. III. Title.
DT6.7.B56 2011
960—dc23 2011023142

ISBN: 978–0–415–41367–1 (hbk)
ISBN: 978–0–415–41368–8 (pbk)
ISBN: 978–0–203–15349–9 (ebk)

Typeset in Times New Roman and Franklin Gothic
by Keystroke, Station Road, Codsall, Wolverhampton

Printed and bound in Great Britain by
TJ International Ltd, Padstow, Cornwall

Contents

Plates

Figures

Tables

Boxes

Preface

Writing a book on Africa is at the same time both exciting and frustrating. Exciting, because there is just so much going on in this continent of fifty-three countries, and frustrating because it is quite impossible to cover everything and, since the pace of change is so rapid, by the time the book is published it is inevitable that further changes will have taken place. Although one may feel one knows this continent well – and, collectively, we have a total of over seven decades of experience teaching and researching on and in different parts of Africa – there is never a shortage of engaging revelations.

As we write this Preface and prepare to despatch the typescript to the publishers, North Africa is in turmoil as popular democratic uprisings attempt to overthrow dictatorial regimes in Tunisia and Egypt, while in Libya what amounts to a civil war is in progress, as Colonel Muammar Gaddafi attempts to cling on to the power he has held for forty-two years. Elsewhere in the continent democratic elections have been held in Nigeria, Africa's most populous state, and a referendum has been conducted among the population in the south of Sudan, the continent's largest state, which has indicated overwhelming grassroots support for breaking away from the north and forming a separate nation. On the economic and social fronts, while Africa still has many of the world's poorest countries, some African states have experienced rapid economic growth in recent years and have actually made some significant progress towards achieving the Millennium Development Goals and reducing poverty. Meanwhile, South Africa, the continent's 'economic powerhouse', was in April 2011 invited to Beijing to join a meeting of the leaders of Brazil, Russia, India and China in a summit of the so-called BRIC group of rapidly developing nations. We are in no doubt that in the decades ahead Africa will be the continent to watch in terms of economic and social progress, and will hopefully be engaging on more equal terms with the rest of the world.

A book such as this is, we believe, a timely offering, both to dispel stereotypical perceptions and to raise awareness of Africa's considerable diversity and potential. The book has its origins in an earlier book, *Tropical Africa*, written in 1994 by Tony Binns. In this new book, Tony has joined with Alan Dixon and Etienne Nel to consider the entire continent; with a much broader scope and greater depth, this book is very different from the earlier one.

Over the following ten chapters we have attempted to examine many aspects of Africa's diversity, and key issues which play a role in affecting progress and the quality of life experienced by Africa's people. One of the key aims of this book is to draw attention to the complex relationships between poverty and development in Africa, and the various factors that influence this. In formulating an appropriate structure for the book, we decided at the outset not to include separate chapters on such topics as gender and politics. Since we firmly believe that these issues are absolutely crucial in understanding so many aspects of the present situation and future prospects, they are considered in a variety of contexts at different points throughout the book: for example, gender aspects of rural life and health; and politics in relation to historical events, conflict and economics. We also recognize that while individual chapters focus on specific themes – such as the environment, population and health – there are actually many linkages between them.

Chapter 1 examines how Africa has been perceived at different points in time, suggesting that inappropriate perceptions and stereotypes have often obscured realities and interfered with our understanding of patterns and processes. The significance of history is recognized in both shaping present-day Africa and in looking forward to the future development of the continent.

Africa's population is the focus of Chapter 2, in which we explore a range of demographic indicators and official policies towards population growth. Both positive and negative future population scenarios are considered in the context of improving economic growth and enduring problems, such as poverty and the HIV/AIDS pandemic.

Chapter 3 is concerned with Africa's environments, and particularly the important interface between people and environment. Following consideration of the need to adapt to climate change, other aspects of management and adaptation to environment are examined. The main thrust of the chapter is that we need to achieve a more detailed appreciation of people–environment relationships, rather than merely see Africans as passive victims of uniformly harsh environments.

Africa's rural sector is where the largest number of the continent's people still live and work, and it is the subject of Chapter 4. Following a consideration of land tenure, different types of rural livelihoods are then examined. Issues such as rural diversification, marketing and food security are discussed and examples of rural development strategies are critically reviewed.

Although still predominantly rural, Africa is rapidly urbanizing. In Chapter 5 we consider Africa's towns and cities, their origins and recent rapid growth, and the implications of this growth for the provision of shelter, food security and employment. The changing nature of urban environments is discussed, with a focus on the problems of delivering basic services, such as water, sanitation and power.

Health is a key influence on the quality of life, and in Chapter 6 we review the status of human health and health systems in Africa and show that women and children are disproportionately disadvantaged. Various aspects of environmental health are examined and the incidence of communicable diseases is evaluated, particularly HIV/AIDS and malaria. The chapter concludes with an investigation of the quality of health systems in light of limited financial expenditure in many African countries.

Meaningful progress in achieving development is dependent upon stable and transparent governance, and conflict has been a feature of many African countries, peaking in the 1990s. Chapter 7 investigates the causes of conflict, including the possible relationships between poverty and conflict. The impacts of conflict on individuals, communities, economy and infrastructure are investigated, and the challenges of the post-conflict period in reconstructing livelihoods and achieving sustainable peace are articulated.

Chapter 8 provides an overview of the state of Africa's economies. Following an examination of their relative size and strength, discussion then focuses on change and development in the different sectors – agriculture, industry and services. The vulnerability of certain African countries which are dependent on 'one-product economies' is then considered, and the potential for future diversification and development is examined in the context of selected countries.

Chapter 9 considers the contested meaning of the concept of development, and then undertakes a chronological review of different phases of theory and practice since the Second World War as they relate to Africa. Attention then turns to examining development theory and practice in the twenty-first century, with emphasis on locally based and regional development as well as the achievement of the Millennium Development Goals. The important issues of aid, trade and debt are then examined.

In conclusion, Chapter 10 looks to the future of Africa and its people, taking stock of key issues and possible directions for progress in four main areas – social, environmental, economic and political. Many countries have a long way to go in delivering basic services and uplifting the lives of all their citizens, but there is much optimism and an impressive sense of resilience and resourcefulness that will hopefully lead to significant progress in the decades ahead.

We hope that this book will go some way towards dispelling popular myths and media stereotypes about Africa. After getting to know large parts of the continent ourselves and making enduring friendships through working at grassroots level in urban and rural communities, we have no hesitation in saying that there are many positive things happening, and we are firmly convinced that everyone should be giving greater attention to Africa.

We would particularly like to acknowledge the help we have received from Andrew Mould and Faye Leerink at Routledge while compiling this book. Our thanks are also due to Tracy Connolly, who drew the figures, and to Jerram Bateman, for his help with referencing. Most of all, we would like to express our deep appreciation to the many friends and colleagues in Africa with whom we have collaborated, in some cases for almost forty years. Through working together in universities, development agencies and, especially, the field, we have shared some rich experiences and learned so much, and we firmly believe that we understand things much better as a result. This book is a tribute to our cherished friendships and collaboration.

Tony Binns, Alan Dixon and Etienne Nel
May 2011

1 Africa: continuity and change

1.1 Images of Africa

In the early twenty-first century, Africa is widely perceived as the world's poorest continent with a seemingly endless agenda of development priorities. Yet, in many African countries, solid progress is being made (ODI, 2010), and Africa deserves to have a stronger voice, such that both its problems and its potential are placed 'centre stage' in world economic and social development forums. A once popular image of Africa was that of 'the dark continent', as it was first portrayed by nineteenth-century explorers such as Stanley and Livingstone. In the past, Africa has been regarded as being 'off the map', a mysterious *terra incognita*, populated by wild animals and characterized by harsh environments such as vast deserts and impenetrable forests. In the last two or three decades, however, Africa has become more synonymous with famine, drought, poverty and diseases such as malaria and HIV/AIDS. It is a continent where poor governance and political instability are often seen as the norm rather than the exception, and where seemingly little progress has been made in achieving economic, social or cultural development.

Africa is also still seen as a predominantly rural continent, where, it is often suggested, an inability to feed its growing population is due to inefficient and outdated farming systems, operated by an inadequately trained and poorly motivated workforce who are reluctant to adopt modern methods. Another feature of the post-independence period is that many African countries have at times become dependent on large-scale imports of food, together with a multitude of aid and development programmes sponsored by international agencies, governments and NGOs (non-governmental organizations). But the situation is certainly changing. Although only 39 per cent of Africa's population was urban in 2007, the United Nations Population Fund estimated an urban growth rate (2005–2010) of 3.2 per cent, the highest rate among the world's major regions, and considerably above the world average urban growth rate of 2.0 per cent (UNFPA, 2007). Already, Cairo has an estimated 12 million people, with Lagos (9.8 million), Kinshasa (8.2 million) and Johannesburg (3.5 million) all growing rapidly. Some 50 per cent of Nigeria's 140 million people are already urban-based (see Chapter 5).

Such generalized images and stereotypes unfortunately ignore the great physical and human diversity of the African continent and also fail to appreciate the complex historical processes which underlie this diversity (see Figure 1.1a). Africa is a vast continent, second in size only to Asia, stretching 8320 km from Tangier (Morocco) in the north to Cape Agulhas on South Africa's southern coast, and 7360 km from Cap Vert near Dakar (Senegal) in the west to Cape Guardafui, the easternmost point of the Horn of Africa in Somalia (see Figure 1.1b). The continent and surrounding islands now comprise fifty-three countries. Once the cradle of the world's earliest civilizations, Africa now has over a billion people, comprising a wide range of ethnic, language and religious groups. The continent's pre-colonial history was rich, varied and often highly sophisticated. It is only since the

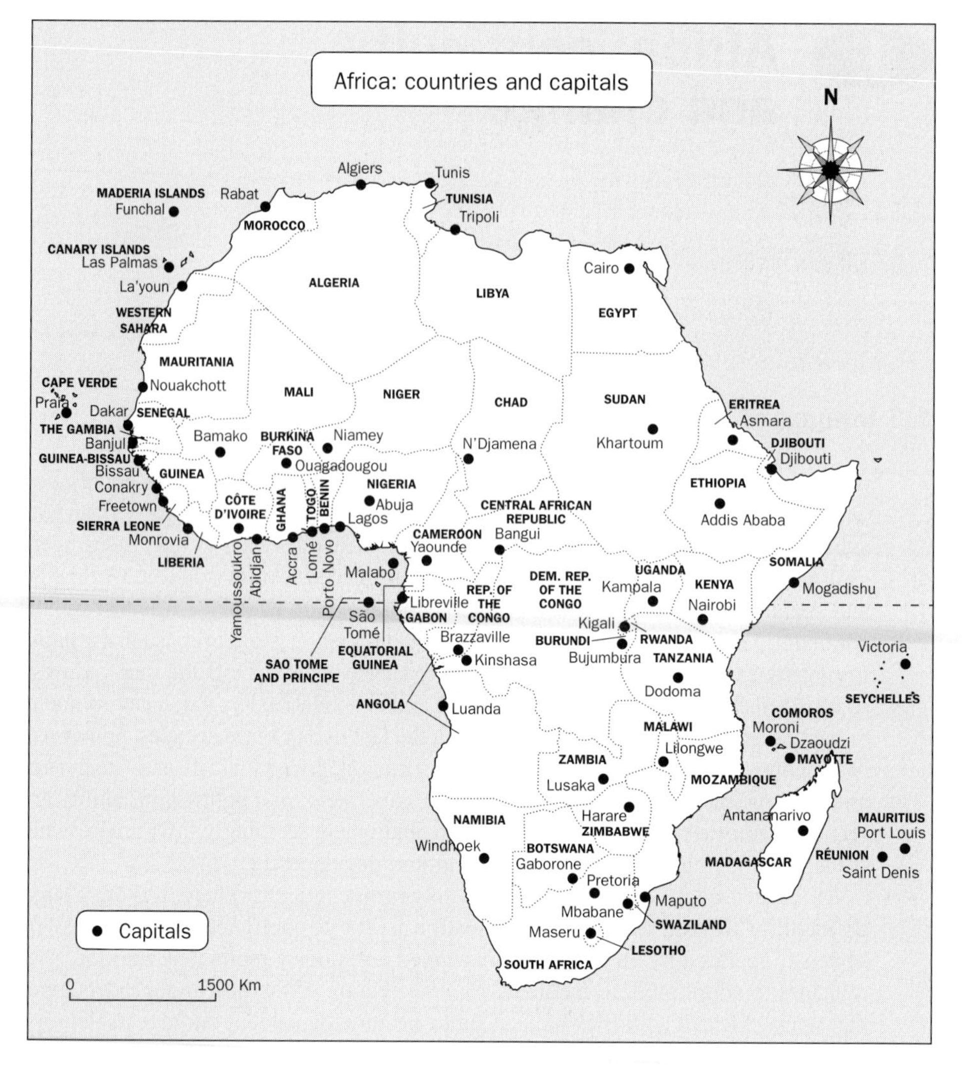

Figure 1.1a Africa today: countries and capitals.

Source: Adapted from Binns, 1994

nineteenth century that colonialism has significantly transformed economies and societies and has pulled Africa, sometimes unwillingly, into the world economic system through trade in crops, minerals and other resources.

1.2 Understanding and misunderstanding Africa

Longstanding myths and stereotypes about Africa, built up since the first Europeans set foot on the continent, and continually portrayed in the media, are difficult to eradicate. These perceptions, which are often founded upon an inadequate understanding of African environments, societies, cultures and economies, have sometimes, directly or indirectly, compounded Africa's problems. There are many examples of this, such as in the shape of countries, alignment of boundaries and ethnic composition of African states. The 'great powers', meeting in Berlin in 1884–1885 to divide up the African 'cake', showed little concern for the future viability, governance and development of African countries and

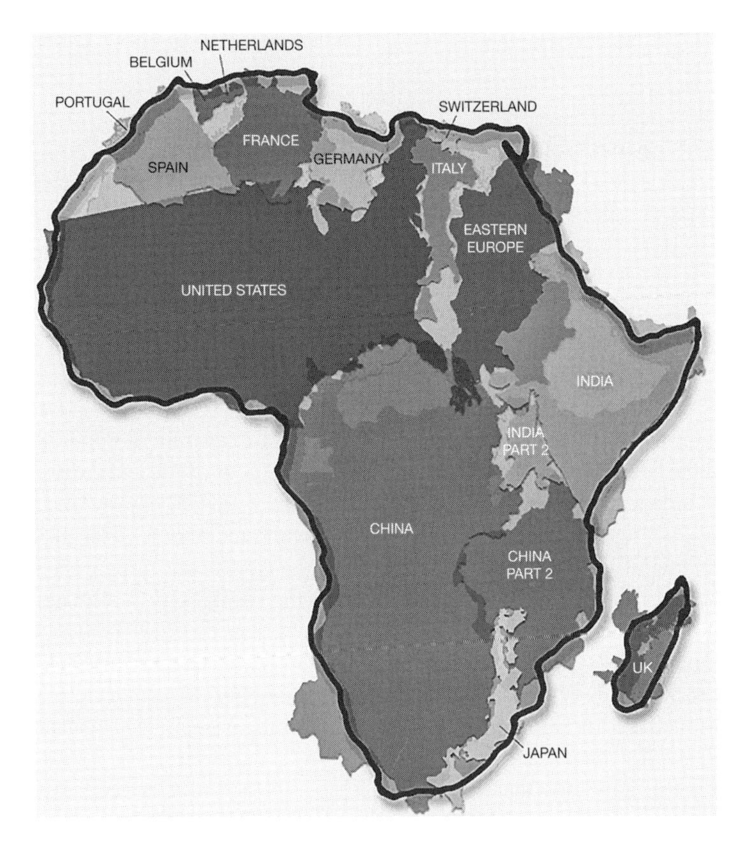

Figure 1.1b Africa today: the true size of Africa.

peoples. The Europeans merely staked their claims during the 'scramble for Africa', with little consideration for the consequences (Griffiths, 1995).

The result was the formation of strangely shaped countries such as the British colony of The Gambia (see Box 1.1). In some cases the colonial powers established boundaries, often straight lines, which showed little regard for topography or traditional ethnic homelands (see Box 1.2). The boundary between Ghana and Togo in West Africa, for example, divides the homeland of the Ewe people. Algeria, Libya, Namibia and Egypt each have at least one straight-line boundary, while in the case of Tanzania and Kenya, the mostly straight-line border bends around Mount Kilimanjaro to include the mountain in Tanzania. A popular, yet somewhat contentious, myth suggests that Queen Victoria gave Kilimanjaro to her German grandson Kaiser Wilhelm II as a birthday present. However, a more reliable account suggests that Carl Peters, founder of the German East Africa Company, persuaded local Chagga chiefs to sign treaties ceding territory to his Society for German Colonization.

One of the most ridiculous borders in Africa is the so-called 'Caprivi Strip', in southern Africa, where Namibian territory stretches eastwards in a narrow 450-km-long 'pan-handle', with Angola and Zambia to the north and Botswana to the south. Acquired in 1890 after negotiations with Germany's main rival, Britain, the strip of land is named after the German Chancellor Leo von Caprivi, who was keen to ensure that German South-West Africa should have access to the River Zambezi and beyond to the east coast of the continent (see Box 1.2).

In the carving-up of Africa among the Europeans in the late nineteenth century, states were created virtually overnight and disparate peoples thrown together. This has led to

Box 1.1

The Gambia: colonial legacy

With an area of only 10,380 sq km (smaller than Northern Ireland or the US state of Connecticut) and a population in 2005 of 1.5 million, the West African state of The Gambia is Africa's smallest mainland country and one of the poorest (see Figure 1.2). With a per capita Gross Domestic Product of $1921, and over 50 per cent of the population living off less than $1 a day, The Gambia was ranked 155 out of 177 countries in 2007 according to the UNDP Human Development Index (HDI) (UNDP, 2007; see Tables 1.1 and 1.2)

The Gambia is a long, thin country, 350 km from west to east but no more than 48 km wide, at the heart of which is the River Gambia, the focus of life and economic activity. Until independence in 1965, The Gambia was a British colony and the official language is English. The administrative framework, together with the health, education and judicial systems, were put in place by the British during more than a century of colonial rule. Banjul (formerly Bathurst), the capital city, and once the hub of the colonial state, had a population of only 34,828 at the 2003 census, but the population of the Greater Banjul conurbation was 357,238. Many of the features of a colonial town are still evident in the layout and style of buildings in Banjul, and government ministries, foreign consulates and development agencies are located there (see Plate 1.1).

It is something of a miracle that The Gambia has managed to survive as a state in its own right for well over a century. It is surrounded on three sides by the French-speaking state of Senegal, which is more than seventeen times larger. The Gambia has been described as 'a hot dog in a Senegalese roll' and, as Leopold Senghor, a

Plate 1.1 Colonial house, Banjul, The Gambia (Tony Binns).

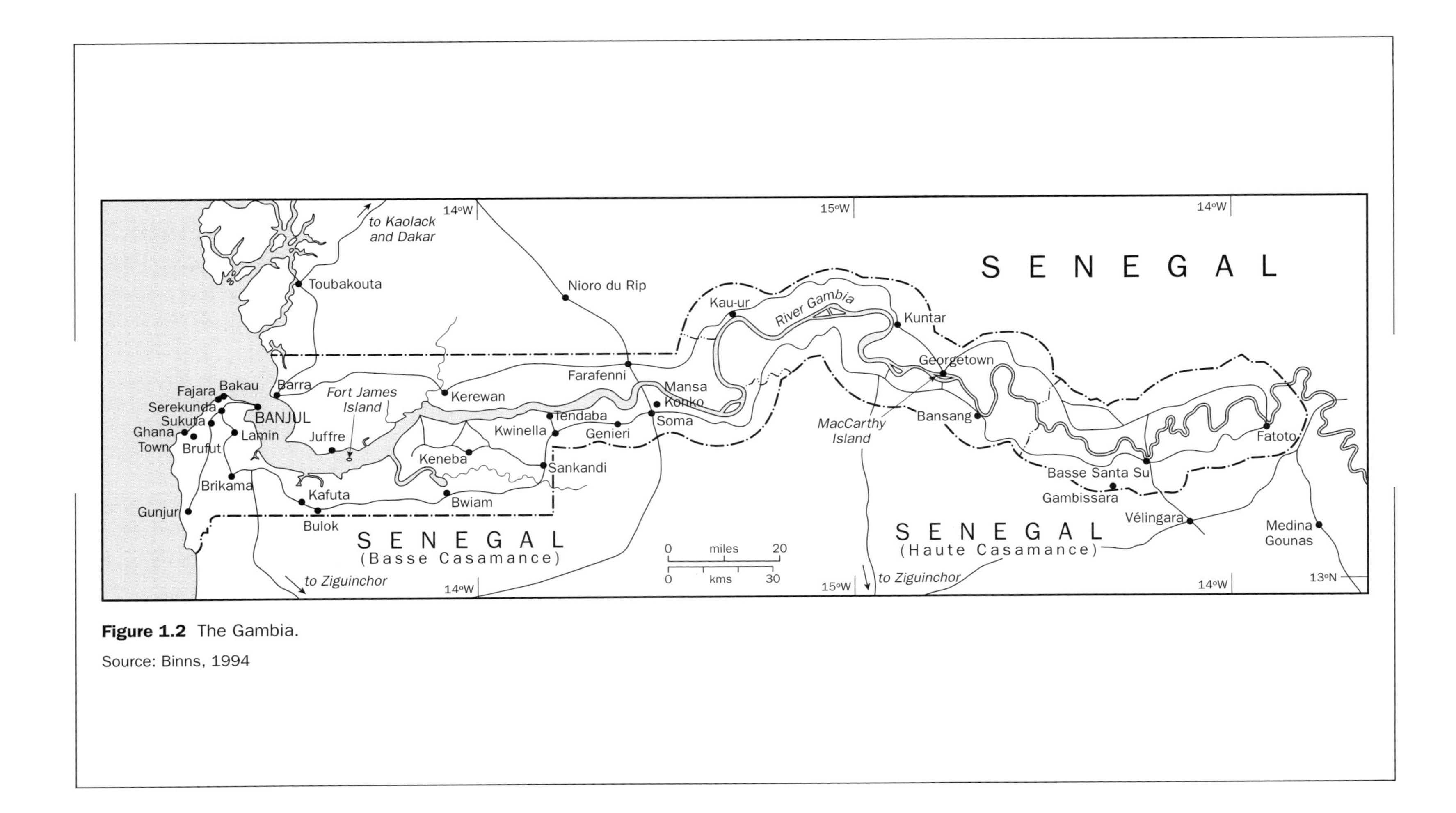

Figure 1.2 The Gambia.

Source: Binns, 1994

former president of Senegal, said in 1961, 'pointed like a revolver in the very stomach of Senegal'. Senegal's administrative structure is based on that introduced by the French colonial rulers before independence in 1960. Although the bureaucracies of The Gambia and Senegal are very different, the social and cultural features of their peoples share strong similarities and a common disregard for the artificial international border which cuts through their homelands. Most of The Gambia's people are Muslim, with the Mandinka and Wolof comprising the main cultural groups. In Senegal, the same cultural groups are also important, and there is constant movement of people and goods across the border dividing the two countries. Many families in The Gambia have relatives in Senegal and visit them regularly. Farm workers from Senegal have for many years migrated into The Gambia to assist with the harvest. Some 150 km up the navigable River Gambia, at Mansa Konko, the important Trans-Gambia Highway crosses the country, linking northern Senegal and Dakar, its capital, with the southern region of Senegal, known as Casamance. With an international border of over 1000 km, it has, understandably, always been difficult to control the movement of people and goods between the two countries, and smuggling has been a constant drain on the two economies (Swindell, 1981).

For a short period in the 1980s it did seem that The Gambia and Senegal might be coming closer together in the 'Confederation of Senegambia'. The confederation was a direct result of a Libyan-backed coup in The Gambia on 30 July 1981, while the country's president, Sir Dawda Jawara, was attending the wedding of Prince Charles and Lady Diana Spencer in London. During the brief uprising, 500 people were killed and over £10 million worth of damage was done, mainly in and around Banjul. In 1981, The Gambia had no national army and, to everyone's surprise, Jawara invited Senegalese troops to enter his country and put down the rebellion. Jawara returned to his liberated country on 2 August 1981, and within six months he and the Senegalese president, Abdou Diouf, had proclaimed the Confederation of Senegambia. Some progress was made on trade agreements and the establishment of a monetary and customs union. However, in August 1989 the confederation collapsed when Senegal pulled out after Jawara suggested that its presidency should rotate between Diouf and himself. After the failure of this brief and somewhat unexpected experiment, there is today a widespread feeling that both countries need each other and it remains to be seen whether a further period of close cooperation develops.

The Gambia had the distinction under Jawara of being one of the oldest multi-party democracies in Africa. But in July 1994, the Armed Forces Provisional Ruling Council (AFRC), under the leadership of Yahya A.J.J. Jammeh, deposed the Jawara government and banned opposition political activity. Since then, Jammeh has continued to hold power. In September 1996, as a civilian presidential candidate, he won 56 per cent of the vote; and in October 2001 he beat four opposition parties to win almost 53 per cent of the vote. A third election in 2006 returned Jammeh to power, in what was generally regarded as a free and fair campaign, though in the lead-up to the poll Jammeh controversially asserted that he would develop the areas that voted for him, while those who did not should expect nothing.

One of the legacies of the colonial period is that The Gambia is a one-product economy dominated by groundnuts (peanuts). Groundnut cultivation was encouraged by the British, with the crop first exported to The Gambia in the 1830s. Agriculture, which employs about 70 per cent of the Gambian labour force, accounts for about 30 per cent of GDP, while in 2005 groundnut production represented only about 7 per cent of GDP. A striking feature of many Gambian villages is that women comprise over 70 per cent of the agricultural labour force and produce more than 80 per cent of domestic food requirements. When groundnuts were introduced, it was the men who mainly adopted the new cash crop, leaving the women to grow rice, millet, maize, sorghum and a large variety of vegetables. Cash obtained from groundnut sales is still a key element in most families' income and any change in the price paid for groundnuts can have widespread repercussions in rural communities. In the late 1980s, groundnut production was relatively more significant in the country's economy, accounting for 80 per cent of domestic exports and about 60 per cent of total crop land, and contributing around 35 per cent of the country's GDP. These figures have gradually declined in recent years due to the increasing importance of manufacturing, commerce, services and particularly tourism.

The Gambia's tourism website, 'Visit the Gambia' (http://visitthegambia.gm/), welcomes readers to 'the smiling coast of Africa' and 'the leading destination in responsible tourism'. The Gambia has provided an ideal location for the development of tourism, catering for Europeans who want to exchange the cold and gloom of winter for guaranteed sunshine and temperatures around 30 degrees centigrade. Since the first group of international tourists arrived from Sweden in December 1965, tourist arrivals in The Gambia have grown steadily, with the exceptions of 1981 and 1995, due to political instability and negative international travel advice. Some 90,000 international tourists visited the country in 2004 – a relatively small number when compared with other African countries such as South Africa, Egypt, Morocco and Tunisia. But tourism receipts in that year represented 13 per cent of The Gambia's GDP, and tourism provided 20 per cent of all private sector jobs. It was estimated that in 2006 there were 4126 full-time equivalent jobs generated by tourism, some 2700 of which were in hotels, where the average monthly earnings were £46. Other tourism-generated jobs include taxi drivers, craft retailers, fruit sellers and ground handlers associated with the various tourism companies. Wages in all these sectors were higher than for hotel workers, with taxi drivers earning an average of £103 per month in 2006 and fruit sellers £89. Twenty large hotels provide almost 90 per cent of the total 3000 rooms and 7000 beds, but in April 2006 a further ten hotels were under construction with over 1000 additional beds (Mitchell and Faal, 2006).

In recent years, some 55 per cent of all tourists have come from Britain and the remainder from Scandinavia (notably Sweden), Netherlands and Germany. The present and potential contribution of tourism to Gambian development must be considered carefully. The extent to which it might help to reduce poverty is a key consideration, for example by sourcing more food supplies from within the country rather than relying on imported food. There has been some progress here, in that in 1986 it was estimated that 65 per cent of hotel foodstuffs were imported, whereas by 2006 this figure had fallen to 45 per cent, although there is still considerable variability between hotels in terms of their food-sourcing policies (Mitchell and Faal, 2006). But there is a danger that in a country as small as The Gambia, tourism could easily swamp the local economy and society and cause an increase in theft, begging and prostitution.

Tourism still affects only a relatively small proportion of The Gambia's land and people, and for over 70 per cent of the population traditional forms of agriculture and pastoralism provide their main livelihood. In much of rural Gambia methods of food and livestock production have probably changed very little over hundreds of years. Relative to the tourists, many of the local people are very poor indeed. Life expectancy in The Gambia is only 58 years (compared with 79 years in UK) and some 60 per cent of the population is illiterate, with a significantly higher illiteracy rate among women. The statistic which perhaps most effectively highlights the level of poverty is the mortality rate for the under-fives, which in 2005 was 137 per 1000, (compared with 6 per 1000 in UK) (UNDP, 2007). With a population growth rate of around 2.6 per cent per annum, and a density of over 150 people per sq km, The Gambia is already one of the most densely settled countries in West Africa, and more than twice as densely settled as neighbouring Senegal.

Clearly, The Gambia has major problems to contend with in its future development. Its size, shape and the dominance of groundnuts in the economy are all legacies of the colonial period. While tourism has helped to diversify the economy and provide much-needed income and jobs, for the bulk of the country's people there has been little improvement in rural living standards since independence in 1965.

Box 1.2

Africa's boundaries

The present-day political map of Africa reveals some unusual boundaries: for example, there are many straight-line boundaries, in North Africa between Algeria and its neighbours, Mauritania, Mali and Niger, and between Egypt, Libya and Sudan; and in southern Africa between Namibia and Botswana. The small West African country of The Gambia is surrounded on three sides by the much larger Senegal, and the borders at the eastern end of The Gambia are delimited as arcs of circles. It has been suggested (though with little historical substance) that the arcs were drawn from the points where shells landed when fired from a British gunboat on the River Gambia. The small southern African country of Lesotho is entirely surrounded by South Africa, while the present-day country of Equatorial Guinea (formerly the colony of Spanish Guinea) is rather fragmented and includes territory on the continent (Rio Muni), the island of Bioko (formerly Fernando Po) with the capital Malabo, and other islands such as Annobon and Corisco (Griffiths, 1994, 1995).

The African territorial 'cake' was divided at the Berlin Conference of 1884–1885, and most of the international boundaries were fixed over the next thirty years. The boundaries show little regard for the topography and cultural homelands of African peoples, and in some cases have led to protracted disputes and even military conflict. The process of delimiting a boundary and then demarcating it with boundary pillars was fraught with difficulty. Griffiths raises some interesting questions in relation to boundary delimitation:

> A watershed seems a precise enough concept until an attempt is made to delimit one, such as that between the Congo and the Zambezi, on a wide, almost level, plateau surface. A river is a river, but is the boundary line a thalweg (the longitudinal outline of a river bed from source to mouth), or the median line or bank? The status of islands is unambiguous with thalweg or bank, but what if the main channel of the river changes? Does a river (or lake) bank boundary preclude a country from riparian, fishing or navigational rights?
>
> (Griffiths, 1994: 68)

Some significant international boundary changes were made after the First World War, when German territories were allocated to other European powers under League of Nations mandates. France was given most of Togo and Kamerun, while the western parts of these countries were allocated to Britain and were subsequently administered under Gold Coast (later Ghana) and Nigeria, respectively. German East Africa (Tanganyika) became British, but the state of Ruanda-Urundi was taken from Tanganyika and given to Belgium under a League of Nations mandate. German South-West Africa (now Namibia) was given to South Africa. Other boundary changes were made after the First World War when, for example, Portugal took the 345 sq km Kionga triangle from Tanganyika. Jubaland in the south-western part of Somalia, bordering on Kenya, was ceded to Italy by the British in 1925, apparently as a reward for joining the Allies in the First World War, and Kenya was extended westwards at the expense of Uganda. In 1935, France gave part of Chad to Libya.

The position of international boundaries has caused many problems across the continent. The straight-line boundaries in North Africa are in most cases in inhospitable desert areas and in some cases were never properly delineated when large parts of the region were under French colonial rule. This has led to boundary disputes in various places and at various points in time. In 1964, at a time when many African states were gaining independence, the Organization of African Unity (OAU) made an important statement that, 'Considering that border problems constitute a grave and permanent factor of dissension . . . all Member States pledge themselves to respect the borders existing on their achievement of national independence' (quoted in Griffiths, 1994: 68).

Africa has almost 80,000 km of international land boundaries and many of these cut through cultural homelands. For example, some 1.6 million Ewe people are split by the international border between south-eastern Ghana and Togo. The Nigeria–Cameroon border splits fourteen cultural groups, while the boundaries of Burkina Faso cross twenty-one cultural areas (Griffiths, 1994). After many years of diplomatic wrangling, and at times violent disputes, in June 2006 Nigeria finally recognized the ruling made in 2002 by the International Court of Justice that granted the oil-rich Bakassi Peninsula to Cameroon, citing a 1913 agreement between Britain and Germany over the border. The UN-sponsored Cameroon–Nigeria Mixed Commission (CNMC), chaired by the UN Special Representative for West Africa, Ahmedou Ould-Abdallah, worked for several years to try to resolve disputes relating to the Nigeria–Cameroon border. These have their origin in the demarcation of territory between Britain and Germany in the late nineteenth and early twentieth centuries.

Where a river marks an international boundary, there can be problems when the river changes its course. The Semliki River between Uganda and the Democratic Republic of the Congo (DRC) has changed course several times in the last fifty years, causing disputes in an area which it is now known is rich in oil. Melting ice caps in the Ruwenzori mountains since about 1987 have led to increasing water flow in the river, widening of about ten metres, erosion of the river banks and changes in the direction of its course (*Independent*, 2009).

In southern Africa, the so-called 'Caprivi Strip' is another strange boundary anomaly associated with the demarcation of territories by the European powers in the late nineteenth century. The strip consists of a narrow protrusion of Namibia eastwards for about 450 km to the Zambezi River, between Botswana to the south and Angola and Zambia to the north. It varies in width from 105 km to only 32 km and takes its name from the German Chancellor Leo von Caprivi, who negotiated acquisition of the land with the UK in 1890 as an annex to what was then German South-West Africa. The strip gave the Germans access to the Zambezi, and a route to the east coast of Africa, where the German colony of Tanganyika (now Tanzania) was located. More recently, it was of military importance during the Rhodesian Bush War (1970–1979), during South Africa's involvement in the Angolan Civil War, and when the ANC (African National Congress) and SWAPO (South West Africa People's Organization) were conducting operations against the apartheid regime (1965–1994). Further instability in the Caprivi Strip has been associated with the activities of the Caprivi Liberation Army (CLA), a secessionist movement formed in 1994, with the goal of uniting the Lozi people who are currently separated by the border and live in three neighbouring countries, Botswana, Zambia and Angola. Finally, a longstanding boundary dispute between Namibia and Botswana, associated with the course of the Chobe River, was eventually resolved by the International Court of Justice in 1999, which ruled that the main river channel, and therefore the international boundary, was further north, and so the contested Seddudu Island was deemed to be part of Botswana.

frequent problems in both colonial and post-colonial periods, as in the case of Nigeria, Africa's most populous state. With some140 million people and 400 language groups, the Hausa, Fulani, Yoruba and Igbo are the largest tribal groups. While northern Nigeria is dominated by the Hausa and Fulani and is predominantly Muslim, the south, dominated by the Yoruba and Igbo, is mainly Christian. Since independence from Britain in 1960, Nigeria has had seven military *coups d'état* and a bloody civil war (1967–1970), and only since 1999 has the country had a reasonably stable civilian government (see Box 1.3). Independent Nigeria has experienced great difficulty in holding important population censuses and surveys, although a much-needed national census was undertaken in November 1991, followed by another in March 2006. The country's recurring political, religious and tribal rivalries have sometimes led political commentators to ask whether such a heterogeneous country will ever be truly governable.

Box 1.3

Leadership in post-independent Nigeria

Period	*Leader*	*Remarks*
1 October 1960–15 January 1966	Prime Minister Sir Abubakar Tafawa Balewa	**Elected**. Balewa was assassinated, together with other leaders of the northern region
1 October 1963–16 January 1966 COUP	President Nnamdi Azikiwe	Removed from power
16 January–29 July1966 COUP	General Ironsi	He was assassinated and thereafter civil war erupted
1 August 1966–29 July 1975 COUP	General Yakubu Gowon	He was toppled and relocated to the UK
29 July 1975–3 February 1976 COUP	Major General Murtala Mohammed	He was assassinated, but the regime continued under Major General Olusegun Obasanjo
13 February 1976–1 October 1979 RESIGNED	Major General Olusegun Obasanjo	He handed over successfully to the first democratically elected president
1 October 1979–31 December 1983 COUP	Alhaji Shehu Shagari	**Elected**. He was toppled three months into his second term in office
31 December 1983–27 August 1985 COUP	General Muhammadu Buhari	He and his deputy were removed from office
27 August 1985–26 August 1993 RESIGNED	General Ibrahim Badamasi Babangida	He 'stepped aside' after annulling a free and fair election held on 12 June 1993, handing over to Chief Ernest Shonekan
26 August–17 November 1993 COUP	Chief Ernest Shonekan	He was 'forced' to hand over to General Abacha by the military
17 November 1993–8 June 1998 DIED IN OFFICE	General Sani Abacha	He died after 'eating an apple'
8 June 1998–29 May 1999 RESIGNED	General Abdulsalami Abubakar	He handed over to a democratically elected government

Period	Leader	Remarks
29 May 1999– 29 May 2007 END OF TERM OF OFFICE	Chief Olusegun Obasanjo	**Elected**. He had successful first and second terms, and conducted a second general election in 2007
29 May 2007– 5 May 2010 DIED IN OFFICE	Umaru Yar'Adua	**Elected**. There was some dispute over the election results
6 May 2010– present	Goodluck Jonathan	Became President on death of Umaru Yar'Adua and was then **elected** in April 2011

Misunderstandings of a rather different nature are apparent in the ways that traditional farming and pastoral systems have been viewed and treated in both colonial times and since independence. Traditional African farming systems, such as shifting cultivation and rotational bush fallow, have often been accused of having low productivity because of their utilization of primitive methods, and of being environmentally damaging and unresponsive to market demands. African farmers are commonly characterized as behaving irrationally and being unaware of economic trends because they are more concerned to satisfy family food requirements rather than maximize agricultural productivity. However, such observations are frequently based on partial understanding, perhaps through examining production systems only at certain times of year, when little farm work is being undertaken, or neglecting to talk with the farmers themselves. Commonly in such surveys, more remote villages, poorer families and women are ignored, because it is generally easier to communicate with literate male farmers living near a main road and close to a town. As Chambers (1983) has shown, such biases must be recognized and avoided, wherever possible.

Between the 1960s and early 1990s, agricultural policies in African countries were dominated by a 'technological transformation' approach, with large-scale development schemes, heavy use of machinery, fertilizers, pesticides and high-yielding crop varieties. As a result, the continent is littered with broken tractors and failed development schemes. In recent years there has been a notable shift in the approach to raising the productivity of African farming systems, through harnessing valuable local knowledge and introducing smaller community-based projects which are more carefully attuned to the needs and aspirations of individuals, households and communities (Chapter 4).

Similar misunderstandings surround pastoral systems in marginal arid and semi-arid areas. At least 40 per cent of Africa's land area is devoted to pastoralism, but pastoralists represent only about 15 per cent of the continent's population. With the exception of a few countries, such as Mauritania, pastoralists are in a minority in national populations. African governments have often been critical about regular migrations of people and livestock in search of pasture and water, since such movements can lead to tax evasion, the spread of disease, and clashes with settled cultivators. Pastoralists are also often criticized for overgrazing and using such practices as burning to stimulate new grass growth, which is seen as environmentally destructive and is often linked with desertification. Consequently, many African governments have adopted a negative attitude towards pastoralists, believing that the quality of livestock and pasture can be improved only with sedentarization (settlement) and the development of commercial livestock farming modelled on ranch beef production systems in such places as the Americas and Australia (Scoones, 1995a). Such schemes have had limited success in Africa, however, suggesting that there is a need to understand traditional pastoral systems better and to work together with herdsmen to enhance the productivity and sustainability of these systems.

It should be appreciated that mobility is an important way of coping with the ever-present threat of drought in marginal and non-equilibrium environments. The restriction of pastoralists' mobility and the loss of land to the cultivation of cash crops such as groundnuts and cotton have pushed herders into even more marginal areas where vulnerability to drought is much greater. Research has shown that African pastoralists have a wealth of environmental knowledge and their lifestyles may actually be the most effective and sustainable in light of the harshness of the climate and the poverty of environmental resources (see Chapter 4).

Some government officials and so-called development 'experts' in both the colonial and post-colonial periods have shown a remarkable lack of understanding of the people with whom they are dealing. If meaningful and sustainable development is to occur in Africa, then there is a need for greater empathy with poor people and a better appreciation of livelihood systems and the wealth of knowledge which ordinary Africans possess. A dictatorial 'transformation approach' to development initiatives must be replaced with a more democratic form of development which has Africans, both rural and urban, as its main focus and genuinely seeks to fulfil their needs and aspirations (see Chapter 9).

1.3 Africa's colonial legacy

The colonial period was a relatively brief interlude in African history, yet its impact was profound and debate continues over the merits and problems of colonialism. The colonial powers demarcated boundaries and introduced administrative, legal, health, education and transport systems modelled on their own. Economically, colonialism programmed African countries to consume what they do not produce and to produce what they do not consume. Africa was brought into the world economic system with the large-scale production and export of agricultural and mineral resources. The marketing of commodities by trading companies and multinational organizations on a global scale developed under colonialism. The production of cash crops for export took priority over domestic food crops, while in the mining and industrial sectors, emphasis was placed on the extraction of minerals and their export in an unprocessed state. Manufacturing industries were discouraged, since their products might compete with European manufactured goods. Within African countries, traditional power structures were subordinated to colonial structures, with expatriate governors and district officers making key decisions that affected people's everyday lives. In many territories, the colonial governor ruled by decree or proclamation, was deferred to by all, and lived in sumptuous surroundings. The perceptions and aspirations of individual Africans were transformed as a result of their contact with foreign people, foreign systems, foreign structures and foreign ways of life.

Even before the massive distribution of land resulting from the Berlin Conference of 1884–1885, some of the continent had already passed under the control of the imperial powers (see Figure1.3a, b and c). Treaties were signed with various African leaders, sometimes with the latter not really understanding the full implications of what they had signed. Traders and missionaries often penetrated the countries first, while the tentacles of formal administration were gradually extended into the remotest areas. In other cases, where African peoples were less willing to cooperate, expensive and often bloody military campaigns were conducted against them, such as the four Anglo-Ashanti wars between 1823

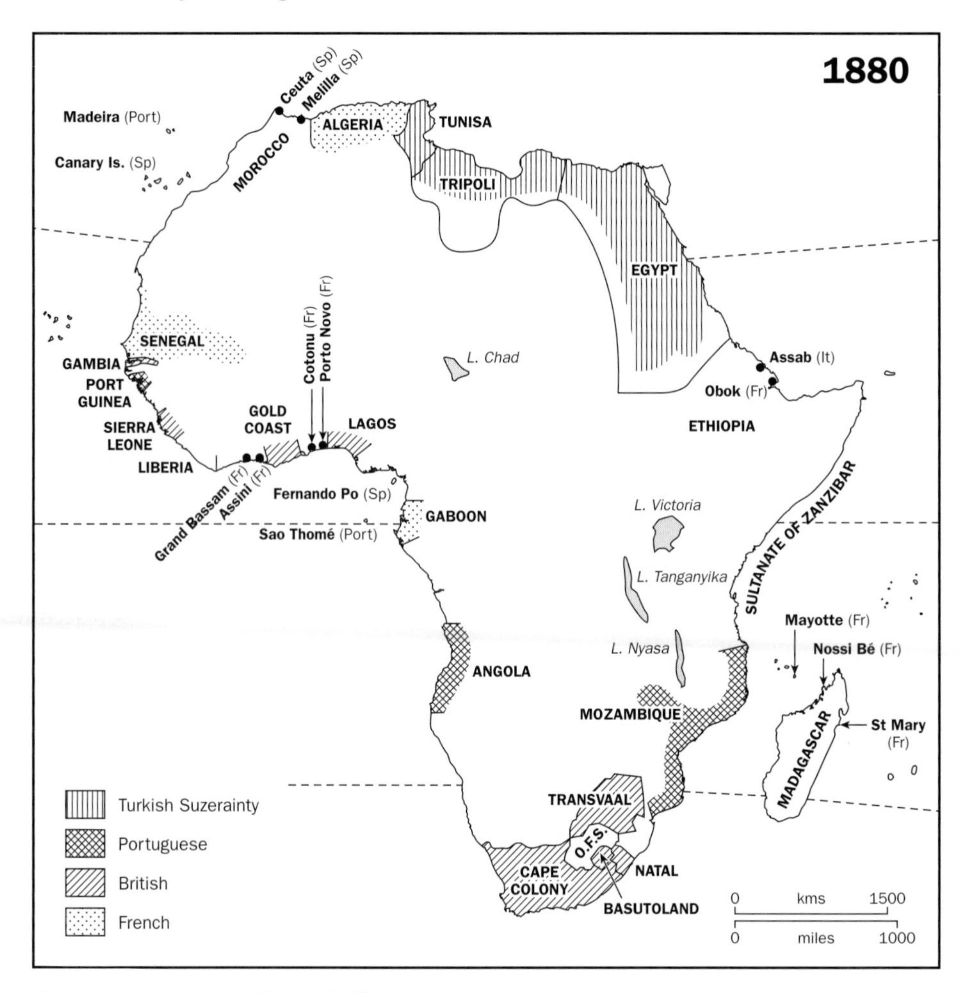

Figure 1.3a Historical Africa: 1880.

Source: Binns, 1994

and 1894 in what is now Ghana. Occupation and pacification were followed by control and administration (Iliffe, 1995; Fage, 2002). A key figure in the acquisition of territories for Britain was the entrepreneur Cecil Rhodes, who envisaged the construction of a railway from Cape Town to Cairo, entirely on British soil (see Plate 1.2).

By 1914, the political map of colonial Africa was complete and the colonial powers were unfolding their various policies and programmes. Approaches to colonial rule differed among the European powers. The British adopted a pragmatic and decentralized approach, the essence of which was 'indirect rule', whereby Britain maintained indigenous cultures and societies by working wherever possible through local social and political systems, such as the emirs in northern Nigeria. The British respected local institutions and feared the problems which might follow their decay. Under indirect rule, large areas could be administered by relatively few officials, though the effectiveness of the system depended on the strength of indigenous institutions. European settlement in countries such as Kenya and Zimbabwe (then Southern Rhodesia) was another feature of British colonial policy, with much of the best land taken over by the settlers, while African farmers were crowded into less fertile reserves, often with disastrous ecological results. In the case of Zimbabwe, the persistence of white settlement and white minority rule led to a considerable delay in

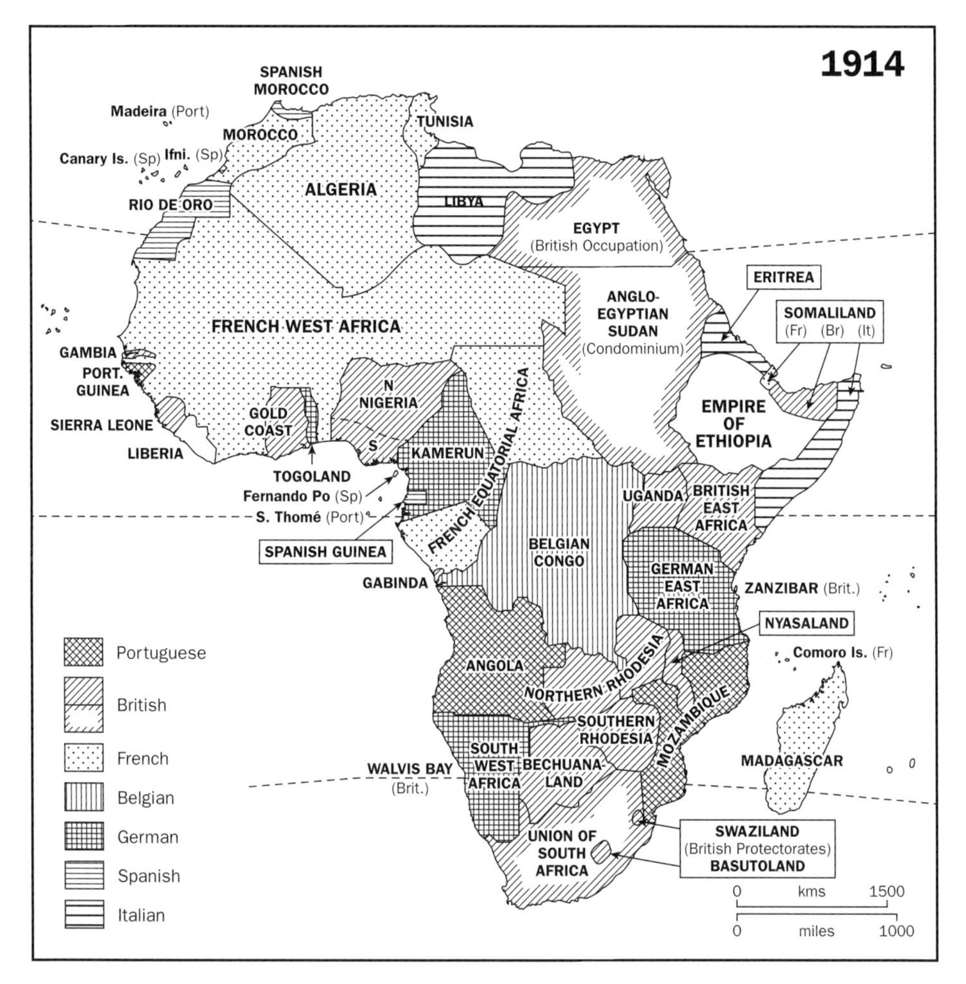

Figure 1.3b Historical Africa: 1914

Source: Binns, 1994

the granting of independence (see Plate 1.3). In British West Africa there was no significant white settle-ment and the ultimate granting of independence to the territories – Nigeria, Ghana (then Gold Coast), Sierra Leone and The Gambia – was much less troublesome.

France's approach to its colonial territories was very different from Britain's. Whereas British colonialism was designed to create 'Africans with British characteristics', French policy was designed to create 'black Frenchmen'. French colonialism was more direct and centralized, with the colonies regarded as parts of France and referred to as 'Overseas France'. French policies were essentially geared towards assimilating the colonies into France and the French way of life. Large numbers of French administrators went into the African colonies and, unlike the British, there was little attempt to work through traditional institutions or leaders. Local cultures and languages were given little encouragement. The French believed in a bureaucratic and hierarchical state in their territories and created small indigenous elites whose members were often educated in France (see Plate 1.4). White settlement was particularly common in the African countries bordering the Mediterranean, but there were also sizeable white minorities in tropical Africa, such as in Dakar (capital of Senegal) and in the plantation areas of Cameroon, Guinea, Ivory Coast and Madagascar. As early as 1688, French Huguenot Protestants, escaping their homeland after the

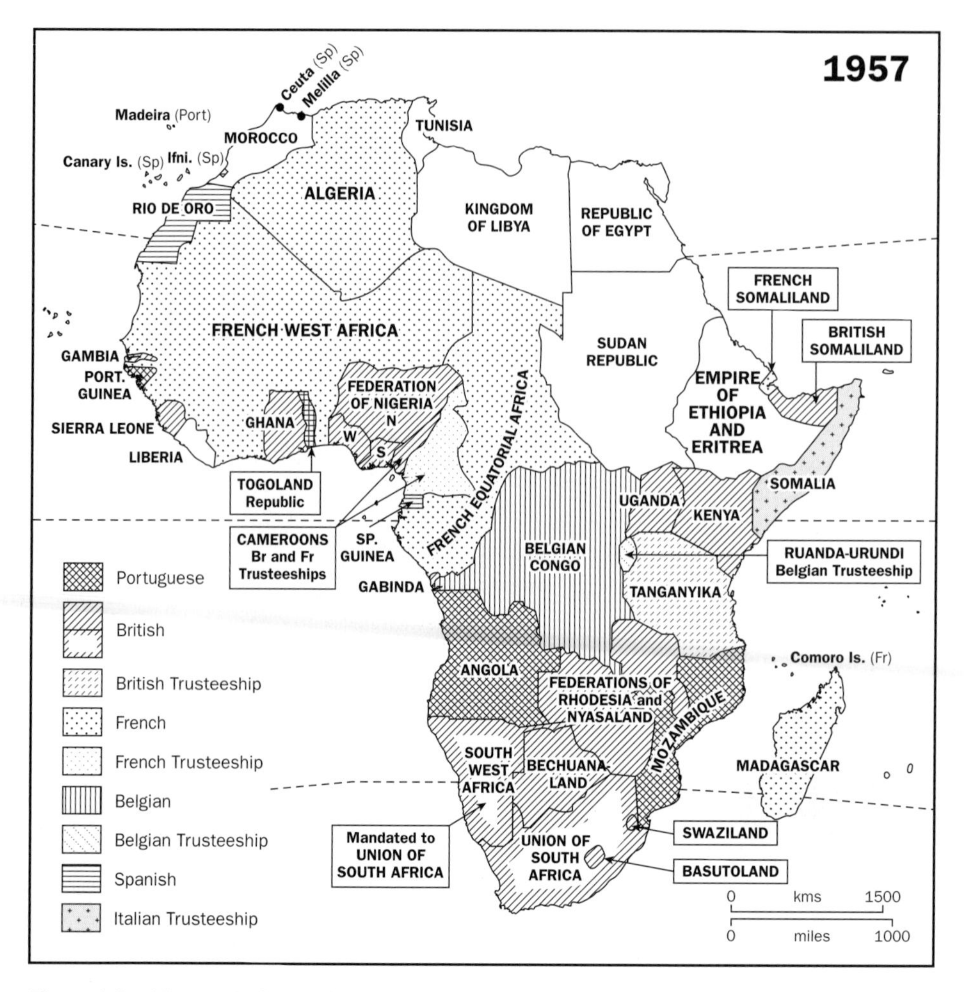

Figure 1.3c Historical Africa: 1957.

Source: Binns, 1994

revocation of the Edict of Nantes, settled in South Africa and played a key role in developing the wine industry in the Franschhoek area of what is now Western Cape Province.

Belgium controlled one vast territory, the Belgian Congo (now the Democratic Republic of the Congo – DRC), initially as the personal possession of King Leopold. The Belgians adopted a policy of assimilation similar to that of the French but, like the British, it was aimed ultimately at independence. Emphasis was placed on economic and social development with a focus on elementary education, so there was no significant growth of an indigenous educated elite. The Belgians argued that Africans needed guidance and tuition before they could take over, and political expression was suppressed until just before independence (Iliffe, 1995).

Portugal was the other major colonial power, and played a key role in making initial contact with the African coast during the voyages of discovery in the fifteenth century. As early as 1482 the Portuguese established fortified trading posts on the coast from Senegal to the Gold Coast and were responsible for naming the territory which became Sierra Leone. In the scramble for Africa, however, Portugal was less successful in gaining territory than either Britain or France (see Box 1.4). By the early twentieth century it controlled Portuguese Guinea (now Guinea-Bissau) in West Africa, Cape Verde, São Tomé and

Plate 1.2 Statue of Cecil Rhodes in Cape Town, South Africa (Etienne Nel).

Príncipe, and its two much larger colonies of Angola and Mozambique. Portuguese policy, like the French, was aimed at assimilation, with all territories being regarded as part of the Portuguese Union. There were also substantial numbers of white settlers in the Portuguese colonies (Fage, 2002; Griffiths, 1995).

Plate 1.3 Bulawayo, Zimbabwe, celebrates independence day (Tony Binns).

Plate 1.4 Colonial post office building in Bamako, Mali (Tony Binns).

Box 1.4

Portuguese decolonization

Portugal's links with Africa began long before the other European powers became interested in the continent. Portuguese exploration of the African coastline started in the early fifteenth century (from around 1419). In 1462, a Portuguese seaman, Pedro da Cintra, sailing off the West African coast, used the term *Serra Lyoa* ('lion range' or 'lion mountain') to describe the peninsular mountains behind the present-day city of Freetown, and the country later became known as Sierra Leone. In 1488, Bartolemeu Dias rounded the Cape of Good Hope and landed in what is now Mossel Bay in South Africa. This voyage was followed by many other Portuguese navigators, most notably Vasco da Gama, who reached India in 1498 by way of the Cape of Good Hope and the Kenyan port of Malindi.

Despite their early navigation of the African coastline, Portugal gained only five territories in Africa: Portuguese Guinea (now Guinea-Bissau), Cape Verde off the coast of West Africa, the islands of São Tomé and Príncipe, and the two larger colonies of Angola and Mozambique. Portugal adopted a policy of assimilation with its African colonies, and in 1951 the Portuguese constitution removed the word 'colony', referring to the African countries as overseas provinces of 'Greater Portugal'. The Portuguese government under the dictator Antonio Salazar proclaimed that 'Portugal is not a small power', stressing that the colonies were twenty-two times the size of Portugal and that both were now a single political entity. In the late 1950s, at a time when Britain and France were reducing their colonial commitments, Portugal strengthened its presence in its African colonies with further large numbers of white settlers continuing to arrive, such that by the end of the 1950s there were 335,000 in Angola and 200,000 in Mozambique. In light of the fact that there were only 7 million Portuguese at the time, these numbers were quite significant.

In Angola, a revolt started in the north of the country in 1961, inspired by earlier independence in the Belgian Congo and also the alienation of land by white settlers for coffee plantations. Although this initial revolt was put down by over 17,000 Portuguese troops, African liberation movements strengthened during the 1960s, and the PAIGC (Partido Africano de Independencia da Guine e Capo Verde) launched an insurrection in Portuguese Guinea in 1962. The PAIGC's leader was Amilcar Cabral, an influential political strategist who was committed to revolutionary democracy and agrarian socialism. In 1972 Cabral held elections, which, according to Wilson (1994: 183), was a unique instance of 'a revolutionary nationalist movement holding elections before independence and setting up structures to maintain accountability'. Cabral saw decolonization as a process rather than a single event, but he was assassinated in January 1973, leading to greater pressure from the people of Portuguese Guinea for independence. By that year the PAIGC controlled most of the country. In September 1974, just five months after a revolution in Portugal itself (see below), Guinea was the first Portuguese African colony to be granted formal independence, becoming Guinea-Bissau.

Meanwhile, in Mozambique, three exiled organizations united in 1962 to become FRELIMO (Frente de Libertaçáo de Mozambique), which started a war in 1964, using Tanzania as a base. FRELIMO's military effectiveness increased after 1970 with the use of bases in Zambia to launch guerrilla attacks on the Trans-Zambesi and Beira railways and the Cabora Bassa hydro-electric project (which had been started in a bid to attract a million Portuguese immigrants).

In Angola, the nationalists remained split between the MPLA (Movimento Popular de Libertaçáo de Angola) and UNITA (União Nacional de Independencia Total de Angola). Despite internal rivalries, the MPLA provided the stronger military challenge to the colonial power. Its leader Agostinho Neto spoke of 'freeing and modernizing our people by a dual revolution – against their traditional structures which can no longer serve them, and against colonial rule' (quoted in Hargreaves, 1996: 232). In countering the liberation movements and guerrilla activities in Africa, Portugal adopted a policy of Africanizing its forces, such that by the end of the various independence struggles about 60 per cent of the troops fighting the nationalists were Africans.

Salazar had been determined to hang on to the colonies when many African countries were gaining their independence in the early 1960s. He argued that losing its overseas territories would mark the end of Portugal's independence, and asserted that he did not wish to hand over the colonies to what he believed were Marxist movements endorsed by the Organization of African Unity (OAU). He was a great supporter of Rhodesian Prime Minister Ian Smith; and, after Rhodesia declared independence from the UK in 1965 until 1975, Portugal supported Rhodesia economically and militarily from its neighbouring colony of Mozambique, including giving Rhodesia important access to Mozambique's seaports.

By 1970, the USA was showing concern about communist influence in some leading African liberation movements, including links between the MPLA and Cuba. Under President Richard Nixon (1969–1974), the US increased its economic and military cooperation with Portugal through the NATO framework. Portugal's earlier founding membership of the European Free Trade Area (from 1960) signalled a commitment to greater European cooperation. Defence spending in the colonies was placing a strain on the Portuguese economy, with military expenditure increasing from 4.55 to 8.3 per cent of GNP between 1960 and 1971, equating to 45.9 per cent of government expenditure in the latter year (Hargreaves, 1996). The military campaigns in Africa dragged on with little prospect of a resolution.

Portuguese colonial policy in Africa ended later and more abruptly than had been the case with the other European colonists. It was precipitated by a left-leaning military coup in April 1974, the so-called 'Carnation Revolution', which ousted Salazar's successor as dictator, Dr Marcello Caetano. This coup was actually inspired by the pro-independence guerrillas whom Portugal had been fighting in its African territories. General Spinola, leader of the post-coup military junta, had recently returned from commanding the troops in Portuguese Guinea, and complained that the army there was 'frustrated by the lack of any progressive political design for the empire' (Hargreaves, 1996: 233).

Independence for the African states came quickly after the downfall of Caetano. As has been said, Guinea-Bissau was the first to gain independence, followed by Mozambique in June 1975, Cape Verde and São Tomé and Príncipe in July 1975, and Angola in November 1975. While the island states had a relatively smooth transition to multi-party democratic systems, Guinea-Bissau endured a brief civil war and transition to civilian rule in the late 1990s, while both Angola and Mozambique had protracted civil wars. The speed of the Portuguese decolonization process inevitably meant that African political leaders and civil servants were often ill prepared for assuming control, whereas in the French and British territories the transition from colonial to independent government was generally rather more measured and better planned.

The other colonial powers in Africa included Spain, which had long held the two enclaves of Ceuta and Melilla in Morocco (see Box 1.5), and gained Western Sahara and Spanish Guinea (now Equatorial Guinea). Meanwhile, Germany controlled Togoland (now Togo) and Kamerun (Cameroon) in West Africa, South-West Africa (Namibia), and Tanganyika (Tanzania), and Ruanda-Urundi (Rwanda and Burundi) in East Africa. However, Germany's hold on African territory was short-lived, since after the First World War its colonies were mandated to other European powers – Britain, France and Belgium.

Box 1.5

Ceuta and Melilla: Spanish enclaves in Africa

While there has been an ongoing international debate between Britain and Spain about whether the former should continue to hold claim to Gibraltar at the southern tip of Spain, the two Spanish-owned city states of Ceuta, on the northern tip of Africa opposite Gibraltar, and Melilla, some 300 km further east along the North African coast, are situated within the boundaries of present-day Morocco (Griffiths, 1994, 1995).

Both cities have been important ports for centuries, with Ceuta founded by the Carthaginians and Melilla by the Phoenicians. Spain captured Melilla from the Kingdom of Fes in 1497 and acquired Ceuta from Portugal in 1580 (Portugal had captured the city from the Kingdom of Fes in 1415). When Spain recognized the independence of Spanish Morocco in 1956, Ceuta and Melilla remained under Spanish rule. In both cases, Morocco now claims the two autonomous city states, along with various offshore Spanish islands.

Ceuta is the larger of the two enclaves, with an area of 28 sq km and a population of 78,320 in 2006, while Melilla is 20 sq km, with a population of 72,000. Ceuta is culturally regarded as part of the Spanish region of Andalusia and was formerly attached to Cadiz Province, while Melilla was part of Malaga Province until 1995. Both states were granted autonomous status on 14 March 1995, and are now regarded as parts of the European Union – making them the only EU territories in mainland Africa. Both use the euro as their currency.

Ceuta and Melilla have important port functions, with sizeable fishing fleets and a considerable amount of cross-Mediterranean traffic from Spain. Ceuta is only ninety minutes by ferry from Spain, and there are ferry services from both cities to Malaga, Algeciras and Almeria. While Melilla has an airport, Ceuta does not. In the 1980s, over 2.5 million passengers used the port of Ceuta annually. Significant industrial and retail centres have developed in both cities in connection with these activities. An estimated 36,000 Moroccans cross the border daily into Melilla for work, shopping and to trade goods. But in 2007 the two states had among the highest rates of unemployment in the EU (Ceuta, 20.3 per cent; Melilla, 18.2 per cent; BBC, 2008a).

Both enclaves are surrounded by fences to deter illegal immigrants from Morocco and beyond, notably from the sub-Saharan West African countries. The two port cities are seen as important stepping-stones to Spain and from there to the rest of Europe. In October 2005, with a wave of migrants attempting to enter Melilla, both Spain and Morocco agreed to deploy extra troops to try to secure the borders. There were strong protests from King Mohammed VI and the Moroccan government when King Juan Carlos of Spain visited the territories in November 2007. Despite strong historical claims on the two territories from the Moroccan government, it seems likely that Ceuta and Melilla will remain part of Spain for the foreseeable future.

One significant and very unpopular element in the establishment of administrative frameworks in the African colonies was the introduction of monetary taxes, to be paid at all levels and at particular times. For many Africans, taxation of any kind was a complete innovation, while others had previously paid only indirect taxes. In order to obtain money to pay these taxes, cash crops had to be grown. Additionally, many people, usually men, had no alternative but to leave their villages and go to the rapidly growing towns and plantations where wage labour was needed. There are many instances of African people demonstrating their displeasure at colonial taxation – for example, the so-called Hut Tax War in Sierra Leone, initiated in 1898 by Temne chief Bai Bureh in response to the British imposition of taxes on all homes (Hopkins, 1973; Iliffe, 1995). But tax evasion was brutally discouraged and frequently led to harsh punishment and forced labour. In fact, the imposition of colonial rule in general was often harsh, and any form of resistance to the colonial state was severely suppressed.

The economic policies pursued by the colonial powers had a profound effect on the African people. The widespread cultivation of cash crops for export was one such policy. In French West Africa, northern Nigeria and The Gambia the groundnut (*arachis hypogaea*) became the principal cash crop, grown primarily to provide oil for growing industries in France and Britain. The cultivation of cash crops often meant that food crops were neglected and the nutrition of rural people suffered. The situation became particularly severe during drought periods, when many people died of starvation. In other parts of Africa cash crops were encouraged not so much to meet demands in Europe as to generate revenue. In Kenya, for example, sisal, coffee and maize made the colony self-supporting.

As each colony was viewed not as a separate economic unit but as a cog in the vast imperial machine, there was a tendency towards specialization in some countries and regions. This is the origin of the 'one-product economies' that are still to be found in many African countries today: for example, copper from Zambia, groundnuts from The Gambia and Senegal, cocoa from Ghana, and diamonds from Sierra Leone. As a result of such specialization, these and other countries were, and remain, vulnerable to world price fluctuations and dependent on the colonial powers and industrial countries for markets, capital and technology. There was little trade between colonies and any profits went to the colonial power rather than being invested in the colony.

The period leading up to independence was often characterized by a lack of preparation for the event and its responsibilities. In the French and British territories the preparation for taking over the legislative, executive and administrative roles and structures by Africans began only after the Second World War. Few chiefs in British Africa were allowed any initiative in the administration before 1945. Meanwhile, in Guinea, the French tried to destroy the fabric of the state before they departed, even removing the books from the law library of the Ministry of Justice. However, the Portuguese territories and the Belgian Congo probably provide the most notorious examples of the lack of preparation for the transfer of the institutions of state (see Box 1.4, above).

A question that is often asked is: to what extent can colonialism be blamed for Africa's current problems, given that independence was gained fifty or more years ago? Indeed, the colonial experience and its aftermath remain major topics of debate among historians and there are many conflicting viewpoints. Suffice it to say here that colonialism initiated a type and pace of change that were unprecedented in Africa. Colonial policy was generally exploitative and often coercive and confrontational. Although it is important to evaluate the many other factors which may have contributed to the present condition of African countries, such as political instability and economic mismanagement, it would be inappropriate to ignore the profound effects of colonial policies and their legacy.

1.4 Challenges, prospects and opportunities?

During the last two decades, development indicators have deteriorated rather than improved in some African countries. Statistics for life expectancy at birth, for example, have changed a great deal over this period. In the early 1980s, the lowest life expectancy figures were associated with a number of Sahelian West African countries, such as Burkina Faso, Mali and Niger. But in the early twenty-first century, the countries with the lowest life expectancy are Botswana, Lesotho and Zimbabwe in southern Africa, a region with the dubious distinction of having the world's highest incidence of HIV/AIDS. However, in economic terms, these countries have above-average Gross Domestic Product (GDP) per capita when compared with other African countries (see Table 1.1).

Some of Africa's fifty-three states are among the poorest in the world. Mali, in the Sahel region of West Africa, for example, is more than five times the size of the United Kingdom, but in 2005 it had only 13.5 million people, compared with 61 million in the UK (Table 1.1). Mali had a per capita GDP in 2005 of only $1033, compared with a figure of $33,238 in the UK. (Here and throughout the book, figures given in dollars refer to US dollars, unless otherwise stated.) Life expectancy at birth in Mali was 48.6 years, compared with 79.0 years in UK (Table 1.2). Mali's situation, like that of so many other African countries, could deteriorate further in the next decade, with a relatively high projected population growth rate of 3.0 per cent per annum, compared with only 0.2 per cent in the UK. Furthermore, Mali is one of Africa's fourteen landlocked states and is situated in a zone where rainfall is notoriously unreliable and drought a regular occurrence (see Box 2.3, below).

But not all African countries are like Mali, and it is important to be aware of the great diversity across the continent. Within mainland Africa, Botswana had the highest per capita GDP of $12,387 in 2005 (Table 1.1). But such national statistics conceal major variations within individual countries. It is common in Africa for people in urban areas to be relatively better off than those living in rural areas. In fact, the differential between the 'haves' and the 'have-nots' in African countries is generally far greater and frequently more visible than in most developed countries of Euro-America. Fleets of air-conditioned Mercedes Benz cars, sumptuous residences guarded with high gates, electrified fences and barbed wire, modern hotels and banks are now common features of most African capitals. But there is poverty in African cities, too, and often in close proximity to the symbols of wealth. Wooden (or even cardboard) shacks, with no running water supply, proper sanitation or electricity, can be found in many large African cities; and it is likely that there will be more of them in the future (see Chapter 5).

African countries face many problems and challenges. Most have suffered greatly since the early 1970s with the knock-on effects of the oil price rise in 1974–1975 and the subsequent world recession. From 2007, African countries in different ways both benefited and suffered from soaring oil prices, high food prices and worldwide recession.

Table 1.1 Africa: key statistics

	Population (millions) 2005	*Land area (,000 sq km) 2005*	*Population density (persons per sq km) 2005*	*Gross Domestic Product per capita (purchasing power parity, $) 2005*	*Population living off less than $1 a day (%) 1990–2005*
Algeria	32.9	2,382	13.8	7,062	<2
Angola	15.9	1,247	12.7	2,335	–
Benin	8.4	111	75.7	1,141	30.9
Botswana	1.8	567	3.2	12,387	28.0
Burkina Faso	13.2	274	48.2	1,213	27.2
Burundi	7.5	26	288.5	699	54.6
Cameroon	16.3	465	35.1	2,299	17.1
Cape Verde	0.5	4	125.0	5,803	–
Central African Repuplic	4.0	623	6.4	1,224	66.6
Chad	9.7	1,259	7.7	1,427	–
Comoros	0.6	2	300	1,993	–
Congo, Rep.	4.0	342	11.7	1,262	–
Congo, Dem. Rep.	57.5	2,267	25.4	714	–
Djibouti	0.8	23	34.8	2,178	–
Egypt	74.0	995	74.4	4,337	3.1
Equatorial Guinea	0.5	28	17.9	7,874	–
Eritrea	4.4	101	43.6	1,109	–
Ethiopia	71.3	1,000	71.3	1,055	23.0
Gabon	1.4	258	5.43	6,954	–
Gambia, The	1.5	10	150.0	1,921	59.3
Ghana	22.1	228	97.0	2,480	44.8
Guinea Bissau	1.6	28	57.1	827	–
Guinea	9.4	246	38.2	2,316	–
Ivory Coast	18.2	318	57.2	1,648	14.8
Kenya	34.3	569	60.3	1,240	22.8
Lesotho	1.8	30	60.0	3,335	36.4
Liberia	3.3	96	34.4	–	–
Libya	5.9	1,760	3.4	10,335	–
Madagascar	18.6	582	32.0	923	61.0
Malawi	12.9	94	137.2	667	20.8
Mali	13.5	1,220	11.1	1,033	36.1
Mauritania	3.1	1,025	3.0	2,234	25.9
Mauritius	1.2	2	600.0	12,715	–
Morocco	30.2	446	67.7	4,555	<2
Mozambique	19.8	784	25.3	1,242	36.2
Namibia	2.0	823	2.4	7,586	34.9
Niger	14.0	1,267	11.0	781	60.6
Nigeria	131.5	911	144.3	1,128	70.8
Rwanda	9.0	25	360.0	1,206	60.3
São Tomé and Príncipe	0.2	1	200.0	2,178	–
Senegal	11.7	193	60.6	1,792	17.0

Table 1.1 Continued

	Population (millions) 2005	*Land area (,000 sq km) 2005*	*Population density (persons per sq km) 2005*	*Gross Domestic Product per capita (purchasing power parity, $) 2005*	*Population living off less than $1 a day (%) 1990–2005*
Seychelles	0.1	0.5	193.1	16,106	–
Sierra Leone	5.5	72	76.4	806	57.0
Somalia	8.2	627	13.1	–	–
South Africa	46.9	1,214	38.6	11,110	10.7
Sudan	36.2	2,505	15.2	2,083	–
Swaziland	1.1	17	64.7	4,824	47.7
Tanzania	38.3	884	43.3	744	57.8
Togo	6.1	54	113.0	1,506	–
Tunisia	10.0	155	64.5	8,371	<2
Uganda	28.8	197	146.2	1,454	–
Zambia	11.7	743	15.7	1,023	63.8
Zimbabwe	13.0	387	33.6	2,038	56.1
NORTH AFRICA	152.9	5,738	44.8		
SUB-SAHARAN AFRICA	743.7	23,619	31.5		
ALL AFRICA	**896.6**	**29,358**	**38.2**		

Sources: World Bank, 2008a; UNFPA, 2007; UNDP, 2007

Table 1.2 Africa: key statistics

	Life expectancy at birth (years) 2005	*Population annual growth rate (%) 2005*	*Child mortality (under-five per 1000) 2005*	*HIV/AIDS prevalence rate (% adults aged 15–49) 2005*	*Malnutrition (% children under 5 underweight) 2000–2005*
Algeria	71.7	1.5	39	0.1	10.4
Angola	41.4	2.9	260	3.7	30.5
Benin	55.0	3.1	150	1.8	30.0
Botswana	35.0	–0.2	120	24.1	12.5
Burkina Faso	48.5	3.1	191	2.0	37.7
Burundi	44.6	3.6	190	3.3	45.1
Cameroon	46.1	1.8	149	5.4	18.1
Cape Verde	70.7	2.3	35	–	–
Central African Rep.	39.4	1.3	193	10.7	24.3
Chad	44.0	3.1	208	3.5	36.7
Comoros	62.6	2.1	71	0.1	25.4
Congo, Rep.	52.8	2.9	108	5.3	–
Congo, Dem. Rep.	44.0	3.0	205	3.2	31.0
Djibouti	53.4	1.8	133	3.1	26.8
Egypt	70.5	1.9	33	0.1	8.6

Table 1.2 Continued

	Life expectancy at birth (years) 2005	*Population annual growth rate (%) 2005*	*Child mortality (under-five per 1000) 2005*	*HIV/AIDS prevalence rate (% adults aged 15–49) 2005*	*Malnutrition (% children under 5 underweight) 2000–2005*
Equatorial Guinea	42.3	2.3	205	3.2	18.6
Eritrea	54.9	3.9	78	2.4	39.6
Ethiopia	42.7	1.8	127	–	38.4
Gabon	53.8	1.6	91	7.9	11.9
Gambia, The	56.8	2.6	137	2.4	17.2
Ghana	57.5	2.0	112	2.3	22.1
Guinea Bissau	45.1	3.0	200	3.8	25.0
Guinea	54.1	1.9	160	1.5	32.7
Ivory Coast	46.2	1.6	195	7.1	17.2
Kenya	49.0	2.3	120	6.1	19.9
Lesotho	35.2	–0.2	132	23.2	18.0
Liberia	42.5	1.3	235	–	26.5
Libya	74.4	2.0	19	–	–
Madagascar	55.8	2.7	119	0.5	41.9
Malawi	40.5	2.2	125	14.1	21.9
Mali	48.6	3.0	218	1.7	33.2
Mauritania	53.7	2.9	125	0.7	31.8
Mauritius	73.0	1.1	15	0.6	–
Morocco	70.4	1.0	40	0.1	10.2
Mozambique	41.8	1.9	145	16.1	23.7
Namibia	46.9	1.1	62	19.6	24.0
Niger	44.9	3.3	256	1.1	40.1
Nigeria	43.8	2.4	194	3.9	28.7
Rwanda	44.1	1.7	203	3.1	22.5
São Tomé and Príncipe	63.5	2.3	118	–	12.9
Senegal	56.5	2.4	119	0.9	22.7
Seychelles	73.0	1.0	13	–	–
Sierra Leone	41.4	3.5	282	1.6	27.2
Somalia	47.7	3.3	225	0.9	25.8
South Africa	47.7	1.1	68	18.8	–
Sudan	56.7	2.0	90	1.6	40.7
Swaziland	41.5	1.0	160	33.4	10.3
Tanzania	46.3	2.6	122	6.5	21.8
Togo	55.1	2.6	139	3.2	–
Tunisia	73.5	1.0	24	0.1	4.0
Uganda	50.0	3.5	136	6.7	22.9
Zambia	38.4	1.6	182	17.0	23.0
Zimbabwe	37.3	0.6	132	20.1	–
NORTH AFRICA	71.1	1.5	35	0.1	8.3
SUB-SAHARAN AFRICA	46.7	2.2	163	6.1	26.3
ALL AFRICA	**50.8**	**1.9**	**149**	**3.1**	24.7

Sources: World Bank, 2008a; UNFPA, 2007; UNDP, 2007

Mounting debt and frequent shortages of basic commodities due to a lack of foreign exchange are commonplace. Non-oil producers experience regular fuel shortages, preventing the movement of goods and people and further compounding already serious economic problems. In cities, electric power cuts occur regularly, caused by the fuel shortages or inadequate maintenance of ageing generating plants and power lines. The effects of prolonged power cuts on industry can be devastating. In Nigeria, Africa's most populous country and one of the world's largest exporters of crude petroleum, supplies of refined petroleum to filling stations are highly unreliable due to the inadequate capacity of the country's refineries and massive profiteering from fuel imports (FT.com, 2009).

It is in the rural areas, however, where most of Africa's population still live and work (see Chapter 4). Some would argue that the towns have already received too much of the wealth and investment, and that there is a need for a genuine and concerted effort to raise rural living standards and make agriculture more attractive and profitable. In most African countries, over 60 per cent of the labour force works in agriculture, mainly producing traditional food crops and using methods that have been handed down from generation to generation with little change. In some countries – such as Rwanda and Burundi, Niger, Burkina Faso, Mali and Tanzania – a very high proportion of the labour force (at least 70 per cent) is employed in farming and pastoralism. Although the proportion working in agriculture is declining slowly, it is likely to remain high for the foreseeable future. By contrast, less than 2 per cent of the UK's labour force was employed in agriculture in 2009.

For those rural young people with some education, agriculture is often much less attractive than the prospects of urban employment. African governments need to restore faith in agriculture by helping farmers to raise production and by paying them good prices for their crops. Greater food production in Africa should reduce the need for food imports which, together with possible food exports, would improve the foreign exchange position. But such policies are inextricably linked with the world trade in food commodities, such that local producer food subsidies in North America and the European Union mitigate against food that might be imported from African countries (see Chapter 9).

As farming systems become more efficient, they may gradually shed surplus labour to the growing industrial sector, which is usually located in or near the towns. Improving Africa's agriculture will take time and must be done carefully. Above all, there is a need to learn from the many failures of the past, and for governments and agricultural advisers to work together with farmers rather than merely instruct them which crops they must grow and which techniques they should use. African leaders are fond of talking about the importance of agriculture, but very often little is done to improve it, as their main concern is to keep food prices low in the towns so that the better-organized, and politically more threatening, urban population stays content.

A further priority for the future will be to improve the transport systems of African countries, which in many cases were laid down in the colonial period. Roads with crumbling surfaces and pot-holes in the rainy season, and railways with worn-out track and rolling stock, are all too common. If agriculture is to become more productive and the benefits of this more widely distributed, then an efficient marketing and transport system is absolutely vital. Much investment is needed, but many African governments cannot afford the financial outlay.

Africa's future is both exciting and uncertain. There is great potential, but the means, predominantly financial, of achieving this potential are lacking. Many African countries are now relatively worse off than they were at independence more than forty years ago. World recession, indebtedness and a shortage of foreign exchange, coupled with limited achievements in agricultural development, poor health and welfare, rapid population growth and crumbling infrastructures, have all contributed to the poverty in many African countries today. Furthermore, economic instability has often bred political instability and vice versa. One-party states, military regimes and *coups d' état* are common features in Africa, making potential foreign investors cautious about investing in unpredictable locations. One of the great post-independence capitalist 'success stories' in Africa was Ivory Coast (Côte d'Ivoire), which benefited from a relatively long period of political stability during the 1960s and 1970s under President Félix Houphouët-Boigny, but then the economic situation steadily deteriorated from the 1980s. Following Houphouët-Boigny's death in 1993, the country experienced continuous political instability, leading to *coups d'état* in 1999 and 2001, and in April 2011 to the ousting of incumbent president Laurent Gbagbo and installation of a democratically elected president, Alassane Ouattara.

Many other African countries, such as Chad, Ethiopia, Mozambique, Somalia and Uganda, have suffered long periods of political instability and even internal wars. In early 2011, the North African countries of Tunisia, Egypt and Libya experienced pro-democracy uprisings leading to the overthrow of longstanding authoritarian leaders Zine El Abidine Ben Ali (Tunisia, 1987–2011) and Hosni Mubarak (1981–2011). Meanwhile, neighbouring Libya was involved in civil war from April 2011 as Colonel Muammar Gaddafi, the Libyan leader for forty-two years, attempted to hold on to power.

Economic stability and progress are more likely to be fostered if African countries could only settle their internal differences of opinion more calmly and generate an atmosphere of greater political stability. Possibly only then will significant advances in economic and social development be possible.

1.5 Perspectives on Africa's future development

The 1970s and 1980s were a particularly difficult time for Africa, with some countries experiencing negative economic growth rates. Mozambique, for example, had an average annual GDP growth rate of –0.7 per cent between 1980 and 1990, while Niger's rate was –1.3 per cent, among the lowest rates in the world. At the other end of the economic spectrum, the Congo Republic had an annual growth rate of 3.6 per cent between 1980 and 1990, and Kenya 4.2 per cent. More recently, between 2000 and 2005, many African countries had positive annual GDP growth rates, with Mozambique experiencing 5.3 per cent growth, while oil-rich Equatorial Guinea achieved a staggering 21.3 per cent annual growth rate during this period (see Chapter 8).

In 1960, W. W. Rostow, in his Stages of Economic Growth Model, suggested that economic progress in particular countries could be categorized into certain phases, of which the most important was 'take-off' – a period associated with increased industrial investment such as occurred during the British industrial revolution in the late eighteenth and early nineteenth centuries (Rostow, 1960). Rostow's model implies that all countries proceed through the same five phases of growth, that industrialization is the key to growth, and that growth leads to economic and perhaps social improvement (see Chapter 9). The model has, however, been heavily criticized in recent years on the following grounds:

- its unilinear approach, implying that 'things get better' over time;
- its eurocentricity, which implies that all countries will imitate the Euro-American experience;
- confusion about the definition and character of growth and development (Binns, 2008).

Some writers have drawn attention to the existence in certain developing countries of 'growth without development' where, despite high rates of economic growth (as measured by increases in GDP), the bulk of the population does not seem to have benefited as one might have expected (see Box 8.2, below). Dudley Seers (and others; see Chapter 9) draws a clear distinction between broader conceptualizations of development and economic growth (Seers, 1969, 1979). Development, Seers suggests, should involve a reduction in poverty, unemployment and inequality, while individuals and groups should also be able to fulfil their ambitions. In many cases where economic growth has occurred, only a small minority of the – usually urban-based – population has benefited. This is what happened in Liberia in the 1960s, where the small Americo-Liberian ruling elite benefited from the country's rapid economic growth largely based on mining and rubber production. The benefits of such growth should ideally be fairly distributed and penetrate into the remotest areas of countries, rather than being 'siphoned off' by a relatively small number of people located at the nation's political and economic hub. But how can this be achieved, and how can any progress be monitored effectively?

In most African countries it is difficult to obtain reliable statistics on poverty, inequality and unemployment. Furthermore, it would be necessary to define each of these terms carefully in relation to the situation being examined. For example, poverty in the UK and other West European countries may be very different in character and magnitude from poverty in Burkina Faso or Zambia. While in Western Europe reference would probably be made to minimum wage levels and receipt of state benefits, in many African countries similar social security measures do not exist. In an African context, some reference to nutritional and health status as well as to income would perhaps be more meaningful (UNDP, 2007).

Similarly, measuring inequality and unemployment in most African countries would be difficult without reliable census and other socio-economic data. Registers of unemployed people are usually kept only, if at all, in African cities, and even there they are unlikely to reflect the situation accurately. Employment is more seasonal in rural areas, being related to cycles of cultivation, with the heaviest work often coming during the rainy season. The dry season may be a period of relative unemployment, or rather under-employment, when other jobs such as house-building or craft work are undertaken in rural villages. Alternatively, for those persons living close to large towns, there may be an opportunity to take up seasonal paid work in the town, perhaps as a carpenter, mechanic or in some other trade. Such off-farm dry season work can make a significant contribution to household income, as in the region around the city of Kano in northern Nigeria.

While we might agree with Seers that definitions of development should include reference to a series of non-economic as well as economic variables or indicators, it is often difficult to measure these in many African states. Seers (1979) suggests that once the magnitude of poverty, unemployment and inequality is recognized, governments and development agencies should then introduce programmes to fulfil basic human needs and reduce these three key indicators of development. The poorest groups should be the main focus of development planning, with particular attention directed towards provision of adequate food, water, shelter, health care, education and employment – the basic needs of all human beings.

It has taken a long time for African governments to take on board these ideas, and some remain wedded to the much-criticized Rostow model. In the last twenty years, however, there has been an increasing move towards smaller-scale, community-based projects which are aimed at providing basic needs to the poorest groups (see Chapter 9). Whereas earlier projects and programmes for promoting economic growth usually focused on wealth generation, more recent development projects in Africa are increasingly 'pro-poor' and focus on poverty alleviation among the poorest individuals and groups. This is illustrated by the Reconstruction and Development Programme (RDP) introduced by Nelson Mandela in a highly unequal South Africa after the country's first democratic elections in 1994 (see Box 1.6). In fulfilling basic needs, it is argued that people will be able to perform more efficiently and live longer, more fruitful lives, which in turn will raise their economic output in the long term. Thus, fulfilling basic needs will eventually lead to economic growth as well as development. In later chapters of this book, more detailed reference will be made to specific examples of development schemes and programmes and their impact on reducing poverty, inequality and unemployment.

Box 1.6

South Africa: reducing inequality in the 'rainbow nation'

Democratic South Africa, or the 'rainbow nation' as it was proudly proclaimed, which emerged following the elections of April 1994, faces an enormous, some would say insurmountable, task of 'reconstruction and development' in an effort to raise living standards among so-called 'historically disadvantaged' households and communities who constitute a large proportion of the national population. The Reconstruction and Development Programme (RDP), launched as an ANC policy document before the elections, and then formalized in September 1994 as the government's 'White Paper on Reconstruction and Development', aimed to address the extreme social and spatial inequalities engendered by years of apartheid. As the policy document states, 'Every aspect of South African life is deeply marked by minority domination and privilege' (ANC, 1994: 119).

South Africa is characterized by levels of social and spatial inequality which place it among the ten most unequal countries in the world, according to income distribution (UNDP, 2007). Before the end of apartheid, in 1993, South Africa's white population had a Human Development Index of 0.901, similar to those of many Western European countries. However, the HDI was 0.836 for the Indian population, 0.663 for 'coloureds' (people of mixed race) and only 0.5 for the majority black population, making the latter comparable with such countries as Papua New Guinea and Cameroon (Lester *et al.*, 2000). Stark spatial inequalities are also apparent within South Africa, such that 'sophisticated urban centres with their established business corporations, which are indistinguishable from those in the Western world, can be viewed from the same spot as the crime-ridden squatter camps, typical of those in South American cities' (Lester *et al.*, 2000: 235). On a broader regional basis, those provinces which have incorporated the former 'black homelands' (the largely rural, black racial reserves) are generally much poorer than others. Eastern Cape Province, for example, which in 1994 incorporated the homelands of Transkei and Ciskei, had the highest 'official' unemployment rate, the highest infant mortality rate and the second-lowest life expectancy in South Africa in 1996 (Binns, 1998; DBSA, 2000; Statistics South Africa, 1999).

The Reconstruction and Development Programme, which was designed to tackle such inequality and poverty, identified the need for 'an integrated and sustainable programme', which was motivated by the recognition that

> The legacy of apartheid cannot be overcome with piecemeal and unco-ordinated policies. The RDP brings together strategies to harness all our resources in a coherent and purposeful effort that can be sustained into the future. These strategies will be implemented at national, provincial and

> local levels by government, parastatals and organizations within civil society working in the framework of the RDP.
>
> (ANC, 1994: 4–5)

The RDP was promoted essentially as a 'people-driven process', focusing 'on our people's most immediate needs, and [relying] in turn, on their energies to drive the process of meeting these needs' (ANC, 1994: 5). The RDP also placed much emphasis on grassroots empowerment, suggesting that 'development is not about the delivery of goods to a passive citizenry. It is about active involvement and growing empowerment' (ANC, 1994: 5).

The ANC government believed these objectives could be achieved by giving much more responsibility for development to local government, which is viewed as the primary level of democratic representation, through what is referred to as 'developmental local government', a term which seems to be peculiar to South Africa (Nel and Binns, 2003a). As the RDP stressed, 'The democratic government will reduce the burden of implementation which falls upon its shoulders through the appropriate allocation of powers and responsibilities to lower levels of government, and through the active involvement of organizations of civil society' (ANC, 1994: 140). While these are understandable and admirable objectives, they have placed a considerable burden of responsibility on the local tier of government, a situation aggravated by the very real human and financial capacity constraints experienced by many of the smaller local authorities (Nel, 2001). For example, many non-white councillors, often elected for the first time, have been deprived of a proper education, and many small towns have been unable to raise an adequate income through rates and local taxation.

In order to achieve 'developmental local government', local authorities were expected to maximize both social development and economic growth and to help ensure that local economic and social conditions are conducive for the creation of employment opportunities (Nel and Binns, 2001). In addition, local government was required to take a leadership role, involving citizens and stakeholder groups in the development process, to build social capital and to generate a sense of common purpose in finding local solutions for sustainability. Local municipalities thus have a crucial role to play as policy-makers, and as institutions of local democracy, and they are urged to become more strategic, visionary and ultimately influential in the way they operate.

The wide-ranging National Constitution further reinforces the place of local government in society, requiring it to 'encourage the involvement of communities and community organizations in matters of local government' (RSA, 1996: 81).

However, over fifteen years since the end of apartheid, it is only really in the largest cities, with their greater financial and human capacity, that significant progress, in terms of infrastructure provision and economic support, has been achieved. With a national unemployment rate of 22.7 per cent in September 2007, and local unemployment rates approaching 80 per cent in some areas, together with an adult HIV/AIDS infection rate of 18.8 per cent (one of the highest in the world), South Africa still has a long way to go in improving the quality of life of its people and eradicating the legacies of apartheid.

Summary

1. Longstanding myths and stereotypical views of Africa are difficult to eradicate and frequently inhibit accurate understanding.
2. Although the colonial period was a relatively short phase in Africa's history, the effects of colonialism have been far reaching.
3. Inadequate understanding of Africa's peoples and environments has often led to inappropriate development strategies and projects which have sometimes caused more problems than they have solved.
4. Considerable social and economic inequalities exist between and within African countries.
5. In countries where positive economic growth has been achieved, development and the fulfilment of basic needs have often not occurred.

Discussion questions

1. Examine some of the common images of Africa and suggest how popular myths and stereotypes might be modified.
2. 'Colonialism programmed African countries to consume what they do not produce and to produce what they do not consume.' Examine this statement with reference to particular colonial policies and countries.
3. Why are so many countries in Africa worse off now than they were at independence?
4. Explain the difference between 'growth' and 'development'.

Further reading

Griffiths, I.L. (1994) *An Atlas of African Affairs* (2nd edn), London: Routledge.
Griffiths, I.L. (1995) *The African Inheritance*, London: Routledge.
Iliffe, J. (1995) *Africans: The History of a Continent*, Cambridge: Cambridge University Press.
ODI (Overseas Development Institute) (2010) *Millennium Development Goals Report Card: Learning from Progress*, London: ODI.
Wilson, H.S. (1994) *African Decolonization*, London: Edward Arnold.

Useful website

UNFPA (United Nations Population Fund) (2007) *State of World Population 2007*. http://www.unfpa.org/swp/2007/english/introduction.html (accessed 4 September 2008).

2.1 Introduction

In 2009, Africa's population exceeded 1 billion people for the first time and, given the reality that Africa has some of the fastest growth rates in the world, the continent's population could double by the year 2050 (PRB, 2009). The current average African population growth rate of 2.4 per cent per annum is double the world average and this has significant implications for a variety of the key themes examined in this book, including: access to land, employment, social services and housing. The fact that population growth rates exceed economic growth rates in many parts of the continent, and have done so for much of the post-independence era, is a continuing cause for concern, as are higher rates of infant mortality, malnutrition and HIV/AIDS infection than are experienced in other continents. In addition, significant population relocation has taken place, either in search of employment, such as in southern Africa, or forced relocation in the case of refugee movements from political flash-points, such as Darfur and the Great Lakes region. A clear link exists between key demographic trends, population growth rates and overall levels and prospects of economic well-being and development (UNECA, 2001). Over and above this, significant regional differences exist between tropical Africa and North and southern Africa, not only in terms of key demographic, nutritional and economic indicators, but also in terms of such issues as religion and cultural diversity.

Africa undoubtedly suffers from the stereotyping of its people that characterizes much of its image in Western popular media. While poverty, low life expectancy and ethnic tensions are undoubtedly endemic to the continent, these are extremely variable in space and over time. A visit to South Africa, Egypt or even Ethiopia will reveal distinct differences in people's development status within society. Historically, Africa was regarded as being a continent short of people rather than land (Binns, 1994). The presence of domestic slavery long before the colonial powers arrived, and continuing in some places until independence, testifies to the importance of owning labour (Hopkins, 1973). However, in a study of Africa's population, there is as much variation across the continent today as there was in the past. The homelands of the Ibo and Yoruba peoples in southern Nigeria, the Ethiopian Highlands and the two tiny states of Rwanda and Burundi, west of Lake Victoria, have long been densely populated regions. By contrast, countries such as Niger and Chad, bordering the Sahara Desert, together with Namibia, Botswana, the Central African Republic and the Congo Basin have always been sparsely settled. Taken as a whole, Africa's population is increasing rapidly and pressure on land and other resources is growing in some cases to dangerous levels. Two differing interpretations emerge from this reality – one which argues that 'Africa is sitting on a time bomb' because of its population growth (allAfrica.com, 2010), and a second which argues that Africa may in fact be under-populated and reap a 'demographic dividend' as economic prospects improve (*The Economist*, 2009). It is important to note that the demographic situation is not static, and that over the last sixty years there have been noteworthy changes in key demographic indicators. For example, in the 1950s the average woman in Africa was likely to have 6.7 children in her lifetime. That has now fallen to 4.8, although this is still well ahead of the world average of 2.6 (Africa News, 2009).

In helping to set the context for the rest of this book, this chapter overviews the continent's key demographic indicators and current population trends relative to those occurring globally. The discussion also highlights regional and national differences and investigates some of the key population-related challenges which the continent currently faces as well as their associated developmental implications. The chapter concludes with an overview of Africa's population in terms of the key issues of culture, ethnicity and religion. Text boxes overview such key demographic issues as HIV/AIDS in southern Africa (Box 2.1), the challenges posed by rapid population growth in such countries as Kenya (Box 2.2), and under-population in Mali, where low levels of population density have provided opportunities for enhanced food production and export to support other regions (Box 2.3).

2.2 Population and development

2.2.1 The reliability of census data

According to the Population Reference Bureau, the population of Africa reached 1 billion in 2009 (PRB, 2009). Most commentators would argue, however, that this is a significant under-representation because of the unreliability of the population data itself. Censuses are expensive and demanding of time and labour; they may also be difficult to administer due to the inaccessibility of people in remote rural areas, and high illiteracy levels among the population. Moreover, people may be suspicious of censuses, associating them with taxes and law enforcement, so they either avoid completing them or give misleading information. Census data can also be manipulated for political advantage, particularly when it is used as leverage for development assistance, or where it seeks to under- or over-represent different ethnic groups within a population (Potter *et al.*, 2008). For example, the results of the December 1974 census in Sierra Leone were withheld for many years for fear of disturbances that might be caused if one ethnic group were seen to have more members than another. The fractious political process in Zimbabwe over the last fifteen years, where election-related tensions have led to allegations of vote rigging to favour a particular tribal/political grouping, is another case in point.

Twenty-six out of the fifty-three African countries undertook their last national census prior to 2000 (Angola's last census took place in 1970). So it is perhaps not surprising that the true population of Africa is believed to be much larger (US Census Bureau, 2009) than a billion. But why should this be an issue for development? In short, regular and reliable data on the demographic characteristics of the population, trends in migration, and the spatial distribution of populations are essential prerequisites for developing appropriate and sustainable infrastructure, such as schools, hospitals, roads, water supplies and sanitation. Census data can draw attention to existing or potential development issues, such as the lack of access to healthcare or education – statistics which, as highlighted later in this chapter, are central to attempts to quantify development using the Human Development Index.

2.2.2 Population size and growth

Table 2.1 highlights the most recent population estimates (2009) for each country, along with details about rates of natural increase, the projected population in 2025, the infant mortality rate and the total fertility rate. As the table indicates, Africa's total population comprises just under one-seventh of the global population. However, it is growing at twice the global rate of natural increase (2.4 per cent), as is reflected in Africa's total fertility rate (TFR) of 4.8, which is nearly double the global average of 2.6 (TFR indicates the average number of children that a woman is likely to bear in her lifetime). These figures stand in stark contrast to a European rate of natural increase of 0 per cent and TFR of 1.5 (below the estimated population replacement rate of 2.1), and an Asian rate of natural increase of 1.2 per cent and TFR of 2.3.

By far the most populous state in Africa is Nigeria, with 152.6 million people, almost twice as many as the next most populous country, Ethiopia, with 82.8 million. At the other end of the spectrum, the island states of the Seychelles and São Tomé and Príncipe have populations of only 100,000 and 200,000, respectively. Of the mainland countries, those with the smallest populations are Djibouti (900,000), Equatorial Guinea (700,000), The Gambia (1.6 million), Guinea-Bissau (1.6 million) and Swaziland (1.2 million).

Evidence suggests that the total population of Africa was fairly stable at around 100 million for many centuries, probably until the late nineteenth century, when colonial rule commenced. It is estimated that in 1900 Africa's population was about 133 million, and this more than doubled in sixty years to reach 270 million in 1960, on the eve of independence. By 1970, the population had reached 344 million, and in 1985 it was approaching 550 million (Binns, 1994). Since then, the population has nearly doubled.

Table 2.1 Key demographic details about Africa: total population and population change

Country	*Population 2009 (millions)*	*Projected population 2025 (millions)*	*Rate of natural increase (%)*	*Infant mortality rate (per 1000)*	*Total fertility rate*
WORLD	6,810	8,087	1.2	46	2.6
AFRICA TOTAL	999	1,385	2.4	74	4.8
NORTH AFRICA					
Algeria	35.4	43.7	1.9	26	2.3
Egypt	78.6	99.1	1.9	19	3.0
Libya	6.3	8.1	2.0	18	2.7
Morocco	31.5	36.6	1.4	31	2.4
Sudan	42.3	56.7	2.2	81	4.5
Tunisia	10.4	12.2	1.2	19	2.0
Western Sahara	0.5	0.8	1.8	44	3.0
WEST AFRICA					
Benin	8.9	13.8	3.2	89	5.7
Burkina Faso	15.8	24.8	3.2	89	6.0
Cape Verde	0.5	0.7	2.1	29	3.1
Côte d'Ivoire	21.4	29.9	2.4	100	4.9
Gambia, The	1.6	2.3	2.8	93	5.6
Ghana	23.8	32.2	2.1	50	4.0
Guinea	10.1	15.2	2.7	104	5.7
Guinea-Bissau	1.6	2.3	2.6	117	5.9

Table 2.1 Continued

Country	*Population 2009 (millions)*	*Projected population 2025 (millions)*	*Rate of natural increase (%)*	*Infant mortality rate (per 1000)*	*Total fertility rate*
Liberia	4.0	5.9	3.0	99	5.8
Mali	13.0	18.6	2.8	110	6.0
Mauritania	3.3	4.6	2.5	73	5.1
Niger	15.3	27.4	3.9	88	7.4
Nigeria	152.6	207.2	2.6	75	5.7
Senegal	12.5	17.9	2.9	61	5.0
Sierra Leone	5.7	8.1	2.0	89	5.2
Togo	6.6	9.3	2.7	91	5.1
EAST AFRICA					
Burundi	8.3	11.2	2.1	120	5.4
Comoros	0.7	0.9	2.5	53	4.2
Djibouti	0.9	1.1	1.9	67	4.2
Eritrea	5.1	7.4	2.9	58	5.3
Ethiopia	82.8	113.1	2.7	77	5.3
Kenya	39.1	56.5	2.7	67	4.9
Madagascar	19.5	28.4	2.9	70	5.0
Malawi	14.2	21.6	3.1	80	6.3
Mauritius	1.3	1.4	0.7	15.4	1.7
Mayotte	0.2	0.3	3.6	–	4.5
Reunion	0.8	1.0	1.3	8	2.5
Rwanda	9.9	14.5	2.5	62	5.5
Seychelles	0.1	0.1	1.0	12.9	2.2
Somalia	9.1	13.9	3.0	111	6.7
Tanzania	43.7	67.4	2.3	69	5.3
Uganda	30.7	51.8	3.4	76	6.7
CENTRAL AFRICA					
Angola	17.1	26.2	2.7	125	6.6
Cameroon	18.9	25.5	2.3	74	4.7
Central African Republic	4.5	5.5	1.9	106	5.0
Chad	10.3	13.9	2.6	106	6.3
Congo, Rep.	3.7	5.3	2.3	75	5.3
Congo, Dem. Rep.	68.7	109.7	3.1	92	6.5
Equatorial Guinea	0.7	1.0	2.4	102	5.4
Gabon	1.5	1.9	1.8	55	3.6
São Tamo & Príncipe	0.2	0.2	2.6	75	4.1
SOUTHERN AFRICA					
Botswana	2.0	2.3	1.3	48	3.2
Lesotho	2.1	2.4	0.2	83	3.4
Mozambique	22.0	29.9	2.4	–	5.4
Namibia	2.2	2.8	2.1	46	3.6
South Africa	50.7	54.4	0.8	45	2.7
Swaziland	1.2	1.5	1.6	85	3.8
Zambia	12.6	18.3	2.9	70	6.2
Zimbabwe	12.5	16.0	1.4	60	3.8

Sources: PRB, 2009; UNDP, 2009

As these statistics indicate, the post-1945 period has seen a rapid acceleration in the growth of Africa's population, such that today some countries have population growth rates which are among the highest in the world. Growth rates above 3 per cent per annum are not uncommon, as in Benin (3.2 per cent) and Burkina Faso (3.2 per cent), and throughout Africa the population is increasing. This stands in contrast to the situation in much of Europe, where growth is minimal or even negative, such as in Russia, where the growth rate is –0.3 per cent per annum (PRB, 2009). Some African countries, however, do have lower population growth rates than the rest of the continent, notably the island republics of Réunion (1.0 per cent), the Seychelles (0.1 per cent), Mayotte (0.3 per cent) and Mauritius (0.7 per cent). On the mainland, Tunisia has the lowest growth rate at 1.2 per cent, while in AIDS-affected Botswana and Zimbabwe the rate of natural increase has fallen to 1.3 per cent and 1.4 per cent, respectively.

As can be seen in Table 2.1, the TFR varies from 7.4 in Niger, to 6.7 in Somalia and 6.6 in Angola, to 2.3 in Algeria and 2.0 (i.e. below replacement) in Tunisia, which, together with significant differences in the rates of natural increase, indicates that populations are starting to stabilize in North Africa and some of the island states, while rapid growth is still occurring in large parts of western, eastern and central Africa. In southern Africa, rates of natural increase are also stabilizing, but this is happening amid prevailing high TFRs, indicating the negative effects of the AIDS pandemic (see below).

2.2.3 Life prospects and economic well-being

Table 2.2 reflects the poor performance of Africa's population in relation to a range of key indicators. While the world average life expectancy is 69 years, in Africa it stands at a modest 55 years. By comparison, average life expectancy in Asia is 69, in Europe it is 76, and in the Americas it is 75. Africa's highest life expectancies are to be found in North Africa and the island states, with Tunisia having an average life expectancy of 74 years, Réunion 76 years and Mayotte 74 years. By contrast, in most of central and eastern Africa, life expectancy is in the 50s, while in southern Africa it has fallen into the 40s (Lesotho is 40 years and Zimbabwe is 41 years). These appalling rates in southern Africa are due largely to the prevalence of HIV/AIDS, which has cut nearly twenty years from the average life expectancy in the last two decades. This reality is reflected in the percentage of adults (15–49 years) living with HIV/AIDS (see Plate 2.1). While the world average is 0.8 per cent, Europe is 0.3 per cent and Oceania is 0.1 per cent, Africa has an average HIV/AIDS incidence of 4.3 per cent. While figures as low as 0.1 per cent (Tunisia) are noted in North Africa, they rise to over 2 per cent in parts of East Africa (with Kenya at 7.4 per cent and Malawi at 11.95 per cent). But in southern Africa the *lowest* figure is 14.3 per cent, rising to a staggering 26.1 per cent in Swaziland and 23.9 per cent in Botswana. (See Box 2.1 for a more detailed examination of the impact that HIV/AIDS is having on the demography, society and economy of southern Africa.)

Africa's high rates of population growth, coupled with generally poor rates of economic performance, are reflected in the large numbers of people living under the minimum survival line of $2 per day (see Table 2.2). At a global level, 48 per cent of the world's population live under this level, but in Africa the figure is 65 per cent, contrasting starkly with figures of 19 per cent in South America and effectively 0 per cent in most of the Global North. In several African countries, over 90 per cent of the population lives on less than $2 per day: for example, Madagascar, Rwanda, Tanzania and Mozambique. It is only in North Africa that percentage rates under 15 per cent are noted (14 per cent in Morocco and 13 per cent in Tunisia), while even in relatively wealthy South Africa, 43 per cent of the population survive on less than $2 a day. The high poverty levels are further reflected

Table 2.2 Key demographic details about Africa: life and economic prospects

Country	*Life expectancy*	*% adults (15–49) living with AIDS 2007/8*	*% living on under $2 per day 2005*	*GNI per capita (purchasing power parity, $) 2008*	*Net migration rate per 1000*
WORLD	69	0.8	48	10,090	–
AFRICA TOTAL	55	4.3	65	2,660	–1
NORTH AFRICA					
Algeria	72	0.1	24	7,940	–1
Egypt	72	–	18	5,460	–1
Libya	73	–	–	15,630	1
Morocco	71	0.1	14	4,330	–3
Sudan	58	1.4	–	1,930	1
Tunisia	74	0.1	13	7,070	0
Western Sahara	65	–	–	–	20
WEST AFRICA					
Benin	56	1.2	75	1,460	–1
Burkina Faso	57	1.6	81	1,160	–1
Cape Verde	71	–	40	3,450	–5
Côte d'Ivoire	52	3.9	47	1,580	–1
Gambia, The	55	0.9	57	1,280	2
Ghana	59	1.9	54	1,430	–1
Guinea	56	1.6	87	1,190	–6
Guinea-Bissau	46	1.8	78	530	0
Liberia	56	1.7	95	300	5
Mali	48	1.5	77	1,090	–3
Mauritania	57	0.8	44	2,000	–1
Niger	53	0.8	86	680	0
Nigeria	47	3.1	84	1,940	0
Senegal	55	1.0	60	1,760	–2
Sierra Leone	48	1.5	76	750	2
Togo	61	3.3	69	820	0
EAST AFRICA					
Burundi	49	2.0	93	380	5
Comoros	64	<0.1	65	1,170	–3
Djibouti	55	3.1	41	2,330	3
Eritrea	58	1.3	–	630	2
Ethiopia	53	2.1	78	870	–1
Kenya	54	7.4	40	1,580	–1
Madagascar	59	0.1	90	1.040	0
Malawi	46	11.9	90	830	0
Mauritius	72	1.7	–	12,480	0
Mayotte	74	–	–	–	4
Reunion	76	–	–	–	0
Rwanda	48	2.8	90	1,010	0
Seychelles	73	–	–	19,770	6
Somalia	50	0.5	–	–	–6
Tanzania	54	5.7	97	1,230	–1
Uganda	50	5.4	76	1,140	–1

Table 2.2 Continued

Country	*Life expectancy*	*% adults (15–49) living with AIDS 2007/8*	*% living on under $2 per day 2005*	*GNI per capita (purchasing power parity, $) 2008*	*Net migration rate per 1000*
CENTRAL AFRICA					
Angola	46	2.1	70	5,020	1
Cameroon	52	5.1	58	2,180	0
Central African Republic	45	6.3	82	730	0
Chad	47	3.5	83	1,160	–1
Congo, Rep.	53	3.5	74	3,090	–3
Congo, Dem. Rep.	53	1.3	80	290	0
Equatorial Guinea	59	3.4	–	21,700	3
Gabon	59	5.9	20	12,270	1
São Tamo & Príncipe	65	–	–	1,780	–2
SOUTHERN AFRICA					
Botswana	49	23.9	49	13,100	2
Lesotho	40	23.2	62	2,000	–4
Mozambique	43	12.5	90	770	0
Namibia	59	15.3	62	6,270	0
South Africa	52	18.1	43	9,780	3
Swaziland	46	26.1	81	5,010	–1
Zambia	43	14.3	82	1,230	–1
Zimbabwe	41	15.3	–	–	–11

Sources: PRB, 2009; UNDP, 2009

Plate 2.1 HIV/AIDS poster, Sierra Leone (Tony Binns).

in the income levels shown in Table 2.2. Africa's per capita Gross National Income (GNI) in 2008 was $2660, compared with a global average of $10,090 and figures of $6020 in Asia and $25,550 in Europe. Predictably, there are wide variations across the African continent, from $21,700 in Equatorial Guinea and $15,630 in Libya to a mere $290 in the Congo (DRC) and $300 in Liberia. Disparities between the oil-rich countries and those that have suffered from civil war and destruction of economic infrastructure are obvious from these statistics.

Box 2.1

HIV/AIDS in southern Africa

AIDS (Acquired Immune Deficiency Syndrome) has been labelled the 'scourge of Africa' (Potter *et al.*, 1999: 126) – and rightly so, as this crippling disease has had a far more devastating effect on Africa, and southern Africa in particular, than on any other part of the world. In 2009, the UN estimated that there were 33.4 million people living with AIDS globally, with 67 per cent of them resident in sub-Saharan Africa (up from 60 per cent in 2006), half of whom are children. While the world infection rate for adults is 0.8 per cent, that of Africa is 5.2 per cent of the total population (UNAIDS, 2009). Within Africa, however, infection rates vary from as low as 0.1 per cent in much of North Africa to an average of above 20 per cent in southern Africa, with Swaziland having the highest infection rate in the world at 26.1 per cent (PRB, 2009).

In addition to the human tragedy which is being experienced, AIDS is having a catastrophic effect on the demography, society and economy of sub-Saharan Africa. It has now become the leading cause of death in southern Africa and appears to have shaved more than twenty years off the average life expectancy in many parts of that region. Furthermore, as Stock (2004) notes, several countries in this region have experienced 'reversed transition' in terms of the demographic transition model due to rising death rates. For example, in Zimbabwe the death rate per 1000 rose from 8 to 18 between 1990 and 2000; while in South Africa, AIDS is now the leading cause of death. The effects of HIV/AIDS have contributed to the creation of a 'lost generation', where the children of AIDS victims have to be cared for by their grandparents or even their older siblings, while the overall labour force has contracted (see Plate 2.2).

Significant social costs are incurred through the loss of family members and people at what should be the most productive phases of their lives, and the burdens are placed on grandparents and care-givers to support the ailing and the survivors. This is also leading to the breakdown of family structures and a significant increase in the number of orphans and even child-headed households. In parallel, healthcare systems have been significantly stretched, with the care of AIDS patients often accounting for up to 50 per cent of national health budgets, to the detriment of other healthcare requirements in countries where resources are already limited.

In economic terms, AIDS has often removed the key breadwinners from families, as well as limiting the labour force participation of those who have to become care-givers, both in formal employment and on farms (Gould, 2009). In addition, the increased incidence of death has exacerbated the burden which the costs of traditional African funeral rituals place on families: a funeral can often set a family back one to two years in financial terms. At a broader level, employing firms have to cope with high labour-turnover rates, increased illness-related absences, and the serious loss of skills, while at the broader level there is often a diminishing number of potential taxpayers.

However, there is some cause for hope. There is evidence that the HIV infection rate in Zimbabwe is declining, while in East Africa, and especially Uganda, which was the first part of the world to experience the catastrophic effects of the disease at the macro-level, the HIV infection rate fell from 14 per cent to 8 per cent between 1990 and 2000 (News24, 2009). This, however, came only after years of intensive social engagement and training designed to educate the population and reduce the risks posed by practising unsafe sex. National campaigns were led by the state, with support from churches and NGOs.

South Africa still has the highest number of AIDS sufferers (5.7 million) in the world, and it has experienced significant demographic change which the government, until recently, has seemed unable to combat. It is estimated that 250,000 South Africans died as a direct result of AIDS in 2008, while the overall death rate rose from 316,559 in 1997 to 667,184 in 2006 as a direct result. Of these deaths, 41 per cent were in the most vulnerable group – 25–49 years (AVERT, 2010a). Furthermore, the country's life expectancy plummeted from 62 years in 1990 to 50 years in 2007. The degree to which HIV/AIDS is affecting younger generations is shown by the fact that an estimated 33 per cent of women aged between 25 and 29 are HIV positive, while in 2007 280,000 under-15-year-olds were HIV positive. In addition, it is estimated that there are between 1.5 million and 3 million AIDS

Plate 2.2 Grandmother (right) takes care of orphaned grandchildren in South Africa (Etienne Nel).

orphans in South Africa, and it is expected that by 2015 one-third of all children (more than 5 million) will have lost one or both parents as a result of AIDS. There are already 148,000 child-headed households in the country (*Mail and Guardian*, 2009).

For much of the 1990s and the first decade of the twenty-first century, South Africa seemed to be in denial about the scale of the AIDS crisis and how to respond to it, with former President Mbeki courting the support of AIDS sceptics and recommending the use of traditional African foods as a cure. Sadly, Harvard statisticians have attributed the premature deaths of tens of thousands of people directly to these misguided policies and the associated delays in instituting a meaningful anti-retroviral (ARV) treatment campaign to suppress the symptoms of HIV (Timise, 2009). It has only been in the last few years that a far more determined policy has been initiated to increase access to ARVs, to try to reduce HIV infections and to promote public awareness (AVERT, 2010a).

2.2.4 Human development indicators

Quality of life can be evaluated with reference to a wide range of statistics, though not all are either available or reliable for many African countries. Per capita income, literacy levels, life expectancy, infant mortality and daily calorie intake are some of the most useful variables in evaluating quality of life. Very few countries have reliable data on incomes, unemployment registers or minimum wage levels. Since most of Africa's people live in rural areas where they are primarily engaged in food production, the bulk of which does not enter the market, it is difficult to ascertain incomes without detailed household surveys that are both difficult and expensive to undertake. In addition, many of these people are not engaged in 'formal sector' activities, which makes estimation of formal and informal sector employment and unemployment and under-employment difficult to gauge (Potter *et al.*, 2008).

If per capita Gross Domestic Product (GDP) is used as a measure of national wealth, then African countries are among the poorest in the world. In fact, four-fifths of the world's very poorest countries are in Africa. The Congo (DRC), with a per capita GDP in 2007 of only $298, was ranked by the United Nations as the world's poorest country, followed closely by Burundi ($341), Liberia ($362) and Guinea-Bissau ($477). In mainland Africa, the country with the highest per capita GDP in 2007 was Gabon at $15,167, which is relatively low when compared with $53,433 for Norway and $35,130 for the UK (UNDP, 2009). It should be remembered that these figures are averages and therefore conceal considerable variations among the population. In many African countries, it is common for a small proportion of the population to own a large proportion of the wealth. Small elite groups, often based in the capital cities, and with good access to political decision-making, are common in many African countries. Both the Gini-index and the Inequality Adjusted Human Development Index, which measure inequality within societies, show that South Africa has one of the most unequal societies on earth (UNDP, 2009, 2010). Sometimes these elites are concentrated among particular ethnic groups, such as the white minority in Johannesburg, and the two 'settler communities' of Creoles in Freetown (Sierra Leone) and the Americo-Liberians in Monrovia (Liberia), both of whom once played a major role in economic and political affairs in their respective countries.

Standards of health and education are poor in most African countries. Adult illiteracy is widespread especially in Mali (with a rate of 73 per cent), Burkina Faso (71 per cent),

Niger (71 per cent) and Guinea (70.5 per cent). Female illiteracy in these and other countries is even higher. Lesotho and South Africa are unusual in having relatively low illiteracy rates of 17.8 per cent and 12 per cent, respectively, while their female illiteracy rate of 33 per cent in both cases is well below the continental average (UNDP, 2009). Healthcare, particularly in the rural areas, is at best rudimentary and at worst non-existent in many parts of Africa. This is reflected in low life-expectancy and high infant-mortality figures (see Table 2.1). Infant mortality is a particularly useful indicator, since it reveals the state of childcare, the well-being of the mother and more general household conditions. A figure of over 100 deaths per 1000 live births is common in the tropical parts of Africa (see Plate 2.3).

Daily calorie supply is another key statistic, since it affects life expectancy, infant mortality and susceptibility to a wide range of diseases. Calorie supply in the countries of Western Europe and North America is well over 3200 per person per day, while the average for sub-Saharan Africa is 2098 (see Table 2.3), well below the UN recommended figure of 2350 and the global average of 2768. By contrast, the USA has a figure of 3745 and the UK 3397 calories per person per day. In many African states, the figure is below 2000: for example, Burundi, with 1629, and Ethiopia, with 1808. Some countries experienced a fall in calorie supply between 1965 and 1988, most notably Chad (from 2374 to 1852), Mozambique (1704 to 1632) and Ethiopia (1802 to 1658), all countries which suffered from civil war as well as drought (Binns, 1994). Clear regional differences in calorie intake exist, with West Africa averaging 2120 calories per person per day, central Africa 2041, eastern Africa 2045 and southern Africa 2418, while the per capita intake rates in parts of North Africa are above 3000 calories (FAO, 2009c; van Wesenbeeck *et al.*, 2009). These differences reflect the fact that the lowest nutritional levels are to be found in areas experiencing the greatest environmental and economic stress and often the highest levels of population growth, which does not bode well for the future health and well-being of large numbers of people in Africa. It is worrying to note that in 2005 it was established that 131 million people (or 17.3 per cent of the total population) in sub-Saharan Africa

Plate 2.3 Ante-natal clinic in rural Sierra Leone (Tony Binns).

Table 2.3 Key demographic details about Africa: development and settlement

Country	*Human Development Index 2007*	*HDI position out of 182 reporting countries*	*% urban*	*Population per sq km*	*Calories per person per day 2005*
WORLD	0.753	–	50	50	2768
AFRICA TOTAL	0.514	–	38	33	2098*
NORTH AFRICA					
Algeria	0.754	104	63	15	3095
Egypt	0.703	123	43	79	3317
Libya	0.847	55	77	4	3018
Morocco	0.654	130	56	71	3194
Sudan	0.531	150	38	17	2292
Tunisia	0.769	98	66	64	3275
Western Sahara	–	–	81	2	–
WEST AFRICA					
Benin	0.492	161	41	79	2293
Burkina Faso	0.389	177	16	58	3620
Cape Verde	0.708	121	59	126	2382
Côte d'Ivoire	0.484	163	48	66	2516
Gambia, The	0.456	168	54	142	2135
Ghana	0.526	152	48	100	2687
Guinea	0.435	170	33	41	2538
Guinea-Bissau	0.396	173	30	45	2054
Liberia	0.442	169	58	36	2008
Mali	0.371	178	31	10	2566
Mauritania	0.520	154	40	3	2785
Niger	0.340	182	17	12	2137
Nigeria	0.511	158	47	165	2503
Senegal	0.464	166	41	64	2154
Sierra Leone	0.365	180	37	79	1912
Togo	0.499	159	40	117	2018
EAST AFRICA					
Burundi	0.394	174	10	298	1629
Comoros	0.576	139	28	302	1800
Djibouti	0.520	155	87	37	–
Eritrea	0.472	165	21	43	1533
Ethiopia	0.414	171	16	75	1808
Kenya	0.541	147	19	67	2039
Madagascar	0.543	145	30	33	2005
Malawi	0.493	160	17	120	2132
Mauritius	0.804	81	42	625	2885
Mayotte	–	–	28	503	–
Reunion	–	–	92	324	–
Rwanda	0.460	167	18	375	1941
Seychelles	0.845	57	53	191	2385
Somalia	–	–	37	14	1744

Table 2.3 Continued

Country	*Human Development Index 2007*	*HDI position out of 182 reporting countries*	*% urban*	*Population per sq km*	*Calories per person per day 2005*
Tanzania	0.530	151	25	46	2010
Uganda	0.514	157	13	127	2385
CENTRAL AFRICA					
Angola	0.564	143	57	14	1880
Cameroon	0.523	153	57	40	2233
Central African Republic	0.369	179	38	7	1897
Chad	0.392	175	27	8	1980
Congo, Rep.	0.601	136	60	11	2327
Congo, Dem. Rep.	0.389	176	33	29	1500
Equatorial Guinea	0.719	118	39	24	–
Gabon	0.755	103	84	6	2761
São Tamo & Príncipe	0.651	131	58	169	2598
SOUTHERN AFRICA					
Botswana	0.694	125	60	3	2203
Lesotho	0.514	156	24	70	2425
Mozambique	0.402	172	29	27	2070
Namibia	0.686	128	35	3	2294
South Africa	0.683	129	59	42	2900
Swaziland	0.572	142	24	68	2323
Zambia	0.481	164	37	17	1887
Zimbabwe	–	–	37	32	2037

Note: * Sub-Saharan Africa

Sources: FAO, 2009c; PRB, 2009; UNDP, 2009; van Wesenbeeck *et al.*, 2009

were malnourished. The corresponding figures for the rest of the continent are 5.9 per cent of the population (3 million people) in southern Africa, 16.3 per cent in West Africa, 18.2 per cent in central Africa, and 19.6 per cent in East Africa.

Owing to the unreliability of applying financially based criteria to determine levels of development, as reflected in the artificially high economic scores recorded in oil-rich countries, and poor sources of data supply, the Human Development Index (HDI) is often taken to be a more reliable indicator of development. The HDI is a composite figure made up of life-expectancy figures (a surrogate for healthcare), educational attainment and income. The index figure for any country will lie on a continuum from very low – 0 to a possible maximum of 1 (Potter *et al.*, 2008). In the case of Africa, the average HDI is a modest 0.514, compared with a global average of 0.753. In the case of the European Union, the score is 0.937, while that of Latin America is 0.821 and East Asia is 0.770 (PRB, 2009). In this regard Africa clearly performs very poorly, which is a sad indictment of the low scores attained in the key development indicators that constitute HDI, a scenario that is aggravated by the negative impact of the HIV/AIDS pandemic. At a national level, as reflected in Table 2.3, the highest national HDI in Africa is achieved by the Seychelles, which scores 0.845. Scores above 0.7 are common in North Africa, while scores in West Africa tend to lie well below the continental average of 0.514. Sierra Leone, with a score of 0.34, has the lowest score in the world. Details contained in the second column of Table 2.3 reflect the sad reality that Africa currently has some of the least developed countries on the planet.

2.2.5 Urbanization and population density

Rapid rates of urban growth, often approaching 4 per cent per annum, are a feature of many parts of Africa and are a real concern for global organizations, such as the UN, which have drawn attention to the difficulties faced by African cities that are struggling to cope with the challenges posed to services, housing and employment by their rapid growth (see Chapter 5). This is further reflected by the reality that in many major cities 70 per cent or more of the population are already living in informal housing. Rapid urban growth is likely to continue for the foreseeable future, since Africa is still a predominantly rural continent, with only 38 per cent of the population being urbanized (Stock, 2004), compared with a global average of 50 per cent, 77 per cent in South America and 75.6 per cent in the OECD countries. The most urbanized territory in Africa is Réunion, at 92 per cent (see Table 2.3), followed by Western Sahara at 81 per cent and Libya at 77 per cent. Most of western and eastern Africa scores below 40 per cent, while North Africa and parts of southern Africa are above 50 per cent. The least urbanized countries are Burkina Faso (16 per cent), Ethiopia (16 per cent), Niger (17 per cent) and Malawi (17 per cent). It is noteworthy that in West Africa many of the largest settlements are located on or near the coast and have become foci for dense population, such as Dakar, Abidjan, Lagos and Accra. Away from the coast, the regions surrounding Johannesburg in South Africa, Nairobi in Kenya and Kampala and Entebbe in Uganda have also attracted dense populations. The challenges posed by rapid population growth and high levels of density are examined in a case study of Kenya in Box 2.2.

Box 2.2

Overpopulated Kenya

Kenya, with one of the most rapid population growth rates in Africa, and indeed the world, was one of the first countries in Africa to adopt a national family planning programme – in 1967 (Kenya Population Council, 2009). The country's average annual population growth rate between 1980 and 1989 was 3.8 per cent. In 1984 President Daniel Arap Moi held a conference on population at which he reaffirmed and intensified the country's commitment to reducing population growth. The seriousness of the situation was again recognized by the Kenyan government in 1986, when it referred to population growth 'overwhelming the economy's capacity to produce and provide for its people'. The population of 20.6 million in that year was projected to rise to 35 million by 2000 and 56.5 million by 2025. This rapid growth is partly due to infant mortality being halved between 1948 and 1985 and life expectancy increasing from 39 years in 1955 to 59 years for a Kenyan born in 1989.

At 7.7 per cent per annum, the rate of urban population growth has been even higher than the general population growth rate. Large families are often associated with the low status of Kenyan women, and one of the few ways of improving status is by producing a large number of children, something which is not restricted to Kenya. In recent years there has been an increase in surgical contraception, contraceptive distribution by non-professional workers and the setting up of purpose-built family planning clinics in urban areas (Chimbwete *et al.*, 2005; Kenya Population Council, 2009).

Kangundo Village in Machakos District had a population of 54,147 in 10,163 households, according to the 1979 census. The density was 406 people per sq km and the agricultural land only 0.2 hectares per person. The population is still increasing, with people concentrating on more productive areas of land, living on hillsides and suffering from chronic seasonal food shortages. Holdings are getting smaller, cultivation is more intensive and there is much unemployment. Many people are now landless and, with no space to rear cattle, the lack of manure has resulted in the land losing its fertility. Expansion and movement to other rural areas is restricted by different ethnic territories, so rural–urban rather than rural–rural migration is common.

The Greenbelt Movement has encouraged women, coming together under the aegis of the Family Planning Association of Kenya, to plant trees for economic crops and fuelwood. Most women in the community are aware of the need to limit the size of their families, but the average number of children per woman in the late 1980s was still six, falling far short of the Kenyan government's target of four. Each community-based distribution agent is responsible for distributing contraceptives and information on family planning to around 1000 households. The family planning network is having some success, as the 1989 Kenya Demographic and Health Survey found that 40 per cent of married women in rural Machakos District were using some form of contraception. In Kenya as a whole between 1984 and 2009 the total fertility rate fell from 7.7 to 4.9, a reduction of roughly three children per woman. Women are also marrying slightly later, although 50 per cent still marry and become mothers before the age of twenty. The World Bank argues that a major factor behind the change in views is the increasing cost of education, which has changed the balance of benefits and costs in raising children. Other factors, such as the government's commitment to reducing population growth, the increase in family-planning services, HIV/AIDS, the 1984 drought and a reduction in infant and child mortality, have also played their parts. The study also proposes reducing the social costs of family planning, a greater sensitivity to local needs, and an increased role for non-governmental organizations in the delivery of family-planning services (Chimbwete *et al.*, 2005; Kenya Population Council, 2009).

Kenya is making steady, albeit slow, progress in reducing population growth. A more integrated approach is needed, where family-planning workers and agricultural extension officers work together to raise food production, living standards and the status of women, since these and many other factors are interrelated.

Most of Africa's billion people live in rural areas (62 per cent), mainly in scattered villages. By world standards, the continent is not a densely settled region, having an average of thirty-three people per sq km, compared with a world average of fifty. Population density, however, varies dramatically, with the highest densities occurring in small, heavily populated island states such as Mauritius (625 per sq km) and Réunion (324). By contrast, Mauritania, on the southern edge of the Sahara, has a density of only three per sq km, and Gabon, astride the equator, has six per sq km. These figures may be compared with the population densities of other countries, such as UK (233 per sq km), India (253) and the Netherlands (400). Box 2.3 presents a case study of a low-density country, Mali, and examines how it has been able to supply food to other areas.

Box 2.3

Underpopulated Mali: feeding the people

Mali is a vast landlocked state in the semi-arid savanna–sahel zone of West Africa, with an area of 1,240,192 sq km (larger than the combined areas of the UK, France and Germany). Whereas these three European countries have a total population approaching 200 million, Mali's population in 2009 was estimated at only 13 million, with an average population density of just ten people per sq km, one of the lowest in Africa. The vast majority of Mali's people live in the wetter southern half of the country, where the main towns are located, notably Bamako, the capital (658,275 people), Segou (88,135), Mopti (74,771), Sikasso (73,859), Kayes (50,993) and San (30,772). Mali's most famous town, Timbuktu, was once an important trading, religious and educational centre in the pre-colonial period, with an estimated population in the sixteenth century of 25,000. However, its fortunes have since declined, and in the 1960s its population fell to around 7000 (Binns, 1994).

Mali is a very poor country, with over 68 per cent of the population living in rural areas and 85 per cent of the labour force occupied in agriculture. In 2008, it had a per capita GNI of only $1090, life expectancy was 48 years and the country had one of the highest infant-mortality rates in the world at 110 deaths per 1000 live births. The daily calorie intake was only 2566 per capita in 2005, well below the USA's 3745. If statistics such as life expectancy, literacy and GDP per capita were aggregated, Mali had the world's fifth-lowest Human Development Index in 2007 (PRB, 2009).

Undoubtedly, the key to life and the future development of Mali is water. The country's greatest asset is the River Niger, rising in the Fouta Djallon Plateau of Guinea to the south, flowing northwards through Bamako towards Timbuktu, then bending south through Gao to the border with Niger, a total distance within Mali of 1700 km. The river provides a year-round navigable routeway through the country, which has been used for centuries for transport and trade. However, in 1985, at the end of a long period of drought, the Niger almost dried up near Gao. North of the Niger bend, population and vegetation become progressively more sparse, cultivation is limited and pastoralism is the dominant food production system. Further south, and downstream of Bamako, the river loses speed and braids into a series of channels across a vast area known as the Inland Niger Delta, one of the natural wonders of West Africa. With an area of 103,000 sq km, the delta is not far short of the combined areas of Belgium, the Netherlands and Switzerland. During the rainy season from June to October, the plains of the delta are flooded. Ancient settlements such as Mopti and Djenné stand on islands linked by causeways above the flood plain. Land watered by the Niger flood has long been used by Bambara farmers for the cultivation of crops such as rice, millet, sorghum and maize, while the river channels are important fishing grounds for the Bozo and Somono fishermen. Fulani and Tuareg pastoralists also make use of the delta for grazing their herds (Deiemar, 2004; Diarra, 2009; IPPG, 2009).

With such a small and poor population, living in a vast country endowed with the resources of the Inland Niger Delta, one is tempted to ask the question: could the land be used more profitably to raise food production, improve domestic living standards and possibly earn valuable foreign exchange through agricultural exports? This question was considered by the French colonialists who ruled Mali (then called French Soudan) until 1960. They saw great potential in the country, and believed the delta could become 'the rice granary of West Africa'. In the early years of the French colonial period, cash crop production in Mali concentrated on cotton for export rather than rice for domestic consumption, with a small cotton scheme near Segou at the southern end of the delta. After the First World War, in 1919–20, the French proposed that a million hectares of irrigated land, more than half under cotton and some under irrigated rice, should be developed on the left (west) bank of the Niger. As the area was underpopulated, it was estimated that at least 300,000 immigrants would be needed to work on the scheme over a period of twenty-five years. In 1932 the Office du Niger (ODN) was created to coordinate operations and it planned to build two barrages, including a major one at Sansanding, and a series of related canals and irrigated areas. The ODN was also to install rice mills and cotton ginneries, build settlers' villages and provide medical and education services. Work started in 1934 on the Sansanding barrage, together with a navigable bypass and associated canals. Although the canals were completed a year later, the barrage was not finished until 1947. In the meantime, in 1935, 6600 hectares were planted with rice and 8000 hectares with cotton (Deiemar, 2004; Diarra, 2009; IPPG, 2009).

During the Second World War, the French Vichy government allocated money to irrigate a further 200,000 hectares, mainly for cotton production. The ambitious plans even envisaged the building of a railway across the Sahara Desert to provide a direct route to the Mediterranean. However, by 1945, the ODN had little to show for the massive expenditure. Despite the pilot schemes, inadequate knowledge about soil and agronomic conditions had resulted in a general failure of the initial stages of the irrigation project before the Sansanding barrage was even completed. The colonists who had arrived had difficulty tilling the land with the tools provided and many of the settlers were Mossi people, who came from a very different environment where floodland cultivation was not particularly significant. They often brought large numbers of relatives and hired labourers to work on their farms, with the result that overcrowding occurred in some areas. Although much emphasis was placed on cotton production, rice proved to be more profitable, with good prices and markets available in both Mali and neighbouring Senegal. In 1945 it was decided to stop all further expansion of the scheme and undertake more detailed research. By 1957 the popularity of rice had increased further and the crop occupied 60 per cent of the land, while cotton had reduced to only 15 per cent. Despite poor performance on the scheme, investment continued to flow into the ODN. During the period from 1945 to 1957, it received 30 per cent of the colonial agricultural development funds, whereas the rest of the Agriculture Service received less than 4 per cent of the annual colonial budgets (Deiemar, 2004; Diarra, 2009; IPPG, 2009).

Following independence, the ODN was transferred to the Malian government in 1961. By 1962, thirty years after its creation, the scheme's total area was still only 50,000 hectares. Total expenditure by the ODN was estimated to be £36.5 million (at 1962 values) or £730 per hectare. During the 1960s, rice production was affected by marketing problems, infestation by wild rice and attacks by quelea birds. Fertilizer costs escalated because, with the exception of phosphates, they had to be imported. In spite of all these difficulties, though, in 1986 it was felt that the ODN was worthy of yet more financial assistance, with emphasis now being placed on the production of rice and other food crops with the aim of achieving national self-sufficiency. It was also decided that the ODN should cease cotton cultivation altogether. A major rehabilitation programme of the ODN was therefore initiated which, at a cost of about £23 million, aimed to rationalize the organization's management and increase the total cultivable area to more than 100,000 hecatres, of which 46,730 were to be planted with rice and 7700 with sugar cane by 1989. The programme was supported by loans from the International Development Association, France, Germany, the Netherlands and the EU. Output of raw sugar increased from 10,000 tonnes in 1983/4 to 18,100 tonnes in 1986/7, and to an estimated 20,000 tonnes in 1987/8, thus almost meeting domestic demand. However, yields of rice still remained low, largely due to poor rainfall. In 1988 a further programme was inaugurated, with the aim of improving the irrigation network for rice cultivation. The cost of this programme was estimated at £50 million and was supported by the World Bank and European donors.

After the drought years of the early 1980s, Mali's production of rice increased in the second half of the decade from 187,200 tonnes in 1985/6 to 294,000 tonnes in 1990/1, which meant that domestic requirements were finally met. Seed cotton production also increased from 175,100 tonnes in 1985/6 to 255,000 tonnes in 1990/1, largely due to development programmes in the southern regions, which escaped the worst effects of the 1982–1984 drought (Binns, 1994; Deiemar, 2004; Diarra, 2009; IPPG, 2009). Since 2000, key changes have taken place, including moves to privatize the cotton parastatal, CMDT, improve overall marketing and production, allow the building of a Chinese textile factory in Segou and allow the farmers a greater say over farming operations and irrigation management. In the same decade investment by the World Bank, France, Germany and the EU sought to improve irrigation infrastructure in 35,000 hectares of the scheme. As a result of these endeavours, by 2004 paddy yields had increased from 2 tonnes per hectare in the 1980s to 6.5 tonnes, cultivation intensity had risen by some 80 per cent, water was being used more efficiently and incomes had risen commensurately, leading Deiemar (2004) to comment that after forty years of near stagnation the scheme was finally growing. By 2009, the scheme covered 83,900 hectares and had 56,000 small farmers operating on it. It is worth noting that in 2008 international investors started investing in the Niger Delta. For instance, Libya leased 100,000 hectares of land on which it hoped to produce 200,000 tonnes of hybrid rice annually (Diarra, 2009; IPG, 2009).

In spite of massive investment in the ODN over six decades, the performance of irrigated agriculture in Mali has been disappointing. In the early days too much emphasis was placed on cotton production to supply French industries, and not enough attention was given to feeding the Malian population and exporting food surpluses to other food-deficit countries in the region. Post-independence governments have favoured institutional solutions to agricultural and rural development supported by foreign aid, rather than policies designed to promote more self-reliant rural development. Poor management and maintenance of the ODN scheme combined with inadequate incentives to tenant farmers are major factors to blame for its uninspiring record. The Inland Niger Delta still has much untapped potential. Perhaps the focus and style of future development strategies should change, to encourage community-based smallholder schemes rather than large, complex institutions such as the ODN.

Even within particular countries there are densely settled and sparsely settled areas. The coastal region of West Africa from Ivory Coast to Cameroon, the Nile Valley and the area to the north and west of Lake Victoria are the most densely settled regions on the mainland, together with some smaller, but equally dense, pockets around Kano in northern Nigeria and the Witwatersrand in South Africa. Sparsely settled areas include the Sahara Desert and the countries on its southern fringe, the eastern Horn, the Central African Republic, Namibia, Botswana and the Congo Basin (Binns, 1994; Stock, 2004).

There are many reasons why some areas are densely settled while others have very few people. Other than along the Nile, river valleys have not been popular places for settlement in tropical Africa, owing to the presence of such water-borne diseases as bilharzia (schistosomiasis), river blindness (onchocerciasis), malaria and others. People have avoided some areas because of their harsh climates, notably the lack of water in many countries close to the Sahara. In West Africa, particularly Nigeria, a region known as the 'middle belt' sits between densely settled areas in the north and south. Poor soils and the widespread presence of tse-tse flies are problems here, but the sparseness of population is probably due mainly to slave raiding of the eighteenth and nineteenth centuries before the onset of colonial rule. As people were cleared from such areas, the bush grew back over farmland and tse-tse flies became a greater problem, spreading sleeping sickness among the remaining inhabitants (Binns, 1994; Stock, 2004).

Densely settled areas usually form as a result of complex environmental and social factors. Upland areas such as the Ethiopian Highlands and the Jos Plateau in Nigeria were popular as refuges during times of inter-tribal warfare, and inhabitants have developed intensive systems of terrace farming to survive in these difficult environments (Stock, 2004; see Box 3.6, below). Fertile volcanic soils have attracted dense settlement in western Cameroon, Rwanda and Burundi. The high-quality brown soils of the Kano Close-Settled Area in northern Nigeria support one of the densest rural populations in tropical Africa (Binns, 1994). Other areas of dense settlement include mining centres such as the Shaba–Copperbelt region of the Congo (DRC) and Zambia, and the Witwatersrand region of South Africa. Cities located at higher altitudes, where risk of disease is reduced – such as Nairobi, Harare and Lusaka – have also developed as logical nodes of settlement.

2.3 Population change and demographic structure

Since 1945, most African countries have moved into the second stage of the so-called 'demographic transition', lagging behind many other parts of the world where populations have progressed from Stage 1 (high birth and death rates) to Stage 4 (low birth and death rates), and in some cases, such as Eastern Europe, to Stage 5, when population decline begins (see Figure 2.1). From a situation of high birth rates and high death rates in Africa, death rates have declined in recent decades through medical improvements, but birth rates have generally remained high, falling in only a few countries. In Mauritius, for example, a vigorous family-planning programme helped to reduce the birth rate from around 47.4 per 1000 people in the years 1949–1953 to 26.7 in 1970 and just 20 in 1985. The net effect of lower death rates and continuing high birth rates in most of Africa has been a rapid growth of population, as reflected in the Demographic Transition Model (Figure 2.1; Stock, 2004).

Demographic structure is best seen in a population pyramid that shows the proportions of males and females in different age groups. A broad base indicates a high birth rate, while a narrow peak shows a high death rate and low life expectancy. In many African countries the population pyramid has a broad base with a large proportion of the total population being in the lower age groups (see Figure 2.2). The pyramid tapers, with relatively few people in the upper age groups. In most African countries life expectancy is well below the seventy

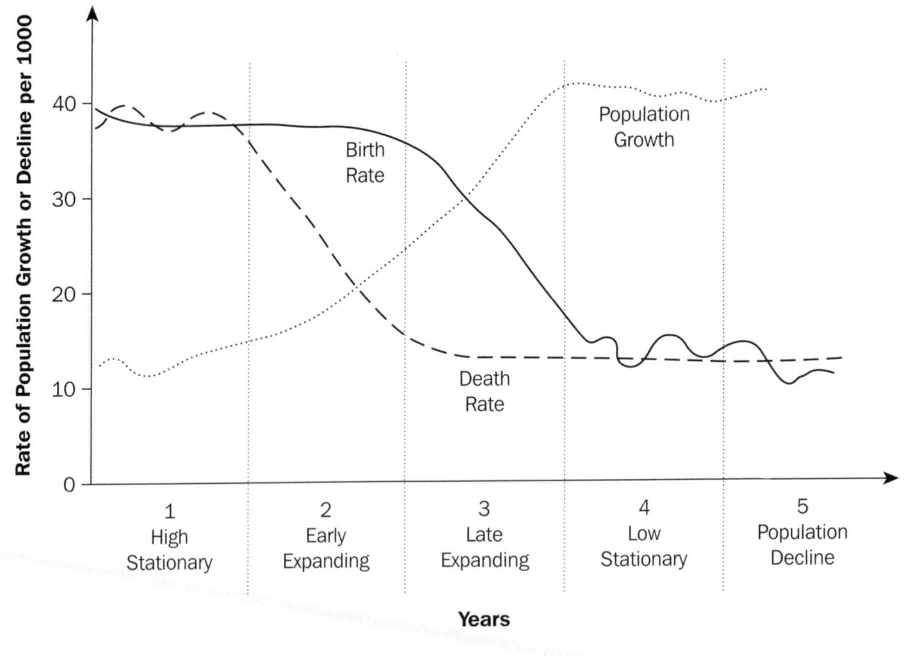

Figure 2.1 The Demographic Transition Model.

Source: Adapted from Potter *et al.*, 2008: 193

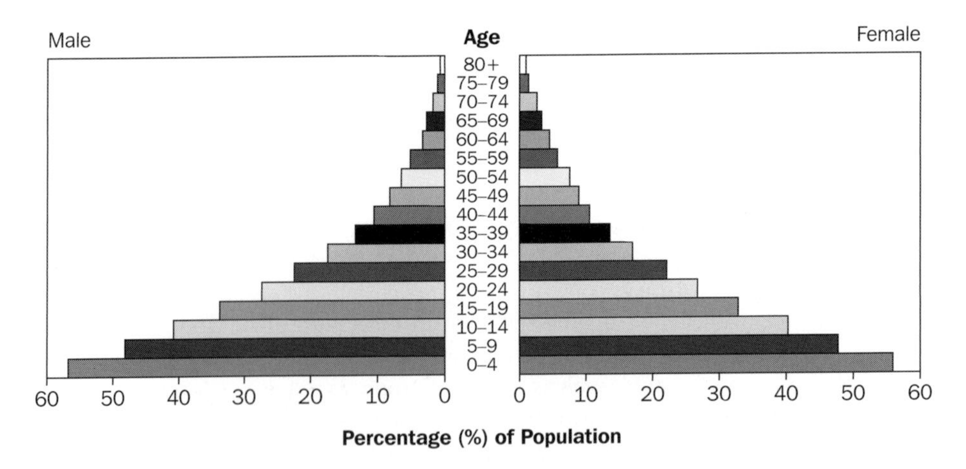

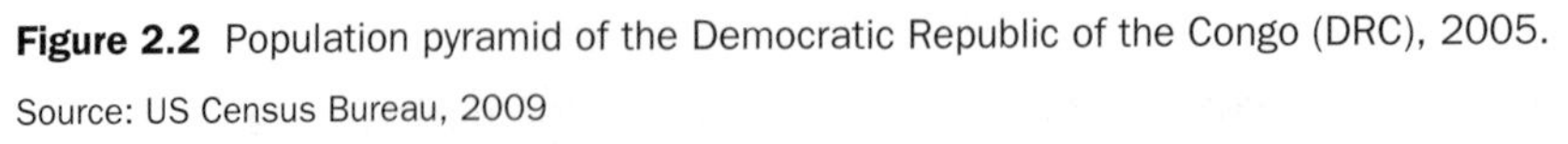

Figure 2.2 Population pyramid of the Democratic Republic of the Congo (DRC), 2005.

Source: US Census Bureau, 2009

or more years that people in Western Europe and America can expect to live (US Census Bureau, 2003). In fact, in much of Africa, the majority of people can expect to live only fifty years or even less. For example, life expectancy at birth is only forty-six years in Guinea-Bissau. In some crowded island states, however, people can expect to live much longer. Life expectancies in Mauritius, the Seychelles and Réunion, for example, are 72, 73 and 76 years, respectively. However, these three countries have also managed to reduce their birth rates significantly, resulting in lower-than-average population growth rates.

It is interesting to compare the population pyramids of two countries such as the Democratic Republic of the Congo (DRC) and Mauritius (see Figures 2.2 and 2.3). The latter, with a low birth rate, has a narrower base to its pyramid and a small proportion of its population in the lower age groups. With a high life expectancy, a good proportion of both males and females live well beyond the age of fifty. By contrast, the population pyramid for the DRC has a broad base, reflecting a high birth rate and a greater proportion of the population in the lower age groups. Moreover, the peak of the pyramid is very sharp, as relatively few people live over fifty. The high percentage of the population under twenty translates into a high level of dependency, and additional burdens are placed on the state to supply adequate levels of education and future employment opportunities as a direct result. A similar situation prevails in many other African countries, such as Sierra Leone (see Plate 2.4).

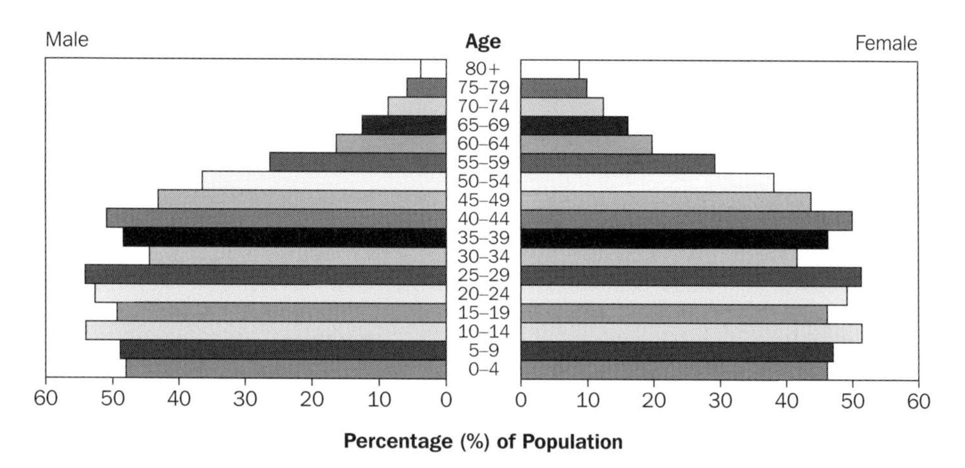

Figure 2.3 Population pyramid of Mauritius, 2005.

Source: US Census Bureau, 2009

Plate 2.4 Children in Freetown, Sierra Leone (Tony Binns).

2.4 Population movement and redistribution

Apart from natural increase in population, the other main factor affecting the size and distribution of population is redistribution, mainly through population movements of one sort or another. Movements of people within and between the countries of Africa have in some cases been occurring for centuries. We can distinguish between two main types of movement – migration and circulation. Migrations are usually permanent or irregular, but involve a lengthy change of residence, often longer than a year. Circulations are usually shorter, sometimes daily, periodic or seasonal, but can be long term, too (Stock, 2004).

There are many examples of migrations in Africa: some are voluntary, while others are enforced. In West Africa, the movement of Mossi men from Burkina Faso to help with the cocoa harvest in Ghana and Ivory Coast has been going on for many years, and the money they earn is a vital lifeline for their poor villages. Others migrate to towns for education or to attend a hospital, while a large number of young people are attracted by the 'bright lights' and relatively modern facilities to be found in towns, particularly capital cities such as Lagos, Nairobi and Addis Ababa, all of which have experienced rapid growth as a direct result. Economics is the main cause of rural–urban migration in most of Africa (Binns, 1994). For many, a spell in a big city is regarded as an initiation into adulthood and Western culture, as reflected in the experiences of migrant workers in the gold, coal and copper mines of southern Africa.

In central and southern Africa, important historic migrations involved the movement of thousands of people, mainly young males, to the industrial areas of South Africa. At any one time, a large number of men from Lesotho, Zambia, Zimbabwe, Malawi and Mozambique are absent from their homes, and all agricultural and domestic work has to be done by the women who remain behind.

Africa also has many examples of refugees who have migrated to avoid civil war and persecution in their homelands. Angola, Mozambique, Somalia, Rwanda, the DRC, southern Sudan and Ethiopia are just some of the many areas from which refugees have migrated in search of sanctuary. In the 1980s, large numbers of refugees crossed from Ethiopia into eastern Sudan and from Chad and Uganda into western and southern Sudan. In the 1970s and 1980s, large numbers of refugees abandoned their homes in Mozambique, escaping the violent attacks of RENAMO on the civilian population and the general state of famine in that country. Malawi was the main destination of these refugees. More recently, in the 1990s, the civil wars in Liberia and Sierra Leone caused many refugees to cross into neighbouring countries, such as Guinea. Other enforced migrations include the repatriation of Ghanaians from Nigeria and Mauritanians from Senegal. More recent movements include the flight of refugees from the DRC into Uganda and from southern Sudan and Somalia into Kenya (UNECA, 2006).

One longstanding, long-distance migration involves the movement of pilgrims to the holy city of Mecca in Saudi Arabia. All Muslims aim to make this pilgrimmage – the Hadj – at least once during their lives. Whereas this migration was once done by foot and took many months, today the round trip to Mecca can be done in a few hours by air. At Kano airport in northern Nigeria, for example, a whole section of the terminal is devoted to the movement of thousands of pilgrims each year.

Various types of circulation usually take place over a shorter distance and time-span than migration. Movements between rural and urban areas are particularly common and involve large numbers of people. On a daily basis, the journey to work in town from surrounding rural areas is widespread. In many African countries public transport is at best inadequate and often non-existent. It is common to see people travelling to work in taxis, converted trucks and minibuses, on bicycles or on foot. Large numbers of people in the so-called 'peri-urban' areas are dependent on urban employment for their livelihoods.

Others make the regular journey into town to sell produce grown in market gardens and on farms. In Sierra Leone, for example, villages on the mountain slopes behind the capital, Freetown, conduct intensive market gardening of fruit and vegetables. Each morning before dawn women head-load heavy baskets of produce down the hill to markets in the capital (Binns, 1994). Others hire taxis or trucks to carry their foodstuffs. Around Kano in northern Nigeria, a similar intensive production of vegetables takes place, in this case making use of low-lying depressions known as *fadamas*, which in many cases remain cultivable throughout the long dry season. Similarly, close to Banjul in The Gambia, the intensive growing of vegetables, mainly by women for the urban market and tourist hotels, has increased in recent years. Around Harare in Zimbabwe many families maintain rural plots of land on which they grow crops that help them survive, as wages are often insufficient to meet all household needs. This involves semi-regular movement in and out of the city. In South Africa, daily and weekly commuting was initially enforced by apartheid, which displaced workers to racial reserves distant from their places of employment. In the post-apartheid era, these movements continue due to the established nature of the dispersed settlement pattern. This leads to workers either commuting over significant distances on a daily basis or maintaining a rural home that they may visit only at the weekend and basing themselves in the city during the working week. In northern Nigeria, where there is a prolonged dry season from October to June, some people, mainly men, migrate to the towns for long periods and take up employment there – a long-established practice known as *cin rani*. Meanwhile, movement of people from urban to rural areas is associated with town-based traders taking manufactured items to rural villages, or townspeople visiting relatives in the countryside.

Circulation within Africa's rural areas is surprisingly widespread, in spite of the poor roads and public transport systems. Farmers, for example, will often walk long distances to get to their farms. Their journey might take more than two hours in each direction, so they frequently leave their village before dawn and return after dark. In areas where rotational bush fallowing is practised, the farm may be in a different location each year. Movement between rural villages is also common, particularly where markets attract sellers and buyers from surrounding areas. Rural people may also leave their villages to assist with agricultural work, such as cash crop harvesting, in other rural areas. There are examples within Africa of enforced migration in rural areas, such as the villagization programmes in Tanzania (see Box 9.3, below) and more recently in Ethiopia (Stock, 2004). Rural non-farm employment may also draw people away from their home villages. In the Eastern Province of Sierra Leone, for example, large numbers of male farmers engage in alluvial diamond mining during the dry season, returning to their villages at the onset of the rains to prepare their farms for cultivation. A lucky diamond discovery can make a valuable addition to the income of a rural family.

Population movements, whether voluntary or enforced, short or long distance, permanent or temporary, have had a marked effect on the distribution of population in Africa. This can lead to gender imbalances in various locations, such as where rural areas have a female predominance as a result of temporary male outmigration in search of work. Table 2.2, above, details the net migration rate per 1000 people in Africa. It is apparent from the table that in overall terms the continent is experiencing a population loss of one person per 1000 as a result of migration. At the national levels, economically stressed Zimbabwe has the highest out-migration rate of people in the continent at 20 per 1000 people, followed by Somalia at 6 per 1000 and Lesotho at 4 per 1000. In terms of recipient territories, Western Sahara has a net in-migration rate of 20 per 1000, well ahead of the next highest, the Seychelles, at 6 per 1000, closely followed by Liberia and Burundi, both at 5 per 1000. The table reflects the reality that most countries on the continent have a fluid migration structure that is the direct result of economic, social, environmental and political factors which variously attract or repel residents.

2.5 Population policies

In the Global South, rapid population growth is often seen as the major barrier to development and the associated improvement of living standards. However, there are two important aspects to be considered. First, it is unwise to generalize, since in Africa there is much variation from one country or region to another in terms of population pressure on available resources. Second, the reasons behind rapid population growth must be understood. With such high infant mortality in many African countries, there has long been an incentive for parents to have more children in the hope that some will survive. Another factor is the early age of marriage and first child-bearing, though this should be considered alongside the relatively short life expectancy compared with countries in the North (see above). In some societies polygamy is common and status is attached to males having a number of wives and their offspring. Children also provide important help in farming and domestic duties, such as fetching water and firewood. Finally, in countries where there is no social security system, parents depend on their children to care for them in old age. In most African societies old people are respected members of the community (Binns, 1994).

These facts must be borne in mind when discussing population growth and control. Some would say that African people are poor because they have too many children. Others argue that as living standards and education improve family size will decline, as it has done in the countries of Western Europe and North America since the late nineteenth century. Indeed, there is clear evidence that over the last forty years the population growth rate in Africa has declined, and there is growth in the size of the middle class, suggesting that Africa is finally starting to shake off the shackles of low economic growth.

Although by world standards Africa is sparsely populated, with only 33 persons per sq km in 2009 (see Table 2.3), compared with 170 per sq km in Western Europe and 129 per sq km in Asia, over half of Africa is environmentally fragile and much of the land currently under cultivation is marginal. However, it is estimated that by 2025, given current growth rates, Africa's population density in habitable areas will be the same as that of present-day Europe. There is, therefore, an urgent need for African countries to consider taking steps to reduce population growth, yet many still have only weak or non-existent family-planning programmes. In 1994, the Organization of African Unity (OAU), in its Population and Development Policy Programme, encouraged all states to institute population policies. The fundamental link between population and development has been highlighted by the United Nations Economic Commission for Africa (UNECA, 2001: 11), which argued, 'Today the population issue is a development issue that calls for a long-term and holistic approach in the context of sustainable development . . . Efforts to slow population growth, reduce poverty, achieve economic progress, [and] ensure environmental protection . . . are mutually reinforcing.'

Ghana and Kenya were the first countries to institute population policies, in 1965 and 1967, respectively. Thereafter, surprisingly, there was a twenty-year lag before a limited number of other African countries drafted national population policy statements: namely, Liberia, Nigeria and Senegal (all 1988), Sierra Leone and Zambia (both 1989) and Sudan (1990). In the 1990s, another twenty-five countries initiated policies. According to Sullivan (2006), the first countries did so because of demographic pressures and economic stagnation; however, from the 1990s, policies were instituted as a result of pressure from and loan condition requirements imposed by the World Bank, suggesting probable low levels of commitment to these policies. This is of serious concern, given the urgency of responding to both the challenges of rapid population growth and the HIV/AIDS pandemic. Financial constraints and weak infrastructure and services have been identified as additional barriers to policy implementation (UNECA, 2001). While many countries are in different stages of formulating a population policy, the effects of the HIV/AIDS pandemic over the last twenty years have understandably commanded the attention of the medical profession and planners in Africa, and both merit and exercise a clear influence over current policy thinking. The tension between population and resources, together with a better understanding of demographic mechanisms, has meant that family planning is no longer perceived as a Western practice imposed upon African societies but, on the contrary, as a way to reinstate traditional African birth-spacing patterns. According to allAfrica.com (2009) 'reproduction in Africa is a cultural issue', and 'only cultural institutions have the ability to change people's perceptions', indicating the need to encourage African responses to existing demographic challenges.

2.6 Culture and ethnicity

Different ethnic origins are reflected particularly in language. In the continent as a whole there are some 1000 languages, divided into about 100 language groups. There is considerable language diversity in some West African countries, such as Nigeria, where no fewer than 395 different languages have been identified, and in some cases a single language, such as Yoruba, may have many dialects. Even a country such as Sierra Leone, which is smaller than Scotland, has sixteen indigenous languages. In contrast, in southern and central Africa there is less language diversity, with languages such as Swahili spoken over a wide area. In addition to Swahili, only three other languages – Amharic, Arabic and Hausa – have more than local/national significance. As a result, the languages of the colonial rulers – notably English, French and to a lesser extent Portuguese – have become widespread and in common use for official government business (Binns, 1994; Fellman *et al.*, 2007). In terms of national languages, while it is not uncommon to have more than one national language, South Africa is something of an exception in having thirteen. Understandably, many governments today are keen to preserve traditional languages alongside European ones, and appropriate policy support and research are in place.

Religion is another aspect of the cultural make-up of Africa's people. Traditional religions with ancestral cults, magic and secret societies still exist in many parts of Africa, but these have occasionally been eradicated, and more commonly overlain, by Islam and Christianity. Islam, which entered Egypt in AD 640, spread rapidly westwards but took some time to penetrate south of the Sahara. From the late eighteenth century its influence spread into the Sahel and the savanna zone of West Africa, while on the eastern coast of Africa Islam flourished due to regular contact with Arab traders (Fellmann *et al.*, 2007). Since the Second World War it seems that the spread of Islam has continued, particularly into southern parts of West Africa, which during the early colonial period were strongly influenced by Christian missionaries. In spite of possibly losing some ground to Islam, Christianity in various forms continues to be widespread in southern Africa as well as in Tanzania, Uganda, Zaire and southern Nigeria, notably in the densely settled homelands of the Yoruba and Ibo (Binns, 1994). While indigenous Christianity has existed in north-eastern Africa, and most notably in Ethiopia, for nearly 2000 years, it was only from the arrival of missionaries in the seventeenth century, and especially in the colonial era, that Christianity started to have a dramatic impact on beliefs and religious practices in most parts of the continent. It was estimated in 2001 that 45 per cent of Africa's population were Christian and 40.6 per cent were Muslim (World Christian Encyclopedia, 2001).

Although the term 'tribe' is often associated with the peoples of Africa, many states have tried to play down tribal divisions since independence in an effort to promote national unity. In Zambia, for example, President Kaunda strove to build a non-tribal society with English as the unifying national language. In many African countries, however, ethnic diversity and, in particular, tribal rivalries have caused internal instability and even civil war (see Chapter 7). One of the most tragic expressions of this was the Rwanda genocide of the mid-1990s which saw the deaths of hundreds of thousands of people. Other examples include the Nigerian civil war of 1967–1970, which was caused by the Ibo people in the south-eastern part of the country setting up the separate state of Biafra and demanding retention of valuable revenues from oil produced in the region. In Liberia, the minority Americo-Liberians controlled much of the economy and society until a military coup in April 1980 overthrew William Tolbert and the True Whig Party (Binns, 1994). Master Sergeant Samuel Doe assumed power, but his own overthrow and assassination ten years later was itself the product of intense tribal rivalries as well as Doe's unpopular and harsh dictatorship. In Sudan, conflict between the north and the south has been due largely to the Khartoum-based government imposing the Arabic language and Islamic sharia law on the predominantly non-Muslim population of the south. In neighbouring Ethiopia, the former Italian colony of Eritrea fought a thirty-year war against successive Ethiopian governments before finally gaining its independence in 1993. Meanwhile, in the Ethiopian province of Tigray, the Tigray People's Liberation Front has also been struggling for greater autonomy for many years. As we will see in Chapter 7, such instability and civil wars have had devastating effects on fragile African economies and have caused poverty, misery and homelessness for many innocent people (Maconachie and Binns, 2007a).

2.7 The future of Africa's population

Between 2000 and 2050, the world population will probably increase by 3 billion people, and no less than 40 per cent of that growth will occur in Africa (UNECA, 2001). This means that Africa's population may increase by more than 100 per cent over the next forty years, rising to 2.2 billion people; that Africa will occupy a larger share of the global population; and, in contrast to most other parts of the world, that Africa will continue to have an extremely youthful population and continue to face many of the challenges that it is currently experiencing. Unless living standards and food supply improve, many of Africa's current development challenges will persist.

Views on the future of Africa tend to be either pessimistic or optimistic. According to allAfrica.com (2009), unless something is done to address the continent's 'uncontrolled growth rate', Africa will be left 'sitting on a time bomb'. This article highlights the reality that unemployment is likely to grow faster than employment opportunities and that there is a direct correlation between high rates of population growth and low economic development. For much of the last forty years of the twentieth century, Africa's population grew at above 2.8 per cent per annum, but its economy only grew at 1.9 per cent per annum, creating an ever-increasing wealth gap.

While it is easy to reach negative conclusions about the future of the continent, one must also acknowledge that positive demographic change is slowly taking place. While the population growth for Africa in the period from 1980 to 2000 averaged 2.7 per cent per annum, by 2009 the rate of increase had fallen, significantly, to 2.4 per cent (Stock, 2004; PRB, 2009). Other noteworthy improvements include: a significant decline in the infant mortality rate from 156 per 1000 births in 1975 to 74 in 2009; average life expectancy has risen from 45 to 55 years; the crude death rate has fallen from 19.8 to 12 per 1000; and the crude birth rate has fallen from 46.4 to 36 per 1000. In addition, the percentage of the population under the age of fifteen has fallen from 44 per cent to 41 per cent (Best and de Blij, 1977; PRB, 2009). Further, the total fertility rate has fallen from 6.7 children per woman in 1950 to 5.3 in 2009 (Africa News, 2009). Gradual slowing of population growth rates may well provide an opportunity for the traditionally slower-growing economy to catch up and gradually start to level the playing field. Support for such an argument comes from Ashford (2007), who argues that Africa's currently youthful population will provide a 'demographic dividend' when this group joins the workforce and has to support a declining number of people under the age of fifteen. Falling fertility rates, improving the educational system and reinvigorating family-planning programmes would greatly assist in raising overall levels of income and welfare.

Such considerations have contributed to the arguments of the optimists, such as the Mo Ibrahim Foundation, which argues that Africa is in fact 'under-populated' as it has 13 per cent of the world's population and 20 per cent of its land mass. The foundation points to the continent's rich resources and the reality that its economy is growing at nearly 5 per cent per annum – a faster rate than that of the Global North (average African incomes are also growing faster than average incomes in the North) (Perkins, 2010). On a similar theme, *The Economist* (2009) argues that while Africa is a demographic outlier in the global sense, 'another Africa' – where people live longer, have fewer children and have better economic prospects – is emerging. So, while very significant environmental, political and institutional challenges persist, there are grounds for some optimism about the future. According to the UN Population Division (as reported in *The Economist*, 2009), if Africa's fertility levels had remained at their 1970s levels, the continent's population would have been 8 per cent larger than it currently is. On an equally positive note, the *Guardian* (2010a) reports, 'Africa's untold story of a booming continent and a growing middle class.' This is a direct result of a continental economic growth rate of between 4.5 and 6 per cent experienced since 2003, and the relatively limited impact that the 2008–2010 global recession had on Africa (see Box 10.1, below). This, together with improved political stability, rapidly growing Chinese investment (see Box 10.5, below) and a fourfold increase in investment and trade in the last ten years, has led to a third of the population joining the middle class, which, if the pattern followed elsewhere in the world is repeated, should lead to reduced average family size and improved life prospects. While this view may be overly optimistic, and such benefits may well remain only a distant dream for the foreseeable future for hundreds of millions of people, the fact that such an argument can even be made clearly shows that the status quo is changing in Africa and that both the economy and the demography of the continent are capable of undergoing significant shifts.

In conclusion, it is apparent that the population of Africa is in many ways an 'outlier' relative to the other continents. Above-average scores on most demographic growth indicators and below-average development scores, coupled with high rates of urbanization and poor nutritional levels, are causes for real concern. The slow implementation of population-control policies, the negative effects of HIV/AIDS and poor policy choices have all taken their toll. That said, as information discussed in this section suggests, the situation is not cast in concrete. There is evidence of falling population growth rates and improving economic prospects. If both of these trends continue and are supported, there is hope that demographic and welfare conditions may start to show significant improvements in coming decades. However, if this cannot be achieved, the consequences could be dire.

Summary

1. Africa's population exceeded 1 billion for the first time in 2009.
2. Questions can be raised about the reliability of census data.
3. Regional variations in growth rates exist between the key regions of Africa, while growth rates have also fluctuated over time.
4. Africa performs poorly on a range of key demographic and development indicators, such as the Human Development Index.

5 Urbanization and in particular the growth of the large cities pose particular human development challenges.
6 Population policies cannot be separated from cultural considerations.

Discussion questions

1 Compare and contrast key demographic trends in the different regions of Africa.
2 What are the implications for the future of higher population growth rates and low HDI scores?
3 What effects have technological improvements had on human migration and associated urban and rural lifestyles?
4 Critically evaluate whether Africa is 'under-populated'?

Further reading

Food and Agricultural Organization (FAO) (2009) *FAO Statistical Yearbook*, Rome: FAO.
Potter, R., Binns, T., Elliot, J.A. and Smith, D. (1999) *Geographies of Development* (1st edn), Harlow: Adison Wesley, Longman.
Potter, R.B., Binns, T., Elliot, J.A. and Smith, D. (2008) *Geographies of Development* (3rd edn), Harlow: Pearson.
Stock, R. (2004) *Africa South of the Sahara* (2nd edn), New York: Guilford Press.
Sullivan, R. (2006) 'The Politics of Population Policy Adoption in Sub Saharan Africa', paper presented to the African Sociological Association Conference, Montreal, 10 August.
United Nations Development Programme (UNDP) (2009) *Human Development Report 2009*, New York: Palgrave Macmillan.

Useful websites

Population Reference Bureau: http://www.prb.org/. Provides a useful overview of global and national population trends and key debates.
United Nations Development Programme: http://www.undp.org/. Provides access to the key human development reports and a range of other useful data sets.
US Census Bureau (2009) *Census Data for Countries and Areas of Africa: 1945 to 2014*. http://www.census.gov/ipc/www/cendates/cenafric.html. Provides Africa-specific information.

3 African environments

3.1 Introduction

Environmental issues in Africa have received considerable attention in recent years. For example, the World Bank's Africa Action Plan (World Bank, 2007a) identifies one of its goals as reducing the cost of environmental degradation and improving the management of non-renewable resources. The Action Plan states,

> the Bank has valuable international knowledge on environmental matters, coupled with recent experience in building country systems for environmental safeguards. Using these resources it can help countries to develop strategies to adapt to climate change and assist them in developing action plans for sustainable environment practices in key productive sectors.
>
> (World Bank, 2007a: 11)

Several of the Millennium Development Goals (MDGs) adopted by the United Nations in September 2000 relate to the environment. MDG 7 is concerned specifically with ensuring environmental sustainability, and includes such aspects as the sustainable management of forest resources, preserving biological diversity, ensuring access to clean water and sanitation and the reduction of slum populations.

Africa's environments have frequently been blamed for many of Africa's problems. 'Diverse', 'inhospitable' and 'unpredictable' are all descriptions that are used as much today as they were over a century ago, when the European powers were examining the possibilities of colonizing the so-called 'dark continent' (see Figure 3.1). But such criticism of the environment is unwarranted, since what may be perceived as a purely environmental issue often has complex social, economic and political elements to it. Famine, for example, can rarely be explained in purely environmental terms, since civil war, economic instability and government policy have played greater or lesser roles in different places and at different times.

However, given the low level of technology and development in much of Africa, people are more vulnerable to the vagaries of environment here than they are in richer, more technologically advanced countries. In a continent where well over half the population works on the land, any environmental 'shock', such as a year with below-average rainfall, flooding or a pest attack (for example from locusts or quelea birds), may have a profound effect on food production and livelihoods.

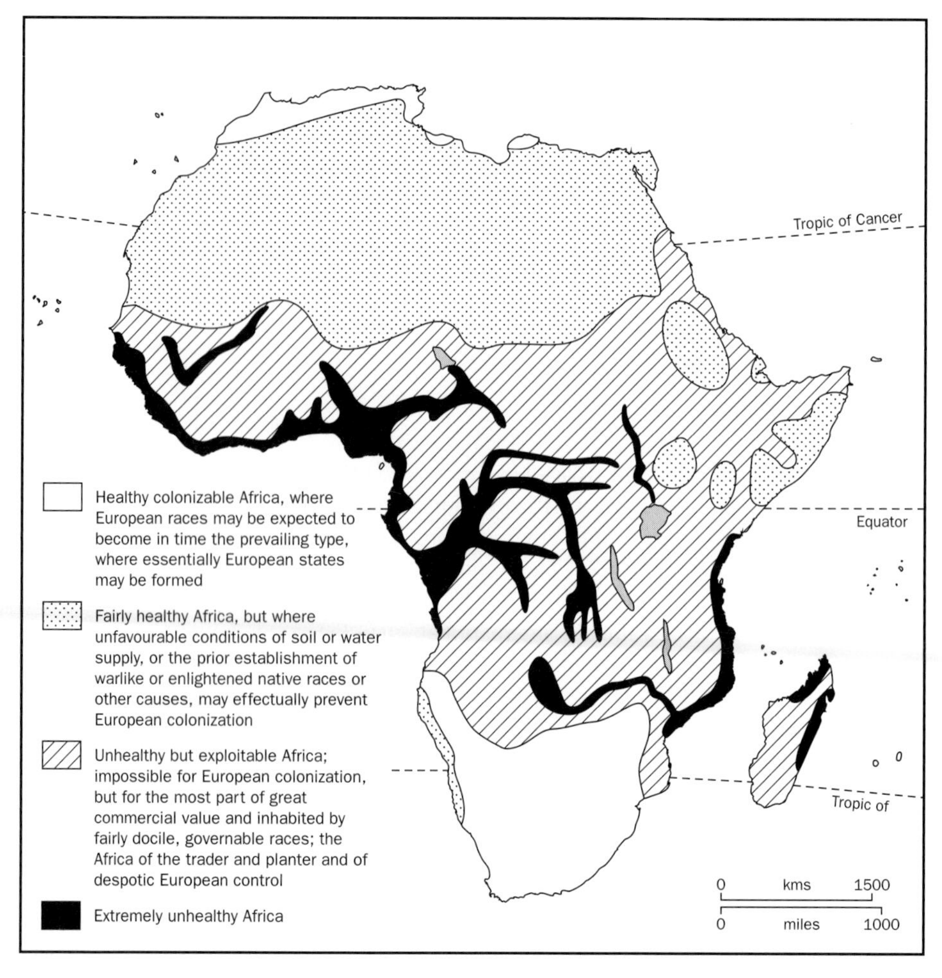

Figure 3.1 Colonizability of Africa.

Source: Adapted from Johnston, 1899

3.2 The diversity of environment

The African continent contains a great diversity of environments, ranging from the great expanse of the Sahara Desert, through the semi-arid areas of the Sahel to the savanna grasslands, rainforest and mangrove swamps. In certain places environment is affected by altitude, such as in the Ethiopian and Kenyan highlands, the Atlas Mountains of Morocco and the Drakensberg of Lesotho and South Africa, where it is cooler and a range of plant types more typical of temperate areas is to be found. Even within a single country, such as Sudan, Africa's largest state, there is an impressive variety of environments – from the moist forest and vast swamps of the Sudd in the south, to semi-arid grasslands and ultimately desert in the far north (see Figure 3.2).

Climate, and particularly rainfall, is the dominant factor affecting environment. Indeed, it might be said that 'water' is the most significant factor affecting much of Africa's survival and future development (see Boxes 3.1, 3.2, 3.3 and 3.4). Proximity to the equator, altitude and distance from the sea are other important factors. Despite the great environmental diversity across the continent, seven major biomes or ecological regions can be identified:

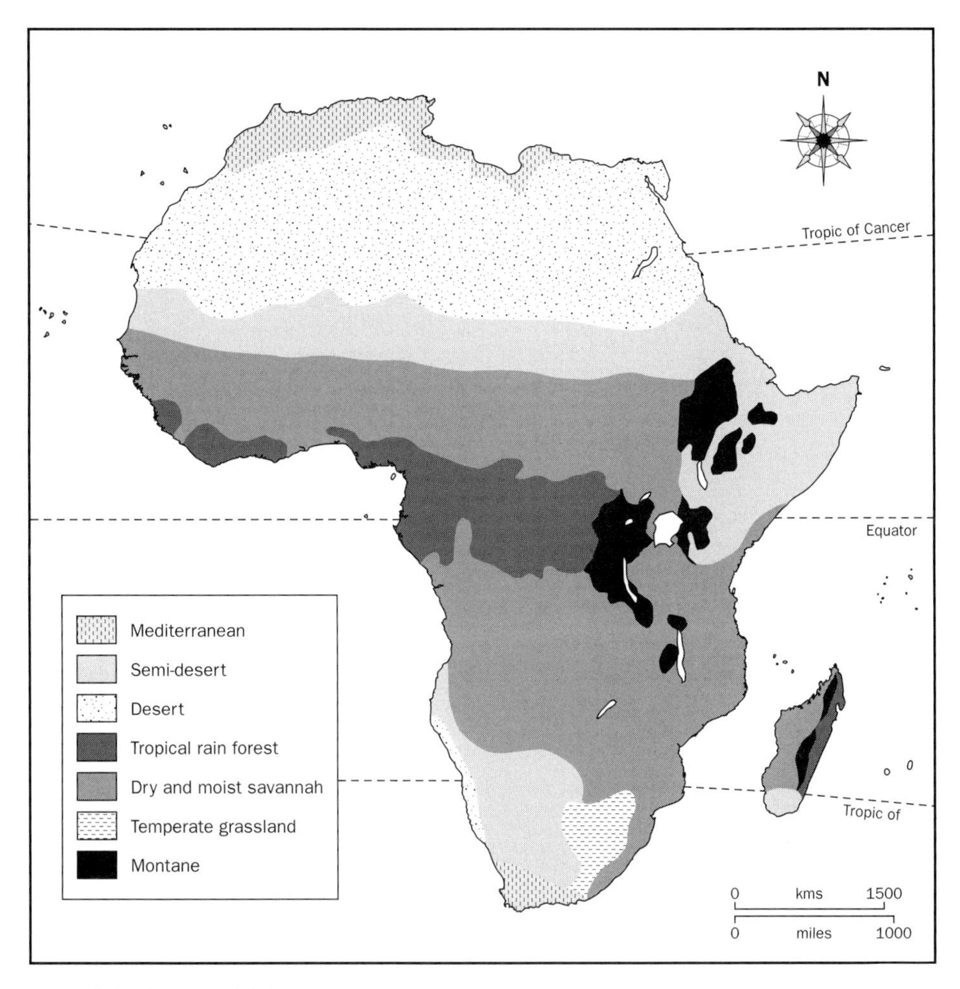

Figure 3.2 Biomes of Africa

Source: Adapted and redrawn from UNEP, 2008

1 *Mediterranean.* This biome is found in North Africa, along the coastal belt from Morocco to Libya, and in southern Africa, along the southern coast of South Africa. The Cape Floral Kingdom in South Africa, whose typical vegetation is known as *fynbos*, is the only one of the world's six floral kingdoms to be restricted to a single country. The region is noted for its considerable biodiversity, with over 6000 endemic plant species (see Box 3.2).
2 *Semi-desert.* This is a region of transition between savanna and desert. In northern Africa it is referred to as the Sahel, while in southern Africa the Karoo and the Kalahari are characteristic areas of this biome.
3 *Dry and moist savanna.* This is the largest environmental region in Africa, and is characterized by a significant dry season, with three to eight months of drought, and heavy rainfall at other times. Tree species and density of tree cover vary according to the amount of precipitation, as well as soils, wildlife, fire and human population. This biome is sometimes divided into two or more sub-regions, according to rainfall levels and species types. Dry savanna, sometimes called 'Sudan savanna', is common in drier areas, while moist or 'Guinea savanna' is more typical closer to the equator. In West Africa, the savanna zone is often subdivided from south to north with decreasing

rainfall into Guinea savanna, Sudan savanna and Sahel savanna, each of which has a distinctive range of plant species.

4 *Tropical rainforest*. This is tall, dense forest, with high biodiversity and characterized by marked layers with a few tall trees emerging above a dense closed canopy. Such forest is found in areas with more than 1400 mm of rainfall and no drought period, mainly on the western side of the continent, in Cameroon and the Congo Basin, and in eastern Madagascar.

5 *Desert*. This region is characterized by sparse vegetation, poor soils, high temperatures and low rainfall. The Sahara, covering over 9 million sq km, is the world's largest hot desert, and stretches across northern Africa from the Atlantic Ocean in the west to the Red Sea in the east. The smaller Namib Desert is situated in south-west Africa. With cold offshore currents in both north-west and south-west Africa, the desert extends right up to the coastline in both regions.

6 *Temperate grassland*. This biome is at its largest extent in southern Africa, where an interior area of high elevation and moderate rainfall produces a landscape that is largely covered by grasses and occasional trees. In South Africa, much of the area is referred to as 'veld' or 'veldt' – an Afrikaans term meaning 'field'.

7 *Montane*. This biome is found in isolated mountainous areas, such as the Ethiopian Highlands and parts of East Africa. Mount Kilimanjaro, Africa's highest mountain (5895 metres), on the Kenyan–Tanzanian border, like other areas of montane vegetation, has a series of marked vegetation zones, with distinctive species that change as elevation increases and temperatures fall.

Box 3.1

Exporting water from Lesotho to South Africa

Large parts of South Africa are semi-arid, and the country has a serious water shortage, with one of the highest water stress indexes in the whole of Africa. Water stress, according to the UNEP (2002), occurs where there is less than 1700 cubic metres of potable water available to each person per annum. With a population of 47 million, and an annual population growth rate of 1.1 per cent (World Bank, 2008a), each person in South Africa has only slightly more than 1000 cubic metres of water per annum. The economic heartland of South Africa, the province of Gauteng, which has some 9.6 million people and includes the capital (Pretoria) and the largest city (Johannesburg), has an insatiable demand for water for drinking, industry and mining.

Less than 400 km to the south of Gauteng is the small (30,355 sq km) independent state of Lesotho, which is completely surrounded by South Africa. Often referred to as 'the mountain kingdom', Lesotho has a population of 1.8 million and, with limited local job opportunities, it has long been a supplier of migrant workers to its larger and considerably more wealthy neighbour. However, with the closure of South African gold and other mines in the last decade, the number of mine migrants from Lesotho in 2005 had fallen to 52,450, almost half of the 1996 figure (Lesotho Bureau of Statistics, 2008). Lesotho also has the dubious distinction of having one of the world's highest incidences of HIV/AIDS, with 23 per cent of adults between 15 and 49 years living with HIV in 2005 (UNAIDS, 2007; see Figure 3.3).

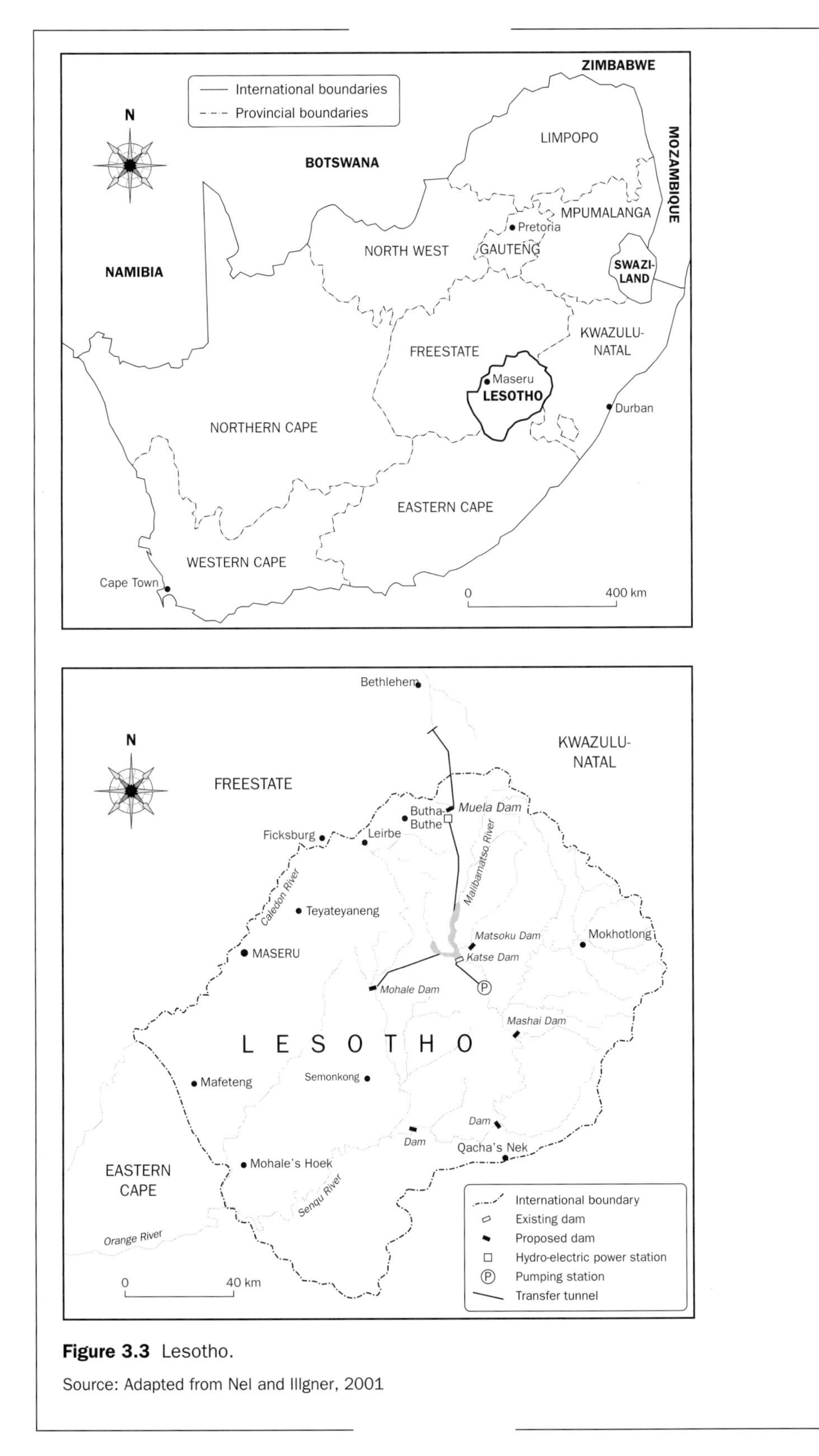

Figure 3.3 Lesotho.

Source: Adapted from Nel and Illgner, 2001

In addition to its migrant labour, Lesotho has another valuable commodity that is desperately needed in neighbouring South Africa: water. Mean annual rainfall in Lesotho ranges from less than 600 mm in the western lowlands to between 1000 and 1900 mm in the mountains that cover 70 per cent of the country and rise to over 3400 metres near the eastern border. The Senqu/Orange River, South Africa's longest river and largest river basin, rises in the Lesotho highlands and flows westward for 2200 km to the Atlantic Ocean.

The possibility of transferring water from Lesotho to South Africa was considered as early as the 1950s, and a detailed feasibility study was undertaken between 1983 and 1986. In October 1986, the foreign ministers of Lesotho and South Africa signed a treaty leading to the development of the Lesotho Highlands Water Project (LHWP, 2008). This $8 billion project is jointly funded by the Development Bank of Southern Africa, the World Bank, the European Union and a number of other sources (Nel and Illgner, 2001). The LHWP, which envisages several phases of development up to 2020, will involve building six dams and two hydro-electric power stations, with a power generation capacity of 276 mw. The Katse Dam (1950 million cubic metres) was first to be completed in January 1998, and the underground Muela power station started generating electricity the following year. In 2002, the Mohale Dam (958 million cubic metres) and a 32-km transfer tunnel between the dam and the Katse reservoir were completed. Over 200 km of water transfer tunnels are planned.

The project provides Lesotho with a source of regular income in exchange for the supply of water to Gauteng Province, and Lesotho is also now self-sufficient in electricity. In the year 2000, the country received $40 million from South Africa for water supply, as well as $400,000 for the export of electricity. The LHWP will involve the construction or upgrading of 650 km of roads, as well as improvements in telecommunications, schools and clinics, and in phase 1A some 35,000–40,000 jobs were created. But the LHWP has also been controversial, as over 20,000 people have been either directly or indirectly affected through the considerable dislocation associated with the flooding of villages, agricultural land and grazing pastures. There have also been numerous complaints about levels of compensation, the lack of local consultation and possible dam-induced seismic activity. Somewhat ironically, while Lesotho is exporting vast quantities of water to South Africa, in July 2008 a news report commented,

> Three parched years in a row have drained Lesotho's water sources and thousands of people that are already facing chronic food security risk losing access to water, the spread of disease and death. According to the Lesotho Department of Rural Water Supplies, 30 per cent of water points [bore holes, wells and springs] in rural areas have dried up, [and] . . . it is estimated that 350,000 people would suffer food deficits over the next six months.
>
> (allAfrica.com, 2008b)

Box 3.2

South Africa's Working for Water Programme

About 70 per cent of South Africa's land area is 'semi-arid', receiving less than 600 mm of rainfall per annum, and about 20 per cent of the country receives less than 200 mm. (The USA defines an 'arid' climate as one that receives less than 250 mm of precipitation per annum; USGS, n.d.) In areas with summer rainfall, high rates of evapo-transpiration are common, reducing the availability of surface water for agriculture and human consumption.

In March 1995, a year after the demise of the apartheid regime, the National Water Conservation Campaign was launched by Kader Asmal, Minister of Water Affairs and Forestry, who argued, 'We could not simply continue to build ever more expensive infrastructure [dams and pipelines] with ever diminishing returns. We will have to find other ways of maximising our water supply' (RDP, 1996: 2). Later in 1995, Asmal launched the innovative Working for Water Programme (WFWP), at a cost of R25 million (£4.3 million). This has involved the mass eradication of alien invasive tree species, which in some areas had displaced indigenous vegetation and were consuming disproportionately large amounts of water (Binns *et al.*, 2001; DWAF, 2008).

Since the arrival of Dutch settlers in the Cape in 1652, over 700 tree species and 8000 shrubs and herbaceous species have been introduced into South Africa, of which 153 are regarded as 'invasive'. Some 45 per cent of the plant species brought from Australia after about 1830 are regarded as invasive, most notably the black wattle (*Acacia mearnsii*). The areas of the country under most stress from invasive alien species are: the Cape Floral Kingdom in Western Cape Province, to the north and south of Cape Town, and extending eastwards into Eastern Cape Province; the Natal Midlands in KwaZulu-Natal Province, north-west of Durban; an area between Ulundi and the coast in KwaZulu-Natal, north-east of Durban; the escarpment and low veld in Mpumalanga Province, east of Johannesburg; and the central bush veld area in the southern part of Northern Province, north of Pretoria.

Since its inauguration, WFWP has impressed a number of international commentators, including the US Secretary for the Interior, Bruce Babbitt, who commented to President Thabo Mbeki in February 2000 that 'the Working for Water programme is the world's leading initiative to combat invading alien plants' (quoted in WFWP, 2000: 3). A key feature of WFWP is that much of the clearance work is undertaken by rural-based, historically disadvantaged communities, which typically have high rates of unemployment and low educational attainment. WFWP also targets potentially marginalized groups within these communities. For example, among each community that is involved, the programme aims to achieve a participation rate of 60 per cent women, 20 per cent youth and 5 per cent disabled. An accredited training programme has been developed, in which all team members and contractors are eligible for a minimum of forty-eight training days in a two-year employment cycle. Those attending training sessions receive 75 per cent of the daily equivalent wage and their transport costs are paid. An orientation course explains the aims and principles of WFWP, the methods used, the nature and conditions of employment, rules and regulations, and such issues as acceptable behaviour and teamwork. Further technical training includes courses in machine (e.g. chain-saw) operation, herbicide application and driver training. Life-skills courses are also provided, including HIV/AIDS and cholera awareness, primary healthcare and nutrition, personal finance, race and gender and literacy and numeracy.

The WFWP has led to a notable improvement in water supply, with local people commenting on the fact that there is now more water in streams, rivers and reservoirs. The livelihoods of poor communities have also been improved through the provision of jobs and training in a wide range of technical and transferable skills (Binns *et al.*, 2001). However, further investigation is needed to evaluate whether community participants in WFWP gain long-term, sustainable benefits from their involvement in the programme, in terms of being able to utilize their knowledge and skills in subsequent occupations that can ensure that they and their families stay out of poverty (Buch and Dixon, 2009).

Box 3.3

Wetlands in Ethiopia and Nigeria

Illubabor Zone, western Ethiopia

Illubabor Zone in western Ethiopia (see Figure 3.4) is one of the most fertile regions of this country. Unlike the semi-arid areas of the north, it has a warm, temperate climate with mean annual temperatures of 20 degrees centigrade and rainfall in excess of 1800 mm per year. The natural environment is dominated by large areas of tropical montane rainforest and land cleared for subsistence agriculture. The climate and undulating topography, ranging between 1400 and 2000 metres above sea level, produce steep-sided river valleys and flat, waterlogged valley bottoms. The accumulation of water in these areas has led to the formation of both permanent and seasonal wetlands.

These wetlands are of critical importance to local communities in terms of their ability to support a range of ecosystem services and socio-economically important products. They provide a clean supply of water for local communities and their animals even during drier periods of the year; they act as sponges and slowly release water, thereby regulating stream flows. The wetland vegetation is dominated by a broad-leafed sedge, known locally as *cheffe*, which is traditionally harvested for use as a construction material in local houses, and for use as a floor covering during religious holidays. Some community members also harvest medicinal plants from the wetlands. One of these, known locally as *balawarante*, is boiled and used as a treatment for skin infections.

In recent years, food shortages in the uplands, increasing population pressure and the gradual influence of the market economy have stimulated much interest in the agricultural potential of the wetlands. In contrast to the uplands, where only one crop per year can be grown, the fertile sediment and moisture retention in the wetlands can facilitate up to three crops each year, provided they are drained and managed efficiently. In Illubabor, wetland drainage and cultivation is now a common sight. Poorer farmers who have no land in the uplands are now growing subsistence crops,

Figure 3.4 Location of Illubabor Zone, Ethiopia.

Source:Dixon (2003)

usually maize, in the wetlands. Richer farmers, who use their upland plots for subsistence crops or coffee growing, are cultivating vegetables in the wetlands, specifically for sale at local markets (Dixon, 2003).

However, there are some concerns regarding the sustainable development of Illubabor's wetlands. In some areas, intensive drainage and cultivation have led to falling water-table levels, resulting in wetland degradation and an inability to support agriculture, traditional uses or environmental functions. This has major implications for food security in Illubabor and also for water resources in the region. Many of these wetlands drain into the Sor River, which has a hydro-electric barrage downstream. As wetlands are destroyed, river flows become less predictable, and their contribution to electricity generation is reduced. Illubabor's wetlands also play a small but significant role in regulating water supply in the Nile Basin. The use of these wetlands by local people in this remote part of Africa can potentially have significant knock-on effects for the politically sensitive topic of water security in downstream Sudan and Egypt (see Box 3.4).

Recent research into local management of these wetlands suggests that widespread degradation is not imminent. This is because communities have developed detailed local knowledge of wetland management, and through a process of trial and error many have learned how to manage the wetlands in a sustainable manner. For example, rather than being completely drained, local people prefer to retain areas of *cheffe* vegetation at the head and outlet of wetlands to help retain the supply of water and to provide *cheffe* for the community. In some areas, cattle are banned from entering wetlands, since they can compact and degrade the soil. Some farmers also use different plants as indicators of the 'environmental health' of wetlands. When certain plants start to appear, farmers see this as a sign to stop cultivation and leave the wetlands to rest (fallow) for one or more seasons (Dixon, 2003).

Such examples of sustainable wetland development are, however, variable throughout Illubabor, and sustainable management practices have not evolved to the same extent in all communities. The reasons for this variability remain unclear, but it has been suggested that environmental, socio-economic, ethnic and demographic differences are possibly the key determinants of sustainable wetland management.

The Hadejia–Nguru wetlands, northern Nigeria

The wetlands of the Hadejia–Nguru floodplain cover an area of about 5100 sq km in the northern Nigerian states of Jigawa, Yobe and Bauchi, and are fed by the Hadejia and Jama'are rivers, which merge to form the Yobe River upstream of the town of Gashua. The Hadejia–Nguru wetlands are of considerable environmental significance in providing a vital source of groundwater recharge for large areas of north-east Nigeria, and parts of the area (Nguru Lake, Marma Channel, Dagona Waterfowl Sanctuary and Baturiya Wetlands Reserve) are listed by the Ramsar Convention (see Figure 3.5). The wetlands also have international ornithological significance, both as a breeding area and as an important resting place for migrating birds in the northern winter, with some 377 recorded bird species (BirdLife International, 2008).

The wetlands are home to over 1.5 million people, whose livelihoods are based on farming, pastoralism and fishing. There is extensive wet-season rice farming, as well as flood-recession agriculture and dry-season irrigation, and the major agricultural outputs were valued at about $75 million in 1997 (Eaton and Sarch, 1997). Seasonally waterlogged areas, known locally as *fadamas*, provide a vital resource for the dry-season cultivation of a wide range of crops, including such grains as millet, maize and sorghum and a variety of vegetables. Fish resources are also significant, with an estimated annual catch of 4000–6000 tonnes, valued in 1990 at approximately $6 million (Milligan, 2002). Fish catches are sold in local markets and further afield – in Nigerian cities such as Kano, Lagos and Abuja, and in the neighbouring state of Niger. Like so many wetlands, the Hadejia–Nguru region also produces other important resources which support local livelihoods, such as timber (for fuelwood and construction), doum palm (for mat-making), potash and many wild foods.

Fulani herdsmen are well aware that the Hadejia–Nguru wetlands are an important resource for grazing and watering their animals, particularly in the long dry season, from October to June. In a single year, the number of cattle fluctuates between 230,000 and 500,000, depending on the season, while estimated numbers of sheep and goats increase from 299,000 and 322,000, respectively, in the wet season, to 437,000 and 529,000, respectively, in the dry season. In the mid-1990s, the livestock sector in and around the wetlands probably generated over $5.5 million per annum (Milligan, 2002).

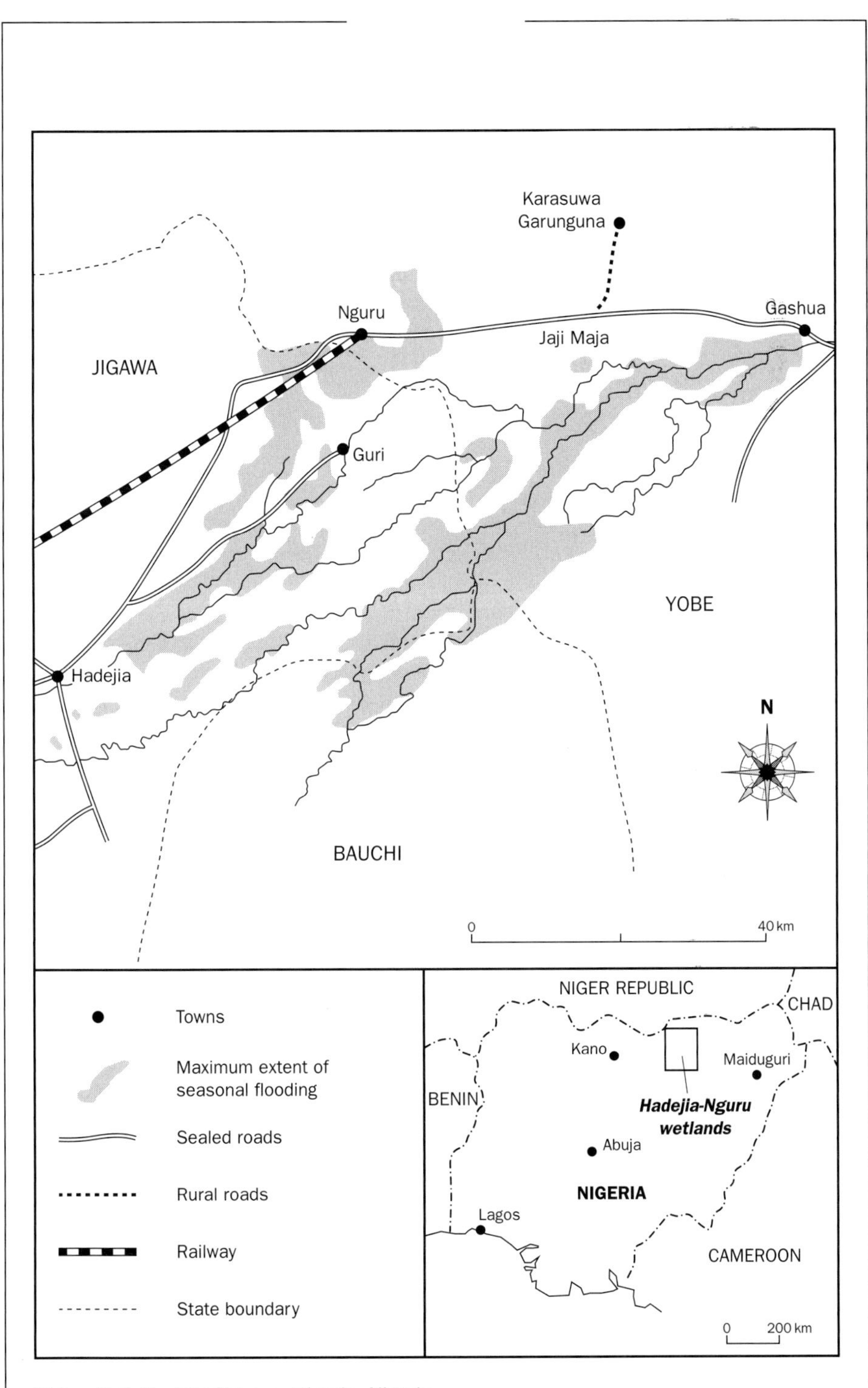

Figure 3.5 Hadejia-Nguru wetlands, Nigeria.

Source: Adapted from Milligan and Binns, 2007

Since the 1970s, a number of large dams have been constructed across semi-arid northern Nigeria, to provide water for growing cities such as Kano, with over 4 million people, and to extend irrigation schemes so that dry-season crops, including wheat, can be grown. The first of these large dams – located at Tiga, 73 km south of Kano – was completed in 1974, and by 1976 it was irrigating wheat, tomatoes and rice in the Kano River Project. Fourteen dams were constructed in Kano State between 1969 and 1978, five more were built between 1979 and 1980, and the Chalawa Gorge Dam was completed in 1992 (Maconachie, 2007).

These dams have undoubtedly had an effect on the wetlands and the sustainability of livelihoods. Water levels in some rivers are now controlled and the absence of an annual flood has had an impact on fish stocks, livestock grazing resources and recessional cultivation in areas which were previously left moist after the retreat of the annual flood. The invasive *typha* grass (a type of reed) has become a major problem in recent years, exacerbated by changes in the water regime, and this has led to increased silting and blocking of channels. Between 2003 and 2008, *typha* invasion in the wetlands increased to over 200 sq km (Wetlands International, 2008). The traditional migratory routes of pastoralists have also been affected, as livestock are no longer welcome where irrigated crops are being grown (Binns and Mortimore, 1989). In recent years, there have been violent clashes between farmers and pastoralists, which in some cases are concerned with competition for scarce resources. This is a complex issue, but changing relationships between people and environment in the Hadejia–Nguru wetlands have undoubtedly had some impact on the sustainability of livelihoods and environments (Milligan and Binns, 2007).

On a more positive note, since 1989 the internationally funded Hadejia–Nguru Wetlands Conservation Project has played a key role in raising awareness of the significance of the wetlands and promoting their sustainable management. In 2008, this initiative was reinvigorated as part of Birdlife International's Wings over Wetlands Project, with support from Wetlands International, the United Nations Environment Programme (UNEP) and other international donors. The new project aims to increase awareness of wetland management, to work with local communities in ecological projects and to reduce *typha* invasion in order to improve waterfowl habitats and floodplain crop production (Wetlands International, 2008).

Box 3.4

The Nile Basin Initiative

The Nile Basin Initiative (NBI) was launched in February 2009 in Dar es Salaam, Tanzania, by the water affairs ministers of the ten countries that share the river. The Nile is the world's longest river at 6695 km, with the White Nile rising in southern Rwanda in the Great Lakes region of central Africa, and the Blue Nile starting at Lake Tana in Ethiopia and flowing north to join the White Nile near Khartoum, the Sudanese capital. The river basin has an area of over 3,349,000 sq km, more than one-tenth of the size of Africa and larger than the Indian sub-continent.

The countries involved in the NBI are: Burundi, the Congo (DRC), Egypt, Ethiopia, Eritrea, Kenya, Rwanda, Sudan, Tanzania and Uganda. The idea of establishing an initiative to promote the development and environmental protection of the Nile Basin originated in 1992 when the water affairs ministers of the DRC, Egypt, Rwanda, Sudan, Tanzania and Uganda formed a Technical Cooperation Committee (TECCONILE). The World Bank has supported the NBI since 1997 and a multi-donor Nile Basin Trust Fund was established to channel these funds efficiently to the NBI (NBI, 2008; World Bank, 2008c). By the end of 2007, multi-donor support to NBI development projects amounted to $700 million. The UK government has given $8.3 million to assist the riparian states in making arrangements for sharing the use and benefits of the Nile and has provided advice on institution building. Additionally, the UK's Department for International Development (DfID) is encouraging grassroots civil society involvement in the NBI through discussion of development priorities via the Nile Basin Discourse (DfID, 2008a).

The agreed vision of the states involved in the NBI is 'to achieve sustainable socio-economic development through the equitable utilization of, and benefit from, the common Nile Basin water resources' (NBI, 2008). A river basin as large as the Nile's has a wide range of associated development issues, including: power, irrigation, catchment and floodplain management, rural incomes, trade, transport and access to basic services.

The NBI focuses specifically on three main areas:

- improving water resource management and development, addressing the implications of climate change and increasing resilience of the states and their people's livelihoods;
- making use of the resources of the Nile to build greater cooperation between countries, rather than allowing it to be a factor in regional tension;
- reducing poverty through increased cooperation on initiatives that lead to economic growth and trade.

The Strategic Action Programme aims: first, to lay the groundwork for cooperative action through a regional programme to build confidence and capacity throughout the basin; and, second, to pursue cooperative development opportunities to realize physical investments and tangible results through sub-basin activities in the Eastern Nile and Nile Equatorial Lakes regions. There is a strong emphasis on cooperation, since action by a single country could threaten the resource base and critical ecosystems in other countries, through downstream effects. Regional cooperation could also be useful in times of stress, such as floods and droughts. Some progress has already been made on the Flood Preparedness and Early Warning Project, which aims to reduce human suffering and damage from flooding. Ethiopia is already sharing flood information with downstream Sudan.

The World Bank suggests that the NBI could well lead to peace dividends through greater regional cooperation. Economies of participating states could become more strongly linked, with cooperation possibly leading to cooperative development in many other areas, such as transport, trade and tourism (World Bank, 2008c). The World Bank's Africa Action Plan, launched in 2005 and updated in 2007, emphasizes the importance of regional integration and specifically mentions the NBI (World Bank, 2007a).

3.3 Rainfall and climatic change

The quantity and timing of rainfall are crucial throughout the continent. A belt of rainfall shifts southwards from equatorial latitudes towards the Tropic of Capricorn between November and April, and northwards towards the Tropic of Cancer between May and October. In January, the highest rainfall totals are to be found in Madagascar and the southern Congo Basin, whereas in July regions to the north of the equator receive most rainfall, such as the Ethiopian Highlands in the East and the coastal states of Liberia, Sierra Leone and Guinea in the West (see Figure 3.6). But as far as human livelihoods are concerned, it is the 'effectiveness' of the rainfall that matters most for agriculture and pastoralism. Whether rainfall is adequate for crop growth depends on the rate at which moisture is taken up by evaporation from the surface and by plant transpiration.

There is also a considerable degree of variability in Africa's rainfall, particularly in desert, semi-desert and savanna areas. For example, between 1968 and 1989 in the Sahelian countries of West Africa, every annual rainfall total was below 'normal' (if a term such as 'normal' can be used). The early twentieth century was also a relatively dry period, with only 1906 and 1909 being 'above normal' in a twenty-one-year period. In Sudan, where relatively good rainfall data exist, two major periods of dryness between 1900 and 1919 and 1965 and 1985 were separated by a generally wet period from 1920 to 1964. Such cycles of wet and dry periods have been detected in many parts of Africa (Binns, 1994; see Plate 3.1).

In recent years, climate change has become widely recognized by the scientific community, though by no means all national governments, as a significant global challenge. In 1988, the First Assessment Report of the Intergovernmental Panel on Climate Change (IPCC) raised concerns about global pollution levels and emissions of greenhouse gases leading to global warming. The IPCCs Second Assessment Report, in 1995, stressed the evidence for ever-increasing emissions, which led to the signing of the Kyoto Protocol in 1997, in which the signatory countries agreed to reduce emissions by 2012. The Third Assessment Report, in 2001, further emphasized the impacts of human-induced climate change and focused on the need for careful adaptation to ameliorate these impacts. It also suggested that climate change was a serious threat to development. More recently, the Stern Review, published in 2006 for the UK Treasury (Stern, 2006), and the IPCC's Fourth Assessment Report, in 2007 (IPCC, 2007), have provided further clear evidence that climate change is happening and is manifested in storm events, such as Hurricane Katrina's devastation of New Orleans in 2005, melting polar glaciers and a variety of other indicators. The IPCC predicts that the average global temperature could increase by about 3 degrees centigrade by the end of the twenty-first century, while sea levels could rise by as much as 59 cm (IPCC, 2007). There is growing concern that while the world's richer countries have been largely responsible for creating atmospheric pollution, the poorest countries are likely to suffer most from its effects.

It is a depressing reality that the 100 countries which are most vulnerable to the effects of climate change, many of which are situated in Africa – the world's poorest continent – have contributed least to world carbon emissions (Huq and Ayers, 2007). The whole of Africa is responsible for only an estimated 4.67 per cent of global emissions, and if South Africa is excluded (it produces 42 per cent of the continent's emissions), then the African contribution to global greenhouse emissions is only 3.2 per cent. Meanwhile, the world's largest polluters – the EU, the USA and China – produce 24.7 per cent, 23.3 per cent and 15.3 per cent, respectively, of global emissions.

While models and predictions for the effects of climate change on Africa are of variable reliability and there is need for much more detailed research, sufficient evidence already exists to indicate that in countries and regions where poverty and vulnerability are already

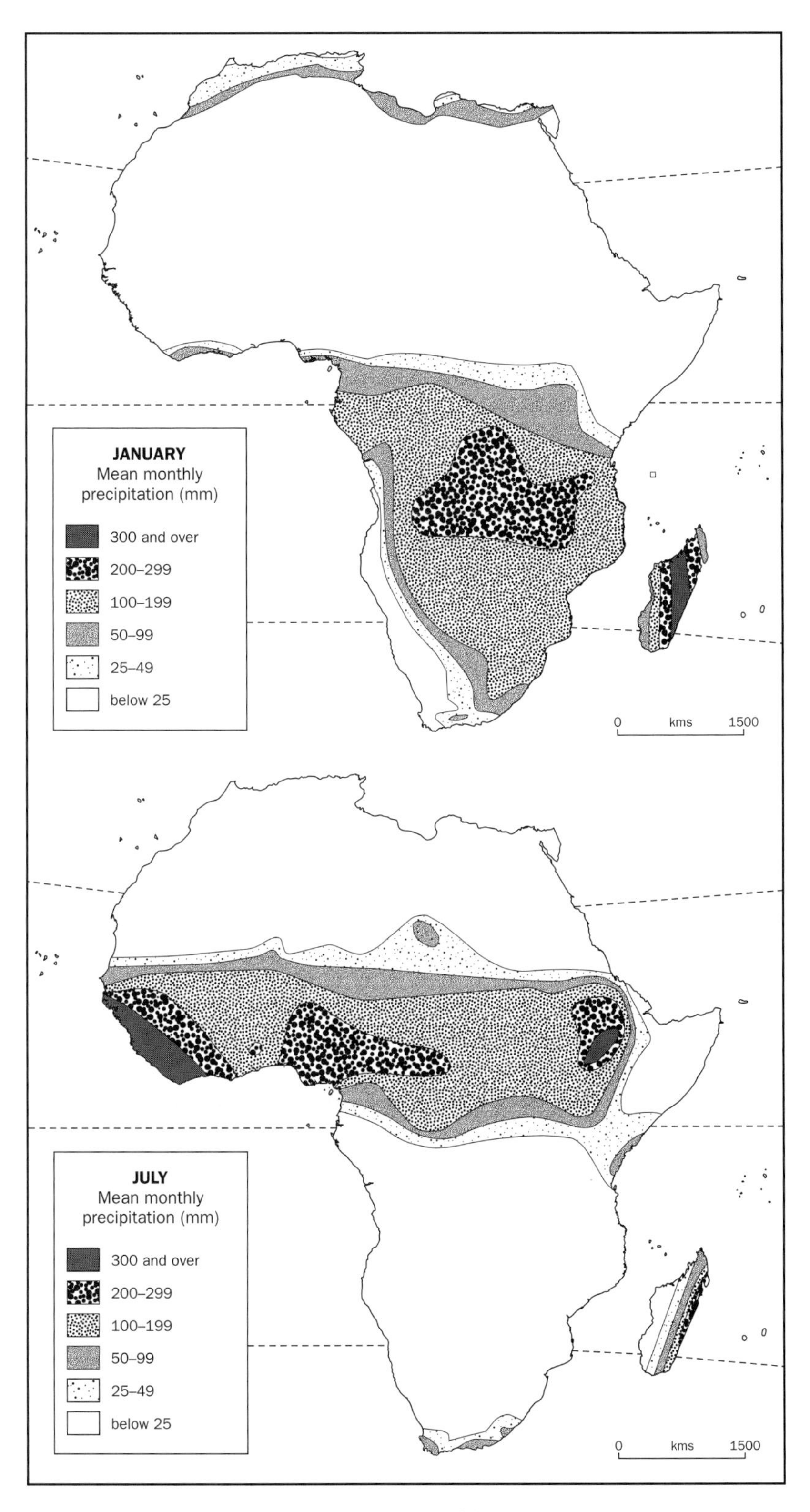

Figure 3.6 Africa, precipitation in January and July.

Source: Binns, 1994

Plate 3.1 Flash flooding on rural road in central Mali (Tony Binns).

features of everyday life, it is likely that climate change will make things worse and could lead to a failure to achieve the Millennium Development Goals by the 2015 target date. While the Fourth Assessment Report shows an awareness of the significance of indigenous knowledge, particularly among poor people living in Africa's marginal environments, there is a need for strategies to be developed at all levels in order to strengthen adaptability to problems associated with climate change and to reduce the vulnerability of households and livelihoods which are likely to be at greatest risk.

More specifically, it is expected that changes in temperature and rainfall regimes in Africa could lead to increasing water stress in North African countries and possible flooding in such areas as the Nile Delta and parts of the East and West African coastlines (see Box 3.4). It is estimated that the area of arid and semi-arid land in Africa could increase by 5–8 per cent (60–90 million hectares), with a decline in maize production from southern Africa and the complete disappearance of wheat production from the continent by the 2080s (IPCC, 2007). The implications of increasing aridity for a country such as South Africa are considerable, and efforts are already being made to maximize the availability of existing water resources by eradicating alien tree and plant species (see Box 3.2), as well as importing water from neighbouring countries, such as Lesotho (see Box 3.1).

The distribution of diseases is also likely to change, with the spread of malaria into previously malaria-free areas such as the East African highlands and a southward spread into South Africa. The economic and social costs will be enormous and the impact of increased malaria in southern Africa, which also has the world's highest incidence of HIV/AIDS, could have a significant effect on levels of life expectancy and mortality (IPCC, 2007). Christian Aid has estimated that by the end of the century some 182 million people could die of disease associated with climate change in sub-Saharan Africa (Christian Aid, 2006).

A number of detailed studies have examined different aspects of the likely impact of climate change in Africa, and are increasingly focusing on developing strategies for adaptation. In Namibia, for example, which contributed less than 0.05 per cent to global carbon dioxide emissions in 1994, temperatures were rising at three times the global average during the twentieth century. It is predicted that rainfall will decline, particularly in the central regions, while higher temperatures will lead to greater evaporation and severe water shortages. The ecological and economic significance of valuable wetland resources such as Etosha and the Okavango Delta in neighbouring Botswana (the world's largest inland delta) could be dramatically affected by climate change. Namibia is particularly dependent on natural resources, and with less plant cover and grassland productivity, it is likely that livestock farmers will be severely affected. It has been estimated that climate change in the next twenty years could result in annual losses to the Namibian economy of between 1 and 6 per cent of GDP – or up to $200 million (Reid *et al.*, 2007).

The scenario which is predicted for Nambia may well be replicated much more widely among pastoral communities in the arid and semi-arid regions of Africa, which are likely to be affected by rising temperatures, decreasing rainfall and more frequent droughts, leading to deterioration in grazing resources and a threat to pastoral livelihoods. Pastoralists may well need to be much more mobile than at present, moving over longer distances, in search of better water and grazing resources for their livestock. But such increased mobility could lead to problems, since increasing competition for scarce resources might fuel conflict between pastoral groups, and between pastoralists and cultivators. There is already growing evidence of such conflict, for example in the Hadejia–Nguru wetlands of northern Nigeria (see Box 3.3), but the situation could become much more serious with deteriorating environmental resource conditions (see, for example, Hesse and Cotula, 2006; Milligan and Binns, 2007).

Rising sea levels, together with more frequent storm surges and flooding, could affect important coastal and riverine habitats, such as mangroves, which are found in West Africa from Senegal to Liberia, and in parts of Mozambique, Tanzania, Kenya and Somalia to the east. Mangroves are salt-tolerant evergreen forests which protect coastal areas from erosion, cyclones and wind, help filter sediments from waterways and can enhance biodiversity in neighbouring habitats, such as coral reefs. They provide valuable nursery areas for fish and shellfish; and, since they boast many species of wildlife and birds, they are important tourism areas, especially in The Gambia and Senegal. Over the last twenty-five years, an estimated 510,000 hectares of mangrove have been cleared in Africa for agriculture, with the timber used for fuel, charcoal and construction purposes (FAO, 2008).

The IPCC's Fourth Assessment Report (2007) also draws attention to the effect of climate change on Africa's urban populations, many of whom live in coastal or delta regions, as in Egypt and West Africa. In the latter, it is predicted that by 2020 a continuous urban megalopolis with 50 million people will stretch 800 km along the coast between Ghana's capital city, Accra, and the Niger Delta in Nigeria. The existence of informal settlements, which commonly lack piped-water supplies, storm drains, sewers and waste collection, could lead to serious problems with rising sea levels, storm surges and flooding. As Reid and Satterthwaite (2007: 1) have indicated, 'many city governments lack the competence and capacity to adapt, and have huge infrastructure backlogs' (see Chapter 5). They advocate the initiation of a range of adaptation measures and working closely with low-income groups to reduce vulnerability, such as community-based settlement upgrading and, where necessary, relocation programmes.

While a continuing reduction in rich-country greenhouse gas emissions and strengthening the Kyoto Protocol beyond 2012 are undoubtedly key items on the climate change agenda, much of the current focus is on the issue of adaptation and, in particular, how poor countries might adapt to the effects of climate change in order to ameliorate its impact on people, environments and economies. There is a burgeoning literature on this topic coming from a variety of organizations and research programmes, such as the UK-based Working Group on Climate Change and Development, which involves collaboration between the New Economics Foundation, the International Institute for Environment and Development (IIED) and several other NGOs, including Oxfam, Friends of the Earth, CAFOD and Greenpeace (WGCCD, 2006). Further reflecting the emphasis on finding ways of adapting to climate change, the Climate Change Adaptation in Africa (CCAA) Research and Capacity Development Programme, a joint initiative launched in 2006 between the International Development Research Centre (IDRC) Canada, and the UK government's Department for International Development (DfID), aims to enhance the capacity of African countries to adapt to climate change in ways that benefit the most vulnerable (IDRC, 2007). The four objectives of the CCAA programme are:

- to strengthen the capacity of African scientists, organizations, decision-makers and others to contribute to adaptation to climate change;
- to support adaptation by rural and urban people, particularly the most vulnerable, through action research;
- to generate a better shared understanding of the findings of scientists and research institutes on climate variability and change;
- to inform policy processes with good-quality science-based knowledge (IDRC, 2007: 6).

The CCAA programme suggests five examples of possible adaptation:

- selecting crop varieties that are tolerant to drought, or that have a shorter growth cycle, to adapt to a shorter growing season in some regions;
- planning urban growth in such a way that new housing is not developed on floodplains;
- developing strategies such as forage resource conservation, and conflict resolution measures to deal with increased competition between farmers and pastoralists, where climate changes affect the movements of nomadic herders;
- adapting fishing practices to changes in marine ecosystems caused by climate change, over-fishing and other environmental drivers;
- adjusting health infrastructure and preventative practice to protect populations in areas where malaria is increasingly prevalent because of rising temperatures (IDRC, 2007: 14).

Tanner and Mitchell (2007) argue that the issue of a pro-poor adaptation agenda should be embedded within wider development debates and practices, with particular emphasis on reducing vulnerability and strengthening resilience among those individuals and groups who might be regarded as chronically poor. With the current widespread interest in devising, embedding and implementing adaptation strategies in Africa, it is important that there is some coordination (rather than duplication) of effort among the many government and non-government agencies involved, and that some tangible benefits are delivered to those who are likely to be most vulnerable.

3.4 Is desertification a myth?

The process of 'desertification' has received much attention in recent years (Binns, 1990; IPCC, 2007). The term was coined in 1949 by a French forester, André Aubréville, but little was heard about it during the 1950s and 1960s, which were generally good rainfall decades in the savanna–Sahel region. During the 1970s and 1980s, however, when different parts of the continent suffered from below-average rainfall in certain years, there was much debate about possible climate change leading to increasing aridity and ultimately the creation of desert-like conditions (see Plate 3.2).

Plate 3.2 Arid landscape, Mali (Tony Binns).

More recently, in the 1990s and early 2000s, the burgeoning literature on climate change has contained numerous references to desertification. For example, the IPCC's Fourth Assessment Report states, 'At present, almost half (46 per cent) of Africa's land area is vulnerable to desertification', and 'Approximately half of the sub-humid and semi-arid parts of the southern African region are at moderate to high risk of desertification' (IPCC, 2007: 442, 439).

In the past, the common image conjured up by desertification was one of an advancing Sahara moving south, smothering villages and destroying farmland and pasture once and for all. It has often been suggested that this process is the result of a long-term decline in rainfall, exacerbated by human practices which are perceived as being unwise and unsustainable, such as overgrazing, burning and deforestation (Binns, 1990).

Reports of an advancing Sahara are by no means new. At a meeting of the Royal Geographical Society on 4 March 1935, Professor E.P. Stebbing, Professor of Forestry at the University of Edinburgh, delivered a much-quoted paper entitled, 'The Encroaching Sahara: The Threat to the West African Colonies' (Stebbing, 1935). Stebbing, reporting on a recent visit he had made to northern Nigeria, spoke of the advancing desert and proposed, as one might expect from a professor of forestry, that two massive forest belts should be planted across West Africa – one through what is now Burkina Faso and northern Nigeria to Lake Chad, and the other further south from Côte d'Ivoire, through Ghana, to Jebba or Minna in Nigeria's middle belt. It was estimated that up to 16 billion trees would have to be planted, though Stebbing encouraged the use of existing mixed deciduous woodland which would be closed and protected from farming, fire and grazing. In a subsequent paper titled, 'The Man-made Desert in Africa: Erosion and Drought', Stebbing (1938) specifically referred to land degradation being caused by human action, notably reducing the length of fallow periods in shifting cultivation systems with increasing population pressure, as well as over-grazing by livestock and the common pastoralists' practice of burning pastures to stimulate new grass growth. However, in 1936–1937, after an unusually wet 1936 rainy season, an Anglo-French Forestry Commission expedition to the area straddling the border of Niger and Nigeria between Niamey and Lake Chad completely refuted many of Stebbing's assertions (Jones, 1938; Stamp, 1940).

More recently, the events which have possibly done most to popularize the use of the term 'desertification' were the two droughts in Africa's savanna–Sahel zone: 1968–1974 and 1979–1984. Following the first drought and associated media coverage, the United Nations Environment Programme organized a World Conference on Desertification (UNCOD) in Nairobi in 1977 to consider the extent and character of the problem and to propose measures to combat it on an international scale (Geist, 2005). UNCOD defined desertification as 'the degradation of land in arid, semi-arid, and dry sub-humid areas. It is caused primarily by human activities and climatic variations' (UNCOD, 1978: 2). A map was produced showing which areas in Africa were thought to be under greatest threat from desertification (see Figure 3.7).

Seven years after UNCOD, a 1984 UNEP report reinforced the gloomy scenario, commenting, 'Desertification threatens 35 per cent of the earth's land surface and 20 per cent of its population; 75 per cent of the threatened area and 60 per cent of the threatened population are already affected through deterioration of the environment and living conditions, and between a quarter and a half of the affected population severely so'

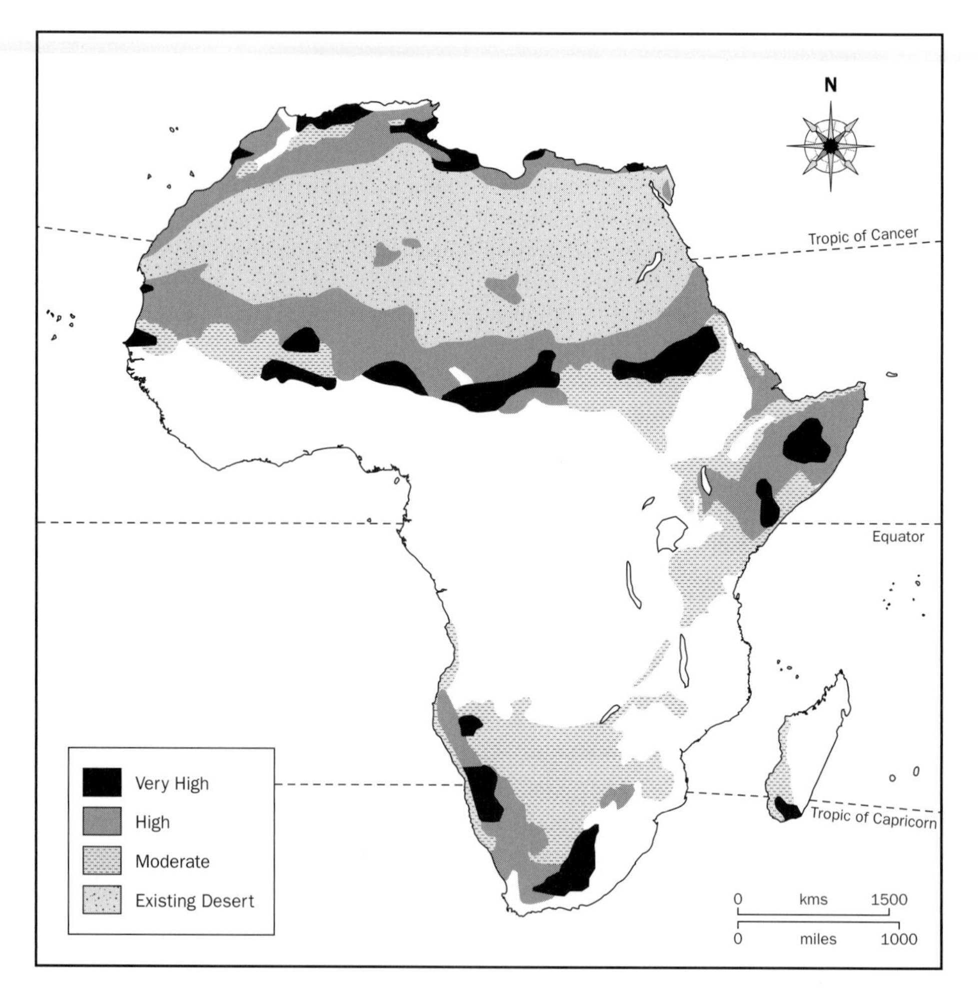

Figure 3.7 Desertification in Africa (UNCOD).

Source: Binns, 1994

(UNEP, 1984: 17). As Swift (1996: 81) suggests, 'This UNEP view became the received wisdom on desertification, scarcely challenged in public policy-making or popular reporting . . . But the received view had never been accepted by many dryland scientists. The data and analysis . . . contain serious flaws and dubious generalizations.'

The Convention to Combat Desertification (CCD), adopted in 1994 by the international community in Paris, further concentrated the focus on desertification and emphasized the role of human activity in soil degradation. The UN declared 2006 International Year of Deserts and Desertification, and predicted that by 2020, as a result of the desertification process, some 60 million people will have abandoned the dryland areas of sub-Saharan Africa and migrated to Europe (UNCCD, n.d.).

But what is the reality of the situation? As Warren and Batterbury (2004: 1) suggest, 'The history of desertification . . . includes many accounts, based on dubious science and sweeping warnings of the imminent demise of dryland environments and societies affected by moving sands, or extreme land degradation.' Evidence from field observations in Africa during the 1980s and early 1990s suggested that desertification was grossly over-emphasized, largely because of insufficient investigation on the ground as to how environments, societies and food production systems respond to periods of drought. A notable UN Environment Department Working Paper written by Ridley Nelson in 1988 stated, 'Contrary to popular belief, the extent of desertification is not at all well known . . . there is extremely little scientific evidence based on field research or remote sensing for the many statements on the global extent of the problem' (Nelson, 1988: 4). As Swift (1996: 73) argues, '"Desertification" is perhaps the best example of a set of ideas about the environment that emerge in a situation of scientific uncertainty and then prove persistent in the face of gradually accumulating evidence that they are not well-founded. This "stickiness" of ideas has important implications for policy.'

Of particular concern is how 'desertification' is actually defined (Binns, 1990). The term has been used widely to raise public awareness, but should probably be used more cautiously to describe land which has forever been removed from productive use. This should be clearly differentiated from areas which have suffered 'land degradation', which, though severe in some cases, could perhaps return to their former status with careful management and two or three good rainy seasons. It is not so much a question of desertification not existing, but rather that its existence over space and time is probably much less than the popular image. The fringe of the Sahara Desert, for example, might be likened to an ebbing and flowing tide, moving southwards during dry periods and northwards when rainfall is more plentiful. Pockets of degraded land may occur at any time for a variety of human and environmental reasons, but the process of degradation is likely to accelerate during dry spells. However, claims that the Sahara is expanding at some horrendous rate are often made, despite the absence of reliable, field-based evidence to support them.

Based on their observations in Burkina Faso and Niger, Warren and Batterbury (2004) suggest that since 1984 the amount of vegetation has actually increased in the northern Sahel region. Reporting on his observations in Niger, Reij (2007) comments that since the mid-1980s about 250,000 hectares of degraded land have been rehabilitated with a considerable expansion of dry-season cultivation. As Reij (2007: 15) observes,

> Since the middle of the 1980s, farmers in the most densely populated parts of Niger have begun to protect and manage young trees and bushes regenerating on their cultivated fields. It has become clear that this is happening on at least 5 million hectares, which is a spectacular scale, unique for the Sahel and probably even unique for Africa.

Whereas twenty years ago all trees belonged to the state, farmers now believe they have exclusive rights to the trees on their fields, and air photos and satellite images of some villages show there are between ten and twenty times more trees now than in 1975. More productive farming systems have been achieved by integrating agriculture, forestry and livestock. Reij (2007: 15) argues, 'Some of the doom and gloom stories about Africa's drylands are not based on facts: they reveal a disconnect between field realities and the development bureaucracies.'

Swift's conclusions concur with those of Reij, suggesting that in the 1970s and 1980s, 'the received narrative of desertification provided a convenient point of convergence for the interests of three main constituencies: national governments in Africa, international aid bureaucracies, especially United Nations agencies and some major bilateral donors, and some groups of scientists' (Swift, 1996: 86). He is reassured by the emergence in the 1990s of a counter-narrative which derives from direct engagement with dryland farmers and herders, focusing on their adaptations in the face of dynamic and often unpredictable environments (see Box 3.5). Swift (1996: 90) concludes, 'Researchers, who come out of the story of desertification with tattered banners, have a particular responsibility this time to get the science right and to ensure that the policy outcomes reflect a more just and efficient distribution of rights and responsibilities.'

Box 3.5

Coping with a marginal environment in Burkina Faso

The name Burkina Faso means 'Land of Noble People'. In the 1970s and 1980s, the people of Burkina (known as Burkinabé) had to cope with the effects of two major droughts, and in certain parts of the country, with high population densities and intensive land use, land degradation required urgent remedial action. The mobilization of rural communities in generating greater environmental awareness and implementing a variety of conservation techniques has been impressive.

Before the change of name in 1984, the country was called Haute Volta (Upper Volta), a name chosen by the French colonialists who ruled the country until independence in August 1960. Like other countries in the savanna–Sahel zone of West Africa, Burkina is very poor, with a per capita Gross Domestic Product of only $1213 in 2005, life expectancy of 51 years and very high infant mortality – 96 deaths per 1000 live births. Its Human Development Index in 2005 was the second lowest in the world, with only Sierra Leone below Burkina (UNDP, 2007).

Burkina – which at 274,000 sq km is slightly larger than the UK – had an estimated population of 13.2 million in 2005, giving an average population density of about 48 per sq km. But in terms of the number of people in relation to available arable land, the rural population density is over 211 per sq km (World Bank, 2008b). More than 80 per cent of the population live in rural areas, with 87 per cent of the labour force employed in agriculture. The capital city and administrative centre, Ouagadougou, is located in the middle of the country and has a population of about 1.4 million. The second major city and economic capital, Bobo-Dioulasso (approximately 450,000 people), is located in a productive agricultural region close to the Côte d'Ivoire (Ivory Coast) border and on the railway that links Burkina with the sea at Abidjan.

French colonial rule created the main structure of the economy, with heavy emphasis on cash-cropping, notably cotton and groundnuts. The French regime was also responsible for initiating the large-scale migration of labour (mainly male) to neighbouring states, notably Côte d'Ivoire and Ghana, where money could be earned in coffee and cocoa plantations or in the large towns. Over 3 million Burkinabé are today working elsewhere in West Africa or overseas, sending vital remittances back to their families in poor rural villages. Political instability in Côte d'Ivoire and the closure of the border with Burkina in 2002 and 2003 seriously affected the movement of people and goods between the two countries, with the latter often re-routed through other West African states, such as Togo, Ghana and Benin. The French colonial regime integrated Burkina into a system of regional trade and transport, with strong dependence on the more prosperous Côte d'Ivoire. Burkina produced cotton and groundnuts, while manufactured and processed goods were imported from France. The lack of colonial investment in infrastructure is reflected in the fact that at independence there were no tarred roads in Burkina.

Yatenga, in north-western Burkina Faso, close to the border with Mali, is one of the country's poorest regions, with a harsh climate and short rainy season, from June to September. Rainfall has decreased in the last thirty years from the long-term average of 720 mm per annum to below 500 mm. Soils are poor, other than in a few river valleys, and much of the natural vegetation has been cleared for cultivation, causing rapid run-off and erosion during heavy downpours. Droughts have been recorded in Yatenga in 1832–1839, 1879–1884, 1907–1913, 1925–1926, 1929–1934, 1940–1942, 1966–1973 and 1979–1985. These droughts often had a disastrous effect on food production and livelihoods, as for example in October 1984, when after four bad harvests the region had a gross annual cereal deficit of 149 kg per person.

Yatenga is the homeland of the Mossi people, Burkina's largest ethnic group, who comprise over 40 per cent of the country's population. Another significant population group are the Fulani, who are predominantly pastoralists and comprise about 10 per cent of the population. Yatenga is heavily populated, with densities over 100 per sq km in the central area (Ouedraogo and Kaboré, 1996). Pressure on the land, poverty of village life and attractions of wage labour elsewhere have caused many Mossi to leave the region since the late nineteenth century.

Those Mossi still in Yatenga grow millet, sorghum and maize, and in the colonial period the French insisted that all farmers should also grow cotton and groundnuts. Intensively farmed vegetable gardens, fertilized with kitchen waste, surround the villages. In many villages, most of the land is under continuous cultivation with little possibility of expanding the agricultural area. Although tree cover has declined considerably, important economic trees such as the shea, dawadawa, kapok, tamarind and baobab are carefully tended and harvested. Unlike the Fulani, the Mossi traditionally kept few animals, but today cattle are seen as a form of investment and insurance against disaster. After the harvest, cattle graze crop stubble, adding valuable manure to the soil, which otherwise receives little fertilization, since chemical fertilizers are too expensive or unobtainable.

Since independence, a number of development initiatives in Yatenga have attempted to enhance environmental management. For example, the Naam Movement was founded in 1967 by Bernard Ledea Ouedraogo, a teacher who moved into rural development work in the early 1960s. Ouedraogo's goal was 'to make the village responsible for its own development, developing without destroying, starting from the peasant: what he is, what he knows, what he knows how to do, how he lives and what he wants'. The movement derived its name from the traditional Mossi village organization (Naam) – a grouping of young men and women formed each rainy season to help with planting and harvesting. It is a traditional self-help cooperative movement, based on equality and featuring elected leaders. Wherever possible, Naam activities use low-cost local tools and materials. In 1976, Ouedraogo set up an umbrella organization, 'Six S', to provide technical and financial help for Naam groups and to raise funds internationally for medicines, cement, pumps and better tools. By 1985, there were 1350 Naam groups and the idea had spread to neighbouring countries.

The Naam group in the village of Somiaga built a pharmacy and a mill, dug several wells for drinking water, and established a tree nursery. A cereal bank was started in 1983 to improve food security in the village and to avoid major seasonal price fluctuations. Cereal stocks are bought cheaply at harvest time and stored until needed, thus avoiding expensive purchases in the late dry season or early rainy season, when market stocks are dwindling. A large dam – 180 metres long and 4 metres deep – was also built and now irrigates up to 60 hectares of rice, vegetables and fruit trees. The dam also raised the water level in local wells, made it easier to water tree saplings and provided a useful source of fish. Upstream from the dam, smaller check dams were built across gullies to reduce the flow and trap the soil to prevent silting up of the main dam.

Projet Agro-Forestier (Agroforestry Project; PAF) is another example of a community-based venture to improve soil and water conservation in Yatenga (Critchley, 1991). Started with support from Oxfam in 1979 as a tree-planting project, PAF has since diversified into promoting water harvesting and anti-erosion measures. Much of Yatenga has a gently undulating landscape, and a key innovation was use of the simple plastic water-tube level to determine contour lines for the construction of rock bunds. The latter are a traditional method of water harvesting and are preferred to earth structures as they are not easily damaged by run-off and therefore require little maintenance. Also, since the bunds are permeable, crops planted in front of them and behind them benefit from run-off. Water retained by the bunds gradually permeates the soil, much of which is covered by a hard crust that would otherwise have prevented water infiltration. Trees and grass planted along the bunds further help to stabilize the soil and in time produce barrier hedges.

Rock bund construction has enabled an expansion of the cultivated area through the reclamation of abandoned land. Since 1983, the number of hectares treated has doubled every year, reaching a total of 8000 hectares in over 400 villages by 1989. PAF encouraged collective treatment of individual fields, and since many young men migrate in the dry season, women play a major role in rock bund construction. PAF provides training in the various techniques and supplies farmers with water-tube levels, shovels, pickaxes, wheelbarrows and donkey carts. Oxfam has helped with the purchase of a tipper lorry to make stone gathering easier (Critchley, 1991).

PAF has also promoted another technique known as *zai*, which can be used in conjunction with the rock bunds. *Zai* are wide and deep planting holes with a diameter of 20–30 cm and a depth of 10–15 cm. With a spacing of 80 cm, the number of *zai* per hectare can reach 15,000. Consequently, the digging of a single hectare of *zai* is a laborious task which can take about sixty work-days at an average of five hours per day. *Zai* are used to rehabilitate lateritic soils and enable cultivation on sterile and hard-pan surfaces (Ouedraogo and Kaboré, 1996). The *zai* collect and concentrate run-off water and, when filled with manure and compost, greatly improve crop yields. The use of bunds and *zai* together produces yields of millet or sorghum of up to 1000 kg per hectare, compared with the average on untreated plots of about 600 kg per hectare. These simple but sustainable environmental management techniques have provided greater food security to households in Yatenga by reducing the impact of poor crop yields in years with low rainfall (Ouedraogo and Kaboré, 1996).

From the late 1980s, PAF broadened its outlook to promote a more integrated approach to village land-use management. Committees were set up to decide on the priorities for improving the area and then to coordinate various conservation strategies, such as stone bunding, compost pits, the enclosure of sheep and goats in the homestead and village fodder plots. Communal land has been protected from overgrazing, tree nurseries started and a wide range of trees planted in the fields. PAF has had considerable success in the sustainable management and improvement of the environment in Yatenga due to its flexible, low-technology and community-based approach, which has built upon a range of indigenous techniques.

Mortimore's valuable longitudinal field-based work in the savanna–Sahel zone of north-east Nigeria serves to endorse Swift's perspective:

> The Sahel is a diverse environment, and the matching diversity of solutions sought by Sahelian households to its challenge has not been adequately recognised in the past. The Sahel is also a disequilibrial environment, and it is necessary to recognise that Sahelians have evolved adaptive strategies to deal with its consequences, and will not be helped by developmental promotions that push them towards specialisation or new forms of dependency.
>
> (Mortimore and Adams, 1999: 196)

3.5 Carrying capacity and land degradation

Land degradation of one sort or another is a feature of many environments throughout the world. But in some of Africa's more marginal environments, features such as soil erosion, gullying and deforestation have received much attention. Frequently, local farmers and pastoralists are blamed for mismanaging their land and taking too much out of it. Overgrazing, the burning of vegetation to stimulate new growth, and the use of timber as fuelwood are among the practices identified as leading to land degradation in Africa. This is the case in South Africa's Karoo region, a vast 400,000 sq km semi-arid interior plateau to the north and east of Cape Town, which is considerably larger than the UK and only slightly smaller than California. As in the savanna–Sahel region, land degradation in the Karoo was frequently blamed on poor management and over-stocking, although, as Dean *et al.* (1995: 258) suggest, the debate is 'characterised by a plethora of opinions and a dearth of published and reliable scientific evidence' (see Plate 3.3).

It is often suggested that the existence of land degradation indicates that there are too many people and/or animals on an area of land. William Allan, writing in 1949, suggested that for every area of land where people are growing crops or rearing livestock, there is a human and animal population limit which cannot be exceeded without setting in motion the process of land degradation. Allan called this limit the 'critical population' or 'carrying capacity' for that system of land usage (Allan, 1949; see also Street, 1969). He suggested that expressing carrying capacity numerically allows a much more precise approach to

Plate 3.3 Erosion gully in northern Nigeria (Tony Binns).

the problems of African agriculture and pastoralism. Allan argued that the calculation of carrying capacities for traditional African food production systems could be achieved, but this required a detailed knowledge of both the nature of the land itself and the management systems being used. To obtain the 'magic' figure, Allan fed the following data into his calculation: data on vegetation and soil type; the total area of each of these types; the cultivation and fallow periods for each of these types based on traditional land-use practices; the cultivable percentages allocated to each type for the system of production; and the total area of land required per head of population. He was then able to arrive at a figure for the carrying capacity of specific areas.

However, would it be feasible to calculate such a figure for every location in Africa? And even if this could be done, would it be useful? Allan recognized that estimates are strictly relative to the physical characteristics of specific areas and to the management systems being used on them, and that the concept of carrying capacity assumes a more or less ideal distribution of population in relation to soil characteristics. He admitted that the two main weaknesses of methods used in calculating carrying capacity are: first, the problem of obtaining accurate figures for the percentage of cultivable land; and, second, the amount of labour needed in surveying and sampling farms. In reality, such data are not available in many African countries, and detailed field surveys would be both prohibitively expensive and practically impossible. Even if carrying capacity figures were obtainable, each value would be appropriate only to a particular area and at a specific point in time.

As we have seen, environmental factors, notably climate, change frequently, and traditional food production systems are also highly dynamic, particularly in arid and semi-arid areas. Scoones (1995b) refers to these as 'non-equilibrium environments'. Indeed, with a system of nomadic or semi-nomadic pastoralism, the movement of people and animals in search of pasture and water would surely make a figure of carrying capacity worthless. Even with more sophisticated technology than Allan had at his disposal, such as air photography and satellite imagery, it is difficult to see the value of calculating carrying capacities with such a dynamic set of variables and a lack of detailed ground-level surveys. Scoones (1995b: 25) concludes:

> The dynamics of many ecological systems are such that equilibrium assumptions do not apply – temporal variability [fluctuations and episodic events], spatial heterogeneity and complex dynamic interactions between species may all mean that non-equilibrium dynamics apply much of the time, especially in dryland areas. In this case, the very notion of carrying capacity becomes meaningless.

However, although the concept of carrying capacity may be of limited practical use for African conditions, the idea has nevertheless provoked considerable debate, and has probably led us to a much better understanding of the relationships between pastoral communities and environmental management.

3.6 Adapting to environment

In light of the prominence given to the need for adaptation to climate change, there is now possibly a greater need than ever before to understand people–environment relationships if such threats as famine and land degradation are to be reduced in future development programmes (Binns, 1995). Such strategies must aim to reduce vulnerability and strengthen resilience among poor African communities. We now have a good amount of accumulated evidence to show how pastoral and farming communities in marginal semi-arid and arid regions have devised, often over many generations, a range of mechanisms to cope with harsh environmental episodes, such as drought, which is endemic in these regions and is frequently a key element in indigenous folklore and oral histories. Many African food production systems display a high degree of structural resilience and a sensitive adaptation to environment (see Boxes 3.5 and 3.6).

Box 3.6

People and environment on the Jos Plateau, Nigeria

The traditional homeland of the Kofyar people is a territory of some 500 sq km on the southern fringe of the Jos Plateau in central northern Nigeria. Research undertaken by Robert Netting in the early 1960s (Netting, 1968) showed that the Kofyar economy and society were skilfully adapted to the rugged topography and provided convincing case study evidence of the processes of intensification and increasing productivity detailed by Ester Boserup in her important book *The Conditions of Agricultural Growth* (Boserup, 1965). More recent field-based research has shown that further adaptations have occurred, taking advantage of new economic opportunities and an easier lifestyle on the Benue plains below (Stone, 1996, 1998; see Figure 3.8).

The Jos Plateau is a distinctive, mainly granitic upland area of about 8600 sq km (80 × 110 km), with an average altitude of 1200 metres, rising to 1350 metres on its southern margin, where there is a steep 300–600-metre escarpment overlooking the plains on the northern edge of the Benue Valley. Old volcanic cones, volcanic

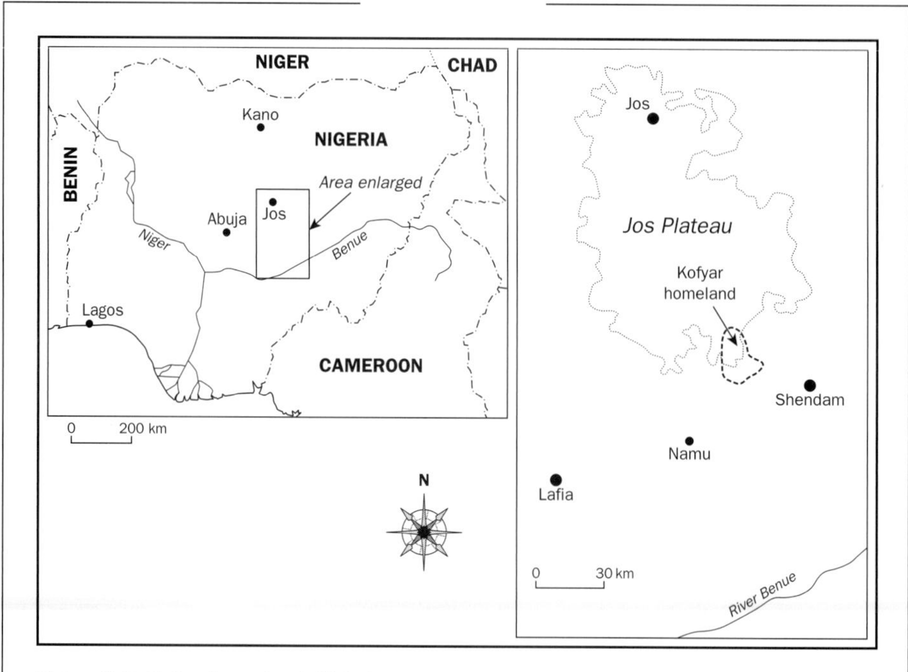

Figure 3.8 Kofyar homeland, Nigeria.

Source: Binns, 1994

boulders and lava flows are features of the area, on which immature and stony soils have developed, in places with good fertility. Rainfall on the plateau is greater than on the plains, with most of the 1000 1500 mm falling between April and October, peaking in August. Temperatures range between 32 degrees centigrade in February and about 25 degrees during the rainy season, though night-time temperatures of under 10 degrees are not uncommon in the hills.

Kofyar country was settled as a defensive site in pre-colonial times when inter-tribal warfare was common, and includes a heavily dissected ridge 24 km long and 8 km wide, and an area of plain-land at the base of the plateau escarpment. Following the British conquest in 1909 and the establishment of colonial rule, more peaceful times encouraged movement out of the hills and on to the plains, but this migration was not significant until the 1950s. At the 1963 census, the total Kofyar population was almost 73,000, with densities reaching 750 per sq km, a very high figure for an African farming community. Such dense settlement is supported in the hills by the intensive and continuous use of skilfully constructed terraced fields, where soil fertility must be carefully maintained to produce successive food crops.

Kofyar hill cultivation reflects an impressive understanding of the local environment's potential and problems, and gullying and erosion are rare. On steep valley sides terraces have been constructed with stone-built retaining walls usually placed at right angles to the slope. The terraces provide a level surface with deep soil for cultivation and help reduce run-off and erosion. On the fields of the homestead farm close to the village, ridges of soil are made in a rectangular formation. Crops are planted on the ridges and their roots are watered from the water collected in the basin, an irrigation technique known as 'basin listing'. Groups of ridges are separated by drainage ditches to prevent waterlogging of the plots.

Unlike other African farming communities, the Kofyar do not experience a 'hungry season' before the harvest. They grow a wide range of grain, leguminous and tree crops that are harvested at different times of the year, ensuring a continuous supply of food. The staples of sorghum, millet and cowpeas are usually intercropped, and other crops include *acha* ('hungry rice'), maize, yams, groundnuts, sweet potatoes and rice in moist areas. Kitchen gardens close to the houses provide herbs, peppers and tomatoes for sauces. Tree crops – such as oil and fan palm, locust bean, pawpaw and mango – are also gathered, further enriching the diet. Sometimes these crops are exchanged for other commodities. The key to the maintenance of such a productive farming system is the regular application of compost, wood ash and animal manure. During the growing season, goats are staked in corrals close to the homestead and at night they are locked into a hut. Their manure and trampled fodder provide vital additions to the soil. On more distant bush fields, which unlike the permanently cultivated homestead farms are cultivated for only six to nine years, Fulani herders are encouraged to walk their cattle over the land, leaving valuable manure.

Technology is simple but effective on the small farms, consisting of hoes, axes and sickles. Homestead farms average 0.62 hectares, whereas the less intensively cultivated and more distant bush farms are larger, averaging 1.2 hectares. The average land requirement for a homestead is about 1.01 hectares. In the past, the main problems facing the hill-farming areas were over-population, limited farm size, poor services, difficult communications and limited access to markets. It was these factors that encouraged the Kofyar, quite spontaneously, to start migrating down the escarpment to establish farms on the Benue plains. Permanent colonization of the plains, close to the town of Namu, began in 1951. By 1960, the Kofyar had built over 100 compounds on the plains and the process increased steadily throughout the decade. Initially, there was regular movement between the plains and the hill farms, but in recent years many Kofyar have found it more convenient and profitable to live on the plains. Those remaining in the hills are often elderly or retired. A survey in 1984 showed that 74 per cent of those interviewed had migrated and over 50 per cent had transferred their sole residence to the plains, abandoning their homesteads and farms in the hills. By the mid-1980s, some 25,000–30,000 Kofyar occupied the Benue Valley lowlands south of their original homeland (Stone, 1996).

In migrating to the plains, the Kofyar have taken their knowledge and farming techniques with them, but certain significant adaptations have also been made. Plains farms are larger – an average of 4 hectares, compared with less than 1 hectare on the homestead farms. Household size also increased in the 1960s – from 4.17 among non-migrants to 6.44 for migrants. By 1984, average household size had reached 8.38 people. Nevertheless, many migrant Kofyar are able to produce sizeable marketable surpluses. Itinerant merchants and lorry drivers visit their farms, many of which are within easy reach of the surfaced road, opened in 1982 and linking the main production areas with key settlements and markets.

Migrant Kofyar household income has increased impressively in spite of little investment in new technology or hired labour. Traditional tools are still used, and extra food production has been generated by farmers and their families working harder and longer hours in the fields. Stone (1998) has shown that Kofyar adult farmers work an average of 1599 hours each year, considerably more than the 500–1000 hours recorded in similar food production systems elsewhere in Africa.

This hard work is reflected in the way Kofyar farmers have extended the farming season, done away with slack periods and mobilized men and women equally in farm-related work.

The Kofyar migrants have adopted yams and rice as cash crops, and have increased production and sale of traditional crops, such as millet and groundnuts. A new variety of yam from Iboland which is of better quality and stores without shrinkage has also been adopted. Fewer goats are kept on the plains, but almost every Kofyar farmer now raises pigs, which were entirely unknown in the area as recently as 1960. Other innovations include growing cassava and bananas for the market and the application of chemical fertilizers. Money earned from crop sales is spent on clothes, taxes, bridal dowries, bicycles, motor cars, school fees, medical care and chemical fertilizers.

The steady depopulation of the mountain homeland of the Kofyar has been occurring since the early 1960s. As early as 1985, the village of Gonkun, for example, had only twelve active farms, compared with twenty-six in 1961. But, as Stone has observed, 'The homeland still contained viable communities in 1985, largely because of deliberate strategies to counter the pull toward the frontier' (Stone, 1998: 247). Stone suggests that the enduring nature of the homeland villages in the face of much out-migration has been due to strong cultural factors. Migrants have been keen to maintain strong links with their home villages and many have relatives staying there, sometimes on a rotational arrangement. Traditional ceremonies and funeral rituals, 'attracted crowds in the hundreds or sometimes thousands from the frontier and the city, and they were part of the rationale for successful urban Kofyar maintaining vacation houses in the homeland' (Stone, 1998: 249). Schools have also continued to be built in the homeland since the 1960s, and they have played a key role in preserving the mountain communities. As Stone comments, 'Sending children to homeland schools was considered a civic obligation, even for families living entirely on the frontier' (Stone, 1998: 250). Another key factor in preserving the homeland communities has been retirement, with the homeland village being the preferred place to die and considered the proper final resting place for heads of households. Despite out-migration leading to the decline of some terraced farming systems and an increase in pests, such as monkeys, the Kofyar have managed to preserve their homeland villages and have achieved some success in securing state funding for school construction and sinking a bore hole with a pump to provide a reliable water supply.

The Kofyar provide a good illustration of how African farmers have adapted to their physical and economic environment. Opportunities were recognized and migrants have colonized new land and established profitable businesses, but they have continued to recognize and support their traditional homeland communities in the mountains.

As the work of Mortimore (1989) and Richards (1985) has shown, African farmers are generally capable of solving many of their own technical and economic challenges. Richards, for example, with reference mainly to Sierra Leone rice-farming systems, shows the capacity of poor farmers to innovate and adapt in light of a wealth of locally appropriate knowledge which development agents have often ignored or misunderstood (Richards, 1985, 1986). Mortimore's work in both the drylands of West Africa and in Machakos, Kenya, also reveals strong evidence of innovation and adaptation, and, as Boserup (1965) has shown, the ability of rural communities to intensify production as population pressure increases (see also Mortimore, 1989; Tiffen *et al.*, 1994; Mortimore and Adams, 1999).

The district of Machakos is situated some 50 km east of the Kenyan capital, Nairobi. Through detailed examination of the historical record, Mortimore and Tiffen (1995) show how the local Akamba people have transformed the landscape in response to a steadily increasing population density and growing market opportunities for crops such as coffee, fruit and vegetables. The construction of terraces, tree-planting, soil-conservation methods, including the use of compost and manure, and the introduction of new technologies such as the ox-plough have all played roles in increasing land productivity in Machakos. As Mortimore and Tiffen (1995: 84) conclude, 'contrary to the expectations expressed in the 1930s, the Akamba of Machakos have put land degradation into reverse, conserved and improved their trees, invested in their farms, and sustained an improvement in overall productivity'. The Machakos story endorses the Boserupian perspective on the positive relationships between people and environment, and counters the more negative Malthusian perspective.

Mortimore's detailed long-term study of communities in the far north of Nigeria reveals an equally impressive range of adaptive mechanisms for coping with the vagaries of the environment (Mortimore, 1989). Productive systems remained resilient, the communities survived, and during the 1980s they even managed to export grain to the market. As Mortimore comments, 'The astonishing densities of population in the Kano Close-Settled Zone (KCSA) [over 400 per sq km in places if the non-farming population is included], have not brought the collapse of sustainable intensification that some expected because the economic base of rural livelihoods has broadened correspondingly' (Mortimore 2003: 235; see Plate 3.4).

There are countless examples of indigenous methods of soil and water conservation in Africa (Reij *et al.*, 1996). For example, in the Ader Doutchi Maggia region of Niger, where rainfall is between 250 and 450 mm, lines of stones have been laid out by people living on the barren plateaux to conserve water and trap windblown sand. After five or six years, sufficient sand has usually been deposited to start cultivation, and the addition of millet

Plate 3.4 Boy and donkey with urban refuse for peri-urban fields, Kano, Nigeria (Tony Binns).

stalks and manure helps to raise fertility in the new plots. Stone lines are found elsewhere in the Sahel region of West Africa. In Yatenga, north-western Burkina Faso, rock barriers, bundles of stalks and branches placed across the direction of water flow help to prevent run-off and soil erosion. Small stone dykes, or 'diguettes', about 30–40 cm high and built along the contours, are even more effective in stabilizing the soil (Atampugre, 1993; see Box 3.5).

Batterbury (2005: 267) observes that his extensive fieldwork in Bam Province on the central plateau of Burkina Faso

> revealed a staggering range of traditional agricultural practices and erosion control techniques among Mossi, Yarsé and Peulh/Fulani cultivators . . . [including] the construction of bunds and barriers from sticks, *andropogon* grasses and stones; micro-variations in planting densities and spacing, taking into account soil fertility and water supply; use of millet-stalk or straw mulches to encourage termite activity in hardened soils; and selective cutting of forest areas to encourage natural regeneration of certain woody species.

On the clay plains of eastern Sudan, *teras* cultivation is widely used, notably in Kassala's Border Area, where some 10,000 to 15,000 hectares are under *teras* and support a population of over 200,000. The *teras* is an earthen bund 35–40 cm high with a base of 1.5–2.0 metres, surrounding three sides of a cultivated plot of up to 3 hectares. The bunds impound run-off and nutrient-rich silt from the plains, and the system produces a chessboard landscape of cultivated rectangular plots interspersed with areas of uncultivated plain. Sorghum is the main crop, with water melons planted on the bottom bund of the open rectangle. Soil fertility has been found to be two to four times higher inside the *teras* field, which suggests that such 'nutrient harvesting' may in some cases be a viable alternative to the application of mineral fertilizers (Niemeijer, 1998: 323; see Figure 3.9).

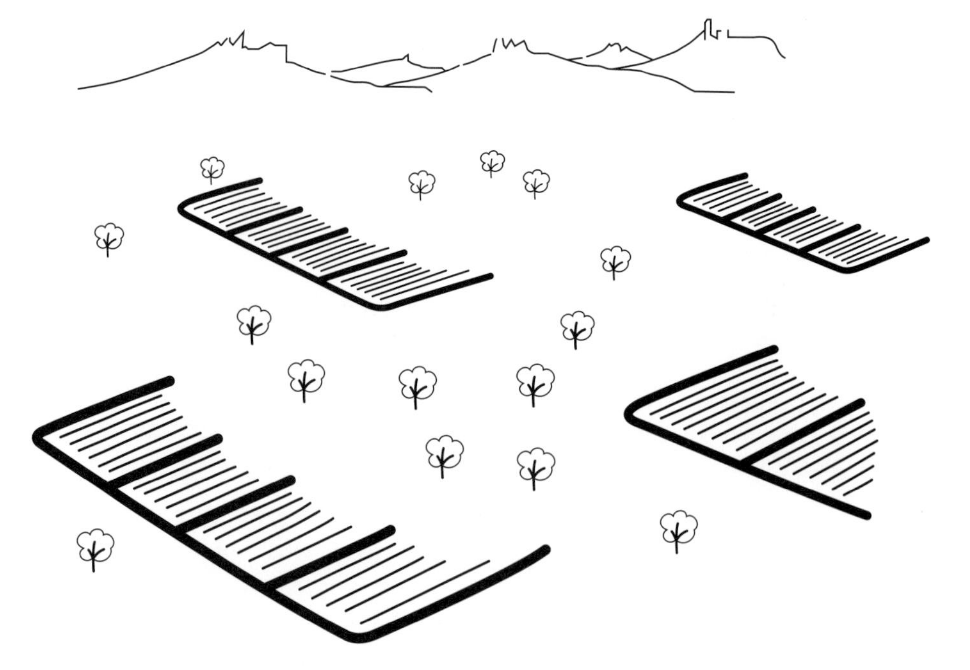

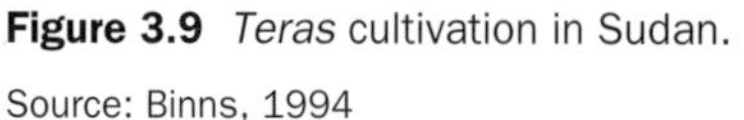

Figure 3.9 *Teras* cultivation in Sudan.

Source: Binns, 1994

The Dogon people in the central plateau region of Mali, south of Mopti, use similar systems of earthen ridges and basins, but unlike the *teras* of Sudan, the Dogon do not leave an open end to the rectangle. Each basin of 1–4 sq metres is made by building an earth ridge in the form of a square. Construction takes place in the dry season before planting, and the ridges are built up during weeding. In fact, the Dogon have developed an impressive range of soil- and water-conservation techniques, including stone lines, stone bunds and earth mounds. The latter, as well as slowing down run-off, act as small compost heaps and help maintain soil fertility (Critchley, 1991).

In Sierra Leone, which has a much higher rainfall than the Sahel, traditional methods of stick and stone bunding in the mountains of the Freetown Peninsula are remarkably effective in preventing soil erosion and require little capital or extra labour input. Bundles of sticks are placed across the slope in continuous lines or bunds and pegged into the ground. Each bund slows run-off and valuable sediment is deposited behind it (Millington, 1984).

Just as in Machakos, many other mountainous areas of Africa employ longstanding and well-maintained terrace systems which are the products of many hours of hard work by local communities. For example, the Mandara Mountains of northern Cameroon have sustained a high-density agricultural population for centuries. As Riddell and Campbell (1986: 90) comment, 'It demonstrates that the successful maintenance of high density populations depends upon the evolution of a production system in which environmental management, social institutions and agricultural practices are intimately linked.' Like so many upland and former defensive sites (see Box 3.6), the Mandara Mountains are home to many different ethnic groups, the largest being the Mafa, who live in one of the most densely populated regions of the Sahel – in places over 400 people per cultivable sq km. The terraced plots require constant attention to prevent soil erosion, while fertility is maintained with crop rotations and the regular application of animal manure as well as household, human and plant waste. Trees are also important to stabilize the soil, with the fallen leaves of the *acacia albida* used as fodder or as green manure in the terraces.

Migration, of one form or another, is an adaptive strategy that is widely used in marginal environments. Pastoralists, for example, move their livestock from one area to another, depending on the availability of water and pasture (see, for example, Adriansen, 2008). Studying one village in south-western Niger, Batterbury and Warren found that the households that sent away the most men cultivated the most eroded fields. As they comment, 'Extending the spatial reach of the household is an effective adaptive strategy for drought' (Batterbury and Warren, 2001: 2).

A survey undertaken in the 1980s of households migrating from famine-affected communities in northern Darfur to the provincial capital, El Fasher, in western Sudan, revealed that some groups were able to stay longer in their villages before moving because they adopted a range of survival strategies. The most popular strategies were the use of alternative income-generating activities, such as the sale of wild foods, transport hire and wage labour. Other survival strategies included changing cropping patterns, cultivating larger areas, multiple cropping, the use of different pastures and sharing of resources (de Waal, 1989).

3.7 Managing forest resources

Africa has an estimated forest area of 635 million hectares, representing about 16 per cent of the world's total forest area. In terms of the amount of forest per person, Africa has an average of 0.7 hectares, which is higher than Asia (0.1 hectares), but lower than Europe (1.3 hectares), North and Central America (1.2 hectares), South America (2.7 hectares) and Oceania (3.2 hectares). In the period 2000–2005, there was a net annual forest loss in Africa of about 4 million hectares (FAO, 2007). Table 3.1 shows key statistics for those African countries with a forest area of more than 10 million hectares.

Central Africa has over 37 per cent of the continent's forests, and the Democratic Republic of the Congo (DRC) dominates both this region and indeed the whole continent, with over 133 million hectares of forest. After the DRC, the largest forest areas are in Sudan (67 million hectares), Angola (59 million hectares), Zambia (42 million hectares) and Tanzania (35 million hectares). West Africa was the region with the highest annual rate of forest loss (–1.17 per cent) between 2000 and 2005, and within this region Nigeria (–3.3 per cent), Benin (–2.5 per cent) and Ghana (–2.0 per cent) were the countries where forest

Table 3.1 Forest statistics for African countries with total forest areas over 10 million hectares, 2005

	Total forest (,000 ha)	*Forest (% of land area)*	*Forest area (ha per capita)*	*Annual change rate (%, 2000–2005)*	*Fuelwood production (,000 cubic metres)*	*Industrial round-wood production (,000 cubic metres)*	*Sawn-wood production (,000 cubic metres)*
Angola	59,104	47.4	4.2	–0.2	3,487	1,096	5
Botswana	11,943	21.1	6.9	–1.0	655	105	0
Cameroon	21,245	45.6	1.3	–1.0	9,407	1,800	702
Central African Republic	22,755	36.5	5.8	–0.1	2,000	832	69
Chad	11,921	9.5	1.4	–0.7	6,362	761	2
Congo, Republic	22,471	65.8	5.8	–0.1	1,219	896	157
Congo (DRC)	133,610	58.9	2.4	–0.2	69,777	3,653	15
Ethiopia	13,000	11.9	0.2	–1.1	93,029	2,928	18
Gabon	21,775	84.5	15.8	n/a	1,070	3,500	133
Côte d'Ivoire	10,405	32.7	0.6	0.1	8,655	1,678	512
Madagascar	12,838	22.1	0.7	–0.3	10,770	183	893
Mali	12,572	10.3	1.1	–0.8	4,965	413	13
Mozambique	19,262	24.6	1.0	–0.3	16,724	1,319	1,246
Nigeria	11,089	12.2	0.1	–3.3	60,852	9,418	2,000
Sudan	67,546	28.4	2.0	–0.8	17,482	2,173	2,171
Tanzania	35,257	39.9	1.0	–1.1	21,505	2,314	2,246
Zambia	42,452	57.1	4.0	–1.0	7,219	834	157
Zimbabwe	17,540	45.3	1.3	–1.7	8,115	992	397
NORTH AFRICA	76,805	8.2	0.4	–0.69	46,371	3,458	**200**
WEST AFRICA	74,312	14.9	0.3	–1.17	145,292	17,128	**3,145**
EAST AFRICA	77,109	18.9	0.4	–0.97	194,818	10,526	**1,296**
CENTRAL AFRICA	236,070	44.6	2.2	–0.28	103,673	12,979	**1,250**
SOUTHERN AFRICA	171,116	29	1.4	–0.66	55,908	26,356	**2,905**
ALL AFRICA	**635,412**	**21.4**	**0.7**	**–0.62**	**546,062**	**70,447**	**8,796**

Source: Adapted from FAO, 2007

depletion was most rapid. For the continent as a whole, the highest annual rates of forest loss during this period occurred in Comoros (–7.4 per cent), Burundi (–5.2 per cent) and Mauritania (–3.4 per cent) (FAO, 2007). With a total forested area of only 5000 hectares, the loss in Comoros was particularly severe. Table 3.1 also shows the scale of production of timber for fuelwood, industrial round-wood and sawn wood in the countries with the largest forest areas. Ethiopia dominates fuelwood production, followed by the DRC, Nigeria, Uganda, Ghana and Kenya.

Ghana has a profitable timber and wood products industry which makes a significant contribution to the country's GDP. In the first nine months of 2008, there was a 7.7 per cent increase in exports compared with the same period in 2007. Trade with European Union countries accounts for 45 per cent of export volume and 33 per cent of value. Yet Ghana has one of the most rapid rates of forest depletion in Africa, with a 2 per cent annual decline in the forest area between 2000 and 2005, and a per capita forest area of 0.3 hectares in 2005. One of the biggest problems facing Ghana, and other countries such as the DRC (see Box 3.7), is illegal logging. Policy measures targeting Ghana's illegal chain-saw operations have not been particularly successful. As a BBC news report in 2007 commented,

> While many communities see little or no benefit from the business of illegal operators – run by organized syndicates – the flood of timber, much of it illegal, on the domestic market means they benefit from cheap prices. Reducing illegal timber harvesting is likely to force mills to close and lead to a shortage of timber for domestic demand. But illegal chain-saw operators need to take the crackdown seriously. In ten years' time, they will lose their jobs because the trees are not there.
>
> (BBC, 2007)

A detailed report in 2005 on the 'Menace of Illegal Logging Operations' states, 'The menace of illegal chain-saw operations to the environment and sustainable agriculture is real and can no longer be taken for granted. And until the government takes a bold and pragmatic decision on illegal chain-saw operations, the country's forest and tree resources face massive degradation and overexploitation' (Sarfo-Mensah, 2005: 18). In concluding his report, Sarfo-Mensah (2005) advocates the introduction of a carefully controlled permit system for a fixed number of registered local groups of timber traders and their chain-saw operators to supply domestic markets.

Box 3.7

The Congo Basin Forest Fund

The rainforest of the Congo Basin, covering some 200 million hectares (twice the size of France), is the second-largest tropical forest in the world, after the Amazon. The forest stretches across parts of Cameroon, the Central African Republic, the Congo (DRC), Equatorial Guinea, Gabon and the Republic of the Congo. It has one-fifth of the world's remaining closed-canopy tropical forest and represents a significant carbon store and reservoir of biodiversity. It is home to over 50 million people, 10,000 plant species, 1000 bird species and 400 species of mammal. Increased logging, population growth, changing agricultural systems and exploration for oil and minerals are leading to some 1.5 million hectares of forest being lost each year, and if this continues over two-thirds of the forest will have been destroyed by 2040.

As early as 1988, discussions in Gabon led to the establishment of the ECOFAC (Ecosystèmes Forestières d'Afrique Centrale) programme, which aims to protect the forest in six central African countries. A number of other initiatives followed, sponsored by the IUCN (International Union for the Conservation of Nature) and WWF (Worldwide Fund for Nature), and in 2002 the Congo Basin Forest Partnership (CBFP) was established at the World Summit on Sustainable Development in Johannesburg. Since then, collaboration in the management and conservation of the Congo Basin rainforest has increased, and in June 2008 the Congo Basin Forest Fund (CBFF) was established to support conservation initiatives. The CBFF is initially being financed by grants of £58 million from the UK government and £50 million from the Norwegian government. It will support project proposals from the Conference of Ministers of Forestry of Central Africa (COMIFAC) to develop the infrastructure and abilities of governments, local communities and civil society to manage their forests and promote livelihood strategies which are compatible with the conservation of forests (CBFF, 2008). Kenyan environmentalist and Nobel Peace laureate Professor Wangari Maathai and former Canadian Prime Minister Paul Martin chair the fund.

At the launch of the CBFF, Jens Stoltenberg, Prime Minister of Norway, said:

> Climate change is one of the defining challenges of our time. The global community will have to find ways to reduce total emissions dramatically over a short time span. Reducing deforestation and forest degradation in developing countries is a main priority for Norway's climate policy, alongside investing in new technologies such as carbon capture and storage. We believe that the Congo Basin Forest Fund is a good example of a mechanism by which developed countries can help shoulder the financial burden of developing countries making significant emissions reductions.
>
> (DfID, 2008b)

Research by the Rainforests Foundation and Forests Monitor suggests that negative environmental and social impacts generally outweigh the economic development benefits of industrial logging in African countries. Typically, local people are excluded from management agreements which are drawn up between logging companies and government forest authorities, while logging companies often manage to avoid paying taxes; corruption is also common. Moreover, armed conflict has been associated with logging, and in both Liberia and the DRC armed rebel groups have been supported by money from the timber trade. Finally, environmental degradation is common in commercial logging areas, with soil erosion, biodiversity loss and pollution (Counsell, *et al.*, 2007).

Forests play a key role in sustaining livelihoods, particularly for poor people, supplying a wide range of products, such as food, medicines, timber for construction and fuelwood (see Chapter 4). But sustainable forest management strategies are needed if this vital resource is to be preserved for future generations. Local people need to be involved in forest management and decision-making. The linkages between deforestation and climate change are also significant. As a 2007 report from the International Institute for Environment and Development comments, 'After several years of dwindling world attention, forests are now set to return to *flavour of the month*' status. Climate change discussions will bring incentives for '*avoided deforestation* to the serious negotiations stage' (Mayers, 2007: 6).

There has been much concern in recent years about the loss of trees during drought periods and the harvesting of firewood in rural and urban communities. Fuelwood is still very important in many of Africa's poorest countries, particularly for cooking. Whereas in the Moroccan cities of Casablanca and Rabat gas is the main fuel used in 99.8 per cent and 98.4 per cent of households, respectively (UN-HABITAT, 2008), in Malawi, Niger, Sierra Leone and Uganda firewood and charcoal are used by 97 per cent of households (World Bank, 2008a). A detailed study undertaken in the 1980s in the cities of Kano (Nigeria) and Freetown (Sierra Leone) found that 47 per cent and 73 per cent of consumers, respectively, used wood for cooking, and the propensity for fuelwood use is greater among poor households (Cline-Cole, 1989). Fuelwood use is generally much higher in rural areas, where alternative fuel supplies are often unavailable. The reduction of woody plants can result in soil erosion and deterioration as the surface is exposed to rain-splash and hard crusts can soon form on bare ground.

It is often assumed that African communities have mismanaged their forest resources, and deforestation is seen as one of the main causes of land degradation and desertification. Deforestation by local communities undoubtedly does occur in certain situations, for example in Freetown during the 1990s, when thousands of desperate refugees abandoned their rural villages in the face of attacking rebels and sought a safe haven close to the capital, where relief organizations were based. Large areas of forest were chopped down in the mountain areas around Freetown to build makeshift homes, to provide fuelwood and to clear areas for growing food crops.

However, as with soil conservation, there is now much evidence to indicate that many communities manage their trees carefully and have long been aware of the need to do so. For example, in the savanna region of northern Nigeria, a study in the 1980s of fuelwood production around the large city of Kano revealed that the area closest to the city, and with the highest rural population densities (up to 500 persons per sq km), and closest to the urban fuelwood market, not only had the highest tree densities (12.3 trees per hectare) but recorded an increase in tree density of 18 per cent between 1972 and 1981. Further from the city, there had been an increase in tree density of 26 per cent between 1972 and 1985. The cutting of living and healthy trees was rare and the volume of timber per hectare was actually greater in farmland than in either shrubland or forest reserves because the trees were much larger. A single silk cotton or kapok tree, for example, may contain as much wood as several hundred trees of a smaller species with a shrubby growth habit. Farmers in the Kano area were well aware of the importance of trees in providing valuable shade and a wide range of economic crops and foodstuffs for people and animals (Cline-Cole *et al.*, 1990; Cline-Cole, 1995).

Research elsewhere in Africa has indicated the local importance attached to managing tree cover. For example, in the Kakamega District of western Kenya, Mearns found that as farm size declines and population density increases, tree cover is actually greater. Furthermore, he suggests, 'the share of the trees and shrubs that are planted is greater on smaller farms and in areas of higher population density' (Mearns, 1995: 106).

Fairhead and Leach's work in Guinea has shed further light on the local management of forest resources. With reference to the savanna areas on the northern edge of the forest zone, they report how scattered forest areas in the savanna have long been regarded as relics of previously more extensive forest cover which 'inhabitants have progressively converted . . . to "derived" savanna by their shifting cultivation and fire-setting practices' (Fairhead and Leach, 1995; 163). Fairhead and Leach identify over 800 'forest islands' in Kissidougou Prefecture alone, most of which have a clearing with a village at their centre.

However, instead of this being a progressively degrading environment, as earlier reports suggested, these forest islands are actually the result of deliberate planting of trees by local communities to protect their villages from bush fires, high winds and excessive heat, to create sources of forest products, and to provide military defence and secrecy for ritual activity. As Fairhead and Leach (1995: 167) comment, 'Comparison of air photographs, documentary sources and oral histories shows that woody cover on the upland slopes and plateaux between Kissidougou's forest islands has generally increased during this [the twentieth] century, and not declined as policy-makers have supposed.' This enrichment of the forest mosaic and increasing agricultural productivity have led to a more productive and sustainable environment, and Fairhead and Leach (1996: 121) argue that environmental agencies should both recognize and support 'the diverse local institutions engaged in resource management'.

The protection of trees, as valuable elements in the environment, has been a feature of many rural communities for generations. For example, in Niger, the sultans of Zinder at the end of the nineteenth century had a policy of executing anyone who cut down certain trees that provided important food supplies. In Mali, there is widespread preservation of tree species that produce valuable fruits for sale or household consumption. The protection of such species is an important example of indigenous silviculture, and clearly indicates that poor farming communities do take initiative in managing their environment through protection and rehabilitation.

3.8 Pests and diseases

Once called 'the white man's grave' by early colonialists, Africa still has many pests and diseases, many of which are in some way environment-related. There is a marked north–south divide within the continent in relation to health and welfare. In those countries north of the Sahara, infant mortality has been steadily declining to levels approaching those of Western European countries, but sub-Saharan Africa still has some of the world's highest levels, with, for example, 165 deaths per 1000 live births in Sierra Leone. Similarly, while life expectancy in North African countries averages 71 years, in sub-Saharan Africa the average is only 46.7, and in some southern African countries (due largely to the HIV/AIDS pandemic) the situation has deteriorated considerably in recent years, with Botswana (by no means one of Africa's poorest states), for example, having an average life expectancy of only 35 years (World Bank, 2008a). (Health issues are considered in more detail in Chapter 6.)

Many of Africa's problems may be summed up with just one word: water. Droughts and floods are frequently in the news, and the absence of a reliable and clean water supply is still a daily reality for millions of African families. In addition, many diseases which are endemic to the continent are water-related (see Chapter 6).

The tse-tse fly, a large brown fly of the *Glossina* genus, is the main vector of trypanosomiasis, causing sleeping sickness in humans and nagana in animals. The disease is transmitted through the bite of an infected fly and initially leads to fever, headaches and itching, followed in the later stages by confusion, poor coordination and disturbance of the sleep cycle (hence the name, 'sleeping sickness'). The disease can be fatal if not treated. Tse-tse flies are found only in sub-Saharan Africa, mainly in sparsely settled savanna woodland. Throughout Africa's long history, the distribution of human settlement and the presence of cattle, horses and many other domestic animals have been severely restricted in many areas through the prevalence of tse-tse fly. There were major epidemics of sleeping sickness between 1896 and 1906 in Uganda and the Congo Basin. By the mid-1960s, the disease had almost disappeared, but then it reappeared due to lack of surveillance. In 1986, the World Health Organization (WHO) estimated that some 70 million people lived in areas where the disease could be a problem, and in 1998 it was estimated that there might be up to 500,000 cases, with many undiagnosed. By 2008, the number of cases had fallen to between 30,000 and 50,000 due to a public–private partnership between the WHO and the pharmaceutical company Sanofi-aventis, with the latter supplying free drugs.

Malaria, transmitted by the female *Anopheles* mosquito, is a major cause of ill-health and death in Africa, particularly among children, with over 90 per cent of the world's annual acute cases of the disease being in Africa (see Chapter 6).

One of the major pests of Africa is the locust, a species of large grasshopper of the *Acrididae* family which congregates in swarms. One of the biblical plagues of Egypt was a swarm of locusts that ate all the crops in the country. It is difficult to predict plagues of locusts. Sometimes they are a major problem after drought periods, but there is also evidence suggesting that plagues are more likely if there is steady rainfall during the insect's breeding season. Swarms can destroy vast amounts of growing crops. In January 2007, for example, locusts threatened to decimate the entire cashew crop of Guinea-Bissau, on which two-thirds of the country's farmers depend, while further west, in Mauritania, there was a severe threat to the nation's staple food crops of maize, millet and sorghum. The Food and Agriculture Organization of the United Nations (FAO) has set up a Desert Locust Information Service (known as Locust Watch) to provide early warnings of possible swarms in different regions. Meanwhile, in 2004, the FAO supported aerial pesticide spraying of over 11 million hectares of land in Africa. However, the use of such chemicals is not without problems for both people and the environment.

Another very significant pest in Africa is the quelea bird, notably the red-billed quelea (*Quelea quelea*), which is mainly found in the savanna grasslands of sub-Saharan Africa. The quelea might well be the world's most abundant wild bird species, with an estimated 1.5 billion breeding adults. The birds breed when rainfall is abundant and a single flock can number in the millions. There are reports of huge flocks taking as long as five hours to fly past and their effect on both wild grasses and grain crops can be devastating. Farmers regard the birds as a serious pest, but there is little that can be done to reduce their numbers (Ezealor and Giles, 1997).

3.9 What future for the environments of Africa?

The experiences of recent years clearly indicate the variable and unpredictable nature of environmental events and the vulnerability of many of the poor people of Africa who live off the land. Recent instances of drought and famine have provoked much concern among major charities and relief organizations, and a considerable amount of detailed investigation has ensued. As a result of the latter, many longstanding and sophisticated coping mechanisms within rural communities in marginal environments to ensure survival through difficult environmental events have finally been acknowledged in the West. Also, the multitude of ways in which local people have adapted to their environment and have taken steps to maintain and strengthen their environmental resource base is impressive (see Boxes 3.5 and 3.6).

It is now also clear that tensions are more likely to develop where food production and social systems have been constrained, for whatever reason, and the efficacy of their adaptive and coping mechanisms reduced, leading to famine and land degradation. Political policies and development programmes have done much to destabilize traditional lifestyles. Drought need not necessarily lead to famine, but the destabilizing effects of civil war on food production in countries such as Ethiopia, Mozambique, Somalia and Sudan have been only too evident and many lives have been lost as a result.

Colonial policies introduced in the late nineteenth and early twentieth centuries also constrained people and their livelihood systems while increasing their vulnerability. For example, the introduction of groundnut cultivation for export by the French in Senegal and by the British in northern Nigeria took little notice of traditional ways of life (Franke and Chasin, 1980). The migratory routes of pastoral groups were disrupted, while farmers were encouraged to grow groundnuts rather than spend their time producing food crops. The introduction of taxation by the colonial powers further encouraged farmers to grow groundnuts for cash and resulted in the large-scale migration of young men to towns and plantations in search of money. The migration of Mossi men from Burkina Faso to the coffee and cocoa plantations of Côte d'Ivoire and Ghana was initiated in the colonial period and continues to this day. While the young men are absent, much of the village work must be done by women. There is no shortage of examples from Africa past and present of government policies and development schemes that have undermined traditional ways of life and caused social and environmental problems.

Of greater value to the rural poor in marginal environments are some of the initiatives that have come from a number of non-governmental organizations working with the rural people at grassroots level and aimed at reducing vulnerability. The establishment of cereal banks in Burkina Faso and Mali by Oxfam since the late 1970s is an example of this. In Niger and Mali, CARE has collaborated with government forestry services in reforestation, soil conservation and agroforestry projects.

Since the disastrous droughts and famines of the mid-1980s, there has been a boom in the business of predicting drought and famines before they occur. This has involved attempts to develop drought and famine early-warning systems, based on a range of environmental and socio-economic data such as rainfall levels, the condition of pasture, migration of people and livestock, level of crop storage and animal sales. However, such data are often unavailable and their interpretation can be time-consuming, requiring a deep understanding of the societies in question. By the time the data have been assessed, the situation may already be serious.

In attempting to reduce vulnerability and improve food security, it is important to investigate and treat each social group and environmental situation in its own right and build upon existing diversification, security and coping strategies. Governments, NGOs and others must work with local people to generate future development strategies that are sustainable. The concept of 'sustainable development' implies a need to manage natural and human resources in such a way that the quality of the resources in the long term is not jeopardized by the desire for short-term benefits. For example, if the intensive cultivation of cash crops is likely to lead to severe land degradation and soil erosion, then it will not be sustainable in the future and will result in the depletion of vital environmental resources. Ensuring environmental sustainability is the seventh Millennium Development Goal (MDG 7). The specific targets include: first, integrating sustainable development principles into country policies and reversing the loss of environmental resources; second, achieving a significant reduction in the rate of biodiversity loss; and, third, halving the proportion of the population without sustainable access to safe drinking water and basic sanitation by 2015 (UN, 2008a).

Much has been learned in recent years about Africa's environments and the ways in which people interact with their environment. Slowly but surely, we seem to be moving away from viewing Africans as passive victims of harsh and unreliable environments and towards an appreciation of people–environment relationships, with attempts being made to build upon these in development strategies that work with the people and are sustainable in the long term. The main obstacles to such progress are likely to come from a failure to incorporate grassroots civil society into decision-making, poor governance and inappropriate and/or untimely political or military intervention.

Summary

1. Africa has a wide range of different environments whose characteristic features are largely determined by latitude, altitude and rainfall.
2. Climate change could have serious implications for Africa's people and environments.
3. Much has been written in recent years about desertification. It is not that desertification does not exist, but rather that its existence over space and time is probably much less than the popular image, and there is a lack of hard scientific evidence.
4. With inadequate data, changing environmental factors and the dynamics of human adaptation to environment, it is difficult to calculate the 'carrying capacity' of an area.
5. African food production systems frequently possess a high degree of structural resilience and indicate a sensitive adaptation to environmental conditions.

Discussion questions

1. In what ways might water be seen as the key to much of Africa's survival and future development?
2. Explain the term 'desertification', and examine the evidence both for and against desertification as a widespread process in Africa.
3. How useful is the concept of 'carrying capacity' in planning for future levels of human and animal populations in relation to resource availability?
4. With reference to specific food production systems, examine aspects of their 'resilience' in coping with difficult environments.
5. What might be the characteristic features of future programmes designed to promote 'sustainable development'?

Further reading

Binns, T. (ed.) (1995) *People and Environment in Africa*, Chichester: John Wiley.

Boserup, E. (1965) *The Conditions of Agricultural Growth*, London: Allen and Unwin (republished in 1993, London: Earthscan).

Fairhead, J. and Leach, M. (1996) Rethinking the forest-savanna mosaic. In Leach, M. and Mearns, R. (eds) *The Lie of the Land: Challenging Received Wisdom on the African Environment*, Oxford: James Currey, pp. 105–121.

Mortimore, M.J. and Adams, W.M. (1999) *Working the Sahel: Environment and Society in Northern Nigeria*, London: Routledge.

Tiffen, M., Mortimore, M.J. and Gichuki, F. (1994) *More People, Less Erosion: Environmental Recovery in Kenya*, Chichester: Wiley.

Useful websites

IDRC (International Development Research Centre), *Climate Change Adaptation in Africa*: http://www.idrc.ca/en/ev. Contains some useful information on climate change adaptation in Africa from Canada's premier development research centre.

IIED (International Institute for Environment and Development): http://www.iied.org. An excellent resource with information and publications on various aspects of people–environment relationships.

UNCCD (United Nations Convention to Combat Desertification): http://www.unccd.int/main.php. A key website which provides a range of perspectives and resources on desertification.

UNEP (United Nations Environment Programme) (2002) *Africa Environment Outlook: Past, Present and Future Perspectives*. Available at: http://www.grida.no/aeo/. Includes some excellent information and downloadable publications concerning environmental issues in Africa.

WHO (World Health Organization): http://www.who.int/en/. An important website with some excellent data and publications concerning health and diseases in Africa.

4 Rural Africa

4.1 Introduction

Despite the increasing trend of urbanization throughout Africa, the majority of the continent's population still live in rural areas, and will continue to do so for the foreseeable future. Recent data from the UN put the rural population of Africa at around 620 million in 2010, which exceeded the urban population by over 200 million. Estimates suggest that the urban population will finally exceed the rural by around 2030 (UN, 2010a; see Figure 4.1). The significance of these statistics from a development perspective lies in the implications that rural life has for the population: a life which, for the vast majority, is characterized by smallholder subsistence agriculture and a high vulnerability to poverty, underdevelopment and food insecurity. Indeed, the scale of the continent, the relative geographical remoteness of many areas and the cultural and ethnic diversity of rural populations have posed significant challenges for governments who have struggled to invest in rural infrastructure, basic needs and livelihood opportunities for rural people.

This chapter examines the ways in which development practitioners and academics have considered development in rural Africa, and how the view of rural populations employing unproductive, unscientific agricultural practices (Trapnell and Clothier, 1937) has given way to one where the cultural dimension of rural development has been widely recognized (Warren *et al.*, 1995), and where individual components of rural livelihoods are analysed in an attempt to understand the nature of rural poverty (Ellis, 2000). In particular, the emergence and adoption of the Rural Livelihoods Approach during the late 1990s has led to a greater and more nuanced recognition of the interactions between natural resources, local labour, culture, credit, the socio-economic and political environment, and all of these interactions' implications for sustainable rural development and poverty reduction (Chambers and Conway, 1992; Carney 1998). We shall outline these characteristics, patterns and processes that dominate rural Africa and influence the lives of the people who live there and the development challenges they face. The chapter begins by addressing the issue of land tenure arrangements, which plays a fundamental role in determining the nature of rural livelihoods, such as agriculture, pastoralism, fishing and forestry. Having outlined these livelihood strategies, we go on to address the importance of food security as a rural livelihood outcome, and look at how approaches to rural development have evolved over time.

4.2 Land tenure in Africa

One of the key aims of this book is to draw attention to the complex relationship between poverty and underdevelopment in Africa, and the various factors that influence this. While one should exercise caution in singling out specific determinants, it can be argued that in

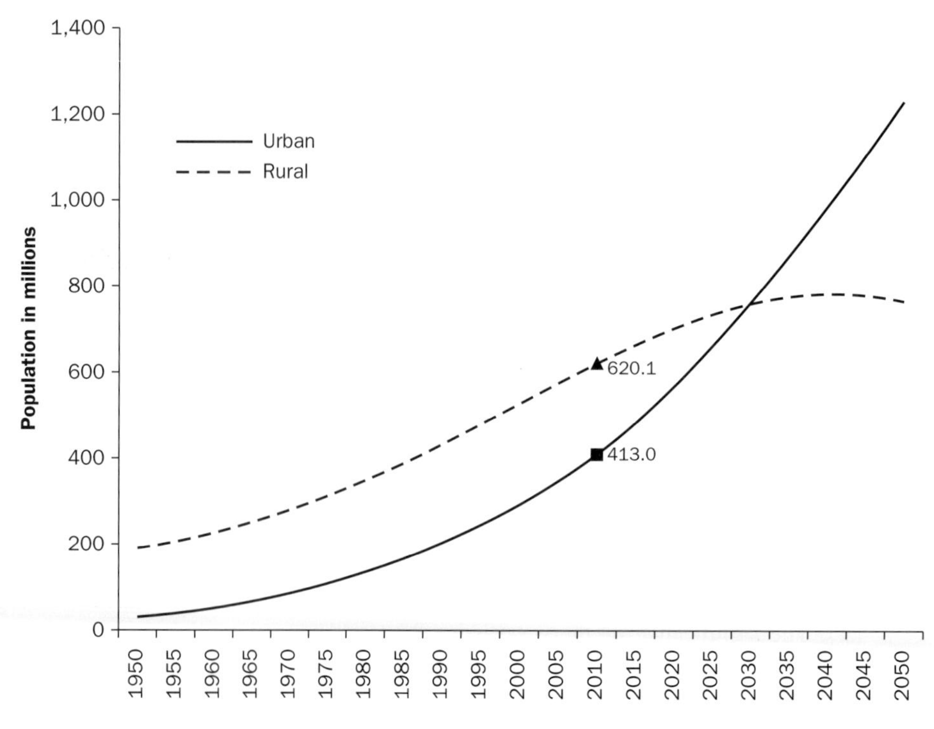

Figure 4.1 Africa's urban and rural populations compared.

Source: UN, 2010a

the context of rural Africa many of the development issues associated with poverty can be traced back to a single issue – the variability of land tenure across the continent. It is fair to say that since the vast majority of people throughout Africa depend on land in order to produce food or sustain a livelihood, it is imperative that they have secure access to land. Though much debated, it is for this reason that many have argued for the need to implement land reform and achieve land tenure security via the granting of freehold property rights that give individuals full ownership and control over their land. Increasingly, security of land tenure has been placed alongside food, water and sanitation as a basic human right (Hellum and Derman, 2004; Moyo, 2006).

The land tenure situation throughout Africa is extremely variable and reflects traditional systems of land accumulation, inheritance patterns, but perhaps most significantly the influence of different colonial governments, which are said to have introduced the concept of freehold tenure and individual land property rights to the continent (ECA, 2005a). Land tenure during the pre-colonial era is usually viewed as being characterized by 'customary systems', although there is little agreement on what this term means, since it was first used by colonial governments to classify and simplify a range of traditional land access arrangements. The relative abundance of land in pre-colonial Africa, however, resulted in a certain degree of flexibility in where people could settle; and, once settled, usufruct rights could be exercised, with land held in trust for the community by village leaders (Pottier, 2005). Land was therefore under communal ownership, with usufruct rights usually being subject to patrilineal inheritance patterns. Another feature of these customary systems was the designation of common land: that is, land that could be used by multiple users for grazing or foraging. In many parts of Africa, these systems evolved further with the emergence of chiefdoms and kingdoms where control over land allocation, and power, became more centralized. Similarly, as occurred with the emergence of the nation state through the

Amhara conquest in Ethiopia, feudal systems in which rulers allocated land to military leaders, the church or the nobility in return for a share of the taxation generated from tenant farmers also operated. Such a system existed in Ethiopia until 1975.

Colonialism had a significant impact on land tenure throughout Africa. Most colonial governments lacked sensitivity to local culture and customs, and land was widely regarded as having to be 'owned' by someone. The imposition of indirect rule, in which chiefs and traditional leaders were handed power by the colonial administration, had the effect of centralizing land ownership and decision-making among those leaders on behalf of the state. Where no leaders existed, they were created as a means of exercising clear authority over land. In other cases, land was simply appropriated for settlers for cash cropping and commercial agriculture. The situation for many countries at the end of the colonial period, therefore, was one where significantly less land was subject to customary land tenure arrangements, and the state became a significant agent involved in granting freehold and leasehold rights. Moreover, these changes to land tenure created a lasting legacy of inequality in land distribution. In Zimbabwe, for example, European settlers were granted freehold ownership of large areas of the most productive land. By the late 1990s, approximately 4500 white commercial farmers still owned 42 per cent of the country's agricultural land (Moyo, 2000).

Post-independence land reforms in many African countries have sought to address these inequalities, although with varying degrees of success. To facilitate a programme of villagization and agricultural reform, the Tanzanian government passed the Freehold Titles Act of 1963, which nationalized approximately 1 million acres of land. However, rather than paving the way for the more equitable distribution of farm land, this policy resulted in compulsory purchases of customary land and widespread disruption to rural livelihoods (ECA, 2005a; see Box 9.3, below). In Zimbabwe, attempts to redistribute land began in the early 1980s with a policy of land acquisition and compensation occurring on a 'willing buyer, willing seller' basis. During the 1990s, the government's power was extended to include the compulsory acquisition of land for resettlement and redistribution, although by 1997 only 71,000 families had been settled on approximately 3.5 million hectares of land, most of which was unsuitable for grazing or cultivation (Human Rights Watch, 2002). A 'fast-track' programme for the resettlement of Zimbabwe's war veterans was subsequently implemented in 2000, which has been characterized by widespread land confiscations, evictions and violence against commercial farmers and their employees. The dismantling of commercial farms, and the lack of support provided by the government for those who have been resettled, has led to the collapse of Zimbabwe's agriculture with disastrous consequences for people's livelihoods and the national economy.

While few would argue against the need for people to have access to good-quality agricultural land in order to make a living, there is a healthy debate among academics and practitioners as to which land tenure arrangements best favour the development of rural livelihoods. One view is that the prevalence of customary and communal land tenure arrangements throughout the continent can be blamed for the lack of development, mainly on the premise that these systems do not provide security of tenure and hence dissuade farmers from innovating and making long-term agricultural investments; customary land, and in some cases land that remains the property of the state, can be taken away and redistributed at the will of community leaders or government. Private ownership, meanwhile, provides security of tenure and creates incentives for long-term investments. An alternative view, however, is that private ownership is costly, difficult to implement in an equitable manner, and ultimately favours the already-wealthier people within society (Sjöstedt, 2007). Moreover, critics of private property rights argue that customary and communal land tenure arrangements can function effectively, particularly where shared natural resources are involved. Indeed, much empirical literature from Africa in recent years has highlighted the key role of local community-level institutions and farmer organizations in managing and deriving sustainable benefits from communal land and water resources (Ostrom, 1990; Agrawal, 2001; Mazzucato and Niemeijer, 2002; Dixon, 2005).

4.3 Rural livelihoods

During the last twenty years, there has arguably been a significant shift in the way development has considered and targeted its interventions in rural Africa. In academic circles, the work of Robert Chambers during the 1980s and 1990s spearheaded a movement away from analysing rural areas in terms of wider socio-economic and political processes to a focus on the ways in which local people make a living, and how this is influenced by such local factors as resource availability, culture and knowledge (Chambers, 1983, 1997). The adoption of so-called 'bottom-up' approaches to development, which allow local people to participate in the analysis of their own circumstances and development decision-making (see Chapter 9), has had a significant impact on our understanding of what rural life means for the majority of Africans. Similarly, the sustainable livelihoods framework (see Box 4.1), which has been widely adopted in Africa's many rural development programmes, has facilitated a deeper understanding of the dynamics of rural livelihoods and the choices that are available to rural people trying to make a living.

Box 4.1

The sustainable rural livelihoods approach

The sustainable livelihoods approach (SLA) was developed during the 1990s at the Institute of Development Studies (IDS, University of Sussex) and the UK Department for International Development (DfID) as a tool for analysing the livelihoods of the rural poor. Drawing particularly on the work of Robert Chambers and Gordon Conway, Scoones (1998: 7) conceptualizes a livelihood as: 'the capabilities, assets (including both material and social resources) and activities required for a means of living. A livelihood is sustainable when it can cope with and recover from stresses and shocks, maintain or enhance its capabilities and assets, while not undermining the natural resource base.'

The sustainable livelihoods framework (SLF), which encapsulates the approach (see Figure 4.2), attempts to illustrate how the livelihood strategies pursued by the rural poor are directly related to their local-level assets, but that the use of these assets are influenced by government policies, market forces, institutions, and a 'vulnerability context' that includes elements of uncertainty such as environmental or socio-economic change. A range of possible goals, or 'livelihood outcomes', is considered, including improvements in well-being (economic and social), reduced vulnerability to shocks and pressures, and the maintenance of the natural resource base. The SLA and SLF have been used extensively throughout Africa in many rural development programmes and research initiatives (Carney, 2002), not least because they allow development agents and researchers to focus on specific components, which can be assessed and subsequently targeted for interventions.

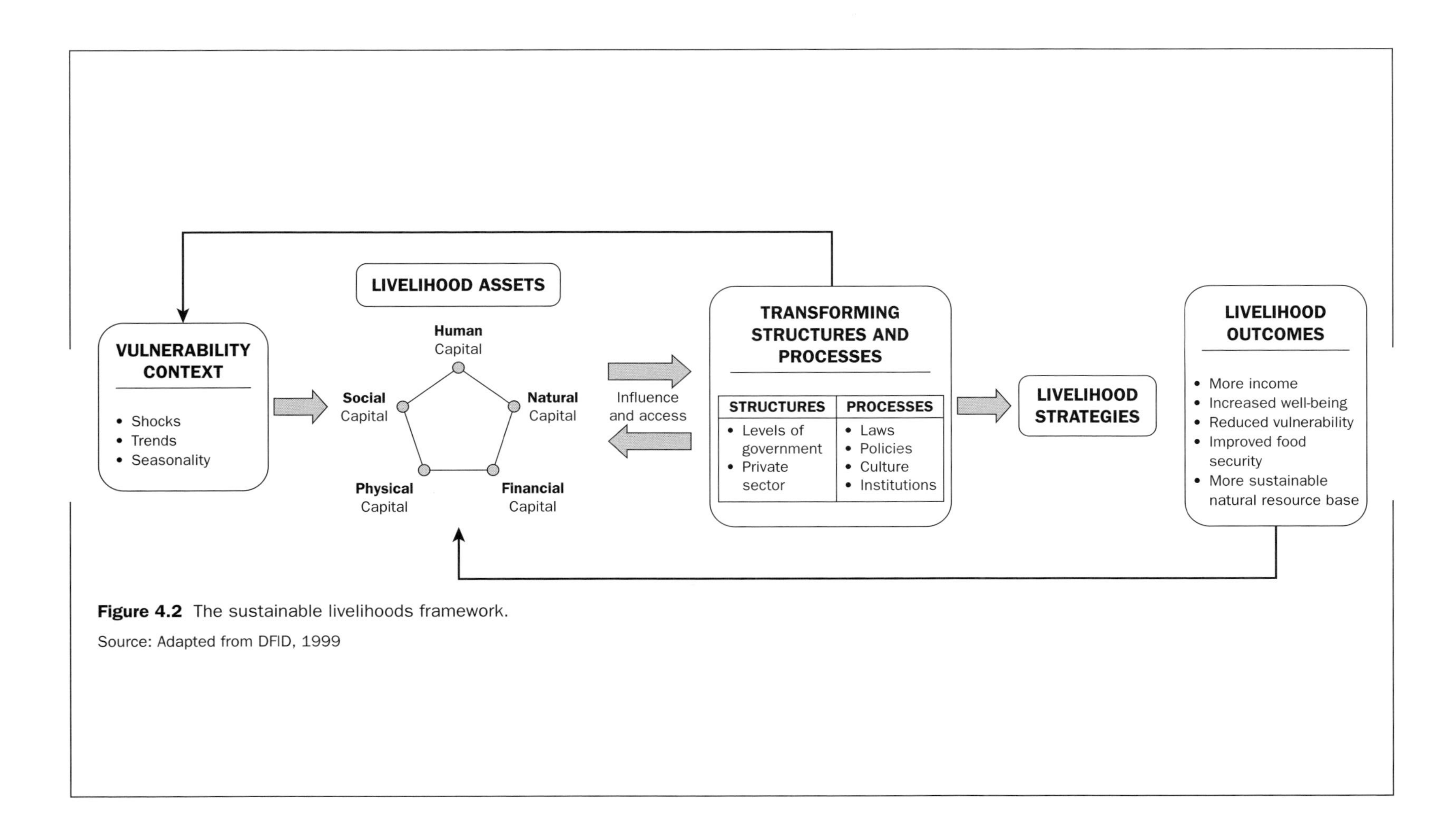

Figure 4.2 The sustainable livelihoods framework.

Source: Adapted from DFID, 1999

Examples of the adoption of the SLA include:

- Tanzania: Ellis (2000) effectively tested the use of the SLA in three villages to compare the rural livelihood strategies of the poor and non-poor;
- Malawi: Orr and Mwale (2001) used the SLA to analyse capital assets and livelihood outcomes for rural households, identifying market liberalization as the cause of higher incomes for tobacco farmers;
- Ethiopia: Babulo *et al.* (2008) adopted the SLA as a means of analysing the relationship between individual household assets and livelihood dependence on forest resources;
- Kenya: Kristjanson *et al.* (2005) identified and mapped capital assets among agro-pastoralist communities to investigate the determinants of poverty;
- West Africa: the SLA has also underpinned West Africa's Sustainable Fisheries Livelihoods Programme, which has led to improved understanding of the problems and challenges facing fishing communities (Allison and Horemans, 2006).

In an examination of the ways in which the SLA has been adopted, incorporated and modified by development and aid organizations in Africa over the last decade, Hussein (2002) also provides case studies of bilateral, multilateral and NGO engagement in Malawi, Uganda, Egypt and Ghana.

Despite its enthusiastic reception during the late 1990s, interest in the SLA appeared to wane during the late 2000s, with critics questioning its utility beyond the household level, its oversimplification of policy and institutional processes, and the lack of attention paid to intra-household dynamics and gender power relations. Nonetheless, the SLA remains a vital component within sectoral programmes of the World Bank, UNDP, FAO and many NGOs, such as Oxfam and Save the Children.

This section reviews the key characteristics and challenges inherent in a range of common livelihood strategies, but it is important to emphasize the spatial and temporal dynamism of rural life. Individual livelihood strategies can change quickly in response to external factors (for example, recent work has focused on the impacts of climate change: Adger *et al.*, 2003; Thomas *et al.*, 2007), and livelihood strategies can be characterized by diversification – a mixture of subsistence and market activities in different sectors. One must therefore be cautious not to generalize; no two rural livelihood strategies are likely to be the same as a result of differences in the resources, circumstances and aspirations of rural people.

4.3.1 Agriculture

At the start of the twenty-first century, rural life in Africa remains dominated by agriculture, which is estimated to contribute to the livelihoods of over 70 per cent of the population, and account for between 25 and 30 per cent of the continent's GDP (ADB, 2010; Foresight, 2011). As shown in Figure 4.3, however, this varies significantly between countries, ranging from 55.5 per cent for the Central African Republic to a mere 2 per cent in the Seychelles. Some countries, such as Rwanda and Burundi, have over 90 per

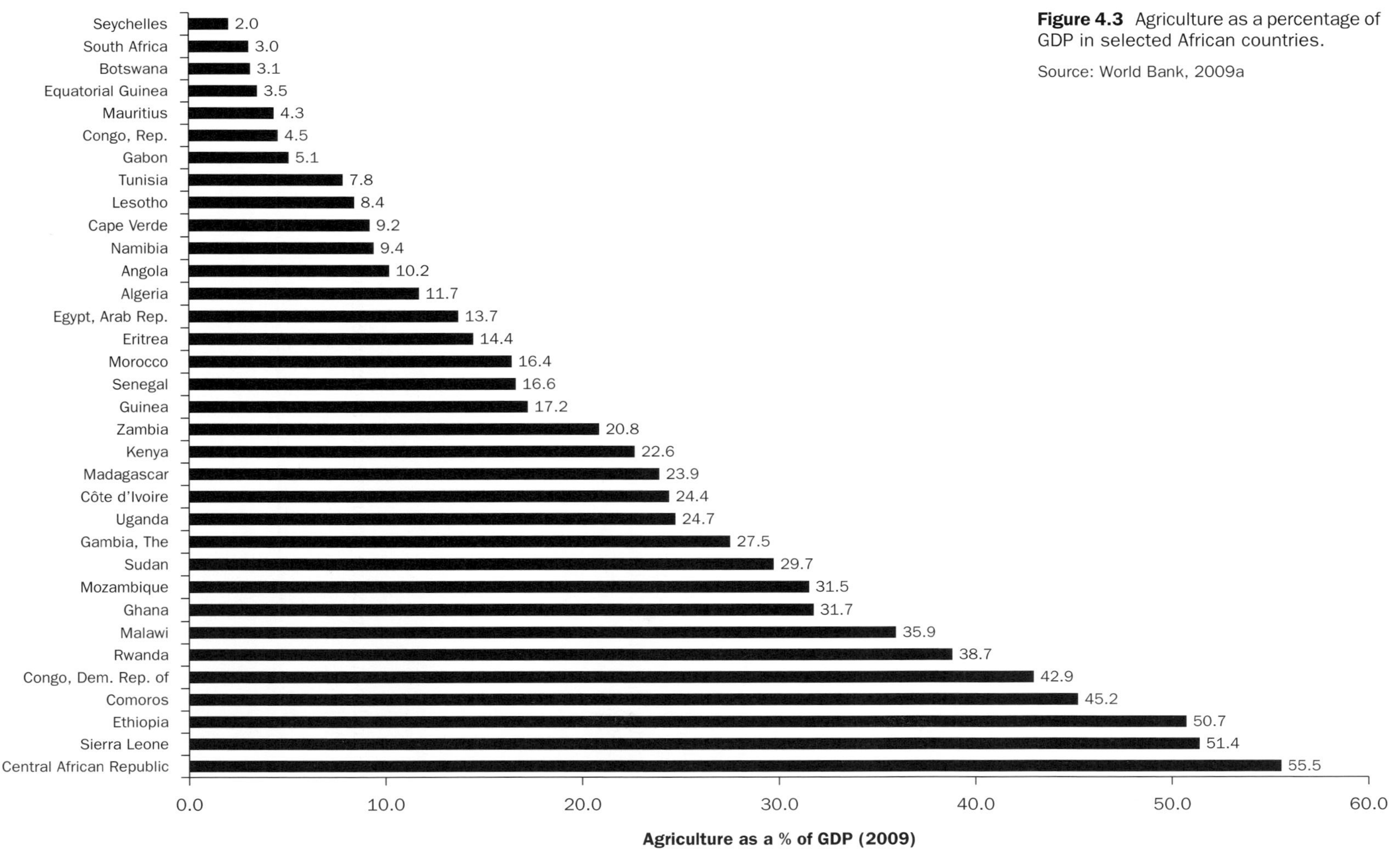

Figure 4.3 Agriculture as a percentage of GDP in selected African countries.

Source: World Bank, 2009a

cent of their populations working in agriculture, while it is only in Africa's small island states and North Africa where this figure falls to below 30 per cent (World Bank, 2009a). In terms of production, while the data in Figure 4.4 shows that yields for most crops have increased steadily over the last fifty years, this has not kept up with the demand that has occurred as a result of Africa's population growth. Statistics reveal that between 1960 and 2009, Africa's crop yields increased from 29,242 kg per hectare to 41,030 kg per hectare (an increase of 40 per cent), while population increased from 285 million to 1.03 billion (a 263 per cent increase) (FAO, 2011). The data in Figure 4.3 also mask the inter-country variation that has occurred as a result of drought, conflict and the HIV/AIDS pandemic.

Agriculture in Africa can be broadly divided into smallholder/indigenous and large-scale/modern (ECA, 2005b). The latter typically employs only a small proportion of those involved in agriculture, and it is more capital intensive, using machinery, fertilizers and irrigation to produce crops mainly for export (see Plate 4.1 and Box 4.2). However, this sector tends to make up agriculture's greater contribution to GDP, and this is set to increase as large areas of rural Africa are leased or sold to foreign-owned agribusinesses. In 2010, for example, the Indian company Karuturi Global leased approximately 740,000 acres of land in the Gambella region of Ethiopia specifically for the cultivation of cereals for export (Parulkar, 2011). In contrast, indigenous smallholder farming (see Plate 4.2 and Box 4.3) constitutes the bulk of the labour force but contributes less to government revenue. Smallholder farming is also labour- rather than capital-intensive, uses indigenous techniques that have evolved over centuries, and, given persistent problems of food insecurity, tends to be subsistence orientated: that is, food is produced for household consumption. Again, however, this grossly oversimplifies the complexity of African farming systems. Many subsistence farmers will produce surpluses beyond their household needs, and will sell these surpluses locally with detailed knowledge of market prices. The production and sale of such surpluses has been going on for centuries in some parts of Africa, and there are long-established marketing systems with flourishing markets linked by trade routes. In pre-colonial times, a variety of currencies, such as cowrie shells and

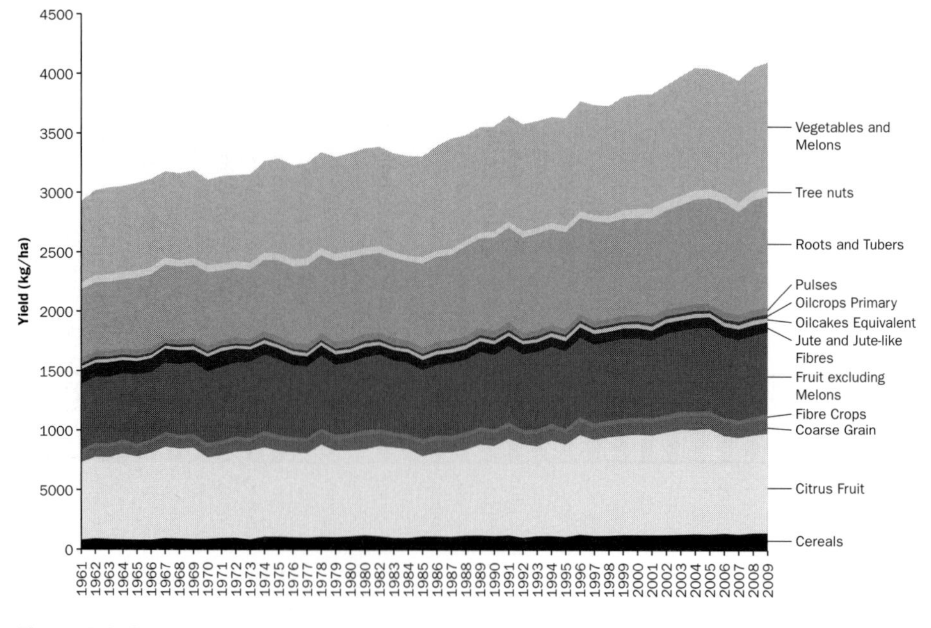

Figure 4.4 Crop yields in Africa, 1961–2009.

Source: FAO, 2011

Plate 4.1 Intensive wheat production outside Lusaka, Zambia (Alan Dixon).

metal bars, were in widespread use, but today these have been replaced by the cash economy (Binns, 1994). The fact that market places and trade routes have flourished over many centuries in parts of Africa suggests that production systems were geared not only to domestic needs but to generate sizeable and regular surpluses for sale.

Box 4.2

Cash crops or food crops?

There is much debate about the merits and problems of growing crops for sale and export rather than for domestic consumption. Although any crop which is traded for money might be called a 'cash crop', the term is usually used to refer to major export crops, such as coffee, cocoa, tea, sisal and groundnuts. Large-scale production of these crops in Africa started during the colonial period, when the European powers were looking for places to produce cheap raw materials for developing food and manufacturing industries. Export crops are basically produced in two ways – on smallholder farms and on large estates or plantations. The latter are changing, but their typical features include:

- expatriate ownership and management, often by a multinational company, with production of a limited range of crops geared to the export market and some processing of the product undertaken on the plantation itself;
- use of labour-intensive production methods, and a relationship between management and labour force which is hierarchical and characterized by inequalities and restrictions; and

- workers living on the plantations and being heavily dependent on them for their livelihood, making them self-contained communities or 'total institutions'.

The establishment of expatriate-owned plantations was discouraged in British West Africa. However, the French had plantations growing coffee, cocoa and bananas in Côte d'Ivoire and bananas in Guinea, the Americans grew rubber in Liberia, and the Germans, as early as 1884, set up rubber, oil-palm and banana plantations in Cameroon. Elsewhere, the Belgians developed rubber, oil-palm and coffee plantations in Zaire. In eastern and southern Africa, white settler farmers established estates in Kenya and Zimbabwe, growing tea, coffee, pyrethrum, tobacco, cotton and sugar. Since independence, some countries have witnessed a retreat of foreign companies from estate ownership, and in certain cases plantations and estates were nationalized, although in Tanzania this was associated with a decline in managerial efficiency, profitability and investment levels. Despite past criticism, there is now more support for plantations among African governments and the movement to nationalize them is no longer quite so fashionable. Kenya has a valuable plantation sector coexisting with smallholder agriculture, and evidence suggests that the poorest rural people may be better off within a plantation environment than if they remain as landless labourers in some of the country's poorest districts. In addition to providing employment, some large Kenyan plantations have healthcare facilities that are superior to those in surrounding areas.

A major concern over cash crop cultivation on smallholder farms relates to problems of food security and the nutritional status of households which may be neglecting food production to cultivate crops for the market. There are many examples of cash crop-growing communities with poorer nutritional status than more subsistence-oriented communities in similar areas. Cash crops are often the

Plate 4.2 Agriculture in Ethiopia using traditional technology (Alan Dixon).

main responsibility of men, while women continue to grow the staple foods. In The Gambia, for example, women grow rice, while men are more concerned with the groundnut crop and upland cereals, such as sorghum, millet and maize. The neglect among men of food crops in favour of cash crops can place a considerable burden on women who are tasked with growing enough food to meet subsistence needs. In some communities, the rewards of cash crop production are kept by the men, and since the women are fully occupied with feeding the family, they have little opportunity to earn any additional income. There is a need for more detailed studies on the relationships between cash crop production and household nutrition.

Box 4.3

Indigenous agricultural systems in Sierra Leone

Sierra Leone, a small (72,335 sq km) West African republic which was a British colony until independence in 1961, has a tropical climate with an annual mean temperature of 26.7 degrees centigrade. Annual rainfall reaches 6700 mm near Freetown, the coastal capital, but decreases inland and eastwards to 2200 mm. Rain falls mainly between May and October, with a dry season from November to April. The cultivation cycles of traditional non-irrigated farming systems are closely related to these seasons, with the main cultivation period corresponding with the rainy season.

Kono District, in north-eastern Sierra Leone, is an area of plateaux and hills separated by river valleys. Primary forest is now found only in protected reserves, much of the present vegetation being secondary bush, the result of repeated clearance and cultivation by farmers. Further north, with declining rainfall, savanna grassland interspersed with woodland is common. Soils are generally light, lateritic and naturally infertile, though they receive a plentiful supply of nutrients from decaying vegetation. The most fertile soils are along river valleys, where there are thick deposits of alluvium, and in inland swamps.

Land in Kono is managed by chiefs on behalf of their communities, and individual farms average 1.25 hectares. Indigenous Kono farmers can clear land where they wish, provided no other person has already started work there. Outsiders, however, are required to approach the paramount chief, who allocates land usually for one cultivation cycle in return for a portion of the harvest as payment. Buying and selling of land are rare away from the main towns of Kono District. The upland rice farm is the most important food production system; although some farmers have swamp farms, kitchen gardens and cash crop gardens, the rice farm is the focus of household and community economic and social activity. A rotational bush fallowing system is used, with periods of one or two years' cultivation separated by longer fallow periods. Settlements are fixed, with most farmers walking from the village to their farms each day during the cultivation season (Binns, 1982a, 1992).

The location of a farm may be chosen several years ahead, with farmers making simple soil tests and examining other natural indicators of fertility. Clearing normally starts in the latter part of the dry season, during January and February, by which time much of the vegetation is dry and brown (see Figure 4.5). The undergrowth is cleared using a cutlass and then trees are felled with an axe at about a metre from the ground, leaving the stump and roots intact. These are difficult to remove with available technology, but tree roots also help to bind the soil together and prevent erosion. When the farm is abandoned, trees and other vegetation quickly grow back, returning valuable nutrients to the soil. Certain economic trees, such as the oil palm and fruit trees, are not felled. The oil palm is fire resistant and provides oil for cooking as well as an intoxicating liquid known as palm wine.

Deciding when to burn the farm can be difficult, but it usually takes place in March or early April, when the first storms occur. If burning is done too soon, the debris may not be thoroughly dry, but if it is left too late, the rain may extinguish the fire, necessitating the laborious regathering and refiring of the debris. The whole household helps to burn the farm, ensuring that the fire neither destroys economic trees nor strays on to neighbouring farms. After burning, timbers are used to build a farmhouse, which serves as the main shelter during the cultivation season and may also be used for storing the harvest.

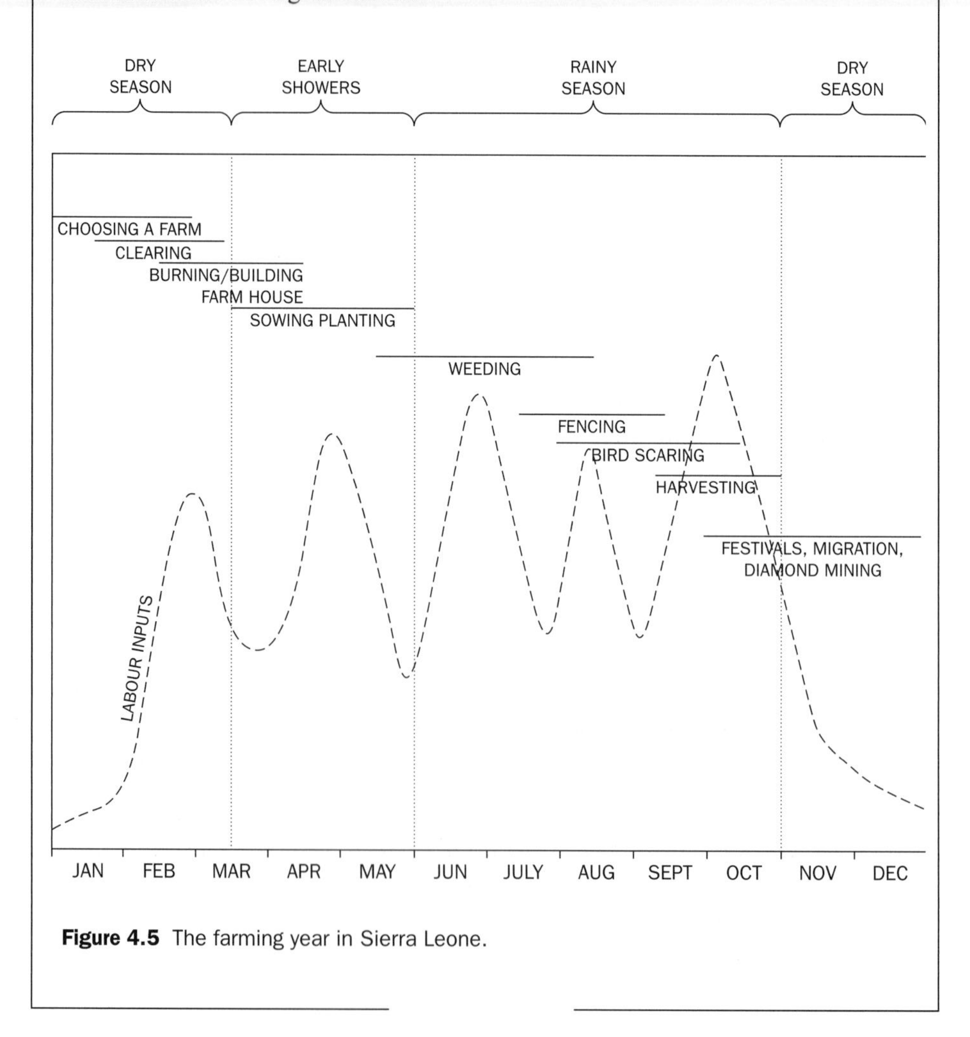

Figure 4.5 The farming year in Sierra Leone.

When rain showers start occurring regularly, in late April or early May, the seeds are sown. A technique known as 'intercropping' is used, in which fifteen or more different crops may be grown on the same farm. In addition to rice, farmers may well plant yams, beans, tomatoes, okra, benniseed, maize, cucumber and cotton, among other crops. This technique was widely criticized during the colonial period, since it was thought that yields of individual crops were lower than in monocropped fields. However, in recent years, the many virtues of intercropping have been recognized. For example, different crops make different demands on the soil and cover the ground, protecting the soil from raindrop impact and possible erosion, and total crop yields may actually be greater than on monocropped fields. Furthermore, intercropping provides the household with a varied diet.

Rice and other seeds are broadcast and quickly covered by men using long-handled hoes. Other crops, such as maize and yams, may have been planted earlier. Maize, for example, is often one of the earliest crops to be harvested, in July or August, and can be a welcome relief during the pre-harvest 'hungry season'. With more frequent rainfall, the crops grow quickly and regular weeding is undertaken by women using short-handled hoes from late May. This work is regarded as too degrading for men. It is back-breaking and time-consuming, and women do it in addition to preparing food, tending vegetable gardens, collecting water and firewood, looking after young children and selling surplus produce in the market. By late June or July, the maturing crops have to be protected from pests, notably the 'cutting grass' or cane rat, a large beaver-like rodent which, in the space of a single night, can flatten and eat large areas of crops. Fencing and trap-making, undertaken by the men, is difficult and unpleasant work since it often has to be done in the rain. August is the wettest month and often a time of food shortage and illness, which may lead to a smaller harvest if farm work is neglected.

Rainfall decreases in September and, as the rice crop swells and ripens, bird-scaring platforms are erected in the fields. Children spend many hours keeping birds away from the crops and they compete with each other in devising ingenious strategies. The rice harvest, between mid-September and November, is the big event of the year and the entire household helps. Labour supply is the main limiting factor on Kono farms, and although a large household with many adults may have the necessary labour force, it also needs to produce a lot of food. Households with a high proportion of very young or old members are often constrained in their farming unless they have access to voluntary help from outside the household or can afford to hire labour. To avoid labour 'bottlenecks' at certain times of the year, farmers may subdivide their land and stagger the operations so that, for example, the entire farm does not have to be weeded or harvested at the same time.

Rice is cut stem by stem using a small knife and gathered into bundles to dry in the sun, after which it is stored in a barn or carried to the village. Most of the rice crop is used for household consumption, but some may be allocated to settle debts or for festivals and ceremonies, such as the initiation of teenage children into secret societies. A good harvest may lead to rice and other crops being sold in the market, though there is nothing more humiliating than selling produce for a low price at harvest time, only to find that by the next rainy season the family is short of food and expensive purchases must be made.

After the harvest, men often engage in various crafts, build and repair houses or try their luck at some alluvial diamond mining. Women continue to visit the farm to gather any remaining vegetables and may even take over a small portion of the farm as a cassava garden for a year or two. But no sooner is the harvest gathered than January arrives and a new farming cycle on a different farm commences.

The Kono upland rice farm is carefully adapted to environment, technology and labour. With low population density, fallow periods of eight to fifteen years are possible and there is no significant shortage of land. However, with greater population pressure and land competition, farming systems will have to adapt further, perhaps with buying and selling of land, fixed farm boundaries and the use of manure and compost to maintain soil fertility.

There have been many attempts to classify African farming systems but, because these systems are so complex and varied, no single classification is entirely appropriate. Farming systems range on a spectrum from extensive systems, such as shifting cultivation, to intensively cultivated and permanent systems, such as horticulture. A useful classification would distinguish between farming systems depending on fallow periods and those with permanent cultivation (see Plate 4.3). Permanent systems may in turn be subdivided into small-scale and large-scale systems. African farms are usually small, with under five hectares being common, and farms and kitchen gardens of less than one hectare are not uncommon, particularly in areas of severe population pressure.

It is likely that all or most African farming systems once involved fallowing or resting part of the crop land for varying periods of time. In the absence of chemical fertilizers and sophisticated machinery, farmers rely on time and environmental processes to restore nutrients to the soil. Fire was, and still is, used to clear plots in many parts of Africa and some farmers apply animal manure and kitchen waste to their farms to improve fertility and yields. Declining crop yields indicate that it is time to rest the land and cultivate

Plate 4.3 Traditional irrigation in West Pokot, Kenya (Tony Binns).

elsewhere. In true shifting cultivation, which is now rare, this would involve movement of the household or even the entire community to another area of forest.

All farming systems are dynamic to a greater or lesser degree, and African systems are no exception. They are affected by a wide range of factors, such as population pressure, proximity to urban areas and a variety of policies introduced by national governments and international development agencies. With increasing population growth and pressure on farmland in recent years, many farming systems have become more fixed, fences have been erected and fallow periods reduced considerably. With more intensive cultivation, farmers are increasingly concerned to maintain fertility and use manure, compost, and chemical fertilizers, when they can afford them and they are available.

The IAC (2004) provides a useful overview of Africa's diverse farming systems based on a classification proposed by Dixon *et al.* (2001). These systems are summarized in Table 4.1. Although pastoralism (see section 4.3.2) accounts for the largest farming land area in both North Africa and sub-Saharan Africa, the majority of Africa's population engage in the 'maize mixed' and 'cereal/root crop mixed' systems. The former, which is arguably the most important food production system in eastern and southern Africa, is characterized by a dry sub-humid to moist sub-humid climate, in which maize is grown as a staple subsistence crop. Sources of cash income include surplus maize, as well as coffee, cotton, tobacco and pulses. This system is extremely labour intensive, relies heavily on cattle for draught power, and has relatively poor crop yields due to shortages of fertilizers and pesticides (both of which are costly). The 'cereal/root crop mixed' system, which accounts for approximately 13 per cent of sub-Saharan Africa's land area, and is practised by 15 per cent of its population, tends to be concentrated across the Sahel and lowland areas of eastern and southern Africa. These areas have high temperatures, are relatively sparsely populated and are host to the tse-tse fly, which spreads trypanosomiasis and thereby severely limits livestock numbers. Maize, sorghum and millet are grown, but in the absence of draught power root crops such as cassava and yams dominate.

Other notable agricultural variations include the 'highland perennial' system of Ethiopia, Burundi, Rwanda and Uganda, where high population densities result in very small (often less than one hectare) land holdings. Here, perennial crops such as banana, ensete and coffee are grown alongside maize, other cereals and a range of vegetables. In less densely populated, semi-arid regions throughout Africa (e.g. Niger and Somalia), sorghum and millet are cultivated as the main sources of food. These 'agro-pastoral millet/sorghum' systems are, however, highly vulnerable to drought, hence agricultural activities are complemented by pastoralism. In some parts of West Africa, particularly in the coastal regions of Sierra Leone and Senegal, there is also a tradition of upland and wetland rice cultivation.

In North Africa, meanwhile, an estimated 30 per cent of the agricultural population engage in a 'highland mixed' system, dominated by the cultivation of such cereals as wheat and barley, along with legumes, fruit and olives. Soil- and water-conservation measures, and small-scale irrigation, are common in Morocco and Egypt, and these systems are often complemented by livestock rearing. This system, however, accounts for only 7 per cent of the total land area. The majority of land in North Africa is classified as 'sparse (arid)', where very little agricultural activity can take place, except in areas surrounding oases.

Farming activities in Africa also reflect the seasonal pattern of rainfall and temperature. Most cultivation is undertaken during the rainy season or shortly afterwards, when there is plenty of residual moisture in the soil; harvest occurs in the dry season. In areas of Ethiopia and Kenya where rainfall is bimodal (two rainy seasons each year), a second period of cultivation may follow the harvest. This has been the case where fast-maturing maize varieties have been adopted, and while this has obvious benefits in terms of food

Table 4.1 Farming systems of Africa

Farming system	*Land area (% of region)*	*Agricultural population (% of region)*	*Principal livelihoods*	*Main areas of production*
Sub-Saharan Africa				
Maize mixed	10	15	Maize, tobacco, cotton, cattle, goats, poultry, off-farm work	East and southern Africa, Cameroon, Nigeria, Zambia, Zimbabwe, Tanzania, Kenya
Cereal/root crop mixed	13	15	Maize, sorghum, millet, cassava, yams, legumes	Sahel region, Mozambique, Angola
Root crop	11	11	Yams, cassava, legumes, off-farm income	Tanzania, DRC, Angola, Central African Republic, Nigeria, Benin, Togo, Ghana, Côte d'Ivoire, Sierra Leone
Agro-pastoral millet/sorghum	8	9	Sorghum, pearl millet, pulses, sesame, cattle, sheep, goats, poultry, off-farm work	Sahel region, Zambia, Namibia, Kenya
Highland perennial	1	8	Banana, plantain, enset, coffee, cassava, sweet potato, beans, cereal, livestock, poultry, off-farm work	Ethiopia, Uganda, Rwanda, Burundi
Forest-based Gabon	11	7	Cassava, maize, beans, cocoyams	DRC, Republic of Congo, Central African Republic
Highland temperate mixed	2	7	Wheat, barley, tef, peas, lentils, broadbeans, rape, potatoes, sheep, goats, poultry, off-farm work	Ethiopia, Lesotho
Pastoral	14	7	Cattle, camels, sheep, goats, remittances	Sahel region, Botswana, Namibia, Ethiopia, Somalia, Kenya

Tree crop	3	6	Cocoa, coffee, oil palm, rubber, yams, maize, off-farm work	Liberia, Côte d'Ivoire, Ghana, Cameroon, Angola
Commercial: large holder and smallholder	5	4	Maize, pulses, sunflowers, cattle, sheep, goats, remittances	South Africa, Namibia
Coastal artisanal fishing	2	3	Marine fish, coconuts, cashew, banana, yams, fruit, goats, poultry, off-farm work	Mozambique, Tanzania, Kenya, Madagascar, West Africa region
Irrigated	1	2	Rice, cotton, vegetables, rain-fed crops, cattle, poultry	Madagascar, Mali, Senegal, Gambia
Rice/tree crop	1	2	Rice, banana, coffee, maize, cassava, legumes, livestock, off-farm work	Madagascar
Sparse agriculture (arid)	18	1	Irrigated maize, vegetables, date palms, cattle, off-farm work	Botswana, South Africa, Namibia, Sudan, Chad, Niger, Mali, Mauritania, Somalia, Djibouti, Eritrea
Urban based	<1	3	Fruit, vegetables, dairy, cattle, goats, poultry, off farm work	All countries
North Africa				
Highland mixed	7	30	Cereal, legumes, sheep, off-farm work	Morocco
Rain-fed mixed	2	18	Tree crops, cereals, legumes, off-farm work	Morocco, Algeria
Irrigated	2	17	Fruits, vegetables, cash crops	Egypt
Dryland mixed	4	14	Cereals, sheep, off-farm work	Morocco, Algeria, Libya, Tunisia
Pastoral	23	9	Sheep, goats, barley, off-farm work	Egypt, Libya, Algeria, Tunisia, Morocco
Urban based	<1	6	Horticulture, poultry, off-farm work	All countries
Sparse (arid)	62	5	Camels, sheep, off-farm work	Egypt, Libya, Algeria, Tunisia, Morocco

Source: Adapted from Dixon *et al.*, 2001

security and income generation, both farmers and agricultural experts have raised concerns about the impact of this double-cropping on soil fertility. Without the capacity to double-crop, however, many subsistence farmers inevitably face a period known as the 'hungry season', which can occur towards the end of the dry season and throughout the rainy season, when food stocks from the harvest are depleted and seasonal conditions preclude further crop cultivation (see Plate 4.4). The cultivation of cassava throughout West and southern Africa is in many ways an adaptive response to this situation. It is a perennial crop, high in carbohydrate and tolerant of low rainfall, which can be harvested when required over a period of two to three years (Hillocks, 2002). Many farmers aim to maximize farm output by planting a number of different crops on the same plot. This practice, known as intercropping, often involves leguminous crops grown alongside staple food crops. It improves soil fertility, increases yields and ultimately minimizes the risk of food insecurity that could occur should one crop be affected by pests or drought.

While it is undoubtedly important to understand the key characteristics and spatial variability of farming systems across Africa, it is also important to understand that this variability is a product of what the sustainable livelihood approach calls 'livelihood assets', at the local level (see Box 4.1). For example, most African agriculture is characterized by low levels of technology and investment, which reflect the shortage of 'financial' and 'physical' capital available to poorer farmers. Instead, agriculture in Africa tends to be heavily reliant on 'human' and 'social' capital, and it is little wonder that development research and practice have increasingly focused on these critical elements. Human capital – especially in the form of labour, but also skills and knowledge – is fundamental to African agriculture. An example of this is shown in Plate 4.5, a field sketch of a farmer-constructed seasonal calendar for a wetland agricultural system in Ethiopia, which highlights the range of labour-intensive tasks (and their duration) needed for the cultivation of a single maize crop during one season. Such activities would typically require the participation of the farmer (sowing, ploughing), his/her family (weeding, pest scaring) and fellow farmers

Plate 4.4 Boy gathering wild fonio in Mali duing the 'hungry season' (Tony Binns).

(ditch clearing, burning). In the absence of this labour, crop production and hence food security would suffer. In this example, there is a clear gender division of labour, with men traditionally undertaking the majority of the agricultural tasks, but this is by no means the norm throughout Africa. Estimates suggest that well over 70 per cent of Africa's food is produced by women; in Sudan, 80 per cent of women are involved in agricultural production; in Morocco, 57 per cent; and in Tanzania, 98 per cent of rural women who are classified as economically active are engaged in agriculture (FAO, 1995). Despite this huge involvement, which typically includes many of the more laborious farm tasks, such as hoeing, weeding and harvesting (undertaken in addition to taking care of young children, collecting firewood and water, and processing and cooking food), women are frequently discriminated against when it comes to securing land tenure rights or receiving agricultural assistance (Joireman, 2008).

In terms of skills and knowledge, those farmers engaged in the wetland agricultural system shown in Plate 4.5 also possess a detailed understanding of the wetland agricultural system that includes knowledge of environmental processes, soil fertility, crop suitability and, critically, sustainable management practices (Dixon, 2003). This 'indigenous' or 'local' knowledge, acquired through generations of experience, innovation and adaptation, is typical of many African farming systems (Reij and Waters-Bayer, 2001), and certainly contrasts with the once firmly entrenched 'Western' view that African agriculture is primitive, disorganized and unproductive. Indeed, far from being disorganized, studies have drawn attention to the links between agricultural sustainability and 'social capital' – that is, the shared norms, values, and networks of trust and reciprocity embedded within communities (Pretty and Ward, 2001). In the Ethiopian example cited here (see also Box 3.3, above), social capital is manifest in local farmer institutions, which set the rules for wetland agriculture and oversee practices that maximize benefit for the community as a whole (Dixon, 2008). Such local institutions are common throughout the continent, and are increasingly regarded as playing a critical role in the adaptation of natural resource management and agricultural systems in the face of social, economic and environmental change (Mazzucato and Niemeijer, 2002).

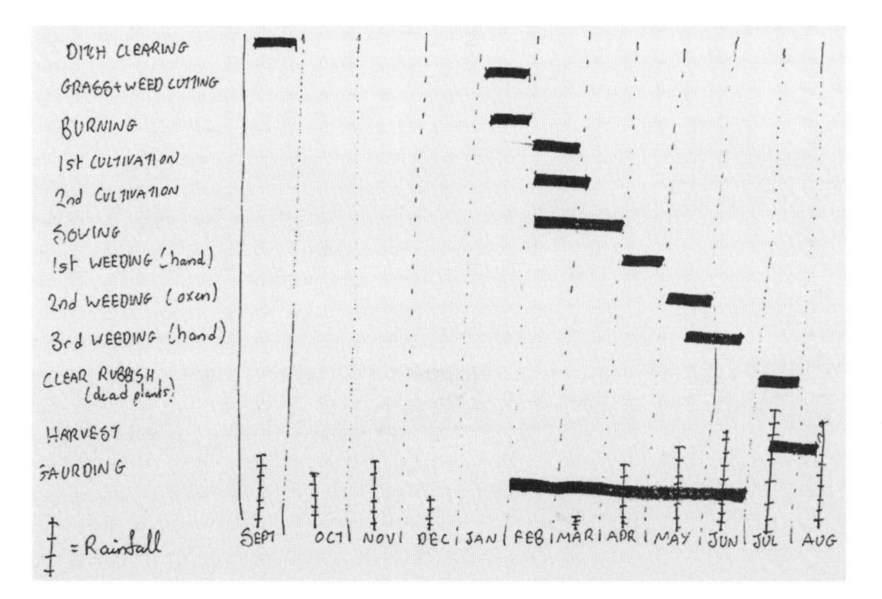

Plate 4.5 Participatory field sketch of a wetland farming calendar from western Ethiopia (Alan Dixon).

4.3.2 Pastoralism

Pastoralism is a farming system in which people 'derive most of their income or sustenance from keeping domestic livestock in conditions where most of the feed that their livestock eat is natural forage rather than cultivated fodders and pastures' (Sandford, 1983: 1). Pastoralism as a livelihood is believed to have developed as an adaptation to medium population densities and the presence of extensive range-lands in semi-arid areas (FAO, 1999b), and in Africa pastoralist communities are found throughout the Sahel (Mali, Niger, Chad), East Africa (Ethiopia, Somalia, Kenya), southern Africa (Botswana, Namibia, South Africa, Zambia), and in parts of North Africa (Algeria, Libya). Recent estimates suggest that 7 per cent and 9 per cent of the agricultural populations in sub-Saharan Africa and North Africa, respectively, are engaged in pastoralism (IAC, 2004), which therefore makes a significant contribution to agricultural output and national economies (Scoones, 1995a).

Livestock reared in Africa's pastoralist systems tend to consist of cattle, camels, sheep and goats, and it is common for a number of different types of animal to be kept, as not all will suffer equally during drought periods; camels, cattle, sheep and goats have different food requirements and varying degrees of resistance to drought and disease. These animals are also multi-functional: meat, milk and hides provide food and income security, but they are also frequently exchanged for goods and services within pastoralist communities, and hence assume significance in terms of engendering values of assistance and reciprocity.

There is also significant variation in the characteristics of pastoralist livelihoods throughout Africa, not least in terms of the degree of movement of people and livestock in range-land areas. Among the Fulani, for example, the Bororo are nomadic pastoralists who migrate throughout northern Nigeria and Cameroon to southern Chad and Niger, following the availability of grazing and watering areas during the wet season (see Plate 4.6). This is in contrast to other transhumant and settled Fulani pastoralist groups in the region (see Box 4.4). Transhumant pastoralism, which is characterized by seasonal migrations of livestock and herders, usually to fixed points, is estimated to involve between

Plate 4.6 Fulani pastoralists watering cattle from a well in northern Nigeria (Tony Binns).

70 and 90 per cent of the Sahel's cattle and 30 to 40 per cent of its sheep and goats (OECD, 2007). It contributes a significant amount to the region economically in terms of meat and milk production, but it is also regarded as having positive environmental impacts because herd migration allows pastures to recover and reduces the risk of land degradation.

Box 4.4

Pastoralists and pressures in northern Nigeria

The savanna lands of northern Nigeria have for centuries been ranged by Fulani pastoralists, migrating with herds of cattle, sheep and goats in search of pasture and water. In recent years, however, the Fulani have had to adapt to new pressures which threaten their pastoral lifestyle.

The Fulani probably arrived in Nigeria in the early fourteenth century after migrating eastwards from what is now Senegal. They were gradually accepted by the resident Hausa population, and strengthened their position as well as that of Islam during the jihad from 1804. The Hausa kings were eventually overthrown and replaced by Fulani emirs. The new ruling class controlled grazing areas, cattle tracks and watering places, and a growing interdependence developed between pastoralists and sedentary farmers, a relationship which has often been called 'symbiotic' – a mutually beneficial partnership between different groups.

This symbiotic relationship continues today, and a visit to the rich agricultural area around the historic city of Kano in February or March, at the height of the dry season, reveals herds of animals wandering across land which is covered with millet, maize, sorghum, beans and groundnuts during the rainy season (April–September). Livestock graze on any vegetation they can find, depositing manure, for which the farmers may sometimes pay the herdsmen. This is the key to a highly productive farming system, where fallow periods are virtually non-existent and the same fields are cultivated year after year.

Not all Fulani are migratory – in fact, a large number of them are now settled, and this process seems to be accelerating. Four major groups of Fulani may be identified according to their level of settlement, migration patterns and amount of land under cultivation. The Bororo remain true nomads, shifting camp regularly and moving south in the dry season and north in the rainy season. Semi-nomadic Fulani have permanent homesteads with adjacent farms where a variety of crops are grown. Elders stay in the homesteads for most of the year, and they are joined by the returning young herdsmen at the start of the rainy season, when the fields must be prepared for planting. Semi-settled Fulani regard cropping and livestock-rearing as equally important. They have fixed settlements and migrate over shorter distances, mainly during January and February. The fourth group, settled stock-holders, graze or corral their herds close to the villages, and young men take animals out to the surrounding area each day. This group increasingly supplements its income by looking after animals owned by wealthy, town-based people who invest in livestock as symbols of wealth and status. Some stock-holders are also engaged in the meat trade, supplying livestock to local markets and arranging transportation of animals to the large urban markets of the south. All four groups of Fulani share a common attachment to their animals and are held in high esteem for their skills at rearing and managing livestock.

The Fulani calendar is closely related to climate, quality of grazing, water resources, and the prevalence of disease. It can be divided into six eco-periods (see Figure 4.6). Pastoralists and cultivators anxiously await the first rain of the *seeto* season. Until then, hot, humid conditions increase contagious livestock diseases and mortality rates rise. Young herdsmen must be prepared to move frequently in search of areas where rain has fallen, but as the rainfall becomes more regular in May they start to move back to their homesteads. Grass usually grows quickly in June, *seeto luginni*, and rain may fall every day. During *ndungu*, the period of heaviest rainfall, grazing quality improves, milk yields increase and female animals give birth. Harvest time, from late September to early December (*yamde*), is when herdsmen plan their routes for the coming dry season. After about three months, when the cold and dusty Harmattan wind blows southwards from the Sahara (*debunde*), southerly winds herald the start of *cheedu*, the hot, dry season, from late February. Morale is low at this time, milk yields decline and if the dry season is protracted (as it was from 1968 to 1973 and again in the early 1980s), animals may suffer and die.

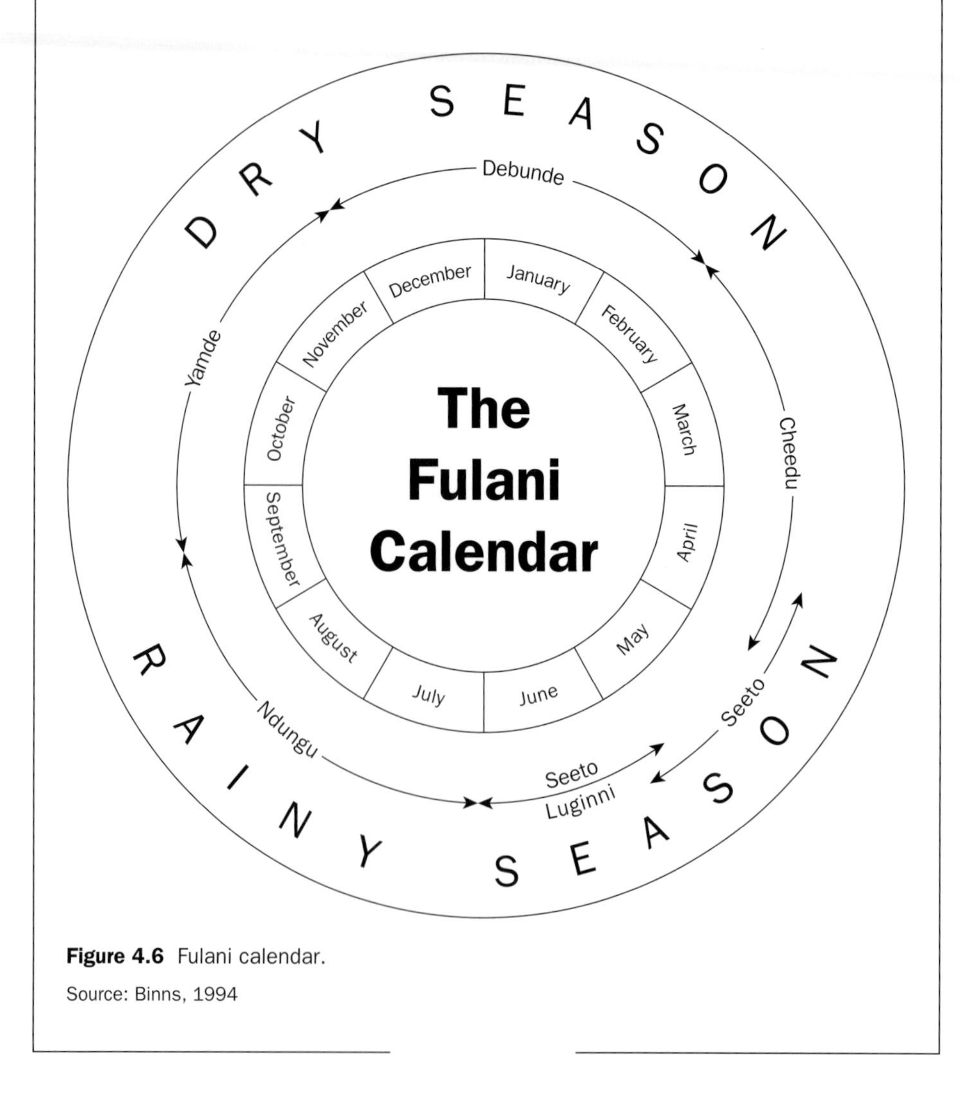

Figure 4.6 Fulani calendar.

Source: Binns, 1994

The provision of adequate supplies of food and water for their animals is the main concern of Fulani herdsmen. After the rainy season, in late September and October, grass grows abundantly, but it then slows down and loses much of its nutritional value as the dry season progresses. Consequently, other sources of fodder must be found. Crop residues, such as sorghum stalks, left in the fields after harvest, may be eaten by the animals, though farmers may guard these carefully for fuel or house and fence construction. Low-lying depressions and river valleys, known in northern Nigeria as *fadama* land, are important dry-season sources of pasture and water, though herdsmen must prevent their animals from damaging crops. Leaves, flowers and fruits of the many large trees in this 'farmed parkland' of the northern savanna are also important sources of fodder.

Although the Fulani lifestyle has evolved over many centuries to encompass a detailed knowledge of environment, rainfall patterns, grazing resources, water availability, pests and diseases, it remains a hard and uncertain existence; and it is becoming even more problematic as the Fulani face rapid changes in the area around Kano. Population pressure is increasing, with densities of more than 200 people per sq km, making the 'Kano Close-Settled Zone' (KCSZ) among the most densely settled land in rural Nigeria and indeed in the whole of tropical Africa. The efficiency of the farming system is maintained by intercropping a variety of grains and legumes on regularly manured land. More than 85 per cent of the total area is now cultivated, and land values and land sales are both increasing. On the outskirts of Kano, every scrap of cultivable land is used to grow crops, and wherever there is well-watered *fadama* land vegetables are intensively cultivated and transported by lorry to the city's markets (Maconachie and Binns, 2006; Maconachie, 2007).

With this intensive use of land, the Fulani are finding it increasingly difficult to source pasture and water. Farmers are concerned about animals damaging their crops and often fence off *fadama* areas. The symbiotic relationship between farmers and herders is not always as harmonious as it is portrayed, and tensions frequently arise (Milligan and Binns, 2007). With increasing population pressure and competition for farmland in the KCSZ, the pastoral, free-range concept of land no longer meets with the approval it once enjoyed. Drought and disease have further troubled the Fulani in recent years. The much-publicized drought of 1968–1973 and the more recent droughts of the early 1980s have depleted herds and created shortages of food and water. The cattle disease rinderpest has killed many animals and a hard-pressed veterinary service has been unable to maintain an effective large-scale immunization programme.

Fulani pastoralists have also been affected by the construction of dams and irrigation schemes, which since the mid-1970s have transformed the landscape of northern Nigeria. Completed in 1976, the Tiga Dam, with a capacity of 1963 million cubic metres of water, irrigates over 60,000 hectares of land in the Kano River Project, where crops such as wheat, rice and tomatoes are grown. East of Kano, in the Hadejia Valley, a further irrigation scheme covers 25,000 hectares associated with a barrage across the river near Hadejia. Fulani interviewed in 1984 spoke of difficult times and declining herd sizes due to severe drought and rinderpest. Whereas the Tiga Dam and Kano River Project had made water more plentiful, pasture and mobility were limited since the large irrigated area had been laid out across traditional migratory routes and crops were now grown throughout the year. Moreover, many trees, which had provided shade and fodder in the dry season, as well as valuable henna hedges, had been removed (Binns and Mortimore, 1989).

Pastoralists faced more acute problems downstream and east of Kano, where, instead of the annual flood, the river regime is now controlled by the Tiga Dam. The river channel has deepened and narrowed and *fadama* land, dependent on the annual flood for water and silt, has been left high and dry above the channel. Fulani in this area complained bitterly about the effects of the Tiga Dam, thus: 'the dam has denied us water in the *fadamas* and has compelled us to reduce the size of our herds to a size manageable by well water'; once perennial streams are now, in the words of one herdsman, 'choked with sand' (Binns, 1984: 642). Many pastoralists, while expressing pride in their herds and lifestyle, gave the impression that they were gradually succumbing to the pressures to settle and adopt a mixed farming economy.

The patterns and processes associated with the general intensification of agriculture around Kano, together with the effects of government intervention in the form of the Tiga Dam and the Kano River Project, have severely constrained the lifestyles of Fulani pastoralists. With the additional impact of drought, it is likely that pastoralists have suffered more than any other group, since their long-established ability to cope with shortages of water and pasture through migration has been progressively restricted. The Fulani way of life and its changing character must be appreciated and provided for in future development plans, to ensure that their living standards do not decline further. However, it is likely that the Nigerian government's ultimate, if unstated, goal is to enforce increasing sedentarization on the pastoralists and their herds (Binns and Mortimore, 1989).

Agro-pastoralist livelihoods are typically practised by settled pastoralists who have acquired land and engage in crop cultivation, while retaining small numbers of livestock (FAO, 1999b). The Gogo of Tanzania, for example, regard themselves first and foremost as livestock herders, but they also undertake a considerable amount of cultivation. Similarly, among the Karamajong of north-eastern Uganda, women live in homesteads in well-watered areas and grow crops, while young men take herds of cattle, sheep and goats in search of pasture. Pastoral groups may develop links with neighbouring farming communities, as in The Gambia and northern Nigeria, where Fulani pastoralists graze their animals on farm stubble after harvest, adding valuable manure to the soil. However, disputes occasionally occur if livestock disturb growing crops (Milligan and Binns, 2007). The exchange of dairy products for grain is also common, and complex systems of loaning animals exist among pastoral groups, sometimes for payment and sometimes as a free gesture in return for help at some future time.

Although such communities have practised agro-pastoralism for generations, evidence suggests that ever more traditional pastoral communities are engaging in sedentary agriculture (Ekaya, 2005). The reasons for this livelihood diversification are numerous and place specific, but explanations have focused on the impact of rising human and animal populations on the carrying capacity of range-lands. Subsequently, over-grazing and land degradation have reduced the availability of pasture, resulting in diminishing livelihood returns for pastoralists who have had to diversify in order to sustain a living. Many would argue that this situation has also been compounded by climatic change, although drought adaptation strategies have arguably always been central to pastoralists' way of life: exchanges of livestock and produce, changes to migratory routes, and the use of wild foods for human and animal consumption are all common adaptive strategies used in difficult times. In reality, pastoralism has come under significant threat from a range of natural and anthropogenic factors. For example, Yemane (2003) attributes the decline of pastoralism and increase in agro-pastoralism in Ethiopia to:

- the expansion of agricultural projects in the form of plantation farms and large-scale irrigated agricultural schemes;
- the expansion of wildlife parks and sanctuaries into range-lands, resulting in the exclusion of pastoralists;
- the expansion of agro-pastoralism in range-land areas due to population pressure and the settlement of refugees;
- the encroachment of unwanted woody plant species as a result of fire bans;
- conflict over range-land resources between and within different pastoralist clans; and
- the prevalence of livestock diseases, such as rinderpest, contagious bovine pleuropneumonia and trypanosomiasis.

Such issues are prevalent in the rest of Africa, too: for instance, the marginalization of Maasai pastoralists in the game parks of East Africa has been well documented (Brockington, 1999; Homewood, 1995), as have ethnic conflicts among the Karamajong in northern Uganda (Inselman, 2003). But also underlying many of these issues are the marginalization and neglect of pastoralism among many developers and planners who fail to appreciate the social, economic and cultural complexity of these livelihood strategies. Pastoralists have been accused of being unhealthy and uneducated, keeping livestock purely for prestige purposes, and being responsible for the environmental degradation of Africa's drylands. Furthermore, their seasonal migrations have been criticized by governments for disregarding international boundaries, creating conflict, and making census-taking and tax-collecting difficult. Consequently, many governments have implemented development strategies that have sought to 'organize' pastoralists and encourage them to take up enclosed systems of livestock rearing that emulate the Western models of the colonial period. As Scoones (1995a: 3) points out, however, such initiatives have been responsible for the 'unremitting failure of livestock development projects across Africa', not least because they have been based on imposing order on non-equilibrium environments – those environments which show extreme spatial and temporal variability in terms of climate, vegetation and human interaction. Instead, Africa's diverse environments require flexible and adaptive livestock rearing strategies that draw on the human and social capital of the pastoralists themselves.

4.3.3 Fishing

Fishing constitutes a significant rural livelihood strategy for millions of people throughout Africa, and it makes an important contribution to food security, employment and GDP. Recent estimates suggest that commercial inland fishing produces around 1.34 million tonnes with an annual value of $750 million. The most productive fishing areas are located in West and central Africa, where an estimated 227,000 people derive a permanent livelihood from fishing inland waters (Molden, 2007). In Malawi and Ghana, fishing is estimated to contribute around 4 and 7 per cent, respectively, of each country's GDP (FAO, 2004); and on the Zambezi floodplain, fishing generates more cash income than cattle rearing and in some cases crop production (see Plate 4.7; Turpie *et al.*, 1999). Equally significant is fishing activity in the inland Niger Delta, which is undertaken by the Bozo and Somono peoples. Small-scale coastal fishing also remains a major component of rural livelihoods in West and southern Africa, where it typically contributes up to 50 per cent of household cash income (FAO, 2004). This is despite a decline in fish stocks caused by poor regulation and, in the case of West Africa, over-fishing by European trawlers (Atta-Mills *et al.*, 2004). Artisanal fishing takes place in the seas off Cape Verde and the Kenyan coast, and by the Fante off the West African coast (Binns, 1992).

Fishing can take a variety of forms, ranging from modern commercial to more widespread but much smaller-scale artisanal activities. For the vast majority of people, however,

Plate 4.7 Fishing – a major source of income on the Zambezi floodplain (Alan Dixon).

small-scale fishing represents just one component of a wider livelihood strategy and hence is undertaken to supplement food supply or household income, often during periods when labour demands for agriculture are low. Typically, these forms of fishing are characterized by little capital investment in materials and equipment, and instead rely on traditional technology and indigenous knowledge of fisheries and fish behaviour. In these traditional systems, it is usually men who catch the fish and maintain equipment, while women play key roles in fish processing and marketing. For many women, this can represent one of few opportunities to earn cash (Molden, 2007).

Although fish stocks in the past have been maintained not least as a result of traditional arrangements that have set rules on net sizes and fishing methods (Geheb and Binns, 1997), increasingly there has been concern that Africa's fishery resources are being degraded. The reasons for this are diverse. Over-fishing as a result of population pressure and the expansion of the commercial fish trade has been well documented with respect to Lake Victoria (Geheb *et al.*, 2008), but environmental pollution resulting in the direct degradation of aquatic ecosystems has also led to the destruction of fishery resources (see Plate 4.8). This has been of particular concern in Nigeria, where the unregulated use of pesticides and the dumping of toxic waste have brought fisheries to a state of near collapse (Adeyemo, 2003). While most African governments have implemented fishery management policies, albeit with mixed results, the key issue, according to Sarch and Allison (2000), lies in recognizing the natural climate-driven variability in fish resources. Local people have an acute awareness of this, and have incorporated it into their fishing strategies for centuries, yet it is often ignored by planners and developers.

4.3.4 Forestry

Trees and forests are multifunctional resources which continue to play a major role in the livelihoods of rural people. Forest and scrub resources are, however, variable across the continent and show considerable diversity in terms of their species composition, which,

Plate 4.8 Luo fishing boat in the Kenyan sector of Lake Victoria (Tony Binns).

to a large extent, determines the manner in which they are used by people. As pointed out in Chapter 3, the majority of rural Africans continue to rely heavily on fuelwood for cooking and heating, yet at the same time people value forests for their provision of edible products, including wild fruits, nuts, vegetables and fungi (Kalaba *et al.*, 2009). This is particularly the case during periods of agricultural food insecurity. Throughout West Africa, tree crop agricultural systems that involve the production of fruits, nuts, cocoa, coffee, palm oil and rubber are major sources of foreign export earnings: 35 and 25 per cent of agricultural exports in Côte D'Ivoire and Ghana, respectively, come from tree crop production (World Bank, 2002a). Although these farming systems are usually associated with commercial plantations, millions of smallholders also derive livelihood benefits through indigenous systems of production, harvesting and marketing of tree crops (NRI, 2004). Facheux *et al.* (2007) estimate that there are around 1.6 million cocoa producers in Ghana, the majority of whom will have smallholdings of less than three hectares, and depend upon family labour for production. In recent years, many of these small farmers have benefited from the development of fair trade marketing arrangements, which has given them an unprecedented degree of livelihood security (see Box 4.5).

Box 4.5

Fair trade for Africa's poor farmers

The last decade has seen unprecedented interest in the application of 'fair trade' principles as a means of improving the livelihoods of the rural smallholder in many parts of Africa. Fair trade, according to those involved in the movement, is 'a trading partnership, based on dialogue, transparency and respect, that seeks greater equity in international trade. It contributes to sustainable development by offering better trading conditions to, and securing the rights of, marginalised producers and workers, especially in the South' (European Parliament, 2006: 14).

Since 1997, the Fair Trade Labelling Organization (FLO) has set international fair trade standards on a range of products, organized support for producers, and regulated use of the Fair Trade logo on products. In effect, it ensures that farmers and producers receive a minimum price for their produce that covers the cost of sustainable production, and it ensures that buyers pay a 'social premium' to the producer or organization that is earmarked for investment in the business or community-based development projects. The guarantee of a minimum price means that many small-scale farmers or cooperatives can plan for the future. In 2010, there were 231 certified producer organizations in 28 African countries, generating a combined social premium of over 12 million euros (FLO, 2010). Businesses range from date production in Tunisia and *rooibos* tea in South Africa to coffee in Kenya and mangoes in Burkina Faso. In Ghana, there has been significant growth and investment in fair trade cocoa production.

Cocoa has been grown in Ghana since the nineteenth century, but until the early 1990s production was controlled by the Ghana Cocoa Board (Cocobod), with the majority of smallholder farmers being dependent on middlemen and suffering under exploitative marketing arrangements. In 1993, however, restructuring of the industry paved the way for the establishment of private companies, such as Kuapa Kokoo (meaning 'good cocoa farmer'), which has grown from an organization of 200 smallholder farmers in 22 village societies to a national cooperative union representing 48,854 cocoa producers. In 2008, Kuapa Kokoo members produced 35,000 tonnes of cocoa beans, representing 5 per cent of Ghana's total export (Fairtrade Foundation, 2011).

Kuapa Kokoo's members typically have three hectares of land under cocoa production, with any remaining areas being used for subsistence crops, such as plantain, cassava and vegetables. The majority also have poor access to healthcare, education, clean water and sanitation. However, farmers have benefited from the guaranteed minimum fair trade price of $1600 per tonne that Kuapa Kokoo receives, particularly in recent years, when the world price of cocoa has risen as high as $3275 per tonne. Income from fair trade sales has funded direct payments to farmers (end-of-year bonuses), the construction of wells, mobile health programmes, HIV/AIDS workshops, non-farm income-generating initiatives for women, and credit schemes. Money has also been invested in agricultural extension activities and business development.

Along with partners Twin Trading and the Body Shop, Kuapa Kokoo launched its own chocolate bar in the UK in 1998, and now holds a 45 per cent stake in the Divine Chocolate company, whose board includes two elected cocoa farmers (Divine Chocolate, 2011). In 2010, Kuapa Kokoo committed to facilitating the spread of fair trade throughout the West African region when it agreed to Divine Chocolate buying a shipment of cocoa from a fair trade-certified cooperative in Sierra Leone.

Agro-forestry practices, which incorporate trees with agriculture, are also widespread throughout sub-Saharan Africa. These systems have been developed by farmers over generations as a risk-management strategy to take advantage of the economic livelihood benefits afforded by forest products, but also the positive impact that trees can have on agricultural production in terms of improving micro-climates and soil fertility. In parts of Malawi and Zambia, for example, farmers incorporate the nitrogen-fixing shrub *Sesbania sesban* into their short fallows, resulting in higher productivity of maize crops (Franzel *et al.*, 2002). In Tanzania, tobacco farmers have been intercropping maize with *Acacia crassicarpa* trees, which are used for fuelwood during the tobacco-curing process. Critically, this has reduced the demand for fuelwood from natural forests (Ramadhani *et al.*, 2002).

Afforestation has been carried out on a commercial basis throughout Africa since the colonial period, but there has been a trend in recent years for smallholder farmers, particularly throughout East Africa, to plant eucalyptus trees on their farms. Eucalyptus is a particularly resilient, fast-growing species and has become an important source of fuelwood for rural communities. It also grows extremely straight and hence is valued as a material for construction in both rural and urban areas (see Plate 4.9). In cities such as Kampala and Addis Ababa, it is also common to see the scaffolding for building construction made of eucalyptus poles. While the sale of eucalyptus is an important source of income for smallholder farmers, concerns have been raised with respect to the environmental impacts associated with the spread of this species, in particular its purported capacity to extract large quantities of water from the soil (Jagger and Pender, 2003; see also Box 3.2, above).

Africa's natural forests, meanwhile, continue to be threatened. Population pressure in the DRC, Zambia, Ghana and Ethiopia have resulted in massive deforestation, with land

Plate 4.9 Construction using eucalyptus poles from the surrounding hillside in Addis Ababa, Ethiopia (Alan Dixon).

being cleared mainly for agriculture and urban development (FAO, 2009a; see also Box 3.7, above). In response, some development strategies have sought to emphasize the important contribution that natural forests can make to rural livelihoods through their capacity to support economically significant non-timber forest products (see Box 4.6). Coffee, for example, is believed to have originated in the forests of south-western Ethiopia, where it still grows wild. Today, many smallholder farmers throughout East Africa retain areas of natural forest specifically because it provides the appropriate shade, temperature and humidity necessary for coffee to grow.

Box 4.6

Non-timber forest products for livelihood development in Ethiopia

There has been increased interest over the last decade in the value of non-timber forest products (NTFPs). Many efforts have been made to develop the value of these products and the trade in them in order to supplement the income of forest-dwelling or forest-fringe communities. In part, this has sought to protect the forests by reducing reliance on timber-based activities. In reality, however, NTFPs are generally not of sufficiently high value to make it worthwhile maintaining forests; as a result, sustainable use of timber and wood products is probably an essential part of a forest-based livelihood which makes it worthwhile for communities to retain their forests and not convert them to farmland.

Nonetheless, NTFPs can form important parts of the livelihoods of people living in and near to forests. This is especially so where NTFPs are suitable for commercialization, as they may then become a core part of the cash income for communities. Utilization of NTFPs in the forests of south-western Ethiopia shows the diverse roles which these products play and their potential in sustainable forest management (NTFP-PFM, 2011).

The highland forests of south-western Ethiopia range from high-altitude cloud forest dominated by bamboo (above 2400 metres) to mid-altitude Afro-montane evergreen forests (between 1200 and 1900 metres). In these various forests more than 300 NTFPs have been identified as being of value to the local communities. The uses of NTFPs are diverse: lianas are used for house construction, herbs for medicinal use, oil for cooking, and tree ferns for termite-resistant fencing. In addition to such domestic uses, commercially valuable NTFPs in the high-altitude forest include honey and Ethiopian cardamom (*Aframomum korarima*), while in the mid-altitude forest wild coffee is the most valuable product.

Over the last five years, a project has been working with communities to increase the value and production of their NTFPs. This is within a participatory forest management initiative which is creating community-managed forests. In some cases, increased production is supported through production initiatives, such as better beehives, while improved harvesting and post-harvest handling techniques are also being developed. However, in most cases, improved market linkages are critical.

Through establishing links with honey-processing companies in Addis Ababa, which buy directly from the forest beekeepers, the price received in the village for honey has increased eightfold in five years. In the case of coffee, while certification as 'organic' and 'forest produced' has helped raise the price received by farmers, global market increases offer most encouragement to forest production. Over recent years, because the price of coffee has remained high, all forest clearance in the coffee-growing area has stopped, as it is more economic for land to be under coffee forest rather than crops.

A survey of the communities in these two forest areas (NTFP-PFM, 2011) showed that 40 per cent obtain income from forest coffee, 26 per cent from forest honey, 18 per cent from forest spices and 6 per cent from bamboo. For the most part, the men are engaged in the commercial NTFPs, and they undertake collection visits deep into the forest for the various products that can be sold. Women mostly collect NTFPs for domestic use from the forest near to their homesteads, although some also collect spices and medicines for sale. Income analysis shows that NTFPs contribute an average of 24 per cent of household income in the higher-altitude areas, and 30 per cent in the mid-altitude forests (because of the high price of coffee). Wealthier households have higher incomes from NTFPs, in part because of their domestic labour resources, but also due to their traditional rights to specific parts of the forest and the NTFPs that grow there. For the poorer households, NTFPs are more important as food supplements and for domestic use.

4.3.5 Rural diversification

One theme that emerges from the preceding sections is that rural livelihoods are not based solely on agriculture. Rather, it has been argued that rural diversification, 'the process by which rural households construct an increasingly diverse portfolio of activities and assets in order to survive and improve their standard of living' (Ellis, 2000: 15), has become the norm throughout Africa since the de-agrarianization brought about by structural adjustment policies in the 1970s and 1980s (Bryceson, 2000). Ellis (1999) reports that between 30 and 50 per cent of rural people in sub-Saharan Africa rely on non-farm income sources, with this figure rising to 90 per cent in some parts of southern Africa.

Although the nature and extent of rural diversification vary significantly from place to place within Africa, Bryceson (2000) proposes a useful typology of non-agricultural activities from which rural people derive an income:

- *local services*, such as handicraft activities (e.g. weaving and metalworking; see Plate 4.10), catering for local markets, brewing, music-making and simply providing labour for others;
- *trade*, in which people may seasonally migrate to urban areas in order to sell agricultural and non-agricultural produce; and
- *transfer payments*, where key sources of income come from remittances and pensions from absent family members.

It is common, however, for households to pursue several income diversification strategies simultaneously, reflecting the availability of opportunities throughout the year, but also household gender divisions. Men, for example, are likely to engage in off-farm income-

Plate 4.10
Rural livelihood diversification in Uganda: women make mats from sedges collected from wetland areas (Alan Dixon).

generating opportunities during quiet periods of the agricultural calendar, while women may generate income all year round from the sale of food snacks, hair plaiting, prostitution, knitting and midwifery (Bryceson, 2000). Although some would argue that women are empowered and benefit from livelihood diversification, it has also been argued that livelihood opportunities for them tend to be more labour intensive and less lucrative than those available to men. Moreover, evidence suggests that livelihood diversification among men may place an increasing burden on women in terms of agricultural and domestic labour, further entrenching them in poverty (Hussein and Nelson, 1998; see Plate 4.11).

Much research in recent years has sought to understand why livelihood diversification has become more common in Africa, and specifically its effects on livelihood security and development. At a fundamental level, it is widely accepted that livelihood diversification represents an adaptive response to uncertainty, and hence constitutes a risk-minimization strategy for rural livelihoods (Chambers, 1997). As highlighted here and in other chapters, uncertainty in Africa takes a variety of forms, including climate variability, changes in economic policy, and conflict and instability, all of which can have profound effects on primary livelihood strategies. Diversification and adaptation have, of course, always been part of the livelihood strategies of people in vulnerable, marginal areas, such as the Sahel, but growing uncertainty throughout Africa offers one explanation for the growth of livelihood diversification strategies. In terms of uncertainty and change in agricultural production, Ellis (2001) also suggests that the successive subdivision of land at inheritance has led to the non-viability of agricultural production for many families. Moreover, agricultural markets have declined in recent years, and agricultural inputs have become scarcer and more expensive. Another explanation for diversification lies in 'pull factors', such as changing terms of trade that favour new markets, and people's perceptions of new opportunities (Hussein and Nelson, 1998). Ellis (2004) notes that wealth and access to opportunities may also be driving diversification in sub-Saharan Africa; poorer households do diversify, but often only in terms of casual work on other farms, whereas wealthier households are more likely to engage in livelihood diversification activities such as trade, transport and shopkeeping.

Plate 4.11 Women harvesting rice in The Gambia (Tony Binns).

In terms of the implications of diversification for livelihood security and development, Ellis (2000) provides a summary of the positives and negatives derived from the empirical literature (Table 4.2), but few today would argue against the view that diversification has had a mainly positive impact on rural livelihoods in Africa. Indeed, rural livelihood diversification and off-farm income-generation strategies have become important components of most rural development strategies implemented by governments and NGOs.

4.3.6 Rural marketing and infrastructure

A major factor influencing livelihood strategies and indeed diversification for many rural Africans is access to rural markets. The presence and permanence of markets indicate that there are economic incentives to produce surpluses over and above household subsistence requirements. Most rural African markets perform a range of functions, including:

- acting as centres of collection and exchange of local foodstuffs, livestock and craft goods;
- provision of services, such as cooked meals, tailoring and repair of such items as shoes, watches and bicycles;
- serving as distribution points for goods coming into the area, either from other areas in the country or from abroad; and
- acting as collection points for goods that will be exported from the local region.

Table 4.2 Positive and negative effects of rural livelihood diversification

Positive effects	*Negative effects*
• *Seasonality* Diversification reduces the effects of the peaks and troughs in production and income caused by agricultural seasonality. • *Risk reduction* The risks for one income source will be different from those for another, hence diversification reduces the risk of livelihood failure. • *Higher income* Diversity can generate more income than farming alone by making better use of skills and resources. • *Asset improvement* Cash from diversification can be invested in different livelihood assets, such as technology or skills development (education). • *Environmental benefits* Cash can be invested in resources which improve the natural resource base, while other activities can reduce dependence on less remunerative natural resource exploitation . • *Gender benefits* Improvement of income-generating capabilities of women, and hence child nutrition.	• *Income distribution* Widens the gap between the poor and wealthier households (the latter have more resources and better access to markets). • *Farm output* Stagnation of home farm agriculture due to labour migration. • *Adverse gender effects* Where male labour is advantaged by diversification opportunities, women's burden is increased.

Source: Adapted from Ellis, 2000

A single market may fulfil all of these roles and perform important social functions. In remote rural areas, without access to newspapers and other media, markets act as meeting places where people can catch up on news and gossip. Although media and fast transport may be lacking, it is surprising how quickly news and information about prices travels through rural Africa.

Some parts of Africa have a longer tradition of market-place trading than others. Some West African countries, for example, have market systems dating from long before the colonial period. Markets in Uganda, however, seem to be a much more recent phenomenon, associated with the sale of manufactured goods and service provision. In the non-Muslim parts of West Africa, such as Ashanti in Ghana and Yorubaland in Nigeria, many of the traders are women. In Muslim areas, such as northern Nigeria and coastal East Africa, men are the dominant traders (Binns, 1994).

A major distinction can be made between permanent markets and temporary or periodic markets. The latter comprise well over three-quarters of all African markets. One of the most intricate systems of periodic markets is to be found in Yorubaland. Five main types of market exist, according to location and timing: urban daily markets in the largest towns; urban night markets; rural daily markets for the sale of fresh produce and meat; rural periodic day markets; and rural periodic night markets. Rural periodic day markets are particularly important, being located about seven miles apart, with no village more than five miles away from a market. By taking place at different sites on different days, a cycle of periodic markets can draw on a larger population of buyers and sellers than a single fixed market. Yoruba periodic markets are arranged in ring systems, normally in four- or eight-day cycles, and successive markets are not normally adjacent. A study of the Akinyele eight-day ring found that most women visited two out of eight markets in the cycle, and in any one day some 10,000 people were on the move, representing 35 per cent of the total population (Binns, 1994).

Most market places in north-western Cameroon also operate on eight-day cycles, though increasing commercialization has led to daily markets becoming more common. Most rural consumers visit only one large market each week to buy commodities such as palm oil, salt, dried fish or meat, and occasionally a small supply of rice or some other commodity not normally grown on their own farm. Until recently, Sierra Leone had no major periodic markets, possibly due to the lack of a basis for interregional trade, since goods produced in one area of the country are similar to goods produced in other areas. However, since the 1960s some large rural periodic markets have developed in the Eastern Province, initiated by local chiefs and satisfying the increasing demand for foodstuffs in the growing urban centres of the diamond-mining region. Much of the produce is transported over long distances to these weekly markets, but local farmers are also benefiting from new trading opportunities and selling increasing amounts of fruit and other foodstuffs (Riddell, 1974).

Certain aspects of rural marketing often used to be under the control of the government until structural adjustment policies led to their dismantling. In some African countries, rural producers were able to sell cash crops to agents or buyers from state-run marketing boards. Commonly, farmers delivered their produce to traders or buying points, though sometimes buyers would purchase crops on the farm. These boards were established after the Second World War in countries such as Nigeria, where they were responsible for buying cocoa, palm products, groundnuts and cotton, and in Ghana, where they purchased cocoa. Marketing boards aimed to stabilize producer prices and foreign exchange earnings and reduce inter-seasonal price fluctuations. When world prices were above producer prices, the surplus earned by the board was put into a stabilization reserve to be used to finance future deficits when world markets fell below producer prices. In theory, the average producer price over a period of years should roughly have equalled the average net price realized in the export market. However, the marketing boards were criticized for failing to stabilize prices and incomes from season to season. It was also argued that the boards had a generally depressive effect on the economy by paying low producer prices, while accumulating large reserves.

While markets are critical components of rural livelihoods throughout Africa, many people are disadvantaged by a lack of access to them. This may be a result of on-farm labour shortages which preclude the transport of goods, but inadequate support services and poor transport and communications infrastructures also contribute to this problem. Poor roads and long supply chains render the transport of perishable produce, particularly fruit and vegetables, impractical for many of Africa's rural farmers, of whom less than 50 per cent live close to an all-season road (World Bank, 2008c). Countless studies have shown that building roads (even non-metalled feeder roads) has a positive impact on agricultural production, food security and economic development (Hanjra *et al.*, 2009).

A particularly significant development in the last ten years has been the massive growth and investment in mobile telecommunications throughout Africa, to the extent that now even small-scale farmers have access to current information on market prices and opportunities (Schuppan, 2009). In Zambia, for example, it is estimated that up to 48 per cent of rural agricultural workers have access to a mobile phone on a weekly basis, enabling them to liaise with local traders to identify the best time to transport their maize, watermelons and vegetables to market. Similarly, in Ghana, an internet and mobile phone-based trading platform known as TradeNet sends text-message alerts to farmers' mobile phones, keeping them updated with the market prices of over eighty commodities from markets across West Africa (de Wulf, 2004).

4.3.7 Food security as a livelihood outcome

Livelihood outcomes can be regarded as the achievements or outputs of various livelihood strategies and can take various forms. According to the DfID (1999), positive livelihood outcomes include:

- the generation of *more income*, which is often used as a proxy for economic development or economic sustainability;
- *increased well-being*, which is a subjective term relating to such factors as improved opportunities and access to health and education, physical security and a sense of control;
- *reduced vulnerability* to external shocks and pressure (e.g. economic, social, environmental and political);
- *improved food security*, which in turn reduces vulnerability and improves well-being and income-generating opportunities; and
- *sustainable use of the natural resource base*, which, arguably, underpins all the above livelihood outcomes.

It has been estimated that over 85 per cent of the poor in sub-Saharan Africa, and 48 per cent of the poor in North Africa, are living in rural areas (IAC, 2004), and a major determinant of rural poverty is an over-dependence of the majority of the population on subsistence farming as a result of persistent food insecurity. In terms of livelihood outcomes, therefore, achieving food security remains the primary concern for the vast majority of rural Africans. Food security, according to the FAO (1996: 2), 'exists when all people, at all times, have physical and economic access to sufficient, safe and nutritious food to meet their dietary needs and food preferences for an active and healthy life'. A more recent analysis by the FAO (2006) suggests that food insecurity and undernourishment are widespread in Africa, and actually increased by around 20 per cent between 1990 and 2002. This is in stark contrast to the rest of the world, where the number of undernourished people has fallen steadily. There is some variation within Africa, however, with the proportion of undernourished people beginning to fall in Angola, Chad, the Congo, Ghana, Malawi, Mozambique and Namibia. In contrast, the FAO (2006) reports that undernourishment in Burundi, the DRC, Eritrea and Sudan has increased significantly in recent years, largely due to conflict-induced famine.

Although it is true that food production in Africa has lagged behind population growth in recent years, the factors influencing food security are diverse and often unique to particular locations. For most subsistence farmers, food security depends on the availability of land, labour, storage, appropriate agricultural conditions and, in many cases, support from agricultural extension agencies (see Plate 4.12). Where people are not directly involved in food production, other livelihood activities must generate enough income to buy food, although here food security is dependent upon production elsewhere and the existence of appropriate marketing and arrangements to guarantee supply. At the macro level, government policy plays a major role in determining what food is grown where, how this is distributed, and to what extent food supplies are imported and exported. According to the FAO (2006), Africa has become increasingly reliant on food imports (particularly cereals) and food aid since the 1980s.

Faced with rapid population growth rates, many would argue that food insecurity is set to worsen throughout Africa in the coming decades. While there is no straightforward solution to improving food security and reducing rural poverty, there is widespread agreement that maintaining conditions of peace and stability which facilitate greater access to food and investment in agriculture should take precedence. Other measures include addressing such constraints as poor governance and weak institutional capacity, while providing security of land tenure is also seen as key to on-farm investments by smallholder farmers. In recent years, there has also been a call for more investment in and promotion of agricultural intensification, with a greater engagement with genetically modified (GM) technologies (see Box 4.7).

Plate 4.12 Traditional granaries, Mali (Tony Binns).

Box 4.7

The role of GM technologies in Africa's rural development

It is estimated that around 200 million people in Africa, of whom a third are children, are undernourished. And, with crop yields gradually declining, the situation is set to worsen. In the absence of green revolution improvements, and a more effective network for distributing and facilitating access to food supplies, there has been much speculation in recent years over the use and adoption of genetically modified (GM) crops as a means of addressing Africa's recurrent food security issues and promoting human development (Scoones, 2006).

Genetic modification allows specific traits, such as appearance, taste, pest resistance, disease resistance and nutritional value, to be modified, enhanced and introduced into crops. To some extent, genetic modification through the selective breeding of plants and animals has always formed part of farming strategies throughout Africa, but what makes GM crops different is the use of biotechnology in the direct manipulation, insertion or 'recombination' of genes (Makoni and Mohamed-Katerere, 2006). Proponents of GM technology suggest that this can result in dramatic increases in high-quality food production, with far-reaching benefits for poor people and their livelihoods. An often-cited example of the benefits of GM crops for Africa is Bt maize, a variety that has been genetically modified to express the *Bacillus thuringiensis* toxin, which is poisonous to the maize stem-borer pest (15 per cent of maize in Kenya is lost to stem-borers). Similarly, drought-resistant maize varieties continue to be trialled in research stations in Uganda and South Africa. With over 300 million people dependent on maize as staple food in sub-Saharan Africa, these GM varieties have massive potential to improve health and well-being (Paarlberg, 2010).

Why, then, has GM technology failed to take off in Africa? The answer to this question is complex. While at least twenty countries – including South Africa, Zimbabwe, Kenya, Egypt, Nigeria, Mali, Mauritius and Uganda – have biotechnology research facilities and are actively engaged in GM research, only South Africa grows GM crops, such as maize, soybean and cotton, commercially. Throughout the rest of Africa, governments, media and public have generally taken a hostile view of GM crops, mainly due to concern over their impact on human health and the environment. Although the adoption of GM crops can reduce dependency on potentially harmful chemical fertilizers, pesticides and herbicides, there are also concerns over the presence of such chemicals within the plants themselves, which may have an effect on human health. Other health concerns relate to the use of potentially pathogenic viruses to insert new DNA into crop genes, and also the potentially harmful effects of vitamin toxicity from nutritionally enhanced crops (Makoni and Mohamed-Katerere, 2006). In terms of environmental impact, critics have argued that 'genetic pollution' through cross-pollination could result in the development of 'super-weeds', leading to a reduction in biodiversity and the enhanced resistance of these weeds to standard herbicides. Crops with engineered pest resistance are also potentially problematic as pests themselves are capable of developing resistance over time, and toxins could have a harmful effect on non-pest species. There are also concerns over the introduction of genetic use restriction technologies (GURTs) – so-called 'terminator technologies' – into crops, which render them sterile. While many fear that this could create an unhealthy dependency on the companies supplying GM seeds, others see it as a direct affront to indigenous systems of seed selection and farmer innovation.

Clearly, GM crops have the potential to improve food security throughout Africa, yet the subject has become highly politicized. While the majority of African governments continue to advocate a precautionary approach to GM in the absence of hard facts, millions of people continue to suffer from hunger and malnutrition. Yet these people rarely have any say in the ongoing debate.

4.4 Rural development

Strategies and approaches to developing Africa's rural areas, reducing poverty and improving food security have changed significantly over the last fifty years, in ways which reflect changing socio-political circumstances but also the evolution of development thinking (see Willis, 2011). In broad terms, the pattern has been one of large-scale development schemes centred on economic growth and technological innovation, giving way to more people-centred, bottom-up forms of development that seek to understand and empower local people and their livelihood assets. Binns (1994) highlights the ill-fated East African groundnut scheme as an example of the former (see Box 9.1, below). This scheme, designed to satisfy an urgent need for edible fats and oils in post-war Britain, originated in 1947 after a rapid (nine-week) survey of East Africa concluded that a vast area of 1.3 million hectares could be cleared and planted with groundnuts, mainly in Tanzania (then Tanganyika), but some also in Zambia (then Northern Rhodesia) and Kenya. The scheme was to be implemented between 1947 and 1952 and involve 1000 Europeans and 60,000 Africans in land-clearing and cultivation. Ex-army personnel and equipment were employed, including using converted Sherman tanks as tractors. The scheme proved to be a disaster and caused a political storm in Britain. The area was very dry, there was little available data on rainfall and soil characteristics, and insect infestation was a problem. No pilot scheme was undertaken and the machinery, which suffered from breakdowns and fuel shortages, proved totally inappropriate for clearing the vegetation. The port of Dar es Salaam also had great difficulty coping with the fourfold increase in tonnage that occurred. The scheme represents a classic example of failure due to inadequate evaluation of natural and human resources. No one thought to talk to the indigenous Wagogo pastoralists about environmental conditions. They were merely recruited on to the scheme as servants and labourers. By the time the scheme was written off in 1951, over £36 million of British taxpayers' money had been spent.

More recently, in 1980, farmers living on the site of the proposed Bakalori Project, near Sokoto in north-western Nigeria, rioted because of dissatisfaction with compensation for their farmlands and homes flooded under a 19-km-long reservoir. Anti-riot police were called in and opened fire, killing fourteen farmers and wounding fifteen.

The results of such projects have often been very disappointing, with poor yields due to unpredictable supplies of inputs (water, seeds, fertilizers, pesticides, mechanized tillage) and poor management (Adams, 1988). Farmers' incomes are low and poorer farmers on the Kano River Project in Nigeria have sold their farms, in some cases seeking work from larger, land-purchasing neighbours who may be wealthy, urban-based, absentee politicians and businessmen. Large dams built across rivers can have profound effects on downstream areas, preventing annual flooding and reducing fish catches, crop production and grazing reserves. A serious flaw in northern Nigeria's strategy of large-scale irrigation schemes is that they have not been integrated with a coherent system of river-basin planning (Binns and Mortimore, 1989).

However, the weaknesses of such large-scale schemes began to be recognized throughout the 1970s and 1980s, as rural development started to embrace emerging ideas about sustainability (economic, environmental and social), and the basic needs approach. In attempting to incorporate these ideas, and embrace a more multidimensional view of rural poverty, many bilateral and multilateral aid agencies sought to implement integrated rural development programmes (IRDPs) in cooperation with African governments. The chief focus of these IRDPs was to increase agricultural productivity through funding and coordinating research, extension, inputs and marketing, but they also addressed infrastructure, health and social welfare issues. Despite attracting significant funding, not least from the World Bank, IRDPs achieved only mixed results, mainly due to a top-down planning process which failed to appreciate or understand the environmental, social and economic complexities of rural Africa (Binns and Funnell, 1983). This experience was reiterated in Robert Chambers's seminal book, *Rural Development: Putting the Last First* (1983), in which he asserted that rural poverty often went unperceived by development agents because of biased research and fundamental misunderstandings about rural people and their livelihoods. Chambers's work was influential in gradually refocusing rural development on people themselves, and in particular their local knowledge and understanding of rural life. Although many African governments were slow to engage with what has subsequently been called the 'farmer first' or 'participatory paradigm' in rural development, these ideas were initially taken forward by rapidly growing numbers of developed world NGOs during the late 1980s and 1990s. The emergence of the sustainable livelihoods framework during the 1990s (Chambers and Conway, 1992; Carney, 1998; Scoones, 1998) further cemented the participatory, household-level focus on rural people (men *and* women), the dynamics of their livelihoods and their development needs.

There is little doubt that the future focus of rural development should be on strengthening smallholder production, reducing vulnerability, increasing food security, and raising living standards. Attention must also be directed towards improving health and education standards in rural communities. Hopefully, such projects will diversify and enhance the productive base of rural areas, making them more attractive places in which to live and work, and thus helping to reduce the drift to towns. Large projects should be replaced by small, community-based and more democratic schemes. All this, however, requires a genuine commitment on the part of African governments, NGOs and aid organizations to moving rural and agricultural issues to the top of their political and financial agendas.

4.5 Conclusion

This chapter has provided a snapshot of the issues and challenges facing Africa's rural population, and it has outlined the ways in which rural livelihoods and development are dependent on spatially variable interactions between environment and society. Indeed, one of the major criticisms of rural development strategies in the past has been the adoption of a 'one-size-fits-all' approach, in which rural diversity, both within and between countries, has often been overlooked. At the same time, however, several decades of bottom-up, local-level rural development initiatives have also failed to reduce rural poverty significantly across the continent.

Clearly, a future challenge for Africa's rural development lies in greater coordination between local, national and international rural development interventions, and between the various stakeholders, including governments, NGOs and civil society. Organizing effective, efficient and sustainable institutions for rural development is therefore one priority area that needs to be addressed. There is also, arguably, a fundamental need for rural development to address the issue of secure rights of access to land and natural resources for the rural population, since this has been shown to have a significant effect on economic and social development. With security of access, sustainable investments in agricultural production and marketing can follow, with the development of the latter potentially providing greater opportunities for the smallholder farmer to access global markets in the way that fair trade and alternative food networks have succeeded in doing in recent years.

Finally, it is perhaps worth reiterating that rural development must be sustainable development; it should be economically, socially, institutionally and environmentally sustainable, inclusive and equitable. Above all, this means listening and responding appropriately to the needs of rural people, and involving them in the development process.

Summary

1. Despite increasing urbanization, the majority of people in Africa still live in rural areas and are heavily reliant on agriculture as a livelihood strategy.
2. Rural livelihood strategies across the continent are, however, extremely varied and reflect the local availability of natural, financial, human and social capital.
3. In recent years, people have diversified their rural livelihood strategies to include non-farm income-generation activities.
4. Insensitive government policies, population pressure and agricultural expansion have led to a decline in pastoralism throughout Africa, with more pastoralists engaging in sedentary agriculture.
5. Security of land tenure is widely regarded as an essential prerequisite to rural development.
6. Food insecurity, caused by under-investment in agriculture and poor infrastructure, is a major determinant of rural poverty.

Discussion questions

1. Why has per capita food production in Africa declined in recent years?
2. What are the advantages and disadvantages of private land ownership in rural Africa?
3. Does pastoralism have a future in Africa?
4. What resources do people need in order to diversify their rural livelihoods?
5. What are the possible effects of climate change on rural livelihoods in Africa?
6. How can rural development initiatives improve food security among Africa's rural population?

Further reading

Chambers, R. (1983) *Rural Development: Putting the Last First*, Harlow: Longman.

Dixon, J., Gulliver, A. and Gibbon, D. (2001) *Farming Systems and Poverty: Improving Farmers' Livelihoods in a Changing World*, Rome and Washington, DC: FAO and the World Bank.

Ellis, F. (2000) *Rural Livelihoods and Diversity in Developing Countries*, Oxford: Oxford University Press.

Reij, C. and Waters-Bayer, A. (2001) *Farmer Innovation in Africa*, London: Earthscan.

World Bank (2008e) *World Development Report 2008: Agriculture for Development*, Washington, DC: World Bank.

Useful websites

FAO (Food and Agriculture Organization of the United Nations): http://www.fao.org. The FAO aims to 'help developing countries and countries in transition modernize and improve agriculture, forestry and fisheries practices and ensure good nutrition for all'. The website holds a repository of reports and current statistics relating to food security, agriculture, pastoralism and fisheries in Africa. The website http://www.fao.org/farmingsystems/SSA_leg_en.htm provides an analysis of different African farming systems.

Pastoralist Communication Initiative: http://www.pastoralists.org/. A website that brings together various interest groups and NGOs concerned with pastoralist development issues in the Horn of Africa.

WFP (World Food Programme): http://www.wfp.org/. The World Food Programme of the United Nations monitors, responds to emergencies and implements strategies to prevent food insecurity and famine throughout the world.

World Agroforestry Centre: http://www.worldagroforestrycentre.org/. The centre supports research into agro-forestry to enhance food security and forest protection. The website provides case studies and resources relating to various initiatives in sub-Saharan Africa.

World Bank: http://www.worldbank.org/. The 'Agriculture' and 'Rural Development' pages of the World Bank's website provide an overview and case studies of key issues affecting Africa, as well as various rural development initiatives. The website also hosts the World Development Indicators statistical database which provides information on rural populations and their livelihoods in Africa.

5 Urban Africa

5.1 Urban and rural: a false dichotomy?

Too often in the past, 'urban' and 'rural' have been seen as quite separate and different. Although we have decided to have separate chapters on rural and urban issues in this book as an organizational aid to dividing up topics, we are very much aware that over the last two decades there has been questioning of whether they should be treated separately, given that urban and rural areas are closely interrelated and often have strong and complex interactions on a daily basis. For example, produce grown in rural areas is sold on urban market stalls; people move to the city for work, shopping, health and education; and people leave the city for recreation in the countryside.

Around many African cities are areas of transition, so-called 'peri-urban zones', interfaces where the city meets the countryside, which are characterized by a mixture of land uses as the city gradually expands and takes in formerly rural settlements and agricultural land (Binns and Maconachie, 2006; Dixon, 1987; Lynch, 2005; Maconachie, 2007; Potter *et al.*, 2008). As Lynch comments, 'One interesting outcome of the research on peri-urban interfaces has been a move away from considering the physical interface between urban and rural, to conceiving of the relationships between them as more important' (Lynch, 2008: 269).

Links between town and countryside are especially strong in Africa, not least because many Africans retain an umbilical link with the village of their birth, wherever they are living. Ceremonies such as christenings, weddings and funerals, as well as various traditional rituals, are good reasons for many urban dwellers to return to their home village for varying lengths of time. Many will also retain traditional rights to land in their village for such purposes as farming and constructing a house. They frequently send money home (remittances) and host visiting relatives from their village. But it is the birth village that is generally the preferred place for eventual retirement (Beauchemin and Bocquier, 2004). In south-eastern Nigeria, migrants who have left the villages are often referred to as 'sons abroad', and are expected to maintain contact with their home and return eventually. It is common for urban-based associations, sometimes called 'improvement unions', to be involved in village affairs. They transmit new ideas and aspirations, are an urban lobby for village interests, and provide advice and finance for village developments, such as roads, bridges, schools, dispensaries and hospitals (Binns, 1994).

So, while this chapter focuses primarily on Africa's towns and cities and the processes and problems associated with them, it is recognized that strong links exist between urban and rural and that these links must be more fully appreciated. The chapter considers the origins of Africa's urban centres, then examines recent rapid urban growth, and some of the implications for peri-urban areas, housing provision, urban environment and employment.

5.2 The urbanization process

Although, as we will see later, some African towns were founded well over a thousand years ago, the rapid growth of towns across the continent has generally been both recent and rapid, with the bulk of the population still living in rural areas. In 2007, only 38.7 per cent of Africa's population lived in towns and cities, making it the only continent with a majority rural population. But population growth rates in Africa exceed those in the world's other major regions, and the urbanization of its population, while lagging behind trends in Latin America and Asia, is now proceeding more rapidly than anywhere else. The population of Africa as a whole is expected to become 45 per cent urbanized by the year 2020, which represents a considerable increase on 1980's figure of 27.9 per cent. This is based upon an annual urban growth rate of 1.4 per cent between 1980 and 1990, slowing to 1.1 per cent from 2000 onwards. The annual urban growth rate within Africa varies considerably from one region to another. Of the four main regions (North, West, East and southern Africa), North Africa is predicted to have the lowest rate of urban growth (0.5 per cent) in the period up to 2020, while East Africa is likely to have the highest (1.5 per cent). This reflects the fact that in 2000 East Africa had the lowest proportion of its population living in urban centres (21.1 per cent), compared with North Africa, which had the highest proportion (51.1 per cent). East Africa is therefore urbanizing later than the other major regions, which have had more large cities for a longer period of time (UN-HABITAT, 2008).

The fact that this urban growth is relatively recent is shown by an examination of the situation in 1960, which for many African countries was the eve of their independence. In that year, West and East Africa had only three cities with over 500,000 inhabitants: Ibadan and Lagos in Nigeria, and Kinshasa in the Democratic Republic of the Congo (DRC). Some fourteen countries in these two vast regions had no city with over 100,000 inhabitants in 1960. In sharp contrast, by 2008, there were over thirty cities in West and East Africa with more than a million inhabitants and over forty in the continent as a whole. These included one mega-city (over 10 million inhabitants) – Cairo, Egypt, with a population of 12.1 million in 2008 (see Box 5.1). Two other prospective mega-cities are Lagos, with 9.8 million, and Kinshasa, with 8.2 million (see Table 5.1).

Box 5.1

Cairo: growth and development in Africa's first mega-city

Cairo is Africa's largest city, already one of the world's twenty-five or so mega-cities, with a population of approximately 12.1 million in 2008. The city grew rapidly after the Second World War, from a population of 2.2 million in 1947 to 9.3 million in 1986, and probably achieved its mega-city status by 1990. However, due to both fertility decline and less rural–urban migration, its annual growth rate has slowed in recent years – from about 2.7 per cent to 1.6 per cent, and 2.1 per cent for the wider region. Sutton and Fahmi (2001) suggest that the slowing of Cairo's growth might also be due to the effects of various restrictive master plans introduced since 1956.

The Greater Cairo Region (GCR), with a population of about 16 million, is the largest metropolitan region in Africa and forms part of what is termed the North

Delta Region (NDR), which includes the coastal cities of Alexandria and Port Said to the north and Suez to the east. In 2007, it was estimated that the NDR was home to 55 million people, representing 76 per cent of the total Egyptian population, and was the base for most of Egypt's industrial capacity and 70 per cent of the nation's agricultural land (UN-HABITAT, 2008).

Cairo's 1970 master-plan adopted a poly-nuclear strategy (as in London and Paris), which aimed to contain the city within its existing built-up area and preserve 'green land' by diverting future growth to satellite towns built on unproductive desert land. To the south of the city, an industrial area was built at Helwan and construction started on four new towns. Emphasis was also placed on transport infrastructure, with the building of an outer ring road and new bridges over the Nile to the north and south of the city. The 1983 master plan aimed to accommodate the growing population with ten new settlements in 'development corridors' to be built with mainly private sector finance on desert land to the east and west of the city. The plan also aimed to redevelop some of the older areas of the city and improve infrastructure through the construction of a new metro and ring road. But concerns were raised about spontaneous urban growth occurring along the ring road, which cuts through agricultural land on the urban periphery. Construction of the metro (the only one in Africa) started in 1982, and in 2008 there were two lines, with planning well advanced on a third line to link the city with the international airport.

During the early 1990s, the 1983 plan was modified with revised upward adjustment to the population forecast for 2000. This led to the cancellation of the green belt project and the merging of three new settlements into what is referred to as 'New Cairo City', to the east of Cairo proper. There was also concern about the preservation of key archaeological sites, and urban growth was largely restricted to the west and east of the city. The inner-city population declined between the 1986 and 1996 censuses, while growth occurred in the outer areas of the city. Unfortunately, the new towns have grown only slowly, failing to attract their anticipated populations, such that by 2020 they could each contain between 150,000 and 200,000 people, rather than 500,000, as originally planned. It seems that a large proportion of the workers employed in industries in the new towns actually commute out from Cairo, as the new towns lack basic social services, such as schools. Denis (1997) comments that many of these 'desert cities' remain empty with poor infrastructure, and in the early 1990s the settlements were largely handed over to private developers to build villa complexes and golf courses.

A particularly significant event in Cairo was the 1992 earthquake, which led to the outwards movement of between 40,000 and 100,000 people from the 800 blocks of flats and 9000 other buildings that were destroyed in densely populated inner districts. This outward movement of mainly lower-class families had been an objective of the master plans, but, on discovering the inadequate services in the peripheral settlements, many of the migrants subsequently returned to inner-city districts.

While the ring road and metro have been successful outcomes of the master plans, the full implementation of the plans was hampered by financial stringencies associated with IMF restructuring from the late 1980s. In particular, it seems that the plans have failed to control unplanned, spontaneous urban growth, typically consisting of modern formal housing and apartment blocks located mainly on agricultural land along the edge of the city and on the fringes of the desert. In the early stages of such development, key services, such as water supply, sanitation and refuse collection, were not available. Sutton and Fahmi (2001: 145) claim that 'the new towns programme has failed to alleviate overcrowded conditions in Cairo through the relocation of small-scale and traditional firms from inner city areas and downtown locations (such as North Gamalia) to these new settlements . . . [I]t has served to divert government money and attention which could have upgraded Greater Cairo's degrading built environment.'

There is an urgent need in Cairo for more low-cost housing stock, yet government policy has encouraged luxury housing construction, and there are a large number of vacant homes. As Singerman (1995: 112) comments, 'as many as 1.79 million completed units remain unused', with the highest vacancy rates in outer and high-income districts. The shortage of affordable dwellings to buy or rent has led to severe overcrowding in middle- and lower-income housing units. There needs to be greater provision of homes for poorer families in areas of the city where they would prefer to live, rather than being forced to resettle in distant outer suburbs with inadequate amenities (Fahmi and Sutton, 2008).

The rapid growth of Cairo has had considerable environmental impacts in terms of encroachment on agricultural land, traffic congestion, air and water pollution and waste management. El Araby (2002) is critical of the lack of coordination among the various urban authorities in dealing with such issues, and advocates the decentralization of political power, resources and responsibilities as the most sustainable way forward.

Table 5.1 Africa's ten largest cities in 2008

City	*Country*	*Total population (,000s) 2008*	*Growth rate (% per year) 2005–2010*	*Total population (,000s) 2020 (estimate)*
Cairo	Egypt	12,098	1.70	14,451
Lagos	Nigeria	9,830	3.74	14,134
Kinshasa	DRC	8,232	4.84	13,875
Khartoum	Sudan	4,886	1.83	7,017
Luanda	Angola	4,252	6.02	7,153
Alexandria	Egypt	4,251	2.03	5,210
Abidjan	Côte d'Ivoire	3,925	3.16	5,432
Johannesburg	South Africa	3,507	2.26	3,916
Algiers	Algeria	3,428	2.22	4,235
Cape Town	South Africa	3,269	1.17	3,627

Source: Adapted from UN-HABITAT, 2008

The growth rate of Africa's largest cities varies considerably, from an estimated 0.8 per cent per annum (2005–2010) in Monrovia (Liberia) and Casablanca (Morocco), to a massive 6.02 per cent per annum in Luanda, Angola. In the case of the Angolan capital, the population is projected almost to double in just twelve years – from 4.2 million in 2008 to 7.1 million in 2020. Kinshasa is projected to grow from 8.2 million in 2008 to 13.8 million by 2020, and by 2025 could have surpassed the populations of Cairo and Lagos. Some of Africa's intermediate cities are growing even more rapidly, such that by

2050 there will be 'more people living in African cities than the combined urban *and* rural populations of the Western hemisphere' (UN-HABITAT, 2008: 4). Some of the implications of this massive rate of urban growth will be examined later.

Urban growth has been most rapid in the capital cities and other large cities. Of the countries with a coastline, the vast majority have coastal capitals, with some notable exceptions being Cairo (Egypt), Kinshasa (DRC), Windhoek (Namibia), Dodoma (Tanzania) and Nairobi (Kenya). Until relatively recently, all the coastal countries of West Africa had coastal capitals, though the pattern changed with the transfer of political power to new capitals at Abuja in Nigeria (1991) and Yamoussoukro in Côte d'Ivoire (1983) (see Figure 5.1).

Many African capitals are also primate cities, with their populations more than twice the size of the second city. In some countries, the urban system is dominated by just one or two cities. For example, in 2000 it was estimated that 66.2 per cent of the Congo Republic's urban population lived in the capital, Brazzaville, while 60 per cent of Angola's urban population lived in Luanda. Some 55 per cent of Guinea's population lived in the

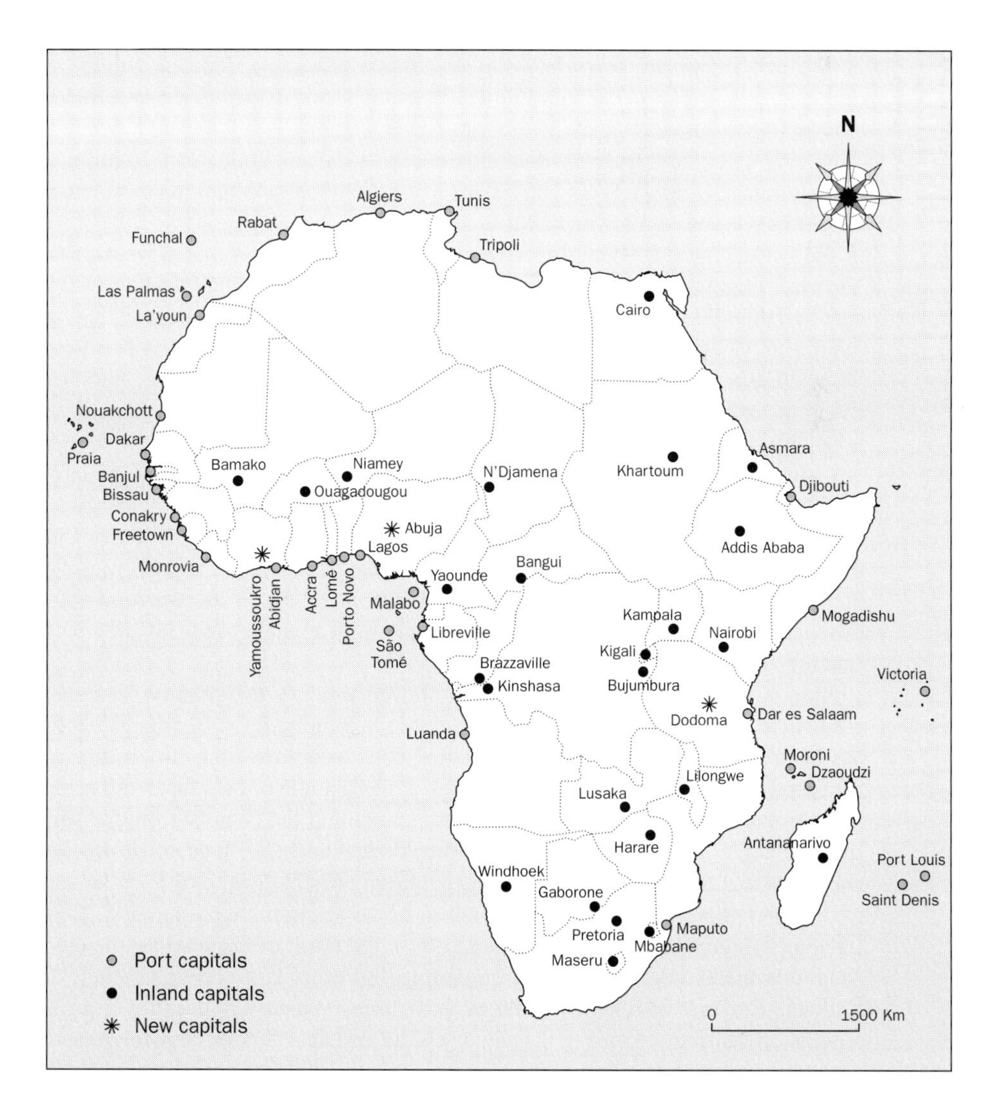

Figure 5.1 Capital cities.

capital, Conakry, but this figure was considerably less than the 84.4 per cent recorded in 1975 (UNPD, 2002). Nigeria is unusual in sub-Saharan Africa in having a well-developed urban system, with only 17 per cent of the urban population in 1980 living in the largest city, Lagos. The 2006 Nigerian census revealed that there were six Nigerian cities with more than a million inhabitants, and no fewer than seventy-three with more than 100,000 inhabitants (Nigerian Government, 2006; see Box 5.2).

Box 5.2

Urban growth in Nigeria

With a total population of 140 million in the 2006 national census, and a projected population of 264 million by 2050, Nigeria currently has a larger population than any other country in Africa. It also has one of the continent's largest economies, with a GDP in 2005 of $99 billion, second only to South Africa ($239.5 billion) (UNDP, 2007), and in recent years has played a leading role in African politics and diplomacy. Nigeria also has some of the largest towns in Africa and it is experiencing a range of pressures associated with rapid urbanization.

Nigeria's population almost doubled between 1980 and 2006, from 71 million to 140 million (UN-HABITAT, 2008). The average annual population growth rate of 2.8 per cent between 1975 and 2005 is looking likely to fall slightly, to 2.2 per cent, in the period from 2005 to 2015. In 1921, it was estimated that 4.8 per cent of Nigeria's population lived in towns and cities, but this figure had risen to 19.2 per cent by 1963. The annual rate of urban population growth was 2.1 per cent between 1980 and 1990, but this is projected to decline to 1.1 per cent by the period 2020–2030. However, it was estimated that 45 per cent of Nigeria's population were living in towns and cities by 2011, meaning that almost 65 million people were urban-based (UN-HABITAT, 2008).

Lagos, with a population of 9.8 million in 2008, overtook Ibadan as the largest city in Nigeria between 1953 and 1963, and has experienced phenomenal growth, especially during the civil war (1967–1970), when it was outside the zone of hostilities and oil companies moved their headquarters there from Port Harcourt. Lagos grew at an annual rate of almost 4 per cent between 2005 and 2010, second only to the Nigerian capital, Abuja, with a staggeringly high annual growth rate of 8.32 per cent (see Figure 5.2).

Nigeria's largest cities are distributed unevenly across the country. The south-west is the most urbanized region, a legacy of the long tradition of urbanism which has been a feature of Yoruba culture dating back to before the tenth century with the founding of Ile-Ife, but reaching a peak by the middle of the nineteenth century at about the same time as the British were penetrating the country from the south. In Yorubaland are the large cities of Abeokuta, Ado-Ekiti, Ibadan, Ile-Ife, Iwo, Ogbomosho, Oshogbo, Oyo and, on the southern coast, Lagos. In the south-east of the country there is another concentration of large cities including Aba, Benin, Calabar, Enugu, Onitsha and Port Harcourt. In the north of the country, the city of Kano dominates, with over 3 million people, but other large centres include Jos, Kaduna, Katsina, Maiduguri, Sokoto and Zaria. What is noticeable from any population distribution map of the country is the absence of many large towns across the centre and in the valleys of the two major rivers, the Niger and Benue. This so-

Figure 5.2 Nigeria: states.

called 'middle belt' has long been a relatively unproductive area of sparse population, due initially to slave raiding and the prevalence of tse-tse fly.

Since independence, national development strategies have generally perpetuated a strong urban bias. Industries, infrastructure and services have been developed in the towns, while the rural areas have been relatively neglected. Many Nigerians view farming as a low priority, not the sort of work to be undertaken by anyone with an education.

Rapid urban growth has caused many problems relating to housing, transport, water supply, the urban environment and urban food supply. Transport is a major problem in many Nigerian cities, and Lagos's traffic congestion is legendary. Elsewhere, the situation is particularly acute in old cities like Kano, where the pre-colonial street pattern was not designed for motor vehicles. Transport provision in many Nigerian cities has been left to the non-subsidized private sector, so buses are generally expensive, numbers of private motor vehicles have increased and there is little provision for pedestrians or bicycles. There is an urgent need in many cities to pedestrianize certain areas and introduce subsidized mass public transport systems to reduce road congestion.

Supplying urban populations with food is another problem. Intensive market gardening areas surround most Nigerian towns. The increasing consumption of bread as a convenience food necessitated enormous imports of wheat and flour in the late 1970s and early 1980s, amounting to 1.5 million tonnes in 1981, with 90 per cent coming from the USA. Import restrictions have been imposed since the late 1980s, together with efforts to increase domestic wheat production, notably in irrigation schemes in the north.

Rapid urban growth and, in particular, the severe congestion in Lagos were key factors in the decision to build a new capital city at Abuja in the sparsely settled and underdeveloped middle belt. At the end of the three-year civil war in 1970, there was also a strong interest in redressing imbalances and establishing a capital city that was closer to the geographic centre of the country. It was felt that Abuja's location might help to reduce the longstanding rivalry between the mainly Muslim north and the Christian south which has plagued Nigerian political history since the country was formed by the British.

The master plan for Abuja and the Federal Capital Territory was worked out from 1973 with experts from the new city of Milton Keynes in Britain and an American consortium, International Planning Associates. A well-known Japanese architect, Kenzo Tange, was involved in detailed design of the central areas (see Figure 5.3).

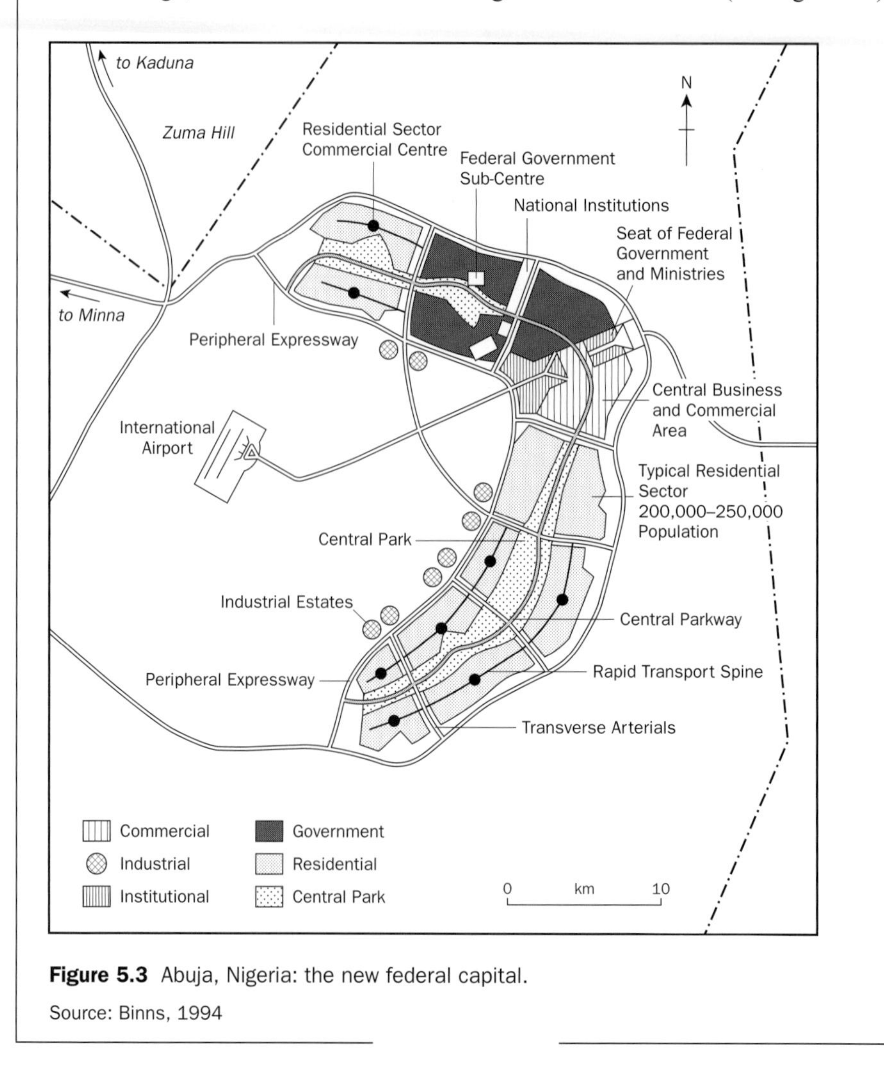

Figure 5.3 Abuja, Nigeria: the new federal capital.

Source: Binns, 1994

The Federal Capital Development Authority was established in 1975 and the plan was eventually published in 1979. The new city has a semicircular arrangement of segregated residential, commercial, industrial and government districts, linked by a network of multi-laned highways. It also has an international airport. Imposing government buildings are located at the centre of the arc; the movement of government ministries from Lagos to Abuja took place before the mid-1990s. Although its construction has been more protracted than expected, the master plan speaks of Abuja serving as a 'symbol of Nigeria's aspirations for unity and greatness, being a central, neutral and healthful place with plenty of room for urban development' (NFCTA, 2008). In recent publicity, the Nigerian Federal Capital Territory Administration speaks of 'Giving Abuja a character' (NFCTA, 2008).

The city's residential districts, each of 40,000–60,000 people, are grouped into 'mini-cities' of 100,000–250,000 people with associated employment facilities. There is more segregation by income in Abuja than in many other Nigerian cities. The city, which thirty years after the publication of the plan was still the country's major prestige project, had 1.7 million people in 2008; but with its average annual growth rate of 8.32 per cent, the population is predicted to reach 2.9 million by 2018. In connection with Nigeria's bid to host the 2014 Commonwealth Games, a series of ambitious new development projects were planned in Abuja, including the construction of two boulevards as the dual spines of the CBD, various new sports facilities, the extension and modernization of the airport, new tourist resorts, more high-rise buildings and a rapid transit rail system. But the Games were awarded to Glasgow (partly because of logistical problems that were exposed during Abuja's hosting of the All-Africa Games in 2003), and it now seems likely that many of the proposals for Abuja will be shelved.

The main factors causing urban growth have been high rates of natural increase of both urban and rural populations, the incorporation of peri-urban villages into sprawling built-up areas, and, to a variable extent, rural–urban migration. Clancy (2008) has argued for more detailed research on the environmental situation in relation to Africa's growing cities and is concerned particularly about their rapidly deteriorating ecological footprints.

5.3 Rural–urban or urban–rural migration?

It is still often the case that, for people living in poor rural areas lacking basic facilities, towns, and particularly large cities, are seen as 'havens of opportunity' where jobs, education, healthcare and consumer goods can be obtained. It has been estimated, for example, that between 1953 and 1963, some 644,000 people migrated to Lagos, accounting for 75 per cent of the city's total population growth during that period. The 1979 Nairobi census revealed that, of the city's 828,000 recorded inhabitants, only 26 per cent had been born there, and most of these were small children. Similarly, in West Africa, the 1970 Ghanaian census showed that only 50 per cent of Accra's total population had been born in the city (O'Connor, 1983).

However, there is evidence to suggest that in the last twenty or thirty years the rate of rural–urban migration has slowed considerably, and some large cities have even experienced a process of increasing return migration to rural areas or to smaller cities. Jamal and Weeks (1994) have shown that the rural–urban income gap in Africa collapsed during the

1970s and 1980s as a result of structural adjustment policies, economic recession and the debt crisis. These factors led to a devastating fall in real urban incomes and the development of a 'new urban poor' in many African cities (see Box 5.3). For example, Jamal and Weeks (1994) estimate that the average non-agricultural wage in Sierra Leone in 1985/6 was 72 per cent less than average rural household incomes, while in Uganda an average farmer's income was 30 per cent higher than that of an urban minimum-wage earner in 1984. Riley (1988: 7) describes Sierra Leone's urban poor in the mid-1980s as 'a deprived group with fewer income or equivalent earning opportunities than the rural poor'. In 1988, it was estimated that 30 per cent of Kenyan urban wage earners could not buy the minimum amount of calories needed to feed a family of five (Jamal and Weeks, 1994). One adaptation to the stringencies of urban life in the 1970s and 1980s was the growing of food in family compounds and vacant urban spaces, a practice known as 'urban agriculture' (see Box 5.4).

Box 5.3

Structural adjustment and urban employment in Ghana

In March 1957, under the leadership of the charismatic Dr Kwame Nkrumah, Ghana was the first country in Africa south of the Sahara to become independent. The country covers an area of 238,537 sq km and in 2005 it had an estimated population of 22.5 million. The predicted annual population growth rate for the period 2005–2015 is 1.9 per cent – significantly less than the 2.6 per cent annual growth recorded between 1975 and 2005 (UNDP, 2007).

Accra, the capital city, had a population of about 1.6 million in 2000, while the second city, Kumasi, had about 1.1 million. These cities are growing more rapidly than the national population, such that their estimated annual growth rates between 2005 and 2010 were, respectively, 3.23 per cent and 3.69 per cent. By 2020, it is likely that Accra's population will have just over 3 million people, while Kumasi will have just under 3 million (UN-HABITAT, 2008; Grant and Yankson, 2003).

Ghana has had a turbulent political history since independence, with fifteen governments, ten of them democratically elected civilian regimes and five military regimes that seized power through *coups d'état*. The economy, which at independence was based on strong exports of cocoa and minerals (notably gold, diamonds and manganese), experienced a sharp decline in the 1970s, caused by inappropriate economic policies, escalating public expenditure and a heavy bias towards the urban sector. Per capita GNP declined at an annual rate of 4.4 per cent between 1973 and 1983, agricultural and industrial output plummeted, and inflation was rampant. The economy reached its lowest ebb in 1983, compounded by the drought of the early 1980s and the need to absorb 1.2 million Ghanaian migrants who were deported from Nigeria. In that year, hunger was widespread and cocoa farmers were chopping down their trees (once the source of the country's wealth) for firewood.

The Provisional National Defence Council (PNDC), led by Flight Lieutenant Jerry Rawlings, seized power on 31 December 1981 from the short-lived civilian regime of Dr Hilla Limann. In the early days, Rawlings managed to generate mass mobilization of the people, with the establishment of local defence, price-control and anti-hoarding committees and a strong attack on profiteering and corruption – known as *kalabule.* 'People's shops' took over the internal distribution of essential commodities.

Rawlings then launched a major International Monetary Fund (IMF)-backed economic recovery programme (ERP) in the budget of April 1983, which reversed many of the previous government's policies with some success. The ERP was implemented in two phases, 1983–1986 and 1987–1989. The first phase devalued and unified exchange rates and aimed to arrest and then reverse economic decline, notably in agriculture. Rehabilitation programmes were launched in the timber and manufacturing industries, in the gold fields, in transport and in the cocoa-producing areas. The second phase focused on increasing economic growth and investment, and rationalized and privatized the management of many state-owned enterprises. As a result of the ERP, 1985 was the first year since 1978 in which real GDP per capita rose, and the improvement continued throughout the late 1980s. The production of food crops more than doubled between 1983 and 1986. The contribution of cocoa increased from 0.7 per cent to 8.7 per cent of GDP between 1982 and 1988, and manufacturing industry's contribution rose impressively from 3.6 per cent to 10.1 per cent in the same period.

In June 1989, the government published its first employment statistics for ten years. In spite of a steadily rising population, recorded employment in Ghana actually fell from 482,100 in 1979 to 207,900 in 1982 before recovering somewhat to 413,700 in 1986. The agriculture, forestry and fishing sector was the most important, providing about 60 per cent of total employment. The second most important sector was commerce – wholesale and retail trade, restaurants and so on – accounting for about 14 per cent of total employment. Although the ERP aimed to increase the efficiency of the public sector, and many employees were made redundant in parastatal organizations, the public sector share of employment actually rose from 74.5 per cent in 1979 to 84.0 per cent in 1986. However, as in other African countries, these statistics relate only to those who were registered with the Employment Service. A survey of employment in Ghana undertaken by the International Labour Organization (ILO) in the late 1980s detected widespread under-employment in towns and an acute shortage of productive employment opportunities due to pruning of the civil service and the rest of the public sector. Real wages and salaries fell by an enormous 83 per cent between 1975 and 1983 and continued to fall up to 1989, forcing many workers to seek additional sources of income. Consumer prices rose by 31 per cent in 1988, 25 per cent in 1989 and 37 per cent in 1990 (ILO, 1995). Grant and Yankson (2003) report that the number of households in poverty in Accra increased from 9 per cent to 23 per cent between 1988 and 1992, and poverty continued to spread during the 1990s, with low-income residents spending a greater share of their incomes (sometimes up to 60 per cent) just on feeding themselves.

The ILO report reached the gloomy conclusion that in the next few years virtually no primary or secondary school leaver could expect formal sector wage or salaried employment. It was suggested that they would have to find jobs in farming and fishing, informal service construction, road building and production activities, or trading. Other than farming, the urban informal sector had to absorb large numbers of school leavers as well as older people made redundant from public sector organizations. Recent evidence suggests that the ILO prediction was reasonably accurate.

Ghana has a dynamic and substantial informal sector, which is characterized by a wide range of micro-enterprises in production and services located in homes, residential areas and along major roads. These include such activities as fast-food operations, retailing in small kiosks, metal- and wood-working, cement block manufacture, urban agriculture and street hawking, with the latter often involving young children. Typically, informal sector enterprises are individually owned and operated, with an average of four employees. For women, 60 per cent of informal sector employment is in trading. In the early 1990s, the informal sector probably accounted for about 22 per cent of the GDP of the total Ghanaian economy (Barwa, 1995).

The Ghana Living Standards Survey undertaken in the late 1990s revealed that only 6.7 per cent of the working-age population (between fifteen and sixty-four years of age) were unemployed, and a very significant 86.3 per cent worked in informal activities (see Table 5.2; GSS, 2000). Additionally, some 17.2 per cent were engaged in 'unpaid family work', with a higher proportion of females than males engaged in both this and informal activities (Table 5.2). Survey results from 1987/8 and 1998/9 indicate a significant increase in self-employment during this period, and a decline in wage employment in both the public and private sectors from 17.3 per cent to 13.7 per cent. As Monk *et al.* (2007: 4) comment, 'the informal sector has done a remarkable job in absorbing the rapid growth in the labour force'.

The urban informal sector is characterized by small enterprises with poor working conditions, using family labour, a well-developed apprenticeship training system and simple technology. There is considerable flexibility in movement between activities and it should be recognized that the informal sector plays an important role in the development of indigenous entrepreneurship. Emphasis in business is often on survival rather than profit, and during difficult times the informal sector plays a key role in alleviating large-scale poverty.

In Kumasi, informal sector activities include trading and manufacturing. The city's central market is one of Africa's largest, operating as a central node in the national food distribution network. Kumasi traders, who are mainly urban-born or

Table 5.2 Ghana: employment and unemployment by gender, 1998–1999

	All (%)	*Male (%)*	*Female (%)*
Total population aged 15+			
Unemployed	6.7	7.2	6.4
Employed	82.1	84.0	80.7
Employed population aged 15–64			
Formal private and public employment	13.7	22.7	6.2
Total informal employment (a+b)	86.3	77.3	93.8
(a) Private informal and non-agricultural self-employment	34.3	21.4	45.1
(b) Self-employment (agricultural)	52.0	55.9	48.7
Sub-category of employed			
Unpaid family work	17.2	10.7	22.7

Source: Adapted from GSS, 2000

urban-raised Ashanti women, have established strong links with a network of small-scale producers, also mainly women, who have been able to expand food production as a result of reliable retail and wholesale outlets paying consistent prices (Lyon, 2000). Entry into trading seems to be easy and there is frequent transfer between different trading roles. The traders show a remarkable degree of flexibility, responding to environmental, economic and political events with changes in their enterprises. When products are in short supply, traders spend long hours visiting supply villages, travelling to ever more remote locations. They liaise with a regular circuit of supply areas, aiming to provide a steady flow of crop deliveries right up to the end of the harvest season, when prices are higher. With the lack of space and perishability of produce, wealthy traders with extra capital that could be used to build up stocks of food for the off-trading season often prefer to loan money to farmers to store produce on the farm rather than take the risk of storing it themselves. Some traders lend money to farmers in the months before harvest to enable the hiring of labour for weeding or harvesting. In return, the lender has the first option to buy the harvested crop from the farmer at the prevailing market price. Professional and salaried women may also engage in part-time trading, but concentrate less on foodstuffs and more on light manufactured goods, especially cosmetics and clothing.

A survey undertaken in the late 1970s found that over 30,000 people were employed in Kumasi's manufacturing and related repair services, with the bulk of them in the informal sector. Enterprises of similar type were clustered in particular parts of the city, usually in poor accommodation and 80 per cent without access to water or electricity. Unlike the food traders, many manufacturing entrepreneurs had migrated into Kumasi, mainly from other parts of Ashanti region (Aryee, 1981). They were relatively well educated, with 60 per cent having more than ten years' schooling, and 90 per cent having gained skills training through an impressive apprenticeship system. On average, 4.5 persons were employed in each enterprise, although the larger businesses, specializing in motor repair and carpentry, had many more apprentices. The average value of investment was $680, and the average amount of capital per worker was generally only a small fraction of that in the formal sector. The average value of output was about $160 per week. Individual earnings were higher than for equivalent work in the modern sector, and foremen – or 'masters' – earned higher wages when they were supervising more apprentices. The latter received token pocket money plus food. Nine-tenths of raw materials and servicing requirements were bought from retailers, usually small shops. Larger informal sector businesses had more links with the formal sector, but three-quarters of sales were to households and individuals. Though direct links with the formal sector were generally weak, indirect links through middlemen were common. For example, carpentry enterprises bought wood from merchants, who in turn had purchased timber from sawmills. Traders gaining contracts for school furniture usually contacted informal businesses to produce the items. Although only 3 per cent of enterprises had received a bank loan, businesses seemed to have few problems in overcoming the barriers to expansion (Aryee, 1981).

The significance of informal sector employment in Ghana has been recognized by the IMF as a key element in the country's poverty reduction strategy: 'Its flexible adaptations to labour market fluctuations and its provision of alternative employment opportunities for alleviating the negative consequences of the structural adjustment policies make the informal sector one of the most crucial in Ghana's development efforts' (IMF, 2003: 73). However, there is a need to address some features of the urban informal sector, such as education and skills training, particularly among women and people with disabilities. The provision of bank credit, the allocation of legally approved sites and the strengthening of enterprise associations, both to overcome disadvantages associated with the small size of businesses and to develop links with the formal sector, might help to strengthen informal sector activities and generate more employment.

Box 5.4

Urban agriculture: ensuring food and household security

In the last two decades, a good deal of empirical research has been undertaken on the phenomenon of 'urban agriculture' in African towns and cities (see, e.g., Mougeot, 2005). This usually involves the growing of crops (grains, vegetables and fruit) within the built-up area, most commonly in 'gardens' within family compounds or on vacant plots. Evidence suggests that urban agriculture can make a significant contribution to ensuring food security, particularly among poor households, as well as providing work in situations where there are high rates of unemployment. Structural adjustment programmes, leading to retrenchment of civil servants and others, have been an important factor in the growth of urban agriculture (see Box 5.2). There is also evidence to show that some of this food is sold in urban markets, helping to satisfy growing consumer demand in Africa's bourgeoning cities, and providing financial reward for producers and sellers (Binns and Lynch, 1998; Ellis and Sumberg, 1998; Lynch *et al.*, 2001; Simatele and Binns, 2008). A survey conducted in the late 1990s in West African cities found that urban agriculture accounted for between 20 and 60 per cent of urban household income and savings (Smith, 2001). In Nouakchott, Mauritania, urban agriculture covers over 150 hectares and is the only source of income for some 6000 people (Cissé *et al.*, 2005). In light of steep increases in the prices of basic foodstuffs in Africa and worldwide since 2007, urban agriculture has become an even more significant activity in many African cities (see Plates 5.1 and 5.2).

A survey undertaken in 1996 in the rapidly growing northern Nigerian city of Kano discovered that considerable amounts of fruit and vegetables were being grown in and around the city. Most cultivators in Kano were men since, according to local Hausa Muslim law, women of child-bearing age must remain in seclusion. While wealthy households and businessmen regarded fruit trees as a form of investment, resource-poor cultivators grew vegetables and fruit, mainly for home consumption and sale. One large area of cultivation in the city was located underneath the Federal Aviation Authority's transmission masts, an area that was opened up in the early 1980s with permission from President Shagari under his 'Green Revolution' initiative. Prospective cultivators first had to seek permission from the Aviation Authority's officers, and land was allocated on a first-come, first-served basis. Vegetables and fruit were generally head-loaded or transported by bicycle to

Plate 5.1 Urban farmer in Kano, Nigeria (Tony Binns).

Plate 5.2 Urban agriculture, central Freetown, Sierra Leone (Tony Binns).

a local market on the southern edge of the production site, though in some cases crops were sold directly to local consumers, market traders and middlemen (Binns and Lynch, 1998; Lynch *et al.*, 2001).

In Lusaka, the capital of Zambia, with a population of 1.4 million, urban agriculture has also become an increasingly important activity among both poor and better-off urban households. Urban cultivation in Lusaka probably dates back to the foundation of the original township in 1929. Nowadays, in addition to the use of urban spaces and back yards, it is being practised widely in other locations, such as 'between railway lines, around industrial areas, along roadsides, in the middle of roundabouts, under power lines, around airports, along rivers or river valleys, on land occupied by educational and administrative institutions, around dams and sewerage installations, and on land which has been officially designated for residential development' (Simatele and Binns, 2008: 8). A recent study has shown that as many as 90 per cent of those engaged in urban agriculture in Lusaka are women (Hampwaye *et al.*, 2007). As in Kano, plot sizes in Lusaka vary in size, but they are generally between 5 and 15 sq metres. Typical crops grown are maize, cabbages, pumpkins, tomatoes, groundnuts, okra, beans, cucumbers and sweet potatoes. Household sustenance and income generation are the main objectives of growing these items. In Chilenje, a planned medium–low-cost housing area, 30 per cent of respondents interviewed between October 2004 and December 2006 reported that urban agriculture contributed 65 per cent of their vegetable requirements. In the poorer, centrally located settlement of Garden Compound, 48 per cent of respondents said that urban agriculture provided 75 per cent of all their vegetable requirements in the rainy season, when these crops are mainly grown. A female respondent commented, 'Life in Lusaka has become difficult. Although my husband and I do not own land, growing our own food has helped us a lot because we are now able to feed ourselves and to save a bit of money for other things' (Simatele and Binns, 2008: 11).

Eastern Cape Province of South Africa is one of the poorest regions in the country, with an estimated 72 per cent of the population living below the poverty line, compared with a national average of 57 per cent (SARPN, 2004). In a field-based study undertaken in 2003–2005, in Grahamstown, Eastern Cape (population approximately 100,000, including a black population of roughly 85,000), Thornton (2008) found 1080 occurrences of urban agriculture, some 947 of which were located in the 'black township' of Rhini, where most of the poorest and unemployed households lived, and where 71 per cent of households were subsisting on social welfare grants. Urban agriculture, on small plots averaging just 1–2 sq metres, was providing subsistence for the 'poorest of the poor', many of whom were attempting to survive on well under $100 per month. On average, the poorest households in Rhini saved up to $25 per month from growing food in their gardens, though social grants, and particularly old-age pensions, provided the majority of poor households with the means to purchase food. These grants were especially important in households with a high proportion of people with HIV/AIDS who were unable to work. South Africa has a national HIV prevalence rate of almost 19 per cent among the 15–49 age group (World Bank, 2008d), and it is likely that the figure in poor townships such as Rhini is considerably higher. But Thornton's overall conclusion was that, despite South Africa's post-apartheid governments encouraging agricultural production by poor households on unused municipal or common land, the existing social security system seems to militate against this. There also seems to be particularly strong resistance to farming among young people who link 'subsistence agriculture to the apartheid legacy of the homeland system' (Thornton, 2008: 258).

While the various studies of urban agriculture in African cities reveal the significance of the activity to a greater or lesser extent, a number of associated problems have also been identified. The availability of a reliable and clean water supply for urban farming is an important issue. For example, in Kano, heavily polluted water is often used on vegetable plots. The city has three large industrial estates producing metal, rubber and plastic goods and a number of tanneries that are now using more polluting chemical dyes rather than traditional vegetable dyes. Both traditional and modern industries discharge their waste water into local water courses, which are then used as sources of water for urban agriculture. As well as polluting rivers and streams in the urban and peri-urban area, the discharge from the Bompai industrial estate affects water quality in shallow hand-dug wells: 'A farmer remarked that he could see that the water quality was not very good at the moment, and his vegetables might get "sick" if he irrigated' (Binns *et al.*, 2003: 438).

Insecurity of tenure is another common problem facing urban farmers. With the rapid growth of many African cities, land which was once used for urban agriculture is increasingly being built upon and can sometimes be appropriated without warning. In Kano, for example, much of the urban farmland was owned by property speculators, who sometimes carefully demarcated their plots with concrete marker posts. But, in a situation where there was a lack of clarity and official documentation associated with land ownership, the farmers continued in their activities, knowing that at some point they could lose the land and a key component of their livelihood (Lynch *et al.*, 2001). Such insecurity of tenure causes considerable stress.

Lusaka's urban farmers were also worried that cultivated land would be seized and developed by titled land owners, government agencies or other developers. They could not have been encouraged by the fact that government legislation in Zambia works strongly against the practice of urban agriculture. The Control of Cultivation Act 1995 (CAP 480, Section 110 of the *Laws of Zambia*) states that 'except with the permission of the Council, no cultivation of any kind will be permitted on un-alienated or unoccupied land within the boundaries of the township' (GRZ, 1995). A number of African governments and urban authorities, including Lusaka, also show concern about possible health effects of growing crops in close proximity to housing. The Public Health Act 1994 (CAP 295 of the *Laws of Zambia*) stipulates that 'a person shall not within a township permit any premises or lands owned or occupied by him or over which he has control, to become overgrown with bush or long grass of such nature as in the opinion of the Medical Officer of Health, is likely to harbour mosquitoes' (GRZ, 1995). There have been occasions when Lusaka City Council workers have been instructed to destroy crops in the urban area, though this has not happened in recent years (Hampwaye *et al.*, 2007).

As Hampwaye *et al.* (2007: 557) suggest in the case of Lusaka, 'urban agriculture has been considered as the antithesis of modernization and indicative of an official failure in the urban development process'. Accordingly, it has been stigmatized as 'backward', 'rural', and 'traditional', an activity that has 'no place in the context of modernizing cities'. Cissé *et al.* (2005) found similar perceptions in their study of francophone West African cities, where the prevailing opinion of the authorities seemed to be that agriculture cannot be an urban activity. By contrast, in Kano, the local authorities generally adopt a permissive attitude to urban agriculture, in effect turning a blind eye to the practice. Meanwhile, in South Africa, official policy has been to encourage urban farming in order to improve household food security in the face of unemployment rates that are as high as 80 per cent in some black townships.

There is an urgent need for governments and urban authorities to look upon urban agriculture more favourably and consider the possibilities of incorporating it into their urban planning strategies. But, as Cissé *et al.* (2005: 153) suggest, 'The recognition of urban agriculture's current or potential importance has not yet been translated into an effective inclusion in the legal and statutory provisions of African countries.' This could be seen as a pro-poor strategy which encourages food production and provides employment. Furthermore, urban agriculture could be part of a 'greening of the city' process, and in Sahelian countries could contribute to combating drought and potential land degradation. Cissé *et al.* (2005) found 3670 palm trees and 1464 fruit trees in Nouakchott's urban farming area. However, in many African countries, urban planning continues to be based on models inherited from the colonial period and used in European countries, while the relatively few planners employed by African local urban authorities have invariably received training that is closely aligned to these models. For example, in Lusaka, the master plan of the late 1970s is still the basis for urban planning over thirty years later. Greater dialogue between planners and various actors involved in urban agriculture is long overdue, such that more appropriate policy and interventions can be devised and better coordinated.

Although at times it is difficult to interpret inadequate and unreliable census data, it seems that Ghana was one of the first African countries to experience serious economic deterioration, and in the period 1970–1984 annual urban growth was only 3.2 per cent. In Uganda, the 1970s were characterized by reverse migration from the major towns, particularly Jinja, although during the 1980s in-migration seemed to resume (Potts, 1995). Evidence from Nigeria in the 1980s indicates increasing difficulties in the large towns and cities leading to migrants moving back to their rural villages (Collier, 1988). Many urban migrants continue to maintain strong linkages with their rural homelands and frequently have land rights there. As Potts (1995: 260) argues, 'If a decision is made to return "home", then the availability of land via systems of customary tenure is regarded as an essential facilitating factor in this process of return migration.'

In a more recent study, Potts has shown that both large and small urban centres in the Zambian Copperbelt, which experienced growth rates of 7 per cent or more in the 1960s, have declined significantly in more recent years: 'Thus both large and small urban centres right across the Copperbelt have been experiencing net out-migration for at least 20 and, in some cases, 30 years, and by the 1990s the rate of such out-migration exceeded natural increase' (Potts, 2005: 592). A Zambian Central Statistics Office report in 2003 specifically mentioned the significance of urban–rural migration: 'The rate of urbanization has continued to decline . . . On the other hand, in-migration to rural provinces has increased in nearly all provinces. This indicates that there is predominance of urban–rural migration over rural–urban migration' (CSO, 2003: 6). Potts concludes that the factors affecting the changing nature of migration are essentially economic, with deepening urban poverty and a decline in public sector support for low-income urban housing and services. In Zambia, economic decline has been associated with fluctuations in the world price of copper, leading to job losses and increasing poverty, while from the late 1980s structural adjustment policies led to further declines in urban living standards (Potts, 2005). The declining rate of urbanization and increasing trends in both return and circular migration are also shown by Beauchemin and Bosquier (2004) in their study of eight francophone West African countries. They comment, 'the role of migration in urban growth seems to be less important than is commonly assumed' (Beauchemin and Bocquier, 2004: 2250).

5.4 Phases of urban growth

5.4.1 Pre-colonial towns

Although Africa has experienced its most rapid period of urban growth since 1960, it would be wrong to ignore the presence of towns in the pre-colonial period. The most important areas of pre-colonial urban development were along the Mediterranean coast, in the Ethiopian Highlands, along the coast of East Africa and in two major zones of West Africa – the forest belt and further north in the savanna region.

Urbanization in North Africa generally occurred earlier than elsewhere in the continent, so the region has a good number of ancient cities, including Fes and Marrakech (Morocco), Algiers (Algeria), Tunis and Kairouan (Tunisia), Tripoli (Libya), Alexandria and Cairo (Egypt). The history of these and other settlements in North Africa was closely linked with the rise and fall of several empires in the region and their associated military and trading activities. For example, it is thought that the Phoenicians – strong and innovative naval merchants from Tyre and neighbouring coastal settlements in what is now Lebanon – arrived in Morocco around 1100 BC. They set up a string of trading centres along the Mediterranean coast and down the Atlantic coast of Morocco, but did not venture far inland. They also established an important purple dye production centre at Mogador (now Essaouira) in Morocco. Elsewhere in Morocco, the city of Fes was established in AD 789 by Idris I, founder of the Idrisid dynasty, and in the following century the Karaouine Mosque and the University of Al- Karaouine were built in the city. Since then, it has been Morocco's capital several times. Further south, the city of Marrakech was founded in 1070 by the Almoravid leader, Abu-Bakr Ibn-Umar, and, like Fes, it was once Morocco's capital (O'Connor, 1983).

In present-day Tunisia, a settlement at Tunes (now Tunis) was founded in the second millennium BC by Berbers, and the city and port of Carthage (now a suburb of Tunis) was founded as a Phoenician colony about 814 BC and settled by people from Tyre (see Plate 5.3). The Carthaginians assumed control of the Phoenician trading empire in the western Mediterranean during the fifth century BC. Carthage became a large and wealthy city until its destruction by the increasingly powerful Roman Empire at the end of the Third Punic War in 146 BC. However, it was subsequently refounded by the Romans, becoming a key city in the Empire until its destruction in the Muslim conquest of AD 698.

Further east, in present-day Egypt, Alexandria was founded in 332 BC by Alexander the Great and was intended to be a crucial link between Greece and the fertile Nile Valley. As an important trading port, the city grew rapidly to become larger than Carthage, and within a century Alexandria was the largest city in the world, and subsequently second in size only to Rome. It formally came under Roman jurisdiction from 80 BC, though there had been a strong Roman influence for over a century before that date. Egypt's capital, Cairo (currently Africa's largest city), was founded in AD 969 by the Fatimid caliphs and soon became an important centre of learning. Situated at the head of the vast and fertile Nile Delta, the city expanded massively from the mid-nineteenth century onwards. The oldest part of Cairo is east of the Nile, where extensive irrigation has allowed the city to spread further into the desert, but it has also spread westwards, swallowing up large areas of former agricultural land (see Box 5.1).

The fact that West Africa was the most urbanized sub-region of sub-Saharan Africa in 1960 testifies to the significance of some large indigenous towns and cities. A series of important empires and kingdoms – such as Ghana, Mali, Songhai, Oyo and Benin – each based on centralized political and economic power located in cities, successively ruled West Africa before the colonial intrusion. In the savanna region south of the Sahara, a line

Plate 5.3 Ruins at Carthage, northern Tunisia (Tony Binns).

of important towns developed at the southern end of a flourishing trans-Saharan trade route. Initially, salt was brought southwards from the Sahara and exchanged with gold, but the trade later diversified, with European and Arab goods exchanged for slaves, skins, gum and spices. The town of Djenné (in present-day Mali) was founded in the ninth century AD, while Gao, Kilwa, Mopti and Kano all date from the late tenth century (see Plate 5.4). Others followed, such as Zaria (1095) and Timbuktu (Tombouctou, 1100). The latter became an important religious and educational centre, with the first university in West Africa. Further east, in what is now northern Nigeria, Kano had at least 75,000 inhabitants in the sixteenth century. However, the coming of the Europeans and the development of the Atlantic slave trade in the eighteenth century switched the trading emphasis in West Africa towards the coast, leading to the decline of trans-Saharan trade and many associated towns (Njoh, 2006).

Other states with important settlements existed in the forest region of West Africa. The Ashanti kingdom, founded on gold mining in what is now south-western Ghana, had its seat of government at Kumasi and exercised considerable power until it was conquered by the British in 1874. Another forest-based state was Benin, in southern Nigeria, famed for its bronze artwork, which flourished in the sixteenth and seventeenth centuries. Perhaps the most famous indigenous towns in West Africa are those of Yorubaland in south-western Nigeria. These walled towns – such as the Yoruba spiritual capital of Ile-Ife, established by the tenth century AD, and the political capital Old Oyo – drew their wealth from productive farming hinterlands and a wide variety of trades and crafts. Wars between Yoruba towns in the early nineteenth century resulted in the creation of new defensive settlements, such as Ibadan in 1830, which by 1865 was the largest Yoruba city.

In East Africa, Lamu – which is at least a thousand years old – claims the distinction of being Kenya's first town. Located on the shore of the Indian Ocean, it was once a major Swahili trading port, shipping ivory, rhino horns and slaves to the Middle East and India.

Plate 5.4 Ancient mosque in Mopti, Mali (Tony Binns)

Further south, the port of Mombasa, currently Kenya's second-largest city (population around 900,000), was probably founded some time before the twelfth century. (In 1151, the Arab explorer Al Idrisi said it was an important trading centre for spices, gold and ivory that had strong links with the Arabian Peninsula, India and even China.)

In southern Africa, the ruins of Great Zimbabwe, located in the south-east of present-day Zimbabwe, cover an area of 7.3 sq km and were built at various stages between the eleventh and fifteenth centuries AD. At its peak, it is thought that some 18,000 people lived in Great Zimbabwe; like the Ashanti kingdom in West Africa, it was known for its gold production and extensive trading network, stretching as far as China. Designated a UNESCO World Heritage site in 1986, the Great Enclosure, with its 11-metre-high walls extending for 250 metres, is one of the largest ancient structures in sub-Saharan Africa.

5.4.2 Urban growth in the colonial period

Though not an indigenous settlement, Cape Town, South Africa's 'mother city' and seat of the national parliament, was established in 1652 by Jan van Riebeeck for the Dutch East India Company. Thereafter, it served as a supply station for trading ships circulating between Europe and the Indian Ocean. Van Riebeeck was instructed to develop amicable relations with the local Khoikhoi, although as the settlement grew relationships between the indigenous people and the colonizers progressively worsened (Lester *et al.*, 2000).

Many of Africa's urban centres are of colonial origin, dating from the late nineteenth and early twentieth centuries. Though they were often initially centres of trade, these towns soon became centres of administration and political control, too. Many were located on the coast, along important lines of communication or near mining centres. Accra, now the capital of Ghana, was initially a number of coastal slave forts, but it grew rapidly from

1877 when the British moved their administration from Cape Coast. Nairobi, Kenya was established almost accidentally as a railway camp in the 1890s in an area without large settlements. The mining towns of Zambia were mostly developed in the late 1920s, when rising copper prices made it worthwhile to invest in mines at Chingola, Kitwe, Luanshya, Mufulira and Ndola.

Colonial towns frequently show a clear distinction in their structure between functional and residential zones, and often segregation of different racial and ethnic groups (O'Connor, 1983). Within residential areas, there were further divisions according to race and class. French towns typically had tree-lined boulevards, parks and pavement cafés. Dakar, capital of Senegal, grew rapidly in the colonial period, acquiring many features of a typical French city and assuming the position of capital of the whole of French West Africa.

In areas of British colonization, the racecourse, golf course and club were key features. This is clearly seen in the city of Kano, where, with the advent of colonialism in 1903, a European residential area (Nassarawa) was established to the east of the old walled city, with wide roads, spacious housing plots and recreational facilities (see Figure 5.4). A quite separate area for non-northern (and mainly non-Muslim) Nigerians, known as Sabon Gari, was laid out on a grid-iron pattern with buildings at a much greater density. A further sector, Fagge, west of the railway station and close to the developing central business district, was settled by Syrian and Lebanese traders. Thus, within Kano, there is a series of clear structural divisions, based on the ethnic and social characteristics of their original inhabitants (Binns, 1994).

In other cities that developed during the colonial period, quite separate settlements built by and for the colonial masters can be seen. For example, in Freetown, Sierra Leone, sometimes referred to as the 'white man's grave' by early settlers, the central area of the city is constructed on a grid plan next to the sea, but the white-settler suburb of Hill Station was built high above the city in a location with a fresh and more tolerable climate. It was linked to Freetown by a railway. Some colonial cities lacked any indigenous element and were established primarily as places for Europeans to live and as administrative centres. For example, the city of Kaduna in central Nigeria was founded in 1913 as a planned city by the British colonizers under Lord Frederick Lugard to serve as the administrative capital of the Protectorate of Northern Nigeria (Binns, 1994).

In East and southern Africa, Nairobi, Harare, Lusaka and Bulawayo were all built on European lines and were established primarily to provide services to European settlers in the surrounding areas. Pretoria, the administrative capital of South Africa, was founded in 1855 and named after Andries Pretorius, who was involved in the victory of the Dutch Voortrekkers over the Zulus in the Battle of Blood River in 1838. It was proclaimed capital of the South African Republic in 1860. The city has many impressive buildings, including the Union Buildings, designed by the British architect Sir Herbert Baker. Completed in 1913, the Union Buildings are today the official seat of the South African government and the offices of the country's president (O'Connor, 1983).

Following independence, urban growth in many countries accelerated. For example, Dar es Salaam, capital of Tanzania, experienced an average annual growth rate of 9.7 per cent between 1965 and 1980, although this rate halved to 4.8 per cent between 1980 and 1988. Meanwhile, Lusaka, Zambia's capital, grew at an annual rate of almost 14 per cent between 1963 and 1969; in the Zambian Copperbelt, Ndola and Chingola grew at nearly 10 per cent each year, and Kitwe at over 8 per cent. But, as occurred in Tanzania, Zambian urban growth rates slowed between 1980 and 1988 – to 6.1 per cent for Lusaka, 4.0 per cent for Ndola, 2.5 per cent for Chingola and 2.4 per cent for Kitwe (Potts, 2005).

In Nigeria there was a desire for a more centrally located capital, so in 1976 the government announced its intention to transfer the capital to Abuja in the middle belt, a relatively

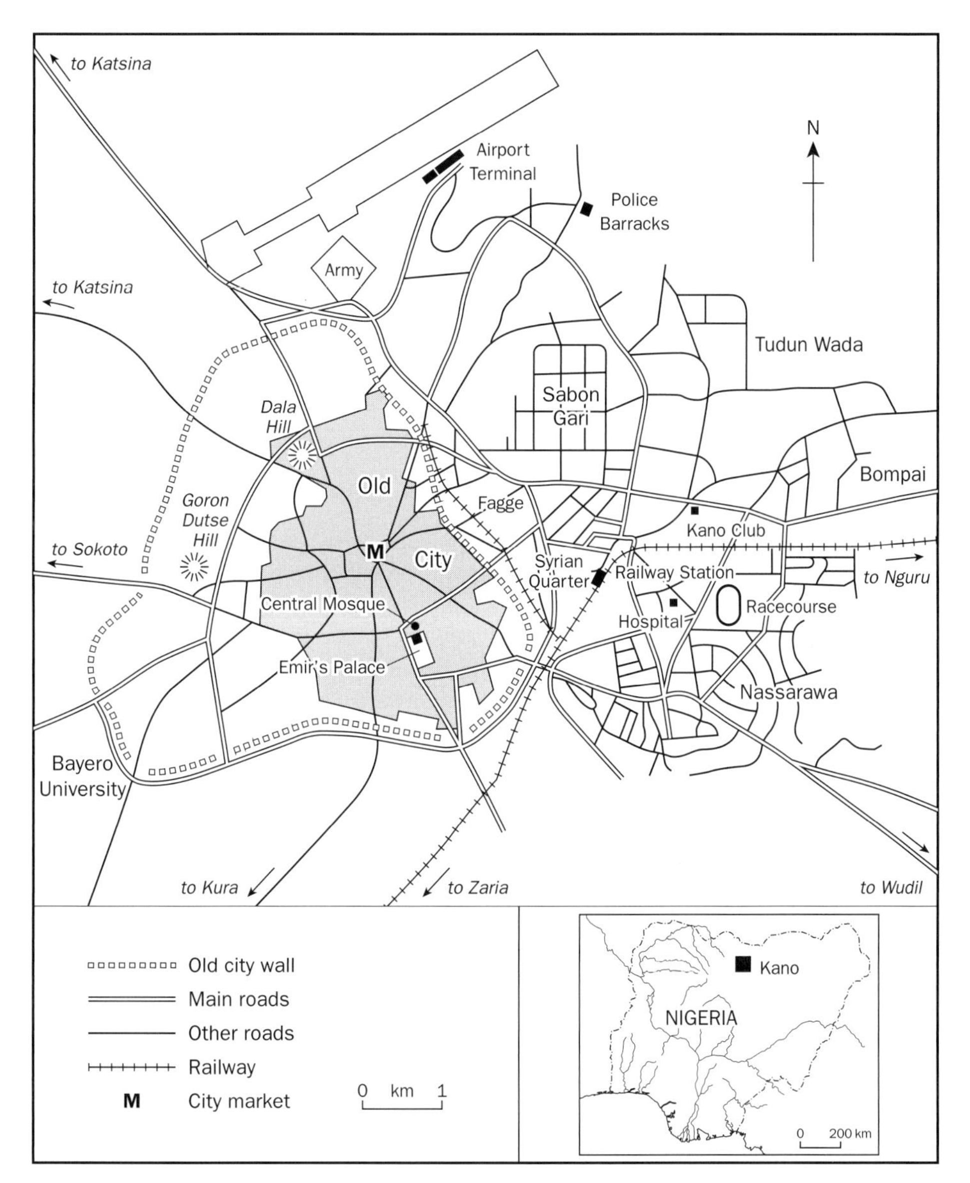

Figure 5.4 Kano, Nigeria.

Source: Binns, 1994

under-populated region. In December 1991, Abuja officially became Nigeria's federal capital and government ministries started relocating from Lagos soon afterwards (see Box 5.5). Also in West Africa, Côte d'Ivoire developed a new national capital at President Houphouët-Boigny's village of Yamoussoukro. As in the case of Abuja, this was a very costly project. The new capital boasts the world's largest place of Christian worship – the Basilica of Our Lady of Peace of Yamoussoukro – which was built between 1985 and 1989 at an estimated cost of $300 million. It was consecrated by Pope John Paul II in 1990.

Box 5.5

Housing and infrastructure in Lagos and Nairobi

Lagos, Nigeria, and Nairobi, Kenya, are respectively the largest cities in West and East Africa. In 2008, the estimated population of Lagos was 9.8 million and that of Nairobi was 3.1 million, with the two cities having very similar annual urban growth rates of 3.74 per cent and 3.76 per cent, respectively (UN-HABITAT, 2008). By examining some of the key issues concerning housing and infrastructure in these two cities, we should gain a clearer picture of the issues facing many large towns and cities in Africa.

Lagos

By the time of publication, Lagos will probably have become Africa's second mega-city (joining Cairo); and by 2020, it will probably be the world's twelfth largest urban agglomeration (UN-HABITAT, 2008). Lagos is already home to some 6.5 per cent of Nigeria's population and almost 14 per cent of the country's urban population. Amazingly, it is also the base for about 70 per cent of all Nigerian industry. It has expanded over 40 km northwards, crossing the state boundary into Ogun State. This vast agglomeration of mainly low-rise developments includes perhaps 200 slums, of which Ajegunle is one of the largest.

To an overseas visitor, Lagos is a chaotic place, such that early editions of the *Rough Guide* and *Lonely Planet* advised tourists to avoid it altogether. In 1991, the UN named it as the dirtiest city in the world. The United Nations Human Settlements Programme comments, 'Lagos is the classic example of a developing country mega-city, combining haphazard, uncontrolled and unrestrained population and spatial growth with little corresponding expansion in housing, infrastructures, services and livelihood opportunities' (UN-HABITAT, 2008: 78). Gandy (2006: 372) sums up the state of the city in the early twenty-first century as follows:

> Over the past 20 years, the city has lost much of its street lighting, its dilapidated road system has become extremely congested, there are no longer regular refuse collections, violent crime has become a determining feature of everyday life and many symbols of civic culture such as libraries and cinemas have largely disappeared. The city's sewerage network is practically non-existent and at least two-thirds of childhood disease is attributable to inadequate access to safe drinking water. In heavy rains, over half of the city's dwellings suffer from routine flooding and a third of households must contend with knee-deep water within their homes.

In short, Lagos might be described as a highly dynamic, yet dysfunctional, city.

Living conditions in many Nigerian cities give much cause for concern, and in Lagos the situation is probably worse than in most. In 2003, it was estimated that only 9.0 per cent of the city's households had access to piped water on their premises and less than half of all households (49.2 per cent) had access to a flush toilet. A large proportion of households (60.8 per cent) were reported to have an insufficient living area (UN-HABITAT, 2008).

The problems now experienced in Lagos can be traced back to the colonial period of the late nineteenth century, when it was seen essentially as an important trading centre in the expanding British Empire. The colonial government showed little concern about the quality of life in the city or the need for adequate services and infrastructure for ordinary Nigerians. Levels of investment in infrastructure development were low and generally directed only towards the wealthy areas. In the 1920s, outbreaks of bubonic plague were symptomatic of a seriously neglected public health environment. A UN survey of the city undertaken soon after independence revealed such serious problems as extreme congestion, a major shortage of housing, high rents, inadequate housing finance, the rapid growth of slums and inadequate sanitation. Furthermore, the newly independent state had very limited expertise in technical and administrative matters (UNTAP, 1964). There has been little improvement in the urban scene since then.

The growing metropolis has extended far beyond its original lagoon location and has 'swallowed up' many former rural villages, such as Olaleye. In 1967, this village was home to 2500 people; by 1983, its population had grown to 20,090. Three-quarters of the people in Olaleye in 1983 were low-income tenants renting on short leases of less than five years from landlords who were mainly self-employed businesspeople, some of them absentee. Reporting on a survey that was conducted in Olaleye in 1983, Aina (1988: 36) commented: 'Large households are predominant with two-thirds of the sample consisting of households of not less than 5 and in certain cases up to 11 persons. Families with 3 or more children constitute a majority, with a significant proportion having 6 or more children. The majority of the sample households, almost 60%, live in single rooms.'

In 1975, it was estimated that an extra 10,000 houses were needed in Lagos each year. Since independence, the private sector has continued to be the main source of housing supply in the city, with the government playing only a marginal role. The national government did little about housing in the 1960s and early 1970s, leaving responsibility for it almost entirely to state governments and offering little, if any, coordination of policies. The first successful attempt to give the housing issue a national focus was the establishment in 1971 of the National Council on Housing, with members drawn from the states. A national housing programme was eventually drafted in 1972, aiming to build 59,000 dwellings throughout the country, with 15,000 in Lagos and 4000 in each of the states.

The Third National Development Plan, launched in 1975, declared that the federal government 'now accepts it as part of its social responsibility to participate actively in the provision of housing for all income groups and will therefore intervene on a large scale in this sector during the plan period' (Nigerian Government, 1975: 10). In 1976, the housing programme was enlarged to 202,000 units, with 46,000 of them in Lagos State. Attention was to be directed towards the provision of basic infrastructure, such as water and electricity, and social amenities, such as schools, health centres and shopping centres. A mortgage bank was established to lend to individuals, state government housing corporations and private developers. Companies with over 500 employees were encouraged to develop housing estates for their workers. However, there was a wide gap between proposals and achievements due to poor planning and economic recession. So, even though some high-profile projects, such as the National Theatre (1977), were completed in Lagos, progress in the provision of housing and infrastructure was slow. By 1980, fewer than one-fifth of the proposed 202,000 new units had been completed.

The Lagos Master Plan of 1980 was a highly ambitious statement on the proposed development of the city, suggesting that all households would be connected to a water supply and sewerage system within two decades. However, in the early 1980s, the global recession and collapse of oil prices had a devastating effect on the Nigerian economy, compounded in 1986 by the introduction of a structural adjustment programme, leading to increased unemployment and poverty and even less investment in public services. As Gandy (2006: 382) comments, 'The failure of the Lagos Master Plan signalled an effective abandonment of attempts to conceptualise the city's problems in any integrated or strategic way, and the rapid urban decline and brutalisation of political life experienced from the 1980s onwards heralded a retreat of policy discourse into the realm of crisis management.'

Progress in dealing with living conditions in Lagos has also been hampered by hostile inter-governmental relations and widespread corruption. It is in this context that Lagos has become a 'self-service city', where survival for many is dependent on ensuring access to clean water, sanitation and, wherever possible, electricity. Water is sold on the streets, while others draw their supplies from bore holes and wells, mobile water tankers and illegal connections to water pipes. Meanwhile, more wealthy citizens can pay for extensions of water and sanitation systems to their homes, and many have their own generators to provide a reliable power supply.

In a bid to combat this very difficult situation, a new non-governmental organization – Metamorphosis Nigeria – was established in 2000 by a group of Lagos professionals with the aim of 'improving the quality of life and opportunities for positive change in urban communities through projects that improve the environment and promote safe sanitary and hygiene practices' (Metamorphosis Nigeria, 2008). This organization has strengthened links between government, civil society, private sector and donor agencies, and has received support from the UK's Department for International Development. Its initiatives are focused particularly on hygiene education and the provision of modern public-convenience facilities in high-density areas.

Since the election in June 2007 of a new governor, Babatunde Fashola, there has been some cautious optimism that the Lagos State government might become more active in cleaning up the city. There is a new target to plant a million trees in three years, while the Lagos Waste Management Authority has been granted new powers, and private sector support, to clean up the city. Litter bins and mechanical road-sweepers have been introduced. Transport is also being given priority. Many of the city's highways and bridges date from the oil-boom period of the 1970s and are in a very poor state. Fashola has initiated the inspection of all bridges, and the Traffic Management Authority has been given greater powers. The Lagos–Badagry Expressway is to be upgraded to a ten-lane highway, and there are plans for a light railway. Work has already started on a mass-transit rail link from the Iddo terminus northwards into Ogun State, a project supported by the Lagos Mega-City Development Agency and backed by the federal government. The Eko Atlantic City, an ambitious development project that aims to emulate the success of Dubai, is planned to be built on reclaimed land on Victoria Island and will include high-rise apartment blocks, shopping malls and conference centres (*New African*, 2008).

Clearly, Lagos is a very dynamic city. However, despite these new initiatives, it seems likely that millions of its inhabitants will see few, if any, improvements to their everyday lives in the foreseeable future. Even the task of ensuring the widespread availability of such basic necessities as clean water, sanitation, housing and power remains truly enormous.

Nairobi

With a population in 2008 of 3.1 million, Nairobi is less than a third of the size of Lagos, but with an annual growth rate of 3.76 per cent between 2005 and 2010, the city is projected to have a total population of 5.8 million by 2025. According to official statistics, 76.6 per cent of Nairobi's households in 2003 had piped water on their premises and 66.5 per cent had access to a flush toilet (UN-HABITAT, 2008). While these figures are better than those of Lagos, there is considerable variability in the quality of life across Nairobi. It is estimated that there are about 200 slums in the city, which are home to almost 50 per cent of its population. The oldest slum settlement is Mathare Valley, with an estimated population approaching 200,000. The slum grew with the influx of people from the rural areas following independence in 1963 (see Plate 5.5). Kibera, with an estimated population of 700,000, is Nairobi's largest slum settlement, situated about 7 km to the south-west of the CBD and covering some 225 hectares. It is split by the Nairobi–Kisumu railway line and is surrounded by middle-income housing estates.

Over 80 per cent of Kibera's inhabitants are tenants in illegal structures, paying an average monthly rent of $12. Some 80 per cent live in single rooms measuring an average of 9.4 sq metres and usually without piped water and sanitation. Water is sold by street vendors and a single pit latrine may be used by up to seventy-five people. As in Lagos, a large proportion of Nairobi's slum dwellers are tenants and, since a third of Kibera's inhabitants earn less than $70 per month, much rental accommodation is unaffordable. A single-room apartment in Umoja Estate, for

Plate 5.5 Informal housing in Mathare Valley, Nairobi, Kenya (Tony Binns).

example, costs about $100 per month to rent (Worldwatch Institute, 2007). With such pressure for cheap rental housing, illegal and poorly constructed multi-storey blocks are being built with shared toilet and washing facilities. Basic services, such as water points, schools and health facilities, are totally inadequate. Although there are fourteen public primary schools within walking distance of Kibera, they can accommodate only one-fifth of the estimated 100,000 children of primary school age living in the area.

The Kenya Slum Upgrading Programme (KENSUP) was formalized by a 'memorandum of understanding' in January 2003, signed by the Kenyan government and the United Nations Human Settlements Programme (UN-HABITAT), to collaborate in the formulation and implementation of a long-term National Slum Upgrading Programme. This collaborative programme, involving the Kenyan government, UN-HABITAT and a number of NGOs, targets the construction and improvement of about 45,000 units annually at a cost of about $440 million. In Kibera, where the provision of water and sanitation has been problematic, the Kibera Integrated Water, Sanitation and Waste Management Project (K-WATSAN) was inaugurated in 2008. This is a pilot project in one of Kibera's thirteen villages, Soweto East, and has been built with support from an NGO – Maji na Ufanisi – and community involvement. The project has provided storm-water drains and seven communal water and sanitation facilities, with each facility having its own management group. Additionally, solid waste management has been improved with small-scale door-to-door waste collection and recycling services.

A non-motorized transport project has been implemented by the NGO Worldbike, in collaboration with the community-based organization Soweto Youth Group. Mobility in the area has been improved with a low-volume road with storm drains and pedestrian walkways. In conjunction with the Kenya Power and Lighting Company, electricity has been extended to 1000 units in Soweto East. A community and youth resource centre has been built with a dispensary, rehabilitation facility for children with disabilities and a one-stop youth centre. Finally, training courses have been arranged to strengthen the institutional and technical capacities of selected target community groups, for example youth training in low-cost construction and carpentry, with funding from a donation by UN Secretary-General Ban Ki-moon, who visited Kibera in January 2007.

The K-WATSAN project is a good example of what Otiso (2003) refers to as a 'tri-sector' development initiative in urban housing and service provision, in which there is a partnership between the state, the private sector and community self-help efforts. He argues that such a partnership is likely to be more successful than projects that are dominated by just one of these three providers, and it represents a significant step forward from earlier 'top-down' development initiatives. Otiso draws on the example of the Mathare 4A Slum Upgrading Project, which covers 18.5 hectares and aimed to improve living conditions by building 8000 rooms, business stalls, kindergartens, a street lighting system, a road and footpath network, a water and sewerage reticulation system, a water drainage system and one 'wet-core' (shower–toilet–washing slab) for every ten households. State involvement was from the Kenyan and German governments. Four private sector participants were involved, too, including consulting and construction firms, while the two major NGO participants were the Catholic Church and Approtec (Appropriate Technology). The project was coordinated by the Catholic Church through an affiliate, the Amani Housing Trust, which encouraged local community involvement by providing money, labour and logistical support. The project resulted in significant improvements to housing and access to basic services, and was able to maintain rents at their pre-upgrade levels. No one was displaced during the improvement process, since new units were constructed before the old ones were demolished. A flexible rent collection programme was introduced and residents were required to sign a legally binding contract to participate in the project before their dwellings were upgraded (Otiso, 2003). The process of slum upgrading in Nairobi will take many years, but at least the process has started and some significant achievements have already been made.

On a much broader scale (covering 3000 sq km), the Kenyan government is preparing an ambitious plan – Nairobi Metro 2030 – which it is hoped will 'spatially redefine the NMR [Nairobi Metropolitan Region] and create a world class city region envisaged to generate sustainable wealth and quality of life for its residents, investors and visitors' (UN-HABITAT, 2008: 129). But formulation of the plan still has a long way to go in defining regulatory arrangements, coordinating activities and strengthening governance in the twelve independent local authorities involved, as well as ensuring participation of key actors, not least the residents of the affected local authorities. In 2007, UN-HABITAT assisted the Kenyan government with the plan, and it was hoped that Kenya's good practice might be replicated elsewhere in Africa (UN-HABITAT, 2008; see Plate 5.6).

Plate 5.6 Central business district, Nairobi, Kenya (Tony Binns)

The capital of Mauritania, Nouakchott, once just a small settlement on a main road along the Atlantic coast, was chosen as the country's new capital three years before Mauritania gained independence from France in 1960. The city's plan reflected the colonial strategy of segregation, with different neighbourhoods for working-class and administrative groups. It was planned as an administrative city for just 8000 residents, but experienced a population growth rate of over 380 per cent between 1955 and 1999. Although this rate slowed after 1990, it is likely that Nouakchott's population is now well over a million.

In Malawi, the decision to relocate the capital to Lilongwe, close to President Banda's birthplace, was taken in 1968, but it did not succeed Zomba as capital until 1974. It was argued that a more central location in an area of good agricultural potential was needed, and with the possibility of locating a new international airport near by. South Africa provided much financial and technical assistance for the project, as well as drawing up a master plan. By 1989, there had been little development in the new capital, aside from government buildings and the airport, though by 2008 the population was well over 600,000, making it the second-largest city in Malawi, after Blantyre.

In 1973, Tanzania decided to move its capital from Dar es Salaam on the coast to Dodoma, 480 km to the west. It was argued that the more central location would stimulate economic and agricultural development in the country's underdeveloped heartland. However, rising construction costs meant that progress on building the new capital was slow, and it was only in 1996 that the National Assembly moved there. Even then, many government departments remained in Dar es Salaam, which continues to be Tanzania's commercial capital.

5.5 Rural–urban links and peri-urban pressures

As anyone who has undertaken field-based research in Africa will appreciate, defining and distinguishing between 'urban', 'peri-urban' and 'rural' in the context of cities and their hinterlands is not easy. For example, there are particular difficulties associated with defining a 'peri-urban' area as a spatial zone around a city, since its size and shape are influenced by a wide range of factors, including topography, density, population distribution, land value and quality, land ownership and transport routes. It is perhaps more useful to consider 'peri-urban' as an environment where the nearby city has a dominant influence on most spheres of activity, and 'where the advantages of combining farm and non-farm work can be maximised' (Swindell, 1988: 98). Taking this further, can we therefore assume that 'rural' areas are those that are too distant to be dominated by urban influence? Not really, because many African cities have longstanding and very strong relationships with rural areas, often extending over considerable distances. The intensification of links between town and country has had the gradual effect of progressively blurring the economic and cultural differences between city and rural areas. This is particularly apparent in peripheral, unplanned areas, where formal infrastructure is often lacking, and many households occupy relatively large plots of land in semi-rural locations (HABITAT, 1996). If researchers and policy-makers are to gain a better understanding of such spatial relationships, then perhaps they should not be constrained by this urban–peri-urban–rural trichotomy (Binns and Lynch, 1998; Lynch, 2005; Maconachie and Binns, 2006).

Rural areas surrounding Africa's rapidly growing towns and cities are gradually being incorporated into the urban system and their character is changing. These peri-urban areas are often zones of intensive market-orientated food production where rural producers and urban consumers are closely juxtaposed. They are areas where once-rural villages have lost their land to urban development, and existing farmland must now be used more intensively to supply profitable urban markets. Where farmland can still be obtained in the peri-urban fringe, urban dwellers may undertake part-time farming to supplement their food supply and income. There is evidence from some towns that capitalist farmers, relying mainly on hired labour, are buying land in peri-urban areas, usually along major roads, to supply urban food markets (Lynch *et al.*, 2001). Commuting from countryside to town in search of paid employment is also common, particularly where dry-season cultivation is not possible. These peri-urban areas show a wide variety of production and exchange relations, with considerable linkage between urban and rural areas expressed in two-way flows of goods, labour and money (Lynch, 2005).

Studies of a number of African cities have revealed intensively cultivated peripheral areas developing in the post-war period, and particularly since independence. Urban dwellers, notably women, grow crops to supplement household food supply, while men often concentrate on cash crop production. For example, women from the Makelekele quarter of Brazzaville, some of them from affluent households, grow cassava, while poorer or unemployed men grow salad crops to sell to wealthier households, hotels and restaurants.

Dar es Salaam, with a population of about 3 million, provides a major commercial opportunity for peri-urban farmers to produce food crops for consumption within the city. Since the 1960s, a large number of people have settled in the peri-urban zone, and in recent years land has acquired more of an investment status, putting pressure on smaller farmers. The peri-urban zone starts at about 20 km from the city centre, and in the early 1980s this agricultural expansion was adopted mainly as a survival strategy when urban incomes and living standards were squeezed. However, since the mid-1980s, the continued expansion has been geared more towards commercial sales rather than satisfying immediate household needs. Over 93 per cent of farmers grow cassava, while other popular crops include bananas, maize, pawpaw (papaya) and a variety of fruits and vegetables. The average size of farms is 1.6 hectares and intercropping is widely used (Binns, 1994).

The mountain villages close to Freetown, Sierra Leone's capital and largest city, have been producing fruit and vegetables for the urban market for many years, but both during and since the civil war of the 1990s, and with the associated massive influx of people into Freetown, these production areas have become increasingly important. Meanwhile, close to Banjul, The Gambia, intensive areas of fruit and vegetable production supply the urban market and the many tourist hotels on the Atlantic coast. Much of this production and marketing is undertaken by women (Binns, 1991).

The city of Kano and the surrounding region in semi-arid northern Nigeria currently support over 5 million people, with population densities between 250 and 500 people per sq km. A survey undertaken by Maconachie during 2001 and 2002 examined the sustainability of Kano's peri-urban area by investigating six study sites on a distance-decay transect along a tarmac road extending north-east from the city through the peri-urban area and into the rural hinterland (Binns and Maconachie, 2006: Maconachie, 2007). Urban development has increased the value and demand for land in peri-urban Kano, and some residents have abandoned farming altogether, selling their plots to urban developers. After a plot has been sold, the developers typically survey the land and lay the cornerstones of a building to indicate their formal ownership and eventual intention to develop the property. Sometimes the land remains unused for many years in the hope that property values will increase (Lynch *et al.*, 2001; Binns and Maconachie, 2006).

Survey respondents in Kano's peri-urban area were concerned about increased levels of resource competition and what was frequently referred to as 'land hunger'. Plot sizes have declined in size and the total size of family land holdings has also decreased in recent years. As farm plots have been required to feed increasing numbers of people, smaller parcels of land have been forced to 'work harder' to ensure household survival. In an area which in the past was characterized by the maintenance of soil fertility through integrated crop production with livestock herding, there is evidence to suggest that inputs of animal manure have declined, and over 60 per cent of households believed that soil fertility was declining each year due to nutrients not being replenished. Increased pressure on land in the peri-urban area has meant that virtually all traditional range-land and cattle paths have disappeared and Fulani herders have sought to diversify their livelihoods. There was also evidence to suggest that the dramatic acceleration of urban growth in Kano had contributed to soil erosion, notably through the highly profitable quarrying of sand and gravel for use in the building industry. Tree cover in the peri-urban area has also decreased, with over 80 per cent of households reporting fewer trees on their land than twenty years before.

Larger quantities of wood were being used for construction and also for cooking and heating water, in the face of the increasing cost and unavailability of alternative fuels, such as kerosene. A further problem in peri-urban Kano was the pollution of water courses from industrial estates situated on the fringe of the urban area. As Binns and Maconachie (2006: 226) comment, 'Although it is acknowledged that some rural–urban linkages have positive implications for local actors, at the same time, physical evidence of land degradation, together with "insider" perceptions of key environmental issues, provide significant warning signs that new stresses are breaking down the sustainability of the system.'

Apart from the effects of urban encroachment on Fulani pastoralists in the Kano peri-urban region, relatively little research in Africa has focused specifically on changing pastoralist lifestyles in peri-urban areas. In Ethiopia, where pastoralists represent about 10 per cent of the population, one of the major pastoral groups – the Borana – live in the south of the country, close to the Kenyan border. Declining returns from livestock herding among the Borana are leading to a diversification of pastoral livelihoods, as in northern Nigeria. For instance, in the peri-urban area around the southern Ethiopian town of Yabello, Borana pastoralists with dwindling herds are increasingly settling and adopting different lifestyles, notably selling firewood, as well as some farming and herding other people's livestock to markets. Earning less than a dollar a day, there is growing competition for agricultural land, water and wood in the increasingly populated peri-urban area and, in desperation, many settled pastoralists must at times resort to illegal logging and charcoal-making. The settled Borana are scorned by those Borana who maintain their pastoral lifestyles. As Aberra (2006: 129) comments, 'Livestock serve not only as symbols of a "pastoral" identity, but also ethnic identity as Borana. Selling firewood is regarded as the most degrading work among Borana pastoralists, indicating complete destitution.'

As evidence from these and other studies shows, the peri-urban areas of Africa's towns and cities are highly dynamic regions that are characterized by a diversity of livelihood activities, which frequently put pressure on key natural resources, such as land, water, trees and soil, thus jeopardizing the sustainability of the livelihood systems themselves. As centres of power and wealth, Africa's towns and cities might be seen as parasitic, dependent on their rural hinterlands for food, labour and other commodities, such as firewood. It has been suggested that, 'For many purposes, it is important to consider the peri-urban zone as an extension of the city rather than as an entirely separate area. This is because the city region functions in a more or less integrated way in terms not only of its ecological footprint, but also of its economic and demographic processes' (Simon *et al.*, 2006: 11). In formulating future development strategies, there is a need for much greater awareness of the complexity and implications of urban–rural linkages, such that neither 'urban' nor 'rural' development is treated in isolation.

5.6 The urban housing challenge

The rapid growth of towns and cities in Africa has put tremendous pressure on housing and service provision. Hard-pressed urban authorities are faced with incoming migrants and existing urban populations that are growing rapidly. A report from the US-based International Housing Coalition in 2007 stated, 'In Zambia, 74% of urban dwellers live in slums; in Nigeria, 80%; in Sudan, 85.7%; in Tanzania, 92.1%; in Madagascar 92.9%; and in Ethiopia, a staggering 99.4%' (International Housing Coalition, 2007: 2).

Residential density in African cities is generally not as great as in some Asian or Latin American cities, due to the existence of spacious, often colonially planned, suburbs and the unpopularity of high-rise housing. However, at least a third of Africa's urban population lives in what might be termed 'slum' or 'squatter' settlements, and these often grow at 15 per cent per annum, meaning they double in size every six years. The term 'slum household' has been defined by the United Nations as: 'a group of individuals living under the same roof that lack one or more of the following conditions: access to safe water; access to sanitation; secure tenure; durability of housing; and sufficient living area' (UN-HABITAT, 2003: 5).

Squatter – or 'spontaneous' – settlements have many of the same characteristics as slums, but with the additional disadvantage that the land is being used illegally. Whereas slums tend to be central, squatter settlements are often peripheral, since squatting in central areas would soon be stopped by the owners, except on marginal land which is of no immediate use. New migrants tend to move to more central slums to be closer to possible job opportunities in the city centre. More established residents who have decided to stay, however, may prefer to squat on the edge of town in their own family houses. In many African cities, some central slums are occupied by longstanding urban residents who continue to live in their family homes, which may have been handed down through several generations. It is not uncommon for particular areas of African cities to be dominated by single extended families or ethnic groups, who may have been urban residents for some time and act as hosts to members of their group who visit from other towns or the home village.

People are forced to live in cramped and unhealthy conditions in many African cities. For example, a survey of Lagos in 1976 revealed an average occupancy rate of 4.1 persons per room. Private rented accommodation, built on plots leased from a municipal or traditional authority or occupied illegally, is very common in Africa. Frequently, the owner lives in the house and rents out rooms to tenants who may or may not be part of his or her own extended family. Rent comprises a large proportion of the weekly expenditure of most households. Housing owned by government (e.g. state or council housing) or by institutions (e.g. housing associations) is much less common (Binns, 1994).

A study of housing conditions in Accra compared survey data from the 1950s with that from the 1990s, and concluded that housing conditions were worse in the latter period (Konadu-Agyemang, 2001). In 45 per cent of the houses surveyed in six high-density, low-income suburbs, conditions were described as 'very poor', and the Town Planning Department concluded that 82 per cent of all buildings in two of these suburbs, Ussher Town and James Town, should be demolished because they were unfit for human habitation. Up to 75 per cent of all households in these suburbs had to share toilet, kitchen and bathroom with either their landlords or co-tenants. In the city as a whole, 85 per cent of households lived in multi-household houses, and it was common for tenants to rent single rooms that were used for living, sleeping and storage. Most rental accommodation in Accra was provided by what might be regarded as 'non-commercial landlords' who lived in the same premises and did not see their property as a commercial investment. Another feature of the Accra housing situation was the very low residential mobility, with 50 per cent of tenants living in the same premises for between seven and fifteen years, and 17 per cent living there for over fifteen years. Houses were much more congested in the 1990s than they had been in the 1950s, with the number of households per house more than tripling, whereas the average number of rooms per house had declined from six to four. Owning a home had become increasingly difficult since house prices had increased more rapidly than average incomes. Konadu-Agyemang (2001: 32) concluded that in the late 1990s, 'Ghana [produced] only 21.4 per cent of the estimated annual need of 140,000 housing units, [could not] adequately maintain existing stock, and [had] a backlog of 300,000 units that [were] needed to reduce congestion in existing units.'

A study of the housing situation in four North African countries – Algeria, Egypt, Morocco and Tunisia – concluded that housing affordability rather than availability was the key problem, with house prices being particularly high in Algeria and Morocco. Meanwhile, low-income groups were increasingly living in informal settlements and unserviced peripheral neighbourhoods. This is already the case in Egypt, but is a growing concern in Algeria and Morocco (Baharoglu *et al.*, 2005).

Improving the urban housing situation represents a daunting task for many African governments. In Luanda, with an estimated population of 6 million, a third of the country's total, in 2008 it was believed that '80 per cent of the urban population is established in compounds considered as precarious or informal settlements, a trend that will remain unaddressed unless measures to reverse the situation are taken' (allAfrica.com, 2008a: 2). In September 2008, the China International Trust and Investment Corporation (CITIC) started constructing a $3.5-billion satellite town for Luanda that will eventually house 200,000 people. The new town will have 20,000 apartments, over 200 shops, 40 kindergartens and schools, and its own power station, water supply and sewerage treatment plants. Some 10,000 local jobs were created during the construction period (allAfrica.com, 2008a).

Improved housing is not only desirable in its own right but can have a significant effect on economic and social development, including the health of the urban poor. One of the major challenges facing the new democratic government in South Africa in 1994 was a massive shortage of housing. The Reconstruction and Development Programme, launched by Nelson Mandela in 1994, endorsed

> the principle that all South Africans have a right to a secure place in which to live in peace and dignity. Housing is a human right. One of the RDPs first priorities is to provide for the homeless . . . A house must include sanitary facilities, storm-water drainage, a household energy supply (whether linked to grid electricity supply or derived from other sources, such as solar energy), and convenient access to clean water. Moreover, it must provide for secure tenure in a variety of forms.
>
> (ANC, 1994: 23)

In 1996, there was an estimated backlog of 1.5 million homes in South Africa, with a further 1 million squatter shacks in the country (see Plate 5.7). Despite a slow start, the government's low-income housing programme eventually achieved some impressive results, with 3 million people (a quarter of those in need) re-housed by 1999. By that stage, the government had built 680,000 houses for low-income residents and had provided 959,000 housing subsidies. It has therefore made some impressive advances in addressing the housing shortfall and improving the livelihoods of millions of people in a relatively short space of time, although some critics have complained about the small size of the housing units, and in certain areas poor building standards and corruption have undercut the general success of the strategy. There has also been impressive progress in providing basic services in South African homes. For example, between 1994 and 1998, the number of houses connected to electricity increased from 31 per cent to 63 per cent, and in the same period the number of telephone lines increased from 150,000 to 386,426. Furthermore, by 1998, 1.7 million additional households had been supplied with water (Lester *et al.*, 2000).

Governments and urban authorities have adopted different attitudes to slums and squatter areas (see Box 5.5). Sometimes it may just be too expensive to do anything about them, since the cost of clearing poor housing and re-housing the inhabitants can be prohibitively high. Even if authority-owned housing can be built, tenants may find it difficult to pay the rents that are needed to bring an adequate return on the investment. In other situations, governments are more hostile to areas of poor housing, seeing them as symbolic of their lack of success in providing for the urban population and modernizing

Plate 5.7 Duncan Village squatter camp, East London, South Africa (Tony Binns).

the built environment. Politicians and wealthy urban elites may consider these areas unsightly and press for their demolition. Alternatively, some urban authorities have helped slum and squatter dwellers to upgrade their housing. As Painter (2006: 5) comments,

> Not only is the destruction of slums fundamentally poor housing policy, it is poor economic policy as well. However appalling a slum may appear, a substantial amount of capital has been invested in its construction. Destroying capital investment . . . cuts off the flow of economic benefits that would have otherwise continued to contribute to economic growth.

Providing security of occupation seems to be crucial if residents are to invest in upgrading their housing. UN-HABITAT has estimated that slum upgrading in Africa generally costs less than $550 per person (International Housing Coalition, 2007).

One such example of upgrading occurred in Chawama, a squatter settlement in Lusaka. By 1974, there were seventeen squatter settlements in the city, accommodating about 40 per cent of its total population. Chawama developed following the cessation of quarrying in the area in 1961, and by 1974 it was the second-largest settlement, with 25,000 inhabitants, the majority living below the poverty line. Initially, the Zambian government took a negative view of the squatter settlements, but from the mid-1960s they were gradually recognized and certain basic services were introduced. In 1974, about a quarter of the families in Chawama had lived there for over eight years, and three-quarters of the working men and women were employed, mainly in construction and other industries. A key move in the long-term development of Chawama had occurred in 1966, when the government bought the land from the white South African land owner. In the same year, the wells in many Lusaka squatter settlements dried up and, after protests by the women, a piped water system was installed, although it soon proved inadequate for the needs of the growing population. A surfaced road was built by the residents in a community self-help project, with tools and materials supplied by the city authorities. Later, in 1969, the city council introduced a bus service from Chawama to the city centre. A market was also built and jointly managed by community and government departments. The only government primary school in a squatter settlement was opened in Chawama in 1969, and in the same year a pre-school group was established.

Improvements in Chawama were gradually achieved by the residents after many protests, but these usually occurred only when it seemed politically expedient for the government to help. Zambia experienced economic difficulties during the 1980s, and between 1978 and 1992 there was a marked decline in urban male, formal-sector employment, as well as a fall in real wage levels. Informal-sector employment in Chawama increased during this period, and a survey undertaken in the early 1990s revealed a number of common household coping strategies, including: greater reliance on extended family support networks and inter-household remittances; increasing hours of women's work; reducing household expenditure; and diversifying income from home-based enterprises, urban agriculture and livestock raising. Whereas the 'top-down' investment in basic infrastructure in Chawama during the 1970s had significantly improved service provision, between 1978 and 1992 there was a marked deterioration in services due to a lack of maintenance, meaning residents had less mobility and were forced to spend more time fetching water and getting to school (Moser and Holland, 1997).

The provision of government housing for low-income groups has been fairly limited in Africa; although some provision has been made in such cities as Harare and Lusaka, it is often a long distance from places of work. High building costs and rents beyond the means of the urban poor have encouraged governments to concentrate on either slum and squatter upgrading or sites and services schemes.

Public–private partnerships for house building have become more common in recent years in a number of African countries. For example, until the late 1970s, the Tunisian government followed a policy of slum demolition and relocation of people to rural areas. But in 1981 it created the Urban Upgrading and Renovation Agency to provide a new impetus to slum upgrading. Since then, Tunisia has significantly reduced the number of people living in slums by encouraging partnerships between the agency, municipalities, service providers and private developers. The Five-Year Economic and Social Development Plans and Municipal Investment Plans reaffirmed the commitment to slum upgrading, and between 1984 and 1994 some 500,000 housing units were built. By 2003, the backlog was said to be only 24,000 units and the government was confident of clearing it over the next few years (International Housing Coalition, 2007).

Sites and services schemes represent another strategy for meeting housing needs in Africa's cities. They also play a role in job creation. Such schemes are usually located on the outskirts of cities, where urban authorities clear and lay out areas with housing plots serviced with access roads, piped water, sewerage and electricity. Households and communities are then encouraged to build their own homes on the plots, often with government assistance in obtaining and purchasing building materials. These schemes are found in many African countries and have received support from the World Bank. They are viewed as more financially viable than schemes that involve the large-scale construction of public housing.

In the 1950s and 1960s, there was an attempt to build public housing in Senegal's capital, Dakar, but the rate of construction was very slow while the waiting list grew rapidly. Meanwhile, there was increasing concern about the development of poor-quality squatter settlements in the centre of the city, such that between 1966 and 1978 the government evicted thousands of people from nine major areas of central Dakar. Concerned that new illegal settlements would emerge, the government adopted the sites and services approach to urban housing provision in its third Four-Year Plan for Economic and Social Development (1969–1973). In 1972, the government signed an agreement with the World

Bank for development of the first sites and services scheme at Pikine, on the outskirts of Dakar. To promote the occupation of lots, settlers were encouraged to build in stages, and start-up loans were made available to purchase building materials.

Many other African countries have also used sites and services schemes as a strategy for alleviating the urban housing problem. The World Bank has been involved in a number of schemes in Lusaka which have provided over 11,500 new housing units. From the early 1970s, Kenya embarked on a number of sites and services projects: for example, in Dandora (Nairobi) and Thika. Between 1970 and 1972, some 3,734 serviced plots were laid out. Although the volume and speed of plot provision were inadequate, a UN-HABITAT report was generally positive about the projects: 'the schemes have made a substantial contribution to accommodating low-income households in the two towns: in Thika, close to 70% of the entire population growth over a ten year period and in Nairobi, about 13%' (UN-HABITAT, 1987: 117).

In Chad, in 2003, President Déby declared that all citizens should be provided with a decent home, and $12 million was allocated from the country's oil revenues for the laying out of 960 serviced plots for families in slum areas, with low-interest loans available for them to build new homes. UN-HABITAT is managing the project and local private sector businesses and civil support groups are also involved. Meanwhile, in Gaborone, Botswana, a self-help housing scheme was introduced in 1992 under the Accelerated Land Servicing and Housing Development Programme, with water-borne sewerage, roads and other services provided, together with access to loans for low-income households.

A common problem of sites and services schemes is that, with their peripheral location, a long and difficult journey to work is often necessary unless employment opportunities can be located near the new settlements. For example, residents in the peripherally located housing scheme of Pikine-Guédiawaye in Dakar complained about the inadequate transport facilities to travel into the city. There was a low frequency of buses and they were often full long before leaving the scheme. In evaluating the relative merits of sites and services schemes, Rondinelli (1990: 161) comments,

> The World Bank's experience with sites-and-services and upgrading projects has been that although these approaches are more effective in providing affordable shelter for the urban poor than slum clearance and public housing programmes, most sites-and-services schemes remain pilot projects serving a relatively small percentage of the poor households which need shelter.

5.7 The urban environment

The speed and unplanned nature of urban growth in Africa have generated many environmental problems in towns and cities, the most serious being a lack of piped water systems for homes and businesses, poor sanitation and sewerage provision, inadequate systems for managing solid waste, and considerable water and air pollution (Hardoy *et al.*, 1992). These issues are all too evident on visiting many African cities and, for those who live there, they are everyday hazards and frustrations that can impact significantly on quality of life. Agenda 21, adopted at the Conference on Environment and Development in Rio de Janeiro in 1992, and reaffirmed a decade later at the World Summit on Sustainable Development in Johannesburg, includes many references to environmental problems that are common in Africa's towns and cities. These include the effects of environmental pollution on urban health, the need to improve the quality of urban infrastructure and the importance of strengthening and expanding waste reuse and recycling systems.

The Millennium Development Goals include a target of halving the share of the population without adequate access to safe water and basic sanitation by the year 2015.

The provision of a clean and reliable piped-water supply varies considerably from one city to another. For example, according to a survey conducted in 2000, Port Said in Egypt had 100 per cent provision of piped water in homes and 90 per cent of households in the city had a flush toilet; whereas a survey undertaken in the same year in Bangui, capital of the Central African Republic, revealed that only 11.3 per cent of households had piped water in their homes and only 4.3 per cent had a flush toilet. Levels of service provision can vary widely even within a single country: in 2003, in Ibadan, Nigeria, only 3 per cent of households had piped water, whereas in Abuja, the capital, the figure was 59.1 per cent. In some cases, household surveys taken at different points in time can reveal a significant deterioration in the situation. For example, a survey conducted in Lagos in 1999 found that 25.6 per cent of households had access to piped water in their homes, but a subsequent survey in 2003 suggested that only 9.0 per cent of households now had piped-water access (UN-HABITAT, 2008). The situation can also change over quite short periods of time: for example, if a major water pipe bursts and maintenance is delayed, neighbourhoods can be left without a piped-water supply. Power outages are also common in African cities. If they are protracted and frequent, they can have a serious impact on life and work, and may adversely affect the water-supply system (Gandy, 2006; see Box 5.5).

A survey of 660 households in Dar es Salaam in 1986/7 revealed that 47 per cent had no piped-water supply either inside or immediately outside their homes, while 32 per cent had a shared piped-water supply. Of the households without piped water, 67 per cent bought water from neighbours, while 26 per cent drew water from public water kiosks or standpipes. Sewage disposal was totally inadequate, with only 13 per cent of dirty water and sewage being regularly removed. In the late 1980s, the municipal sewerage system in Khartoum, Sudan, served only 5 per cent of the urban area and the system regularly broke down, leading to waste discharge into the Nile or on to open land. Meanwhile, in Kinshasa, there was no sewerage system at all in the late 1980s, and about half of the population had no piped-water supply.

For many city dwellers, the daily routine of securing water for their households is fraught with difficulty and uncertainty. As Gandy's enlightening, yet disturbing, study of Lagos showed,

> Inhabitants of slum settlements often face a stark choice between either polluted wells or expensive tanker water distributed by various intermediaries at high and fluctuating prices, making the management of household budgets even more precarious . . . People's daily survival is based on careful distinctions between different kinds of water suitable for drinking, cooking and washing, with much time and expense devoted to securing household water needs. Regulatory authorities also struggle to cope with the proliferation of 'pure water' manufacturers producing small plastic sachets of drinking water sold throughout the city which have been associated with the spread of water-borne disease.
>
> (Gandy, 2006: 383)

Water pollution remains a serious problem in many African cities, despite the introduction of local by-laws that were meant to control pollution levels. Urban farmers in Kano frequently complain about water courses being heavily polluted and the water being unfit for use on their plots (see Box 5.4). The main sources of this pollution are probably the city's tannery and textile industries, which use the largest quantities of water and produce the greatest amounts of waste water. Kano is the centre of Nigeria's tanning industry, and is home to 70 per cent of the country's tanneries. The industry's waste products have high concentrations of the heavy metals chromium and cadmium, and a study that monitored the activities of fifteen tanneries in Kano found that permissible limits for effluent discharge were violated by every one of them. Not only do downstream fish and crops become heavily contaminated by heavy metals, but human health is further threatened in urban and peri-urban Kano because over 60 per cent of local people depend for their water supply on rivers and wells that draw upon groundwater aquifers. As one urban farmer in Kano commented,

> There are three bad colours [of water] that come at different times. The oily red one and the green one will kill the crops, and when we see these colours in the channel, we turn off our pumps immediately. The bluish water is corrosive and causes a red rash when it comes in contact with the skin. We always wash our hands after we come in contact with the blue water.
>
> (Binns *et al.*, 2003: 440)

In Uganda, the Kampala Urban Sanitation Project (KUSP), which was initiated in 2002 with support from the French Development Agency (AFD), has aimed to improve living conditions in thirty-five poor urban communities in the capital by ensuring unpolluted water supplies and providing toilet facilities. The project has installed over 200 toilets and 350 wash points, and has set up a rainwater collection tank in each of the communities. It has also worked to protect the eighty-five natural springs in the city in order to provide a clean water supply to slum communities (Kampala, 2008).

Solid waste is a seemingly insurmountable problem for many urban authorities (see Plate 5.8). Urban farmers in Kano traditionally applied solid waste, usually purchased from urban households, to their plots (Binns *et al.*, 2003). Although composting organic waste has positive effects in improving the soil fertility, and may destroy pathogens if composting temperatures are sufficiently high, there is a danger of introducing heavy metals into the food cycle if composting is carried out improperly. Both sludge from Kano's factories

Plate 5.8 Scavenging on municipal rubbish tip in Freetown, Sierra Leone (Tony Binns).

and untreated raw sewage from toilets were often applied directly on to plots. This material could carry viruses and high levels of faecal coliforms. Solid waste from Kano's tanneries not only emitted appalling odours but provided an excellent breeding ground for disease vectors. The use of unsorted urban waste as fertilizer also caused polythene bags to accumulate on many farm plots. This has now become a major environmental hazard throughout Nigeria and in many other African countries (Binns *et al.*, 2003).

Many urban authorities in Africa still have a long way to go in improving environmental quality and reducing possible health risks caused, particularly, by polluted water and uncollected solid waste. This is an expensive and challenging task, and since the problems are usually greater in the poorer parts of cities, where low-income households are unable to contribute financially to improving the situation, it is often difficult to make any progress.

However, some African cities have achieved a degree of success. For example, in Rabat, the capital of Morocco, the Rabat Environmental Action Plan was formulated in the mid-1990s as a response to Agenda 21, and as part of a national environmental action plan put forward by the Moroccan Ministry of the Environment. The plan tackles pollution in the city through the construction of a sewage treatment plant and the improvement of waste water evacuation, and has set up a monitored disposal site for household waste with composting and incineration facilities. It also includes the development of green space and increasing community awareness of environmental issues through establishing a local environmental observatory and a municipal information bulletin. Gilbert *et al.* (1996), however, are critical about the inadequate financial support and the lack of attention given to such issues as natural habitat, energy efficiency and noise pollution.

The city of Kayes in Mali twinned with the city of Evry in France to produce a Municipal Policy and Participatory Dynamics Project in 1992, aimed at improving living conditions in Kayes. With assistance from colleagues in Evry, the effectiveness of Kayes's city management and resource system has been improved and the community has become involved in household waste collection, with unemployed youths formed into groups to organize house-to-house waste collection. A similar participatory waste management scheme (the Participatory Solid Waste Management Programme) was launched in the mid-1990s by the municipal authorities in Dakar, Senegal, in collaboration with Canada. This involved collaboration between the urban authority, a consortium of waste management companies, local private sector service providers and community-based entrepreneurs. Private sector companies submit competitive three-year tenders to collect and transport waste and are then obliged to subcontract with small, community-based groups of locals called economic interest groups (EIGs). The EIGs are responsible for cleaning the streets, cleaning around rubbish bins, collecting solid waste from inaccessible areas, disseminating information about the waste collection system and educating their communities about the benefits of sanitation and waste management. The waste collection companies provide the EIGs with security equipment, cleaning and storage facilities, oversee their work and pay their wages. An early assessment of the scheme suggested that it was working well, and over 1500 new jobs were created, which it was hoped would lead to reductions in crime and drug use (Gilbert *et al.*, 1996).

Meanwhile, in 2006, Kampala City Council embarked on the Kampala Integrated Environmental Management Project (KIEMP), with support from Belgian Technical Cooperation. The project aimed to improve the environmental conditions in informal settlements by improving drainage and access roads, human waste disposal, access to safe water and solid waste management. It was especially concerned with raising awareness in communities about the significance of personal and environmental hygiene and strengthening local capacity to manage environmental conditions in the city (Kampala, 2008).

A survey of internet sources reveals a plethora of waste management initiatives in Africa, involving urban authorities, local communities and the private sector. One novel initiative, involving the computer company Hewlett-Packard, in collaboration with the Swiss Materials Science and Technology Research Institute and the Global Digital Solidarity Fund, is developing a sustainable e-waste management system in five African countries – Kenya, Morocco, Senegal, South Africa and Tunisia. The project aims to make the recycling of electronic equipment safer, and experiences in China and India suggest that this could lead to significant job creation in the informal sector. The management of e-waste has been identified as a growing worldwide problem by the United Nations Environment Programme, since e-waste contains both valuable materials, such as gold and copper, and highly toxic substances, such as lead and mercury (DSF, 2007).

There has been much debate recently about who should be responsible for providing key services and infrastructure to improve the environment in African towns and cities. The evidence suggests that cash-strapped urban authorities are generally unable to cope logistically and financially with the massive task that they face, and some would argue that in any case both the private sector and local communities should shoulder at least some of the responsibility. There seems little doubt that corruption and poor governance have retarded progress in dealing with such matters, and in charting a way forward these practices must be acknowledged and overcome. In their detailed evaluation of whether the private rather than the public sector might provide better water and sanitation services to the urban poor, McGranahan and Satterthwaite (2006: 2) conclude, 'While there is no single model of good water and sanitation governance, and no reason to favour private providers, good local governance is critical to getting the best out of private as well as public providers.' Gandy (2006: 389) reaches a broadly similar conclusion in response to the chaotic and dysfunctional nature of basic service provision in Lagos:

> the colonial state apparatus and its post-colonial successors never succeeded in building a fully functional metropolis through investment in the built environment or the construction of integrated technological networks. Vast quantities of capital that might have been invested in health care, housing or physical infrastructure were either consumed by political and military elites or transferred to overseas bank accounts with the connivance of Western financial institutions.

In terms of working towards improving the environment in Lagos, and indeed many other African cities, there is a need for long-overdue scrutiny and revision of planning strategies inherited from the colonial period, and the introduction of 'a panoply of institutional reforms ranging across specific areas of law, tax and regulatory intervention which encompass new codes of professional conduct, transparency and accountability' (Gandy, 2006: 390).

5.8 Urban employment: the informal sector

Most rural–urban migrants arrive in the city in search of employment, but in many countries the pace of urban growth is outstripping the capacity of the economy to provide jobs. A major problem in investigating urban employment patterns in Africa is the lack of data and problems with defining such terms as 'employed' and 'unemployed'. Even in certain high-income countries there is an ongoing debate about the accuracy of unemployment statistics. But in many African countries there is no detailed register of unemployed persons, and state unemployment benefits are rare. Those who appear to be unemployed may be more accurately termed 'under-employed', since there are many opportunities for casual paid work in African cities.

Major distinctions are often made between formal and informal sectors of the labour force and between large- and small-scale, wage- and self-employed. Labour markets are much less segregated than in industrialized countries, and it is common for people to have more than one job and to cross the boundaries between different sectors, sometimes on a regular basis. Construction workers, for example, may alternate between working for themselves and for large firms. Trading in different commodities at various levels is undertaken by a wide range of people, sometimes by those who are in full employment, as a source of additional remuneration.

The term 'informal sector' was popularized by research in Ghana (Hart, 1973) and an International Labour Organization report on Kenya in the early 1970s (ILO, 1972; see also Bromley and Gerry, 1979), with the latter seeing the sector as a possible solution to poverty and unemployment. Since then, many people have attempted to define and classify the informal sector and there have been numerous case studies. It was initially argued that informal sector employment is relatively easy to enter with no formal qualifications; that it is small scale and often family based, making use of adapted local technology and resources; and that it is labour intensive. However, the formal–informal dichotomy has been criticized for being too simplistic and rigid. There are many sub-sectors within the informal sector. There is little in common, for example, between the legitimate business of a tailor or metalworker, street hawker and child who guards parked cars, and the illegitimate work of a prostitute, drug-pusher or pickpocket. The simple dichotomy further obscures complex linkages between formal and informal sectors and the ways in which particular enterprises change over time, perhaps moving from being illegitimate to legitimate or from being informal to formal (Binns, 1994).

The formal–informal dichotomy is often seen in a 'dualistic' sense, in which the informal sector is perceived as being traditional, backward, using simple production methods, with low levels of investment and low incomes, whereas the formal sector is regarded as modern, better organized, with modern technology, greater investment and better financial rewards for employees. The implication is that, in time, it would always be desirable for informal sector enterprises to modernize and acquire the attributes of formal sector enterprises. Furthermore, the informal sector is often seen as illegal, clandestine and a possible breeding ground for criminals and political instability.

As Potts (2008) has shown, official attitudes towards informal sector activities have varied considerably over time and space, from being at times positive or benign to being hostile and antagonistic. In some African countries, street traders have been harassed by police, with political leaders suggesting that such activities are untidy and incompatible with achieving a modern and progressive townscape. In Zambia, for example, in 1993 Lusaka City Council attempted to clear away street vendors; then, four years later, it tried and failed to move them into a city market. Similarly, in April 2006, Malawi's government introduced a clean-up campaign and endeavoured to move street hawkers into formal markets. In Nigeria, the Federal Capital Territory authorities announced plans to ban street hawking in Abuja because of the chaotic traffic situation, as hawkers were taking advantage of the slow movement of vehicles to sell their wares at major junctions. The hawkers were also blamed for the large amounts of refuse on roadsides. Relocation to public markets has proved problematic in many African cities, since traders claim that rents are too high and that they sell less if they are based in a formal market rather than on the streets. Meanwhile, in South Africa, the urban informal sector is a relatively new phenomenon, but it has flourished since 1994, particularly street hawking, which was restricted during apartheid (see Box 5.6).

Box 5.6

A new vision for urban development in South Africa

South Africa's cities face tremendous challenges in their difficult quest to address the legacies of apartheid, to achieve growth and to participate in the global economy. The greater power and freedom allocated to local governments have encouraged the larger cities, in particular, to seek defined places for themselves in the world economy through internationally accepted place-marketing and development strategies. It is 'widely accepted that cities in both developed and developing countries now have to compete globally to develop their local economies if they wish to maintain or improve their position' (Jenkinsa and Wilkinson, 2002: 34). Simultaneously, and in response to locally recognized and national government agendas, cities are actively seeking to address the urgent issues of poverty alleviation and inequality. While cities such as Cape Town have obvious natural and economic attractions to offer international investors and tourists, and also have the resources to provide world-class facilities, such as convention centres, high-class shopping malls and international airports, the process of dealing with the massive challenge of urban poverty is a far more daunting prospect and it would be difficult to argue that much tangible progress has been made thus far.

Broad similarities in development goals and objectives exist between the major cities in South Africa. A dual focus on both socially responsible interventions and economic competitiveness is also discernible in smaller cities, such as Kimberley, whose Economic Development Unit has been charged to 'promote economic development through domestic and foreign direct investments, black economic empowerment and small, medium and micro enterprise', in order to 'ensure sustainable job creation, skills transfer and poverty eradication' (Sol Plaatje Municipality, 2001: 12).

The two dominant objectives of poverty relief and economic growth characterize all vision statements and strategies, and the cases of Johannesburg, Cape Town and Durban, examined below, are typical of the broad approach that is being adopted.

Johannesburg

The key challenge facing local government in South Africa in general is succinctly summarized by Bremner (2000: 185) in his study of Johannesburg, when he points out that the task is 'to reinvent a city which could claim a position in the mainstream global economy and . . . become a city which all its citizens could feel part of'.

Following the local government elections of 5 December 2000, the Greater Johannesburg area was created, incorporating thirteen formerly separate local administrative bodies (see Plate 5.9). The new uni-city has since pursued a multi-faceted development programme, known as 'Johannesburg 2030', which incorporates economic and urban regeneration endeavours such as metropolitan marketing, privatization, infrastructural redevelopment and inner-city renewal, a focus on new industrial development and support for the small business sector. The Johannesburg approach to development also includes facilitation of some important private sector initiatives, including support for such consumption-based activities as tourism promotion, casino and convention centre development, the 2002 Earth Summit and the hosting of major sporting events, such as the All-Africa Games (2000) and the Football World Cup (2010), which involved the construction of world-class sporting facilities and associated accommodation.

In terms of addressing the apartheid legacy and poverty-related issues, a range of key projects can be identified. These include the development of the 'Bara-link' economic and transport corridor, linking the centre of Johannesburg to the hitherto separate black township of Soweto. Additionally, there have been various projects to promote township tourism, and concerted efforts to catalyse small and micro-enterprise development through the provision of permanent trading facilities for street hawkers and support services for emerging small businesses. Inner-city regeneration in a series of key precincts – with direct private sector partnerships and a range of flagship projects – has been high on the agenda in recent years (Bremner, 2000).

The Johannesburg CBD Regeneration Programme has seen the redevelopment of the inner-city road–rail transport hub, while the upgrading of infrastructure and the redevelopment of selected urban zones, such as the Newtown Cultural Precinct, are notable initiatives. One of the more impressive initiatives involving Johannesburg and its hinterland is the provincial government's local economic development intervention – known as Blue IQ – a project revolving around government investment and the attraction of private resources into a series of mega-projects focusing on tourism, technology, transport and high-value-added manufacturing, with the aim to make the area a 'smart province'. The provincial investment stands at R1.7 billion (about $170 million) for the first year of implementation and it is likely to increase in subsequent years. The core projects are:

- the Cradle of Humankind World Heritage Site, based on the impressive Sterkfontein caves palaeontological site;

Plate 5.9 Johannesburg CBD, South Africa (Tony Binns).

- Constitution Hill – site of the country's Constitutional Court and an Apartheid Museum complex;
- Newtown – a massive inner-city regeneration initiative, focusing on arts and culture;
- Dinokeng game reserve;
- the Gauteng Automotive Cluster, to boost links between existing motor manufacturers and suppliers;
- Wadeville Alrode Industrial Corridor;
- Innovation Hub – to promote high-tech research and development;
- Gautrain Rapid Rail link – to link the key cities of the region, notably Johannesburg with Pretoria and the international airport (the first section, from the airport to Johannesburg city, opened in June 2010, in time for the Football World Cup);
- Johannesburg airport and IDZ (Industrial Development Zone) – to enhance the capacity of the city's airport and design an export-orientated industrial estate associated with it;
- City Deep Container Depot and IDZ – an inland dry port and industrial facility.

High-priority crime-reduction strategies, involving more visible policing and the introduction of closed-circuit television cameras, have led to a steady reinvestment in Johannesburg's city centre, following a twenty-year period of disinvestment and the progressive abandonment of the CBD by many large businesses in favour of safer suburban locations, for example Sandton in the northern suburbs (Bremner, 2000). Major private sector-funded facilities have been developed outside the CBD, such as the Sandton Convention Centre, the location of the 2002 Earth Summit, and the impressive Montecasino entertainment and hotel complex. The city has also successfully entered into 'twinning' arrangements with some thirty-nine cities internationally and has become the seat of the Gauteng provincial government and the country's Constitutional Court. Less successful, however, have been support programmes for township economic development, where the scale of the task, limited resources and a lack of private sector interest have restricted overall impact.

Cape Town

The vision of the new Cape Town Metropolitan Council is that 'Strong and sustainable economic development in the Cape Metropolitan area will depend on the ability of all stakeholders to work together to address the combined challenge of improving the Cape Metropolitan area's global competitiveness and reducing poverty' (CMC, 1999: 4). In order to achieve this, a dual strategy has been launched. In terms of achieving global competitiveness, issues such as the provision of world-class marketing and services are matched with support for business clusters, infrastructure and capacity development. Meanwhile, the parallel poverty-reduction strategy seeks to provide affordable urban services and infrastructure, to integrate formerly separate areas within the conurbation, to promote community development and job creation, and to establish a 'social safety net' to ensure that the needs of the poorest of the poor are not overlooked (CMC, 1999). Though admirable in intent, it is questionable whether adequate resources exist to meet all of these goals.

Cape Town has worked hard to present itself as 'one of the world's great cities'. While projects are spread across the city, the CBD (or 'city bowl') is clearly the locus of significant regeneration endeavours by both the public and private sector, and a key partnership – the Central City Partnership – is active in this regard. The value of development projects in and around the CBD in 2002 exceeded R5 billion (over $500 million). The largest projects are the Waterfront redevelopment and associated housing construction, which will together cost R2 billion (about $200 million), the international convention centre, completed in 2003 at a cost of R500 million (about $50 million), and the railway station redevelopment, which will cost a further R1 billion (about $100 million) (see Plate 5.10).

One of the city's most innovative interventions is the attempt to develop a major cultural and environmental tourism node at Lookout Hill in the sprawling Khayletisha township. Equally noteworthy is the Noordhoek Valley training centre on the Cape Peninsula, which provides a range of highly successful programmes of skills training, mentorship and job placement for township residents, with the associated development of a tourist craft market adjacent to the main tourist route to Cape Point. Thus far, attempts to promote the city as a centre for 'high-tech' industrial development seem to have been less successful.

At a broader level, support for city partnerships, business attraction, and the encouragement of regeneration in a range of 'business improvement districts' throughout the city and the building of major tourism and retail facilities, such as the up-market Century City shopping and entertainment complex, parallel more modest efforts to improve integration of the poor township areas within the city's economic structure. Much attention has been given to township housing programmes and efforts to establish trading and small business support facilities. The council has actively supported private sector investment in mega-projects, through the provision of bulk infrastructural facilities, road construction and access to land.

Plate 5.10 Cape Town waterfront and Table Mountain (Tony Binns).

Private sector development has, however, been skewed in favour of the more privileged (historically white) areas, while spatial interventions in poorer areas, such as the Philippi metropolitan node and metropolitan activity corridors, have not met with the same degree of success. As a result, it appears as if 'there is a gulf between Cape Town's impoverished townships and its affluent areas, which appears to be widening in important respects. Development trends are tending to reinforce spatial divisions and fragmentation, rather than to assist urban integration' (Turok and Watson, 2001: 136). Indeed, it is both noticeable and regrettable that, despite impressive achievements in terms of the more up-market initiatives, post-apartheid inequalities are increasing, not decreasing (Jenkinsa and Wilkinson, 2002), leading Marks and Bezzoli (2001: 30) to conclude, 'while official discourse seeks the transformation of Cape Town into a more equitable and integrated city, the reality is marked by inadequate investment to alleviate the legacy of apartheid'.

Durban

The city of Durban is seeking to pursue a similar growth and development strategy to that of Cape Town. As the chair of the Economic Development and Planning Standing Committee commented, 'It is our intention to build a globally competitive region in which all communities benefit from economic growth' (Durban Metro, 2000: 2). Specific policies in Durban include:

- promoting Durban as a competitive investment centre;
- tourism promotion;
- encouraging the small business sector;
- economic regeneration of previously disadvantaged communities; and
- provision of information and advisory services (Durban Metro, 2000).

As in other South African cities, the dual focus on major economic interventions and support has been pursued in parallel with poverty relief measures. The city's Long-Term Development Plan, issued in 2002, seeks to improve overall quality of life in the city by meeting basic needs, strengthening the economy and building skills and technology. Five key programmes have been identified within this plan:

- Providing a platform for economic growth based on flagship projects, inner-city regeneration, transport infrastructure, growth nodes and township commercial centres.
- Providing services and support to key customers through business service centres.
- Offering support for tourism, manufacturing, new industry, and research and development sectors.
- Empowerment – procurement and small and medium-sized enterprises support.
- Partnerships and capacity building.

Particularly significant initiatives include the redevelopment of the Point Waterfront, the new King Shaka International Airport (opened in May 2010), the development of major industrial estates and the promotion of the Gateway Shopping Centre, which cost R1.4 billion (about $140 million) and is promoted as the largest shopping complex in the southern hemisphere. The redevelopment of 55 hectares in the Point area at a cost of some R700 million (about $70 million) for recreational purposes, including the construction of what will be one of the seven major Sea-World facilities in the world, is clearly a key project in the urban regeneration of the city (Maharaj and Ramballi, 1998). Support has also been given to a number of defined industrial clusters in which the city has a competitive advantage: for example, petrochemicals and vehicle manufacture. Although Cape Town and Johannesburg (Sandton) both now have international convention centres, the Durban convention centre pre-dates these and has hosted a series of key global events, including the 2001 World Conference on Racism and the launch of the African Union in July 2002.

Earlier efforts to promote township economic development have been less successful. The failure of the private sector to respond with the same enthusiasm and levels of investment as they have towards the development of retail centres in former white areas, together with political difficulties and poor results, have restricted the programme to a focus on the provision of small business and trading support, generally in the CBD rather than in the township areas. One noteworthy exception is the European Union-funded Cato Manor redevelopment initiative, 7 km from Durban's city centre, whereby significant EU resources have been allocated to the economic and social transformation of a largely low-income black-populated area. This was seen as a strategic and visible project designed to address some of the inequalities engendered by apartheid. Hundreds of houses have been built, facilities upgraded and economic growth opportunities for low-income people investigated (CMDA, 2008). This project has been recognized as an example of 'best practice' by UN-HABITAT. It seems that a key challenge for Durban and other cities is 'how to manage the urban process in such a way as to harness the potential of such projects to generate income for the broader public good, while still allowing private development to profit' (Marks and Bezzoli, 2001: 44).

Much more successful have been the promotion of Durban as a major tourism destination and the redevelopment of the beach-front, leading to the city becoming the country's premier destination for domestic tourists. Since Durban is situated within seven hours' drive of Johannesburg, Pretoria and the Gauteng conurbation, this status has been relatively easy to maintain. Another important initiative has been the formation of what is probably the country's most significant public–private partnership. The Durban Chamber of Commerce and the Metropolitan Council have joined forces to establish the Durban Business Partnership, which promotes the development and marketing of Durban. This has also led to the establishment of the Development, Investment and Promotion Agency – a 'one-stop shop' to attract, encourage and streamline investment and development in the city (see Box 9.4).

A significant problem that was noted in discussions with authorities in Durban, which may well be replicated in other South African cities, is the very limited level of public participation in development projects, most of which come to be dominated by business and the local state, despite rhetoric to the contrary (Maharaj and Ramballi, 1998). As a result, one must question 'whether the disadvantaged communities will benefit from local economic development projects' (Maharaj and Ramballi, 1998: 131). In the case of Durban, Maharaj and Ramballi (1998) argue that the major tourism development drive largely ignores the domestic and low-income market, and that inner-city redevelopment projects do not adequately involve surrounding neighbourhoods. In order to achieve true development, it is suggested that 'greater emphasis should be placed on policies that sustain growth through redistribution' (Maharaj and Ramballi, 1998: 46) and improve living standards and the capacity of community residents to become economically self-sufficient.

Other centres

In other centres, similar, albeit less ambitious, development projects have been initiated. In the majority of cases, while there may be a strong commitment to improving the living conditions and economic viability of the townships, it would be difficult to argue that, with the exception of housing and some infrastructural development, any significant progress has been made in terms of job creation and economic development in these poor areas.

In Port Elizabeth (Nelson Mandela Metropolitan Authority), for example, economic development efforts have made a slow start, with future hopes being pinned on a major port and industrial complex to be located east of the main built-up area at Coega. This initiative has received considerable publicity and support from national government. Meanwhile, in East London (Buffalo City Metropolitan Authority), serious efforts to promote the city and encourage township economic development have been attempted, though funding and capacity constraints have impeded progress. The Pretoria (Tshwane) Metropolitan Authority has embarked on a range of projects, including support for local business service centres (trade points), trade promotion, small business development, economic empowerment, tourism, industrial promotion, urban agriculture and place-marketing. While financial and capacity constraints have undoubtedly hindered progress here, too, selling Pretoria as the country's capital city, rivalling the legislative capital of Cape Town, has clearly bolstered the former's image. Several smaller cities, such as Kimberley (Northern Cape), Nelspruit (Mpumalanga) and Polokwane – formerly Pietersburg – (Northern Province), have done remarkably well as a result of their post-1994 acquired status as provincial capitals, which has led to significant government investment and general improvement in city fabric and infrastructure (Rogerson, 1997). In the case of Nelspruit, further growth impetus has come from its position east of Johannesburg, along the so-called Maputo Development Corridor, a major transport and development corridor linking Gauteng to Maputo in neighbouring Mozambique.

Kimberley has initiated one of the most innovative township development programmes. While business attraction and CBD improvement are still getting off the ground, local, provincial and national government resources have been accessed to embark on community-based public works programmes to upgrade infrastructure, emerging-business support programmes and waste-recycling projects.

In summary, urban and economic regeneration in South African cities seems to be revolving around several core themes:

- the promotion of tourism, leisure and cultural activity (Durban, Cape Town, Johannesburg);
- place-marketing (all cities);
- industrial attraction (Port Elizabeth, Durban, Pretoria);
- poverty alleviation, small business development and township development (all cities);

- infrastructure development and housing (all cities);
- the attraction of government investment (Pretoria and the provincial capitals);
- training (Cape Town); and
- inner-city redevelopment (Johannesburg, Durban, Cape Town).

A survey of city practice in promoting developmental local government suggests that while poverty alleviation, particularly in the township areas, might well be a worthy and politically appropriate focus for attention, 'international standard' interventions appear to have achieved much greater success (perhaps understandably). According to Marks and Bezzoli (2001: 27), we are witnessing the transformation of South African cities into archetypal post-modern cities, which have been released from 'the grip of state control and are now at the mercy of that most nebulous of conceits, the free market'.

At a broader level, achieving social justice in South African cities remains a somewhat distant goal. Policy at the national and local levels seeks to attain it, but thus far it has proved elusive to current development practice. To realize this ambition, Beall *et al.* (2000a: 394) suggest that 'pro-poor municipal policy and action need to be devised and executed in a manner that is sensitive to optimizing and supporting the livelihood strategies of the poor'. However, in order to achieve both social justice and pro-poor development, it is also vital that wealth-creation interventions should match endeavours to address inequality (Beall *et al.*, 2000b).

The introduction of structural adjustment programmes from the 1980s often led to massive retrenchments of workers in the public and private sectors, and many found ways of making a living in the urban informal sector or moved to rural areas (see Box 5.3). The informal sector has continued to grow, such that most workers in Africa's towns and cities are now within it. For example, in Kenya, it is estimated that the number of people in the informal sector increased from 105,000 in 1990 to 4.6 million in 2001 (Rogerson, 2001). If nothing else, the informal–formal sector dichotomy has generated much debate and provided a framework for empirical research into employment situations.

During the 1960s and 1970s, Nairobi City Council demolished poor-quality housing in low-income areas which were also frequently informal sector workplaces. For example, in November 1970, some forty-nine shanty settlements with 7000 dwelling units, home to over 40,000 people, were pulled down or burned by police on the instructions of the City Council (see Box 5.5). Since the ILO report on Kenya (1972) revealed the importance of the informal sector, the demolition of homes and enterprises has declined, but as late as 1991 the National Council of Churches of Kenya reported that demolitions in Nairobi during the 1980s led to the loss of 1431 businesses and 4293 jobs, and jeopardized the livelihoods of 34,000 people (NCCK, 1991). Otiso (2002) suggests that many of the demolitions and evictions in Kenya were associated with government attempts to maintain political power, but they were often carried out in the guise of protecting state security, controlling crime and enforcing municipal by-laws.

During the 1990s, the Kenyan government, recognizing the significance of informal sector employment, introduced development policies to support the sector and promote job creation. Market buildings have been constructed in Nairobi and other cities, and in some cases informal sector enterprises have been provided with road access, electricity and financial support. For example, at the hawkers' market in the largely Asian Nairobi suburb of Parklands, in 1991 the local government helped install sewerage and running water, while the Asian Foundation provided funds for the construction of market stalls, kiosks, toilets and a refuse collection area. Stallholders pay an affordable monthly fee for security and cleaning services (Muraya, 2006).

Mathare Valley, with a population approaching 200,000, probably has the largest concentration of informal enterprises in Nairobi, with some 20 per cent of businesses being over twenty years old (see Box 5.5). Many were started with less than fifty dollars, raised from savings, sale of assets or loans from friends and relatives. Four-fifths of enterprises surveyed in 1989 were in trade and commerce, 11 per cent were involved in manufacturing and 9 per cent provided services. Virtually all of the businesses were sole ownerships, with little scope for employing extra labour. Of all operators, 86 per cent were under forty-nine years of age and about half were women, mainly selling fruit and vegetables. Business operators generally had fewer than eight years of formal education and few kept written records. Enterprises mainly sold products directly to customers, from the local area, but some had clients outside Mathare Valley. Average monthly profits compared favourably with average wages in Kenya's private and public sectors (Nelson, 1997).

In addition to profits from these enterprises, almost a third of operators earned additional income from part-time wage employment provided by other operators, spouses and other members of the family; from other businesses, especially the illegal brewing and sale of alcohol; from farming, if they had land; from rents, if they owned property; and directly from friends and relatives. This additional income, when added to the enterprise profits, was adequate for household needs and usually left some for savings. There was considerable interaction between the informal enterprises in Mathare Valley and other enterprises outside the area. Tailors, for example, went outside the area to buy thread, cloth and buttons. Infrastructural facilities were minimal, with poor roads, less than 10 per cent of enterprises connected to electricity, 4.5 per cent to piped water and only 2.5 per cent to a sewerage system. Although the majority of enterprises operated from a fixed location, about 25 per cent worked in the open, while others had temporary corrugated-iron sheds or timber huts, or operated from the house. Operators complained about the lack of infrastructure, problems of obtaining credit, high rents and insecurity of tenure, since many business premises were temporary and rented. Only 11.5 per cent of operators either partially or fully owned their business premises and most were worried that they might be moved by the authorities because of fire risk or other reasons (Nelson, 1997).

More recently, during the late 1990s and early 2000s, the Mathare 4A section of the Mathare Valley settlement was the focus for a squatter rehabilitation project, with land provided by the Kenyan government, and infrastructure and housing units provided by the Federal Republic of Germany (see Box 5.5). In addition to improving housing, water supply and sewerage, residents were encouraged to retain their informal enterprises. Those who operated home-based enterprises before the rehabilitation were given a new home with a workspace, while others requested a workspace. The government relaxed regulations covering licence fees and taxes, such that residents no longer had to fear harassment and possible destruction of their premises by the City Council. Enterprises have benefited from all-weather access roads and footpaths, sewerage, water, street-lighting and refuse disposal areas. The electricity supply enables longer working hours, while the road access is useful for moving goods into and out of the area. As Muraya (2006: 137) comments, 'The new roads allow trucks from the formal sector to supply goods, therefore eliminating the middleman and at the same time helping proprietors save the time and money needed to purchase and transport supplies. These new enterprises have created jobs and now offer a variety of goods and services.'

Many features of Mathare Valley are typical of informal enterprises in other African cities. A wide variety of dynamic and innovative activities frequently play roles in education and skills transmission, generating income and employment, and providing a range of products and services. However, governments and urban planners need to recognize the significance of informal sector enterprises, and consider carefully how such activities can be further supported to enhance their role in generating local and national wealth. The availability of credit and suitable markets are crucial to enterprise success, while the provision of land, infrastructure and eliminating barriers to legality are other vital factors.

5.9 Formal sector employment

It would be incorrect to assume that all formal sector enterprises are necessarily larger than those of the informal sector, but this is often the case. Formal sector enterprises usually employ more people, are characterized by a greater degree of permanence, often have a substantial workplace, are more likely to be recognized and enumerated by the urban authorities, and generally have wider links throughout the country and internationally. But the variety of enterprises and employment opportunities in the formal sector is as great as in the informal sector.

Little modern manufacturing industry was established in Africa during the colonial period, except in South Africa and Zimbabwe. The European powers viewed their colonies primarily as suppliers of raw materials and cheap labour, and industrial development was generally discouraged. However, traditional craft-based industries pre-date the colonial period and still survive in many countries. Unfortunately, they have suffered from external competition and were often neglected by colonial and early independent governments. But, recently, some African governments have attempted to stimulate these traditional craft-based industries, making such items as cloth, wooden carvings and metalwork. These products are primarily for domestic consumption and also for the growing tourist industries. The craft industries of the Yoruba towns in south-western Nigeria are well documented, where the family compound is the main unit of production and emphasis is on use of family labour and local materials. A study of the town of Iseyin in the 1960s showed that 27 per cent of the male population were employed in the weaving industry and numerous other people worked in the related carpentry industry making looms, shuttles and bobbins for the weavers (Bray, 1969). Also in Nigeria, at Ilorin, there is a thriving pottery industry and a narrow-loom weaving industry catering mainly for the luxury market. There are many other examples of traditional craft industries in both towns and rural areas throughout Africa.

Industrialization was seen as an important priority at the time of independence. African countries were keen to emulate their former colonial masters and saw an industrial revolution as key to their future development. In the early post-independence period, industrial development emphasized import substitution, but since much of this industry depended heavily on imported inputs, problems were encountered after 1974 with scarce foreign exchange, falling commodity prices and rising oil prices. Import substitution failed to make full use of domestic resources and de-industrialization occurred in Benin, Congo, Ghana, Madagascar, Tanzania, Togo and Zaire, as well as in Angola and Mozambique, where industrial failure was largely associated with the revolution in Portugal in 1974 and the belated granting of independence in 1975 (see Chapter 8). In contrast, South African cities (see Box 5.6) have experienced significant levels of industrialization and economic development, not only supplying manufactured goods and food products to many parts of Africa but exporting vehicles and a range of advanced sector goods to global markets (see Chapter 8).

5.10 Conclusion

With the rapid growth of Africa's towns and cities, and the seemingly insurmountable problems of providing adequate housing, water, sanitation, electricity and employment for urban dwellers, African governments and urban authorities have a pressing future agenda which has to be matched against other urgent priorities, particularly in rural areas. An important priority is to ensure that urban planning structures and processes are appropriate to meet both present and future needs. In many African cities, planning strategies are still based on legislation that was introduced during the colonial period, with strict by-laws sometimes ruling against, for example, the erection of 'temporary' housing and the practice of urban agriculture. Such a dictatorial 'top-down' approach to urban planning must be replaced by a more democratic 'bottom-up' planning process, in which city dwellers are asked for their views on planning issues, and community-based initiatives are both recognized and incorporated into the process.

Summary

1. Urbanization in Africa has generally been both recent and rapid, though in certain parts of the continent well-established towns existed in the pre-colonial period.
2. The colonial period saw the creation of many new planned towns in which there was often a clear spatial division according to function.
3. 'Rural' and 'urban' have often been seen as quite separate, but in reality there are many complex interrelationships between them.
4. A large proportion of Africa's urban population live in what might be termed 'slum' or 'squatter' settlements and these are growing rapidly. Urban authorities generally lack the funds to replace such housing with low-rental public housing units and instead favour slum upgrading and sites and services schemes.
5. In improving urban environments and providing basic services to households, urban authorities face a daunting task, and they are increasingly looking to the private sector and local communities for support.
6. In understanding urban employment in Africa, the formal–informal sector dichotomy might be criticized for being too simplistic and rigid.
7. Africa's industrial performance since the early 1960s has been disappointing, and its share of world industrial production has declined in the last twenty years.

Discussion questions

1. Compare and contrast the pace and timing of urbanization in Africa with that of other major world regions.
2. Identify the main phases of urban growth in Africa and show how these are reflected in urban structure.
3. Examine the links between 'rural' and 'urban' in Africa. How might an understanding of these links lead to a better appreciation of the development process?
4. Discuss the various strategies which might be adopted to solve the problems of urban housing and service provision.
5. How useful is the formal–informal sector dichotomy in understanding African urban employment?
6. Should industrial development be given greater priority in African countries?

Further reading

Amis, P. and Lloyd, P. (eds) (1990) *Housing Africa's Urban Poor*, Manchester: Manchester University Press.

Hardoy, J.E., Mitlin, D. and Satterthwaite, D. (1992) *Environmental Problems in Third World Cities*, London: Earthscan.

Lynch, K. (2005) *Rural–Urban Interaction in the Developing World*, London: Routledge.

McGregor, D., Simon, D. and Thompson, D. (eds) (2006) *The Peri-urban Interface: Approaches to Sustainable Natural and Human Resource Use*, London: Earthscan.

Mougeot, L.J.A. (ed.) (2005)*Agropolis: The Social, Political and Environmental Dimensions of Urban Agriculture*, London: Earthscan.

Useful websites

ILO (International Labour Office): http://www.ilo.org/global/lang--en/index.htm. UN agency based in Geneva that is concerned with employment issues and a wide range of related issues.

UN-HABITAT (United Nations Human Settlements Programme): http://www.unhabitat.org/categories.asp?catid=9. UN agency for human settlements, based in Nairobi, Kenya. It is mandated by the UN General Assembly to promote socially and environmentally sustainable towns and cities with the goal of providing adequate shelter for all. The website has some excellent downloadable publications and statistics relating to a wide range of urban issues, including urban growth, housing and service provision.

World Bank. Useful publications include *African Development Indicators*: http://web.worldbank.org/WBSITE/EXTERNAL/COUNTRIES/AFRICAEXT/EXTPUBREP/EXTSTATINAFR/0,,contentMDK:21102598~menuPK:3083981~pagePK:64168445~piPK:64168309~theSitePK:824043,00.html, and *The Little Data Book on Africa*: http://web.worldbank.org/WBSITE/EXTERNAL/COUNTRIES/AFRICAEXT/EXTPUBREP/EXTSTATINAFR/0,,contentMDK:21107092~menuPK:3098195~pagePK:64168445~piPK:64168309~theSitePK:824043,00.html.

Worldwatch Institute: http://www.worldwatch.org/node/23. Founded in 1974, the Worldwatch Institute is an independent research organization based in Washington, DC, which aims to generate and promote insights and ideas that empower decision-makers to build an ecologically sustainable society that meets human needs. The website includes resources on such topics as climate change, resource degradation, population growth and meeting human needs in the twenty-first century.

6 Health and development

6.1 Introduction

During May and June 2009, much of the media throughout the developed world was dominated by reports of the emergence of a swine flu pandemic. Originating in Mexico, where the death toll reached 100 by late June, early reports suggested the likelihood of a significant number of fatalities across the globe, and for several weeks TV viewers were presented with maps charting the spread of the disease. Interestingly, one thing that stood out from the maps was how little swine flu had seemingly penetrated the African continent; with the exception of South Africa, and several countries in North Africa, the TV maps showed 'zero cases'. While some would argue that this reflected Africa's relative isolation in a world connected by rapid air transit, a more likely explanation is that swine flu was indeed present, yet most African countries simply lack the infrastructure or capacity to monitor or record these outbreaks to the extent of developed regions. One further interesting observation was the level of media coverage, public concern and government response generated by the outbreak. The UK, for example, began stockpiling readily available antiviral drugs, and ordered enough vaccine to immunize the entire population twice over. By the end of July 2009, 36 deaths had been recorded in the UK and 353 in the USA.

In Africa, over 700,000 people die each year from malaria, over 1.7 million from HIV/AIDS, and over 880,000 from entirely preventable diarrhoeal diseases (see Table 6.1). If you are born in Africa, you are 50 per cent more likely to be malnourished, more likely to be infected with HIV at birth, and more likely to suffer ill health as a result of drought, famine or civil war. If you are a woman, you are more likely to die through childbirth complications than anywhere else in the world. The environment you live in exposes you to a range of debilitating infectious diseases, which are likely to remain untreated, given the lack of basic health services and affordable medicines in most areas. Ill health reduces livelihood opportunities and is a major cause of poverty and underdevelopment, while poverty itself creates further ill health. A key challenge for the majority of African countries, therefore, is how to break out of this vicious cycle of ill health and poverty, which impedes development. In 2000, the Millennium Development Goals (MDGs) recognized the fundamental importance of health in the development equation, with six out of the eight goals relating to specific health issues (see Box 6.1). As of 2011, however, there is little evidence of any of these goals being on target and in some cases there is likely to be a reversal of progress.

Table 6.1 The leading causes of death in Africa (projections for 2008)

	Cause of death	*Number of deaths*	*Percentage of total deaths*
1	HIV/AIDS	1,743,334	14.2
2	Cardiovascular diseases	1,717,379	13.9
3	Lower respiratory infections	1,363,074	11.1
4	Perinatal conditions	1,055,711	8.6
5	Diarrhoeal diseases	885,134	7.2
6	Malaria	723,909	5.9
7	Malignant neoplasms	650,446	5.3
8	Tuberculosis	403,827	3.3
9	Unintentional injuries	352,927	2.9
10	Respiratory diseases	394,850	3.2
11	Digestive diseases	296,246	2.4
12	Road traffic accidents	290,977	2.4
13	Other infectious diseases	277,210	2.3
14	Maternal conditions	231,031	1.9
15	Diabetes mellitus	218,409	1.8
16	Violence	204,026	1.7
17	Measles	166,761	1.4
18	Neuropsychiatric conditions	163,522	1.3
19	Genitourinary diseases	150,591	1.2
20	Nutritional deficiencies	142,312	1.2
21	Meningitis	135,944	1.1
22	Congenital anomalies	109,736	0.9
23	Pertussis	92,358	0.7
24	Endocrine disorders	87,666	0.7
25	War and civil conflict	85,088	0.7
27	Tetanus	63,599	0.5
30	Trypanosomiasis	44,488	0.4

Source: Based on WHO, 2008a

Box 6.1

Health and the Millennium Development Goals in Africa

Six of the eight Millennium Development Goals have targets related to health (see below). Although monitoring progress towards these goals remains problematic, the most recent data suggest that the African region as a whole will not meet any of its goals by 2015. There is, however, significant variation within the continent.

Goal 1: Eradicate extreme poverty and hunger

- *Specific health targets:* Halve the proportion of people who suffer from hunger between 1990 and 2015.

- *Indicators:* 1.8 – the prevalence of underweight children under five years of age; 1.9 – the proportion of the population below the minimum level of dietary energy consumption.
- *Overall progress:* Off track to meet all targets. North Africa has met the goal and fourteen sub-Saharan countries are on track, but rising food and energy prices may hinder progress.
- *Countries likely to achieve goals for child malnutrition:* Botswana, Chad, Egypt, Gambia, Mauritania, Sudan and Tunisia.
- *Countries likely to achieve goals for overall undernourishment:* Algeria, Angola, Egypt, Ghana, Libya, Malawi, Morocco and Tunisia.

Goal 4: Reducing child mortality

- *Specific health targets:* Reduce the under-five mortality rate by two-thirds between 1990 and 2015.
- *Indicators:* 4.1 – Under-five mortality rate; 4.2 – Infant mortality rate; 4.3 – Proportion of one-year-old children immunized against measles.
- *Overall progress:* Off track to meet all targets. Under-five mortality declined a mere 1.8 per cent between 1990 and 2005; nineteen countries experienced no change, while under-five mortality increased in ten countries. Marginal improvements in infant mortality rates (110 to 99 deaths per 1000 births between 1990 and 2005), although more significant improvements in North Africa. Immunization against measles has been variable – 90 per cent coverage rates for Botswana, Tanzania and Tunisia; less than 40 per cent for Nigeria and Somalia.
- *Countries likely to achieve goals:* Malawi, Algeria, Cape Verde, Egypt, Libya, Mauritius, Morocco, Seychelles and Tunisia.

Goal 5: Improving maternal health

- *Specific health targets:* a) Reduce the maternal mortality ratio by three-quarters between 1990 and 2015; b) Achieve universal access to reproductive health by 2015.
- *Indicators:* 5.1 – Maternal mortality ratio; 5.2 – Proportion of births attended by skilled health personnel; 5.3 – Contraceptive prevalence rate; 5.4 – Adolescent birth rate; 5.5 – Ante-natal care coverage (either at least one visit or at least four visits); 5.6 – Unmet need for family planning.
- *Overall progress:* Off track to meet all targets. Estimates suggest a negligible (1.8 per cent) improvement in maternal mortality rates between 1990 and 2005. For attendance of skilled health personnel, no progress has occurred in central, East, southern or West Africa. Small increase in contraceptive use, no decrease in adolescent birth rates, and insufficient increases in antenatal care.
- *Countries likely to achieve goals:* Algeria, Botswana, Cape Verde, Egypt, Gambia, Libya, Mauritius, Morocco and Tunisia.

Goal 6: Combat HIV/AIDS, malaria and other diseases

- *Specific health targets:* a) Halt and begin to reverse the spread of HIV/AIDS by 2015; b) Achieve universal access to treatment for HIV/AIDS for all those who need it by 2010; c) Halt and begin to reverse the incidence of malaria and other major diseases by 2015.

- *Indicators:* 6.1 – HIV prevalence among population aged fifteen–twenty-four years; 6.2 – Condom use at last high-risk sex; 6.3 – Proportion of population aged fifteen–twenty-four years with comprehensive, correct knowledge of HIV/AIDS; 6.4 – Ratio of school attendance of orphans to school attendance of non-orphans aged ten–fourteen years; 6.5 – Proportion of population with advanced HIV infection with access to antiretroviral drugs; 6.6 – Incidence and death rates associated with malaria; 6.7 – Proportion of children under five sleeping under insecticide-treated bednets; 6.8 – Proportion of children under five with fever who are treated with appropriate anti-malarial drugs; 6.9 – Incidence, prevalence and death rates associated with tuberculosis; 6.10 – Proportion of tuberculosis cases detected and cured under directly observed treatment short course.
- *Overall progress:* Off track to meet all targets. The proportion of women infected with HIV/AIDS is increasing. Lower prevalence in North Africa, but insufficient progress to meet the target. The number of people receiving antiretroviral drugs increased significantly after 2003, but demand continues to exceed supply. Sharp increase in the production of insecticide-treated mosquito nets (30 million in 2004 to 95 million in 2007), but less than half of all children sleep under such nets. Proportion of children with a fever who received anti-malarial medication fell between 2000 and 2005. Decline of tuberculosis in North Africa but an increase elsewhere, linked to HIV infection.
- *Countries likely to achieve goals for HIV/AIDS:* Algeria, Botswana, Egypt, Libya, Morocco, Tunisia, Uganda and Zimbabwe.
- *Countries likely to achieve goals for malaria:* Algeria, Benin, Cameroon, Central Africa, Comoros, Egypt, Gambia, Guinea-Bissau, Kenya, Libya, Morocco, Tunisia and Rwanda.
- *Countries likely to achieve goals for tuberculosis:* Algeria, Angola, Egypt, Gabon, Gambia, Libya, Madagascar, Morocco, South Africa, Swaziland, Tunisia and Zambia.

Goal 7: Ensuring environmental sustainability

- *Specific health targets:* Halve the proportion of people without sustainable access to safe drinking water and basic sanitation by 2015.
- *Indicators:* 7.8 – Proportion of population using an improved drinking water source; 7.9 – Proportion of population using an improved sanitation facility.
- *Overall progress:* Off track to meet all targets. Sufficient progress in North Africa, and fifteen sub-Saharan African countries have increased access to clean water in rural areas by 25 per cent, but rural–urban disparities persist. Only a marginal increase in improved sanitation in many areas.
- *Countries likely to achieve goals for access to safe drinking water (rural):* Algeria, Botswana, Burundi, Egypt, Gambia, Ghana, Malawi, Mauritius, Namibia, South Africa and Tanzania.
- *Countries likely to achieve goals for access to sanitation (urban):* Algeria, Egypt, Ghana, Libya, Mauritius, Morocco and Tunisia.

Goal 8: Developing a global partnership for development

- *Specific health targets:* In cooperation with pharmaceutical companies, provide access to affordable, essential drugs in developing countries.
- *Indicator:* 8.13 – Proportion of population with access to affordable, essential drugs on a sustainable basis.
- *Overall progress:* Off track to meet all targets. Despite some progress, patents still restrict the supply of essential drugs to the majority of the poor.
- *Countries likely to achieve goals:* No data.

Note: 'Overall progress' is for the African region as a whole; some individual countries are on track.

Sources: UN, 2001, 2007, 2008a; UNESC/ECA, 2008; ECA, 2005b.

This chapter examines some of these issues in detail, starting with an examination of mortality and morbidity rates within Africa, and moving on to discussions of maternal mortality, environmental health issues and the role of infectious disease. It goes on to examine the emergence of non-communicable diseases, and concludes with a brief review of health services and the way forward for healthcare. One theme that underlies every section in this chapter is the HIV/AIDS epidemic, which is arguably the single most important health issue influencing development across the continent. Since its emergence in the late 1980s, it has lowered life expectancy across the continent, increased infant and maternal mortality, facilitated the spread of other infectious diseases, impoverished people and their livelihoods, and fundamentally stalled development in many nations.

6.2 Looking at mortality rates

Mortality statistics are frequently used to indicate the level of ill health and development throughout the African region. Both life expectancy and infant mortality (the number of infant deaths per 1000 live births) are considered to be indicative of a range of societal health issues, including the prevalence of infectious disease, environmental health problems, and the level of access to health services. Life expectancy is deemed sufficiently important to be used as a key variable in the calculation of the UNDP's Human Development Index.

While the collection of these statistics allows comparisons to be drawn easily between countries, there are some drawbacks with the data. As Bloom *et al.* (2000) point out, those countries with the worst health problems and high mortality rates are less likely to have sufficient resources available to collect and process data accurately. Indeed, the World Health Organization (WHO, 2006a) reports that less than 10 per cent of all deaths in the African region are registered; most statistics tend to be derived from mathematical models. There is therefore likely to be substantial variation in data quality and consistency between countries.

6.2.1 Trends in life expectancy and infant mortality

Table 6.2 provides a breakdown of life expectancy (both male and female) and infant mortality across Africa. The data are ranked on the basis of life expectancy for 2005, with

Table 6.2 Life expectancy and infant mortality rates in Africa

		Average life expectancy						*Infant mortality rate*		
HDI	*Country*	*2005*	*1970–1975*	*Gains*	*Female (2005)*	*Male (2005)*	*Difference*	*1970*	*2005*	*Reduction*
91	Tunisia	73.5	55.6	17.9	75.6	71.5	4.1	135	20	–115
56	Libya	73.4	52.8	20.6	76.3	71.1	5.2	105	18	–87
50	Seychelles	72.7	–	–	–	–	–	46	12	–34
65	Mauritius	72.4	62.9	9.5	75.8	69.1	6.7	64	13	–51
104	Algeria	71.7	54.5	17.2	73	70.4	2.6	143	34	–109
102	Cape Verde	71	57.5	13.5	73.8	67.5	6.3	–	26	–
112	Egypt	70.7	51.1	19.6	73	68.5	4.5	157	28	–129
126	Morocco	70.4	52.9	17.5	72.7	68.3	4.4	119	36	–83
123	São Tomé and Príncipe	64.9	56.5	8.4	66.7	63	3.7	–	75	–
134	Comoros	64.1	48.9	15.2	66.3	62	4.3	159	53	–106
137	Mauritania	63.2	48.4	14.8	65	61.5	3.5	151	78	–73
156	Senegal	62.3	45.8	16.5	64.4	60.4	4	164	77	–87
135	Ghana	59.1	49.9	9.2	59.5	58.7	0.8	111	68	–43
155	Gambia	58.8	38.3	20.5	59.9	57.7	2.2	180	97	–83
143	Madagascar	58.4	44.9	13.5	60.1	56.7	3.4	109	74	–35
152	Togo	57.8	49.8	8	59.6	56	3.6	128	78	–50
147	Sudan	57.4	45.1	12.3	58.9	56	2.9	104	62	–42
157	Eritrea	56.6	44.1	12.5	59	54	5	143	50	–93
119	Gabon	56.2	48.7	7.5	56.9	55.6	1.3	–	60	–
174	Niger	55.8	40.5	15.3	54.9	56.7	–1.8	197	150	–47
163	Benin	55.4	47	8.4	56.5	54.1	2.4	149	89	–60
160	Guinea	54.8	38.3	16.5	56.4	53.2	3.2	197	98	–99
139	Congo, Republic	54	54.9	–0.9	55.2	52.8	2.4	100	81	–19
173	Mali	53.1	40	13.1	55.3	50.8	4.5	225	120	–105
148	Kenya	52.1	53.6	–1.5	53.1	51.1	2	96	79	–17
169	Ethiopia	51.8	43.5	8.3	53.1	50.5	2.6	160	109	–51
125	Namibia	51.6	53.9	–2.3	52.2	50.9	1.3	85	46	–39
176	Burkina Faso	51.4	43.6	7.8	52.9	49.8	3.1	166	96	–70
159	Tanzania	51	47.6	3.4	52	50	2	129	76	–53
121	South Africa	50.8	53.7	–2.9	52	49.5	2.5	–	55	–
170	Chad	50.4	45.6	4.8	51.8	49	2.8	154	124	–30
127	Equatorial Guinea	50.4	40.5	9.9	51.6	49.1	2.5	–	123	–
144	Cameroon	49.8	47	2.8	50.2	49.4	0.8	127	87	–40
154	Uganda	49.7	51	–1.3	50.2	49.1	1.1	100	79	–21
167	Burundi	48.5	44.1	4.4	49.8	47.1	2.7	138	114	–24
124	Botswana	48.1	56	–7.9	48.4	47.6	0.8	99	87	–12
–	Somalia	47.1	41	6.1	48.2	45.9	2.3	–	133	–
166	Côte d'Ivoire	47.4	49.8	–2.4	48.3	46.5	1.8	158	118	–40
158	Nigeria	46.5	42.8	3.7	47.1	46	1.1	140	100	–40
164	Malawi	46.3	41.8	4.5	46.7	46	0.7	204	79	–125
175	Guinea-Bissau	45.8	36.5	9.3	47.5	44.2	3.3	–	124	–

Table 6.2 Continued

		Average life expectancy						*Infant mortality rate*		
HDI	*Country*	*2005*	*1970–1975*	*Gains*	*Female (2005)*	*Male (2005)*	*Difference*	*1970*	*2005*	*Reduction*
168	Congo (DRC)	45.8	46	–0.2	47.1	44.4	2.7	148	129	–19
161	Rwanda	45.2	44.6	0.6	46.7	43.6	3.1	124	118	–6
–	Liberia	44.7	42.6	2.1	45.6	43.8	1.8	180	157	–23
171	Central African Republic	43.7	43.5	0.2	45	42.3	2.7	145	115	–30
172	Mozambique	42.8	40.3	2.5	43.6	42	1.6	168	100	–68
138	Lesotho	42.6	49.8	–7.2	42.9	42.1	0.8	140	102	–38
177	Sierra Leone	41.8	35.4	6.4	43.4	40.2	3.2	206	165	–41
162	Angola	41.7	37.9	3.8	43.3	40.1	3.2	180	154	–26
141	Swaziland	40.9	49.6	–8.7	41.4	40.4	1	132	110	–22
151	Zimbabwe	40.9	55.6	–14.7	40.2	41.4	–1.2	86	81	–5
165	Zambia	40.5	50.1	–9.6	40.6	40.3	0.3	109	102	–7

Source: HDI, 2009

highest at the top and lowest at the bottom. Analysis of this table reveals something about the distribution of health and development across the continent. First, it is clear from Table 6.2 and Figure 6.1 that in 2005 the vast majority of countries had life expectancies of between forty and sixty years (78 per cent of all countries). Indeed, for a relatively significant proportion of countries (39 per cent), life expectancy remains under fifty years. Only the North African and small island states experience life expectancies in excess of sixty-five years. Comparing these data to the average life expectancy during the period 1970 to 1975, it is clear that there have been some significant gains among those countries in the top half of the table, particularly among the North African states. For example, the highest gain within Africa has been in Libya, where life expectancy rose by 20.6 years between 1975 and 2005. At the opposite end of the scale, however, there have been some significant reversals. In Zimbabwe, for example, the HIV epidemic has reduced life expectancy by 14.7 years, compared with its 1975 level.

The difference between the life expectancy of men and women, a factor usually attributed to occupational or lifestyle differences, is also evident in the data. Only in Niger and Zimbabwe does the life expectancy of men exceed that of women. However, this does not necessarily mean that women lead healthier lives than men (WHO, 2007a). As discussed below, maternal mortality remains a significant issue throughout Africa, and those women who survive childbirth are more likely than men to suffer from chronic, debilitating illnesses linked to their household role. With this in mind, 'healthy life expectancy' – defined as 'the equivalent number of years in full health that a newborn can expect to live based on current rates of ill-health and mortality' (WHO, 2004a: 96) – has recently been considered a more informative and useful measure of health and development (Mathers *et al.*, 2001).

With regards to infant mortality, the recent data in Figure 6.1 suggest that very high levels of infant mortality persist across the continent. (It is useful to compare these rates with developed countries, such as the UK and the USA, whose IMRs are 4.8 and 6.3, respectively.) Lower rates are observed for North Africa and the island states, whereas higher rates tend to be located within West Africa, in particular in Sierra Leone, where one in five infants die before their first birthday. It is interesting to note, however, that

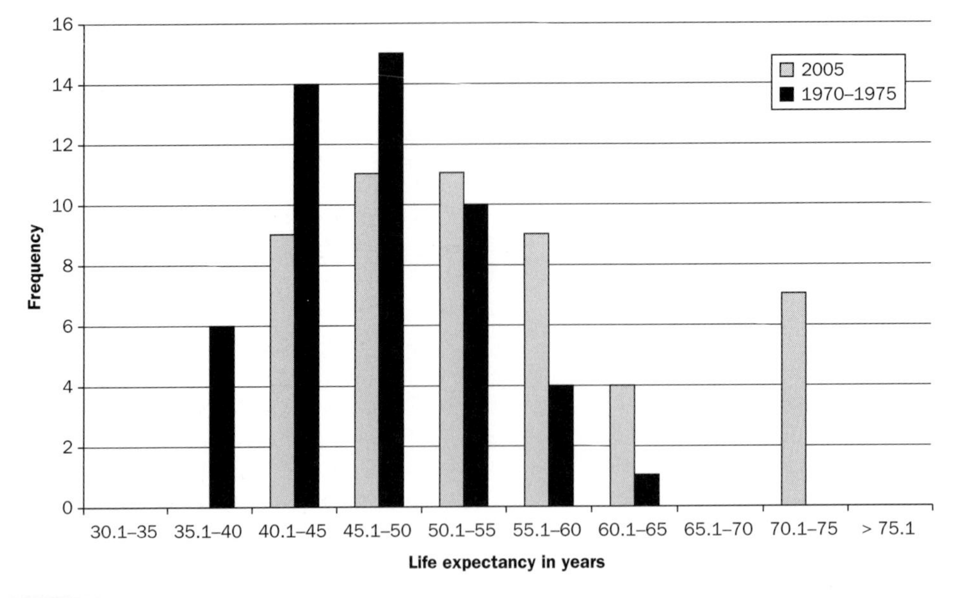

Figure 6.1 Distribution of countries by life expectancy.

Source: HDI, 2009

during the 1970s North Africa experienced similar high levels to the rest of the continent, yet there have since been substantial reductions. In Egypt, the IMR declined from 157 in 1970 to 28 in 2005, largely as a result of improvements in healthcare and a reduction in the fertility rate. While rates have declined throughout sub-Saharan Africa, too, progress has slowed dramatically since the 1990s, again as a result of the HIV epidemic. Nonetheless, Malawi has made significant progress towards reducing IMR and achieving the MDG target on infant mortality, due to an effective government Integrated Management of Childhood Illness Programme, which supports, among other things, the provision of vaccines, vitamin A, oral rehydration therapy and insecticide-treated bednets.

6.3 Maternal health

The World Health Report 2005 outlined the fact that Africa's women experience poorer levels of maternal health and a greater likelihood of death during pregnancy or childbirth than those in any other part of the world (WHO, 2005a). Maternal mortality in Africa (the number of deaths during pregnancy or within forty-two days of termination of pregnancy due to a pregnancy-related cause) totalled 276,000 in 2005 (see Table 6.3), some 51 per cent of the global total (sub-Saharan Africa alone represents 50 per cent) (WHO, 2007b). Within sub-Saharan Africa, maternal mortality rates (per 100,000 of the population) vary significantly from one country to the next: in 2005, Sierra Leone had the highest rate in the continent (2100), while the figure was 880 in Zimbabwe but just 16 in Mauritius.

Although it is often difficult to establish the causes of maternal mortality due to inconsistencies in reporting, the most common recorded causes include post-partum haemorrhage, sepsis, hypertensive disorders, and other indirect causes, such as cardiac or renal disease, which are aggravated by pregnancy (see Figure 6.2). The prevalence of such high levels of maternal mortality has been attributed to a wide range of socio-economic and biological issues that affect women. Rogo *et al.* (2006) suggest five determinants which increase the likelihood of poor maternal health:

Table 6.3 Comparison of 1990 and 2005 maternal mortality rates

Region	*1990*		*2005*			
	MMR	*Maternal deaths*	*MMR*	*Maternal deaths*	*% change in MMR between 1990 and 2005*	*Annual % change in MMR between 1990 and 2005*
WORLD TOTAL	430	576,000	400	536,000	–5.4	–0.4
Developed regions	11	1300	9	960	–23.6	–1.8
AFRICA TOTAL	830	221,000	820	276,000	–0.6	0
North Africa	250	8900	160	5700	–36.3	–3
Sub-Saharan Africa	920	212,000	900	270,000	–1.8	–0.1
Selected countries						
Sierra Leone			2100			
Rwanda			1300			
Malawi			1100			
Zimbabwe			880			
Ethiopia			720			
Kenya			560			
South Africa			400			
Botswana			380			
Tunisia			100			
Mauritius			16			

Source: WHO, 2007b

- *Household and community characteristics.* Women's subordinate role within many African societies and communities means they are less likely to be educated, have legal rights, have the nutritional level of their male counterparts, or have access to money with which to pay for healthcare or medicines. Gender-based violence may be common, while the persistence of harmful cultural practices, such as female genital mutilation, may also have serious long-term health impacts.
- *Biological–demographic variables and risk factors.* The relatively young ages of mothers, combined with a high fertility rate among the adult population, have contributed towards the high maternal mortality rate. Africa has some of the highest adolescent fertility rates in the world, with an average of 119 births for every 1000 women aged 15–19 years (ranging from 199 in Niger – the highest in the world – to 4 in Algeria). This compares with the global average of 48.
- *Malnutrition–infection syndrome.* Pregnancy with concurrent nutrient deficiencies increases vulnerability to diseases such as malaria.
- *Health systems.* The lack of access to a well-resourced maternal health system with skilled workers is a major issue. In Ethiopia, for example, recent estimates suggest that only 6 per cent of births are attended by skilled personnel (WHO, 2009a).
- *National policies and investments.* Support from national policies and investment in health services and infrastructure are fundamental to improving maternal health at the grassroots level.

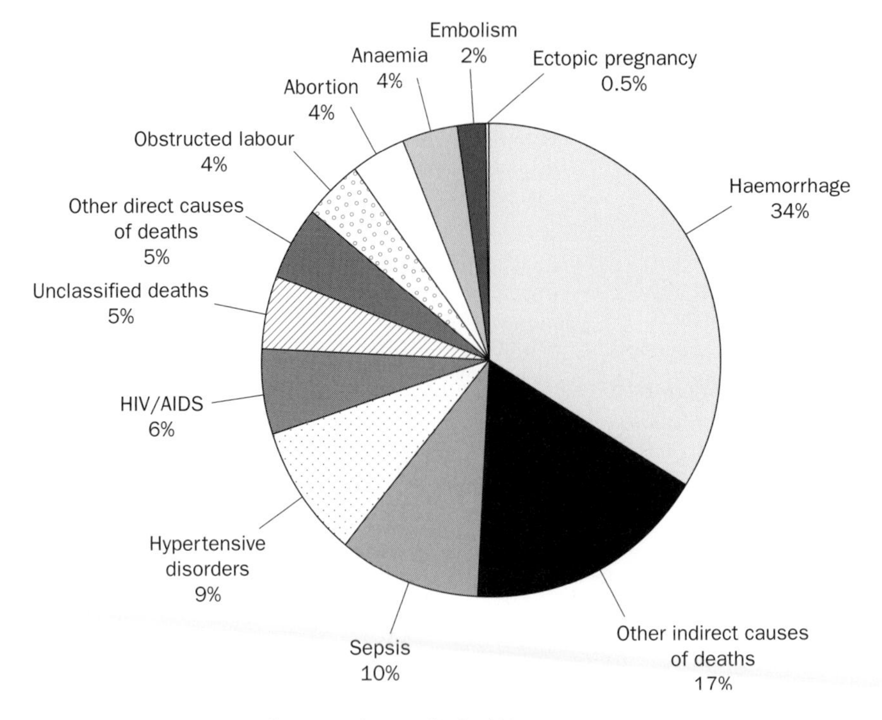

Figure 6.2 The causes of maternal mortality in Africa.

Source: WHO, 2005b

One might add to these points the need for conditions of peace and stability. As is discussed further in Chapter 7, the numerous conflicts that have occurred within Africa during the last forty years have usually been characterized by high levels of sexual violence against women, displacement of people, and the destruction of basic healthcare services. The long-term maternal physical and mental health problems that result have undoubtedly further impoverished women in many countries.

Improving maternal health has therefore become a critical development challenge and one which has been integrated into the Millennium Development Goals (see Box 6.1). Goal 5 aims to reduce maternal mortality and improve access to reproductive healthcare by 2015. The UN road map, which provides guidance on how the MDGs are to be achieved, regards an increase in the number of skilled attendants and health workers at the community level as critical to the achievement of this goal (UN, 2001). However, this will ultimately depend upon the implementation of national policies and the allocation of new financial resources. Perhaps more fundamentally, reducing maternal mortality requires a commitment to recognize and address those determinants discussed above. The empowerment and elevation of women within society, for example, can give them a voice to request family-planning services, ante-natal care and an environment for safe delivery (Hunt and De Mesquita, 2008; Grown *et al.* 2005). It can also give them access to more nutritional food, thereby reducing the likelihood of malnourishment-related complications. Similarly, supporting family planning can help reduce fertility rates, particularly among adolescents, which can lead to long-term improvements in maternal health.

6.4 Environmental health

As mentioned in Chapter 3, the diverse African environment presents various health risks for its population. Bodies of water harbour the insect vector responsible for malaria, as well as such diseases as schistosomiasis and river blindness, while vast areas of savanna woodland host the tse-tse fly, responsible for the transmission of trypanosomiasis. There are, however, a range of other serious environmental risks to health which can be considered to be more anthropogenic in nature: that is, they are in one way or another linked to direct human influence on the natural environment. These risks occur in both rural and urban areas to different degrees, and include air pollution, lack of food hygiene, poor sanitation, inadequate waste disposal, exposure to chemicals, and physical injuries caused by specific environmental hazards. According to the WHO (2006a), environmental risk factors accounted for approximately 23 per cent (2.4 million) of all deaths in Africa during 2002. Inevitably, those members of society who are entrenched in the poverty–poor health nexus are at greater risk from such hazards.

6.4.1 Water supply, sanitation and hygiene

A key target for Millennium Development Goal 7 is to halve the proportion of people without sustainable access to safe drinking water and basic sanitation by 2015 (see Box 6.1). This remains a significant challenge, particularly since an estimated 62 per cent (583 million) of the entire African population lacked access to basic sanitation facilities in 2006, and 36 per cent (341 million) lacked access to safe drinking water (UNICEF, 2006). Despite modest increases in coverage through a range of development efforts since 1990, the increasing demand for water and sanitation due to rapid population growth continues to outstrip supply. With the exception of North Africa and a minority of sub-Saharan countries, most countries will fail to meet their MDG 7's 2015 targets. Figures 6.3 and 6.4 highlight the variation within Africa with regard to access to improved drinking water sources and improved sanitation facilities, respectively – the two indicators used to measure progress towards MDG 7. (Table 6.4 outlines the differences between 'improved' and 'unimproved' water sources and sanitation.) Differences between North Africa and the sub-Saharan countries are evident in both figures: coverage of improved water sources and sanitation is greatest in North Africa and southern Africa, while the Sahel region (Ethiopia, Somalia, Niger and Chad) stands out as particularly problematic.

One trend experienced by all countries is that significantly more people in urban areas continue to have access to improved drinking water and sanitation than those living in rural areas (see Figures 6.5 and 6.6). The lack of infrastructural development means that the majority of people in rural areas rely on natural springs, bodies of water or, at best, communal hand pumps for drinking water. Having to walk, often great distances, to a collection point (which itself becomes a chronic health problem, usually for Africa's women) means that only the minimum amount of water for household daily use tends to be collected. Inevitably, food hygiene suffers as a consequence of this household water scarcity. The lack of sanitation in rural areas also means that many bodies of water from which drinking water is extracted also become polluted with human waste, thereby facilitating the spread of disease. In contrast, urban areas tend to be better supported by a water and sanitation infrastructure, although the data mask the dire situation within the vast areas of informal or slum housing in Africa's cities (as we saw in Chapter 5). It has been argued that these areas lack government investment in water and sanitation infrastructure because of their illegal or temporary nature (Boadi *et al.* 2005), although, as

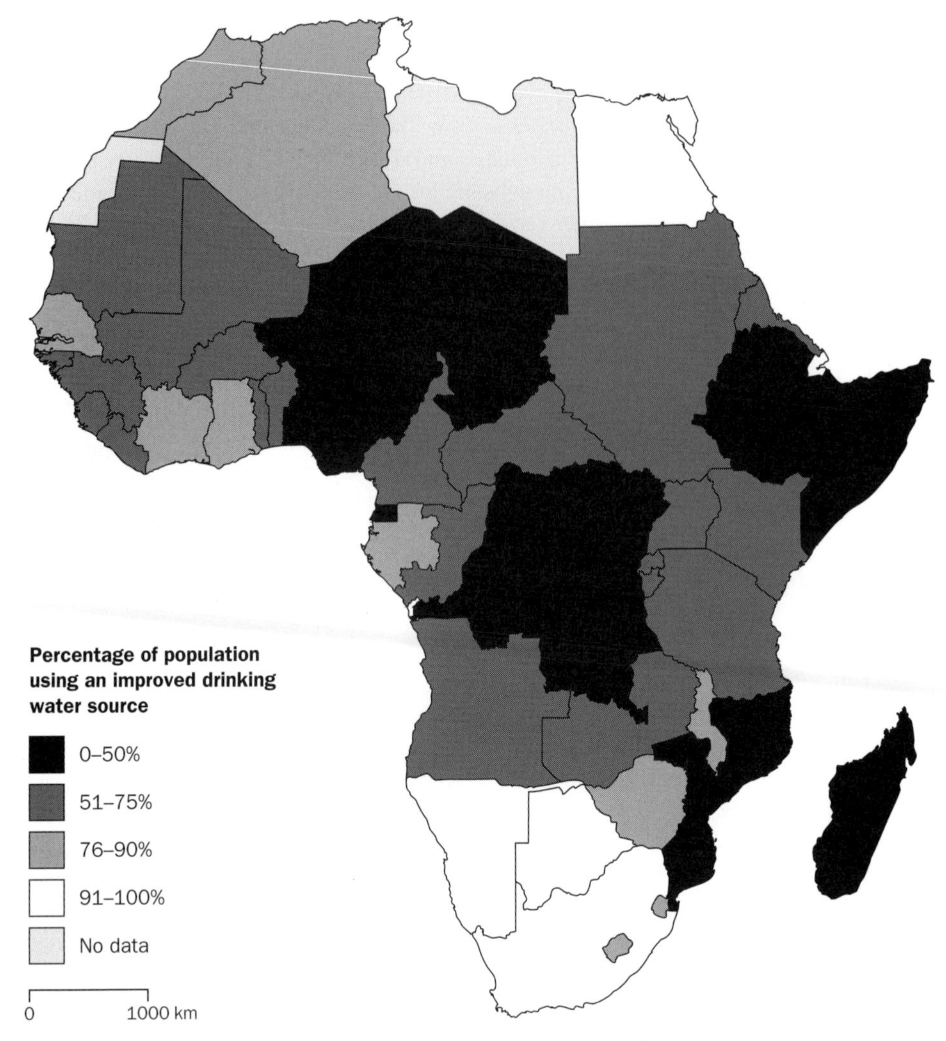

Figure 6.3 Coverage with improved drinking water sources, 2006.

Source: WHO, 2008a

Konteh (2009) points out, it is ultimately a range of social factors, such as poverty, health education and housing, that increases vulnerability to disease in these environments.

Approximately 1.03 million deaths in Africa were linked to inadequate sanitation and the lack of access to clean drinking water during 2003. Of these, the majority occurred as a result of diarrhoeal diseases, caused by the ingestion of such pathogens as rotavirus, salmonella, *Escherichia coli*, cholera and *Giardia* (Boschi-Pinto *et al.*, 2006; Prüss *et al.*, 2008). Cholera, in particular, continues to be a major environmental health issue in sub-Saharan Africa, with 167,000 cases reported in 2007 (WHO, 2008b). The virulent nature of the disease means transmission occurs very quickly in areas with inadequate sanitation where people live in close proximity to each other. For example, following the displacement of people after the Rwandan genocide of 1994, 50,000 people died of cholera in refugee camps in Goma, Tanzania (Goma Epidemiology Group, 1995). Similarly, in 2008, a cholera outbreak in Zimbabwe, which by June 2009 had affected more than 100,000 people, was attributed to the collapse of the urban water and sanitation network after decades of neglect (Mason, 2009).

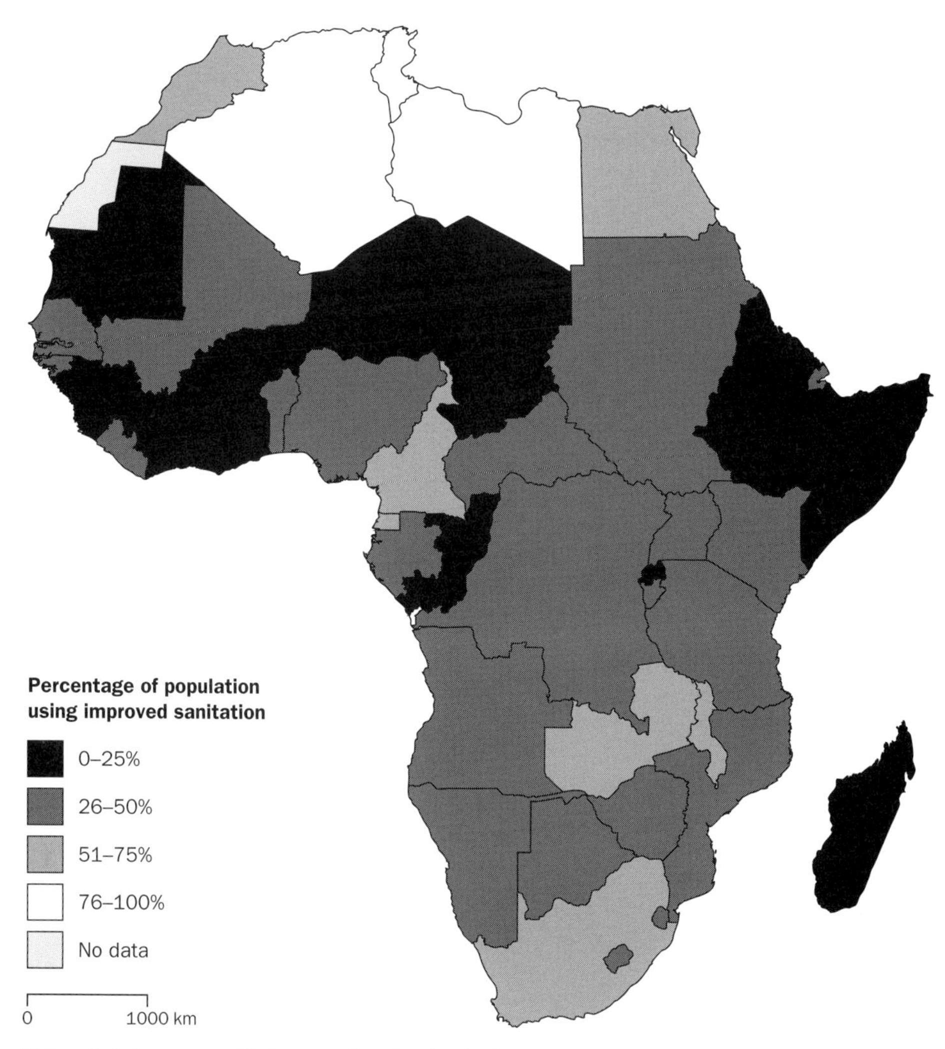

Figure 6.4 Coverage with improved sanitation facilities, 2006.

Source: WHO, 2008a

Other risks associated with inadequate water and sanitation include intestinal nematode infections, such as hookworm. This is transmitted through faecally contaminated soil and leads to severe anaemia, which in pregnant women impairs foetal development and increases the risk of maternal mortality. Brooker *et al.* (2008) estimate that 37.7 million women of reproductive age in sub-Saharan Africa were infected with hookworm in 2005, with a third of pregnancies affected by the disease. Similar prevalence rates among children and adults have a debilitating effect on health, and subsequently the ability to attend school and develop an economically productive livelihood (Brooker *et al.*, 1999; Bethony *et al.*, 2006). Trachoma, an infection of the eye, is the major cause of infectious blindness in Africa. The disease is transmitted from person to person via hands, clothing or flies, and over several years results in scarring of the cornea and eventual blindness. The disease affects around 30 million people, most in the savanna areas of East and central Africa and the Sahel of West Africa (Ethiopia alone accounts for 34 per cent of cases in Africa), even though it is entirely preventable by washing and good hygiene practices (Polack *et al.*, 2005).

Table 6.4 The classification of 'improved' and 'unimproved' water sources and sanitation

Improved drinking water sources	*Unimproved drinking water sources*
• Piped water into dwelling, plot or yard • Public tap/standpipe • Tube well /bore hole • Protected dug well • Protected spring • Rainwater	• Unprotected dug well • Unprotected spring • Small cart with tank/drum • Tanker truck • Surface water (river, dam, lake, pond, stream, channel, irrigation channel)
Improved sanitation facilities	*Unimproved sanitation facilities*

• Flush or pour-flush to:
  • piped sewerage system
  • septic tank
  • pit latrine
• Ventilated improved pit latrine (VIP)
• Pit latrine with slab
• Composting toilet
	• Flush or pour-flush to elsewhere
	• Pit latrine without slab or open pit
	• Bucket
	• Hanging toilet or hanging latrine
	• No facilities or bush or field (open defecation)

Source: UNICEF, 2008

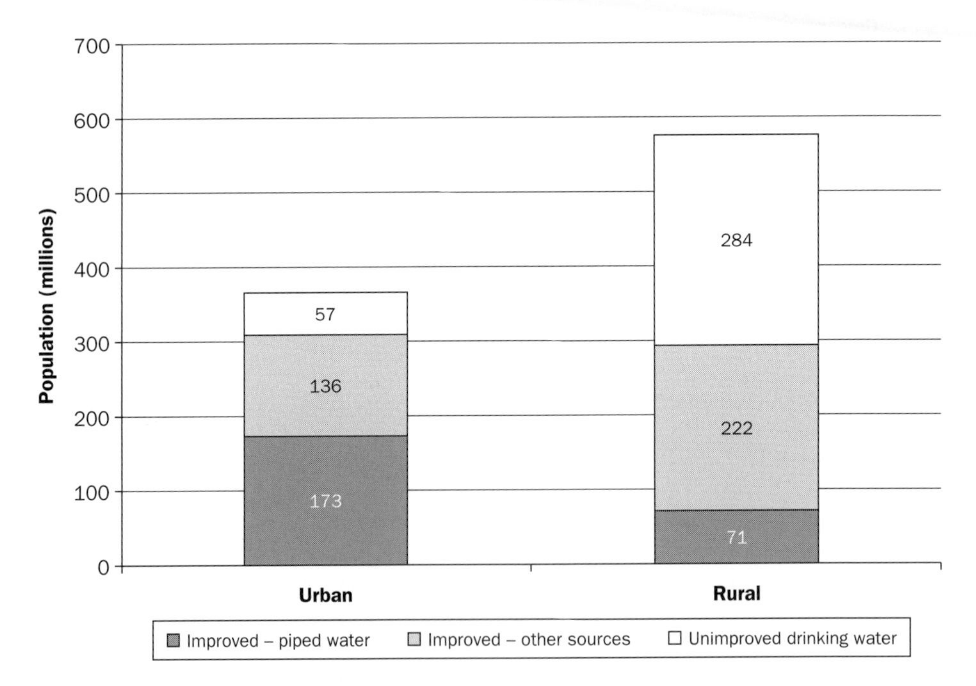

Figure 6.5 Urban and rural differences in access to drinking water.

Source: Adapted from UNICEF, 2008

The traditional approach to addressing such diseases has entailed extending water and sanitation infrastructure in both urban and rural areas. This, however, presents significant challenges in terms of its affordability (for both governments and users) and the logistical feasibility of its installation in remote areas (Paterson *et al.*, 2006). Moreover, provision of such engineered solutions does not necessarily guarantee that they are used or maintained by people (Mara, 2003). In recent years, however, attention has focused more on appropriate

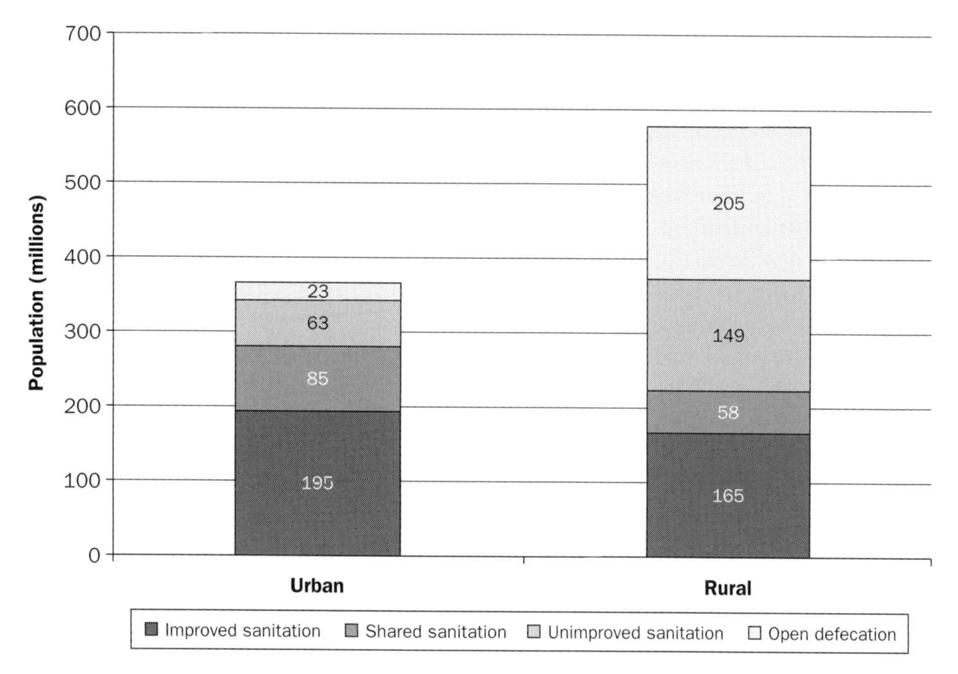

Figure 6.6 Urban and rural differences in sanitation.

Source: Adapted from UNICEF, 2008

technology solutions which are developed in a decentralized manner with the full participation of community members. The WHO's PHAST (Participatory Hygiene And Sanitation Transformation) Initiative was spearheaded in Botswana, Kenya, Uganda and Zimbabwe in 1996, while more recently the community-led total sanitation (CLTS) approach (see Box 6.2) has promoted the use of locally adapted sanitation solutions that are low cost, low maintenance, and employ collective action in their maintenance and monitoring (Kar and Chambers, 2008). Rather than targeting the provision of sanitation 'hardware', the approach focuses on the behavioural change required to ensure changes that are sustainable. As of 2009, CLTS initiatives are being implemented with the support of UNICEF and several NGOs across Africa. Fundamentally, many countries have also adopted the WHO's 'healthy settings' approach, which seeks to integrate environmental health goals within a range of sectors, including transport, planning, education and economic development.

Box 6.2

Community-led total sanitation (CLTS) in Sierra Leone

Over a decade of war between 1991 and 2002 left much of Sierra Leone in crisis and facing critical social, economic and environmental problems. Despite a programme of post-conflict reconstruction and impressive economic growth in recent years, much of the country remains mired in poverty: in 2009, Sierra Leone was ranked 180th out of 182 countries in the Human Development Index. Of particular concern is the high under-five mortality rate, which at 269 per 1000 constitutes the highest in the world. Although the causes of such a high rate are diverse, over 20 per cent of deaths among the under-fives have been attributed to diarrhoeal diseases that are directly linked to the lack of clean water and sanitation (WHO, 2006d). It has been estimated that only 47 per cent of the country's population have access to improved drinking water sources, while only 30 per cent have access to sanitation facilities.

To meet the Millennium Development Goal target of a 66 per cent coverage of improved sanitation facilities by 2015, Sierra Leone needs to build 300,000 latrines. This is particularly challenging, given that many past initiatives which have subsidized the building of improved sanitation facilities have failed to make much of an impact. Since 2008, however, UNICEF, working in collaboration with a range of NGOs, has promoted the adoption of a community-led total sanitation (CLTS) approach as part of its global WASH (water, sanitation and hygiene) initiative. This approach differs from traditional water- and sanitation-improvement initiatives in that it focuses more on the need for community empowerment and mobilization, with the aim of achieving 'open defecation free' (ODF) status for the community as a whole. The first stage in the CLTS approach has trainers and facilitators visiting a community and employing a range of participatory tools which help people analyse their own sanitation situation. These can range from mapping the location of existing resources to participants conducting a 'walk of shame' where defecation sites are identified and attributed to participants. A critical outcome of this initial 'triggering' is the community realizing that in practising open defecation they are effectively eating their own faeces. Subsequently, participants are encouraged to identify ways of addressing the issue of open defecation, and to develop a low-cost, socially sustainable sanitation plan for their village. While communities may draw their own latrine designs, the CLTS facilitators usually suggest a range of latrines that meet acceptable minimum hygiene standards.

In Bauya Old Town, Moyamba District, CLTS activities began in 2009 when a trainer 'triggered' the community into action by demonstrating the links between 'kaka and food' and facilitating 'kaka mapping'. This led to several individuals from the community volunteering to be 'natural leaders', responsible for supervising latrine construction and motivating families to improve their sanitation and hygiene practices. Within months, the community had been declared open defecation free. Alpha Silvalie, one of Bauya Old Town's three natural leaders, suggested: 'The triggering process really helped everyone see the need for change. They didn't need much convincing after that. For us, latrine construction has been a great source of pride and achievement. The community has really come together to make things happen.' According to Ndomawa Banya, the paramount chief: 'Community latrine construction encourages families to use the latrines and take personal responsibility for their upkeep, something which the subsidized latrines of the past failed to do. Now a successful latrine signifies a successful household. It is something to be very proud of' (UNICEF, 2009)

In 2009, a total of 285 communities throughout Sierra Leone were declared ODF, while 899 communities had been 'triggered' through the activities of no fewer than seventeen NGOs, in collaboration with the Ministry of Health and Sanitation (UNICEF, 2009). The CLTS approach is also being rolled out in schools with the aim of influencing community behaviour from an early age (see Plate 6.1).

Plate 6.1 A girl in Sierra Leone participates in UNICEF's community-led total sanitation programme (Emily Bamford, © UNICEF, Sierra Leone 2009, permission obtained).

6.4.2 Air pollution

Air pollution is becoming a serious health issue throughout Africa. The combustion of fossil fuels associated with industrial growth and rapid urbanization has increased the level of pollutants in urban areas to levels that far exceed internationally agreed limits. Of particular concern is the increase in vehicular emissions in urban areas due to a growth in vehicle numbers, poor fuel efficiency among older vehicles, and poor traffic management in congested cities. The rising populations of urban areas mean that ever more people are becoming exposed to harmful particulates and heavy-metal emissions, which can lead to respiratory illness, heart disease and development problems in children. The North African states of Egypt, Sudan and Morocco have the highest levels of urban air pollution on the continent (WHO, 2004b), with Cairo ranking among the world's most polluted cities (Gurjar *et al.*, 2008; see Box 5.1, above). One of the successes in recent years, however, has been the phasing out of leaded petrol in most African countries, coordinated by UNEP's Partnership for Clean Fuels Campaign. Since 2009, leaded petrol has been available only in Egypt, Tunisia and Algeria.

Increasing attention in recent years has focused on the issue of indoor air pollution, which in 2000 was estimated to be directly responsible for the deaths of 392,000 people across the continent (WHO, 2002). Indoor air pollution is a consequence of cooking and heating using solid biomass fuels, such as wood, dung and straw, without adequate ventilation. This results in the emission of a range of health-damaging pollutants, including particulates (soot), carbon monoxide, nitrous oxide, benzene, formaldehyde and carcinogens. Exposure to these pollutants increases the risk of lung cancer, pneumonia, asthma and chronic pulmonary disease (WHO, 2002). The problem is particularly marked in rural areas, where poverty prevents people gaining access to cleaner fuels and where electricity is often absent.

Estimates suggest that 79 per cent of Africa's rural population, and 51 per cent of urban dwellers, are dependent upon solid fuel for cooking or heating (the figures are 87 per cent and 56 per cent, respectively, for sub-Saharan Africa alone). There is a distinct gender dimension to the burden of ill health related to indoor air pollution, with women at greatest risk simply because they spend more time in the home preparing food and breathing in pollutants (up to seven hours per day in some areas). Moreover, women are further disadvantaged by having to collect and transport fuelwood to the home (Plate 6.2).

Plate 6.2 Women disproportionately share the burden of fuelwood collection in Ethiopia (Alan Dixon).

In addressing air pollution, most initiatives have emphasized the need to replace fuelwood with cleaner alternatives. Although solar energy, electricity and gas are more desirable, the more cost-effective and practicable alternatives for the rural poor are kerosene and charcoal. However, while it emits significantly less particulate matter, charcoal production can have a major impact on the environment in terms of deforestation, soil fertility and greenhouse-gas emissions. Alternative solutions have included the development of more efficient wood-burning stoves, installing flues in houses to improve ventilation, and educating people about indoor air pollution hazards. Simply drying fuelwood before burning can improve combustion and decrease smoke production.

6.4.3 Waste management and land degradation

One of the impacts of rapid urbanization across Africa has been the sharp increase in the consumption of resources and an attendant increase in the quantity of waste produced. For example, Lagos, with a population estimated to be in the region of 10 million, generates around 4 million tonnes of solid municipal waste each year (Kofoworola, 2007). As Onibokun and Kumuyi (1999) point out, the problem is not so much the volume of waste, but the lack of capacity among governments and municipal authorities to deal with it. As Table 6.5 shows, few people in some of Africa's major cities have waste-collection services, principally because municipal authorities are under-resourced in terms of human and financial capital (Boadi *et al.*, 2005). As a consequence, it is common to see waste accumulating on roadsides, in bodies of water, or in huge refuse mountains which constitute a health risk and degrade the quality of life for those living near by (Plate 6.3). As highlighted above, the dumping of human waste in the absence of appropriate sewerage can lead directly to outbreaks of diarrhoeal diseases, while the build-up of other types of organic waste can form breeding grounds for disease-carrying rodents and insects. For example, in a study of Monrovia, Liberia, Mensah (2006) reports that indiscriminate waste dumping (some of which occurs directly on to the beach) was directly linked to a series of cholera outbreaks during 2003.

Table 6.5 Percentage of people with access to services and facilities in some African cities, 1993

Country	*City*	*Water*	*Sewerage*	*Waste water treated*	*Per capita solid waste generation*	*Waste collection*
Côte d'Ivoire	Abidjan	62	45	58	1	70
Congo (DRC)	Kinshasa	50	3	3	1.2	0
Ethiopia	Addis Ababa	58	0	–	–	2
Ghana	Accra	46	12	0	0.4	60
Kenya	Nairobi	78	35	90	–	47
Liberia	Monrovia	1	1	0	0.6	<15
Morocco	Rabat	87	95	0	0.6	90
Nigeria	Lagos	65	2	2	1.1	8
Sudan	Khartoum	52	3	45	–	12
Tanzania	Dar es Salaam	22	6	2	1.0	25
Tunisia	Tunis	92	73	87	0.5	61
Uganda	Kampala	30	9	27	6.0	20

Sources: WRI, 1998; Mensah, 2006

Plate 6.3 Pollution of a stream with solid waste in Lilongwe, Malawi (Alan Dixon).

Development interventions to address the problem of urban waste mirror those for water and sanitation, outlined above. Building infrastructural capacity within urban areas, in particular extending the provision of waste bins and collection services, is essential, as is educating urban dwellers on the health risks of indiscriminate dumping (Parrot *et al.*, 2009). In recent years, attention has also focused on the livelihood opportunities presented through non-governmental and informal waste-collection and recycling services. In Cameroon, for example, the Centre International de Promotion de la Récupération (CIPRE) was established in 1996 with the aim of collecting, processing and recycling plastic domestic waste. Informal waste picking, scavenging and collection also present economic opportunities for the poor and marginalized in society, and these activities are becoming widespread throughout Africa (Wilson *et al.*, 2006), although there are significant associated health and safety risks (Cointreau, 2006).

Environmental health in urban areas may be further compounded by the discharge of toxic pollution from industries, particularly those involved in mining or chemical production. Industrial sites are usually located on the peri-urban fringe of major cities, with much of the workforce living within a potentially hazardous distance. Nowhere is this more evident than in the city of Kabwe in Zambia, which has been dubbed the most polluted city in Africa due to its history of lead mining and subsequent concerns over heavy-metal contamination. Environmental pollution is not, however, confined to urban areas. The use of chemical pesticides, herbicides and fertilizers is widespread in rural areas because of attempts to increase agricultural productivity, yet these agrochemicals are often used in an unregulated manner which presents risks to health. According to NEPAD (2003), over 11 million cases of poisoning in Africa each year can be attributed to exposure to harmful pesticides. DDT (dichlorodiphenyltrichloroethane), which has been used for many years as a pesticide for controlling mosquitoes and other insect vectors, is one of several chemicals classed as persistent organic pollutants (POPs), meaning they remain stable in the environment and bioaccumulate in the food chain. Research has suggested that long-term exposure can lead to male fertility problems, an impaired immune system and an increased risk of premature birth. In an examination of pesticide use among cotton growers in Zimbabwe, Maumbe and Swinton (2003) reported that most farmers exhibited symptoms of poisoning, ranging from skin irritation to stomach complaints, yet few sought medical attention. A lack of understanding of the health risks associated with pesticides (characterized by inadequate protective clothing) was attributed to illiteracy among users who failed to understand the toxicity ranking system displayed on containers. While education is clearly an important way forward in reducing exposure to toxic chemicals, attention has also focused on removing the risk altogether.

NEPAD (2003) estimates that over 50,000 tonnes of obsolete pesticide remain scattered throughout Africa, with Botswana, Mali, Morocco, Ethiopia, South Africa and Tanzania each estimated to have more than 1000 tonnes (FAO, 1999a; Haylamicheal and Dalvie, 2009). The Africa Stockpiles Programme (ASP) is one of several international initiatives aiming to dispose of obsolete pesticides safely and prevent further accumulation (ASP, 2009).

6.4.4 Climate change

One issue that seems likely to exacerbate health problems throughout Africa over the next fifty years is climate change. Indeed, despite being the least polluting continent in terms of greenhouse-gas emissions, the impacts of climate change on human health are likely to be greater in sub-Saharan Africa than in any other region on the planet (Ramin and McMichael, 2009; also see Chapter 3). According to the Third and Fourth IPCC Assessment Reports, climate change will extend the geographical range of endemic diseases and increase the frequency of disease outbreaks (Desanker and Magadza 2001; Boko *et al.*, 2007). Malaria is likely to spread into the currently malaria-free highland areas of Ethiopia, Kenya, Rwanda and Burundi, while the increased incidence of flooding in lowland areas is likely to provide new breeding grounds for insect vectors and facilitate malaria outbreaks, as it did in Mozambique in 2000 (Ahern *et al.*, 2005). Lack of genetic resistance among those populations affected will inevitably result in increased mortality. Other diseases that are likely to increase include:

- *Cholera*. Recent research suggests a link between the rise in surface sea temperatures and flooding and the frequency of cholera outbreaks (Colwell, 1996; Mendelsohn and Dawson, 2007).
- *Plague*. Climate variability may also lead to the spread of plague as the rodent population soars and seeks refuge in human dwellings during drought periods.
- *Meningitis*. The range of meningitis is likely to spread beyond existing semi-arid areas of West and central Africa into East Africa.
- *Trypanosomiasis*. Environmental change will increase the breeding range for tse-tse flies and hence the spread of trypanosomiasis.
- *Diarrhoeal diseases*. Flooding will exacerbate sanitation problems and increase the frequency of polluted water supplies.

In addition, water stress is projected to affect between 75 million and 250 million people by 2020 (Boko *et al.*, 2007); hence, it is likely that more people will suffer malnutrition as food production declines as a consequence. Similarly, the predicted increase in flooding in many areas will also lead to crop destruction, food shortages and malnutrition.

6.5 Communicable diseases

As highlighted above and in Chapter 3, Africa's diverse environments are conducive to the spread of infectious, communicable diseases, many of which are entirely preventable through the provision of medicine, vaccinations and environmental health. According to the WHO (2008a), around 6 million people in the African region die each year from communicable diseases, with HIV/AIDS, tuberculosis and malaria accounting for 2.8 million of those deaths (24 per cent of all deaths in Africa in 2008). The implications of these statistics for development are wide ranging in terms of healthy and productive lives, the burden on families, and the capacity of governments to meet healthcare demands. It is therefore unsurprising that the burden of these three diseases was identified and targeted in Millennium Development Goal 6 (see Box 6.1).

6.5.1 HIV/AIDS

In 2007, an estimated 22 million people in sub-Saharan Africa were living with HIV/AIDS, 1.9 million new infections occurred, and 1.5 million people died (UNAIDS, 2008). HIV/AIDS constitutes the most common cause of death and illness in the continent, and hence has become a major social, economic and development issue. AIDS (acquired immune deficiency syndrome), which is caused by the human immunodeficiency virus (HIV), is believed to have originated in central Africa during the 1930s, with the first recorded symptomatic outbreak occurring in Kinshasa (DRC) during the 1970s, before reaching epidemic proportions throughout sub-Saharan Africa during the 1980s. The disease, which is characterized by the destruction of the immune system caused by HIV infection, renders people vulnerable to a range of opportunistic infections that lead progressively to death; tuberculosis has been the most common of these throughout Africa. Transmission of HIV can occur via sexual contact with an infected person, through injection or transfusion with contaminated blood, through the use of contaminated injecting equipment, or from mother to child during pregnancy, birth or breastfeeding. In Africa, however, the major mode of transmission continues to be unprotected heterosexual intercourse (UNAIDS, 2008).

Much research has been undertaken at a variety of different levels on the reasons for the dramatic spread and prevalence of HIV/AIDS throughout the continent, with social, cultural, economic and political factors all playing a role. However, there are several overarching issues. First, the disease has a very long incubation period, so that those who are infected may not be aware that they have the disease for many years. Moreover, the disease has no symptoms before the development of opportunistic infections (which can also occur in the absence of HIV/AIDS), so diagnosis is problematic. Cultural factors include the acceptance within some societies that men can continue to have multiple sexual partners throughout life, especially younger women. However, sexual behaviour also remains a taboo subject, with many people reluctant to discuss it openly, or to undergo testing for HIV for fear of being stigmatized by their family and community. In terms of socio-economic factors, some have argued that poverty itself has been a key driver in the spread of the disease because those living in poverty are more likely to take risks with their sexual health. Poverty drives seasonal labour migration in many areas, which places men, who may be away from their families for long periods, at increased risk of infection from sex workers (Khan *et al.*, 2007). Similarly, the persistence of conflict throughout Africa and the rise in sexual violence that most often characterizes conflict periods have played important roles. Finally, the inadequate response of governments has been identified as a contributing factor to the spread of the disease. During the early stages of the epidemic, in the 1980s and 1990s, many governments either refused to acknowledge its existence or felt uncomfortable prescribing appropriate sexual behaviour in the face of pressure from religious groups (Iyiani *et al.*, 2010).

The nature of the epidemic varies between countries and over time. As highlighted in Figure 6.7, HIV prevalence is very low throughout North Africa, where transmission via contaminated injecting equipment predominates. In contrast, it is much greater in the countries of southern Africa, where approximately 38 per cent of all worldwide AIDS deaths occurred in 2007 (although this may also be due to the prevalence of better monitoring and reporting in this region). Among these countries, Swaziland and Botswana have the highest rates of infection at over 25 per cent and 22 per cent of the adult population, respectively (WHO, 2008c, 2008d), while South Africa has the highest number of people living with HIV in any single country (5.7 million). However, recent data from the UN (UNAIDS, 2008) suggest that in many countries the infection rates have either stabilized or are falling. In Zimbabwe, for example, HIV prevalence among pregnant women declined from 26 per cent to 18 per cent between 2002 and 2006, as did the prevalence

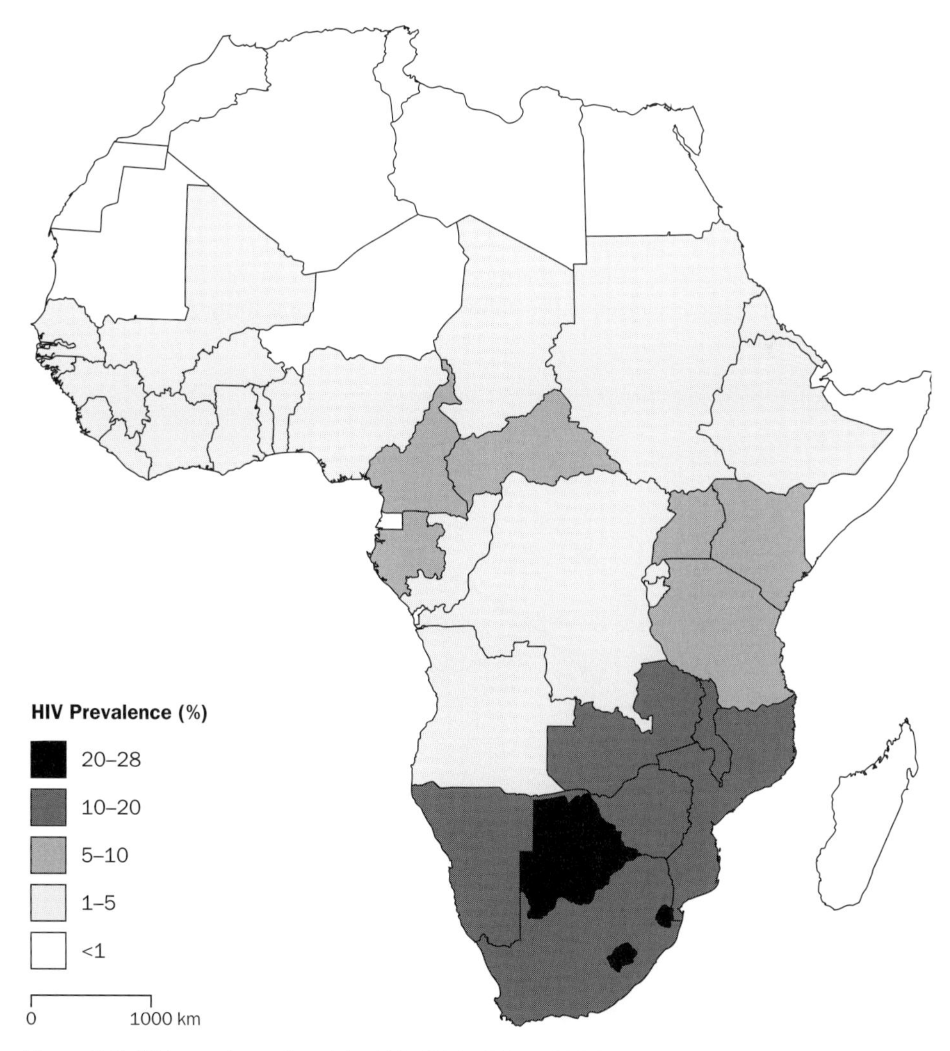

Figure 6.7 HIV prevalence in adults (15–49) in Africa, 2007.

Source: UNAIDS, 2008

among fifteen–nineteen-year-olds in Botswana. Similarly, in East and West Africa there are indications of a levelling out or decline in prevalence. It should be recognized, however, that despite these encouraging signs, the number of people living with HIV/AIDS has continued to rise, due to new infections and the positive effects of antiretroviral (ARV) drugs (see Figure 6.8).

There is a distinct gender dimension to the HIV/AIDS epidemic in Africa, in that women are disproportionately affected in comparison to men (UNAIDS, 2008). Recent estimates suggest that 61 per cent of those living with HIV/AIDS are women, and those in the fifteen–twenty-four age group are up to three times more likely to be HIV positive than men (UNAIDS, 2008; World Bank, 2008d; Figure 6.9) . This vulnerability has been explained in terms of persistent gender inequalities and the subordination of women in society (Marcus, 1993; Baden and Wach, 1998; Buvé *et al.*, 2002). Many cultures demand that women have little sexual experience before marriage, hence they lack the knowledge and power to negotiate safe sex. Similarly, because of their entrenched dependence on men, women are less likely to have access to reproductive healthcare, are more likely to experience sexual violence, and are more vulnerable to HIV infection through sex work. High HIV prevalence among young women, in particular, has affected rates of mother to child transmission of the disease. In 2007, an estimated 1.8 million children were living with HIV, while 11.6 million had been orphaned as a result of deaths from AIDS (UNAIDS, 2008).

The impact of HIV/AIDS has been devastating on many different levels, to the extent that it has become a major development issue. Those who are infected have a short life expectancy in the absence of medical treatment and adequate nutrition. As the immune system deteriorates, the body becomes more vulnerable to illnesses ranging from recurring respiratory infections, chronic diarrhoea and fever to the more serious pulmonary tuberculosis and various cancers. Of course, this is debilitating for those infected, but it also

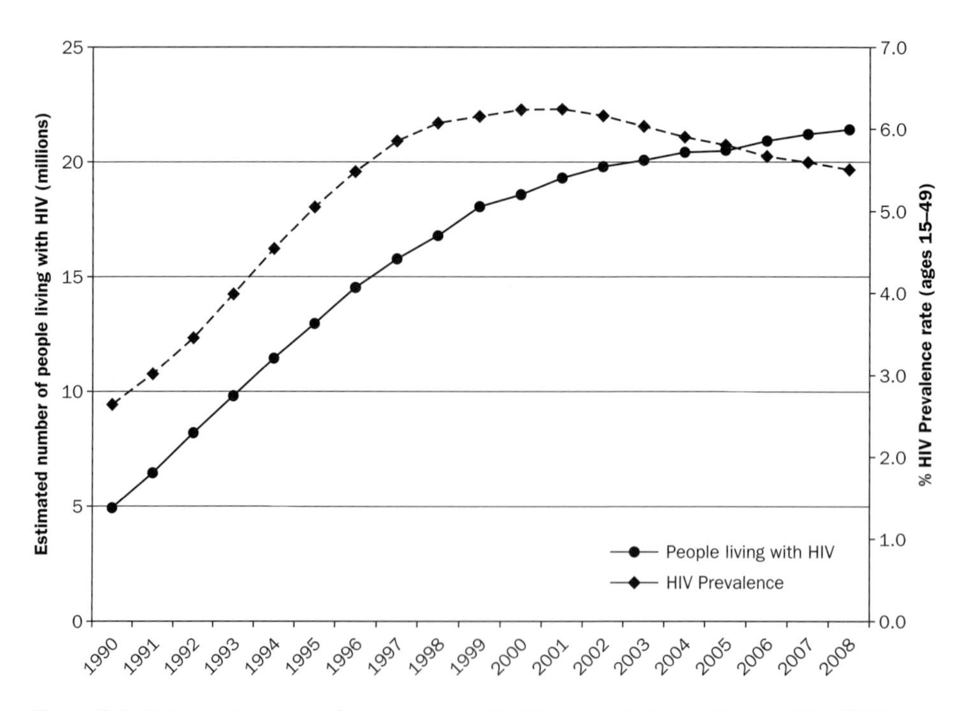

Figure 6.8 Estimated number of people living with HIV in sub-Saharan Africa, 1990–2007.

Source: UNAIDS, 2008

places great pressure on the household, livelihood security, health services and the economy at both local and national levels (Stokes, 2003). One immediate impact at the household level is a shortage of labour (through either death or ill health), which inevitably results in a decline in household income, and debt – a situation made worse by medical bills and/or funeral expenses. Labour shortages in rural areas have an impact on agricultural productivity generally, which can mean less food in local markets and wider food security issues. Despite research suggesting that social capital has been strengthened within many

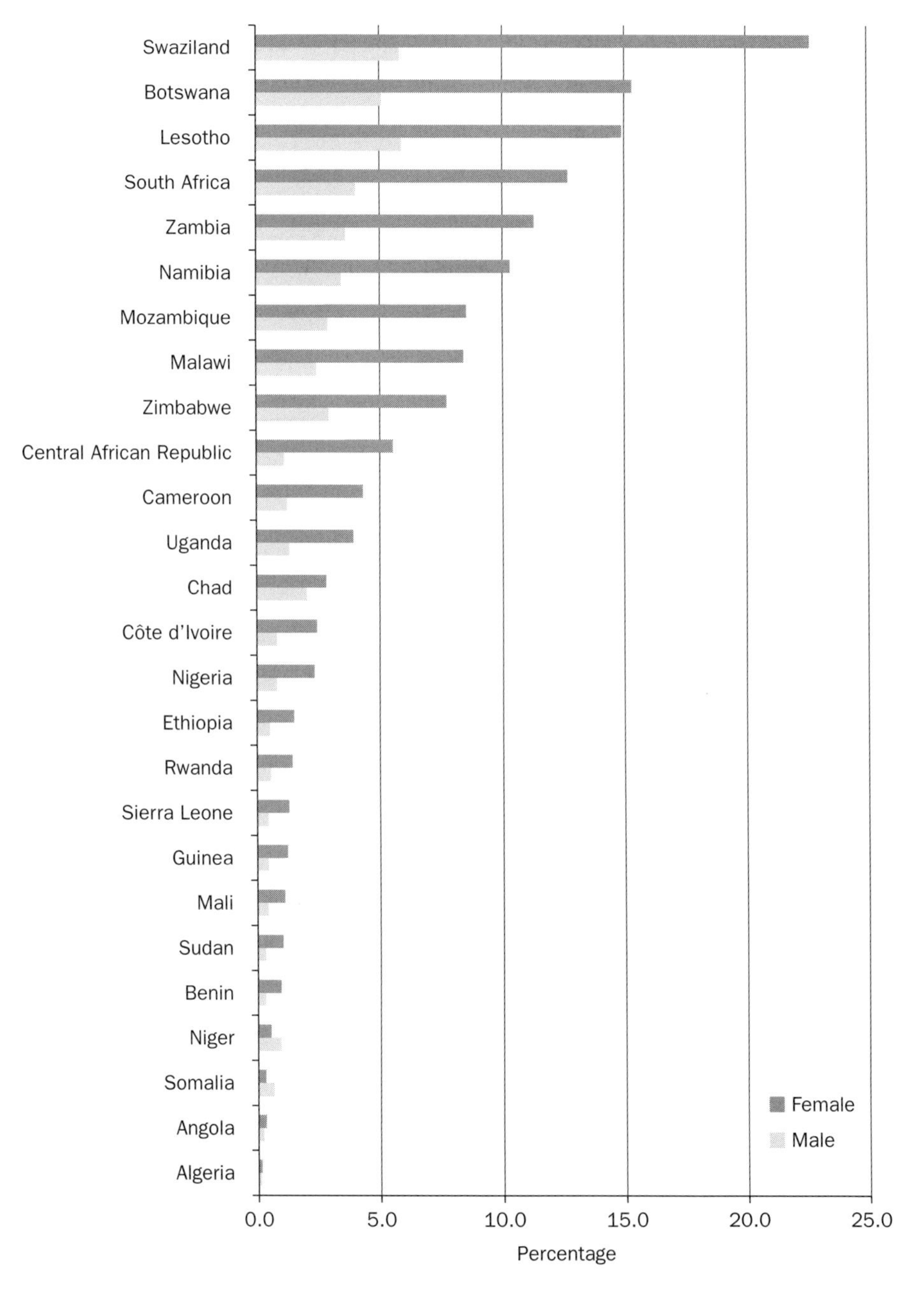

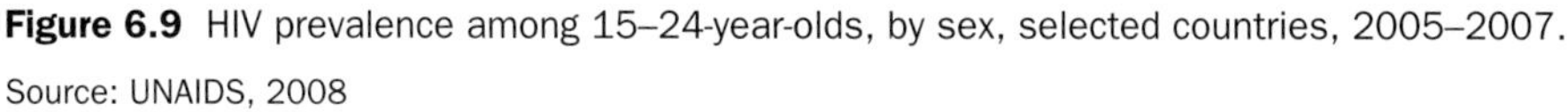

Figure 6.9 HIV prevalence among 15–24-year-olds, by sex, selected countries, 2005–2007.
Source: UNAIDS, 2008

communities as a result of the HIV epidemic, the burden of the sick and orphaned children on communities means that time is diverted from more productive activities. The loss of skilled people with specialist knowledge also influences the functioning and resilience of local communities, and even government departments. For example, the time lost through attendance at funerals has been identified as a major factor influencing the effectiveness of agricultural extension services (Qamar, 2001).

Addressing the HIV/AIDS crisis in Africa is a huge challenge on many levels, not least because of the lack of infrastructure and finance needed to promote and develop prevention, treatment and mitigation strategies (see Plate 6.4). However, significant progress has been made in slowing the spread of the disease in recent years due to the commitment and involvement of governments, international organizations, NGOs and people themselves

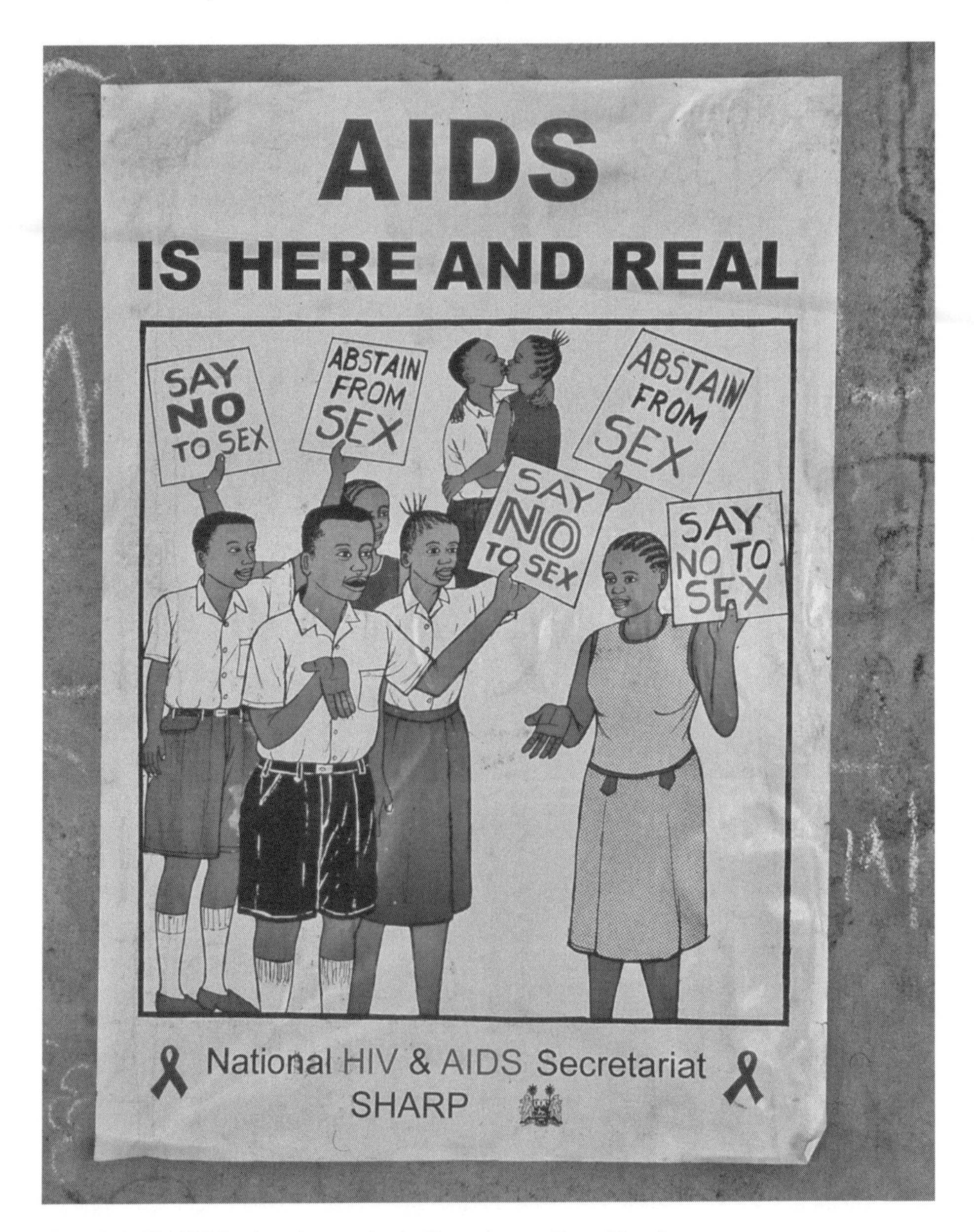

Plate 6.4 HIV/AIDS education poster in Sierra Leone (Tony Binns).

(Iyiani *et al.*, in press). HIV prevention strategies have focused heavily on supporting voluntary testing and encouraging people to alter their sexual behaviour by avoiding sexual contact at an early age, reducing their number of casual relationships and using condoms. In countries such as Uganda and Senegal, early action on HIV prevention has been viewed as the principal cause of the decline in HIV prevalence since the 1990s. It is also important, however, to acknowledge and address the underlying societal causes of HIV infection and risk, in particular poverty and gender inequalities. Increasing women's economic, socio-cultural and political independence has been widely regarded as a fundamental step towards giving women more control over their sexual relationships, and thereby reducing their vulnerability to HIV (Pettifor *et al.*, 2004; UNAIDS, 2008). This is being addressed through strategies targeting universal education for girls, income-generation strategies for women, and political and legal reforms.

There has been significant progress in the treatment of HIV/AIDS during the last decade, especially with respect to the availability of antiretroviral (ARV) drugs. These drugs prevent the HIV virus multiplying to levels which compromise a person's immune system, hence their use increases resistance to the disease and prolongs life. They are also highly effective in preventing transmission of the virus from mother to child during pregnancy and birth. Although ARVs have been used in the developed world since the mid-1990s, their high cost (around $10,000 per person per year) rendered them prohibitive for the majority of people in Africa. However, after sustained pressure on the big pharmaceutical companies from governments, activists and international organizations, including UNAIDS and the WHO, the cost of treatment has fallen dramatically to around $300 per person per year in recent years, not least because of the licensing of cheaper generic versions of the ARVs produced by companies in India and South Africa (WHO, 2006a). In 2006, the UN called for 'universal access to comprehensive prevention programmes, treatment, care and support by 2010' (UN, 2006), and throughout Africa progress has been encouraging. In Botswana, for example, approximately 79 per cent of the population with advanced HIV were receiving ARVs in 2007 (see Figure 6.10), which constitutes an 80 per cent increase in coverage when

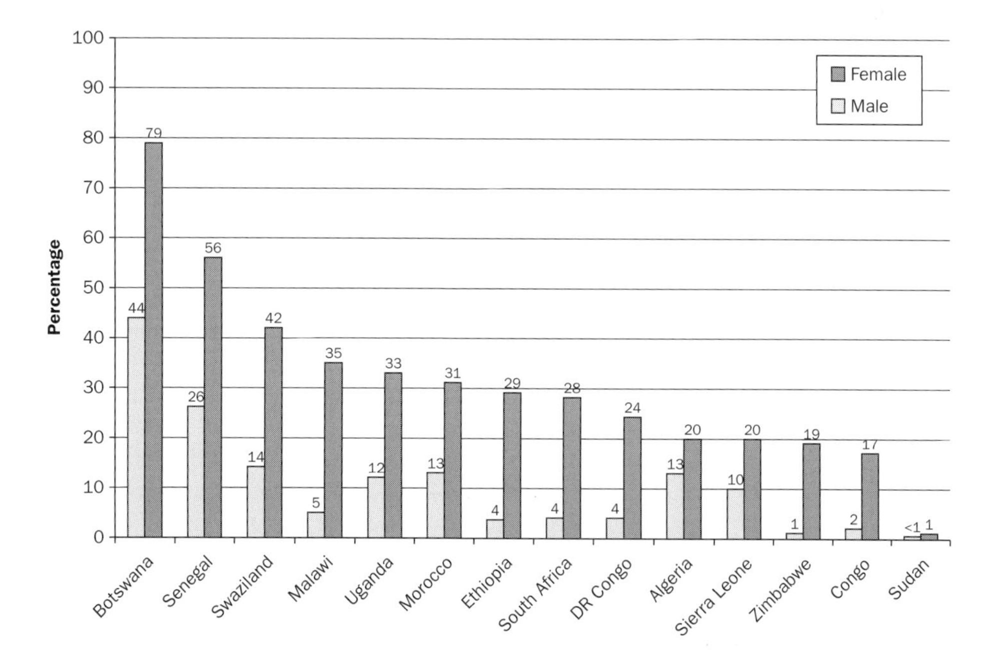

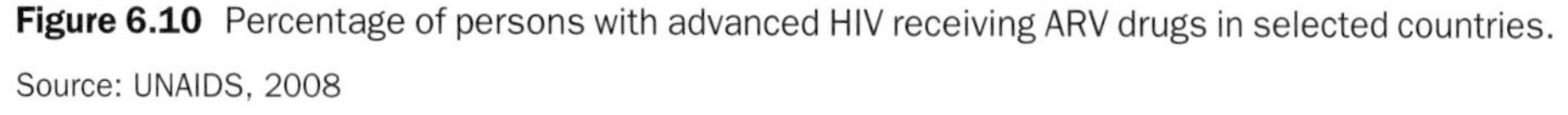
Figure 6.10 Percentage of persons with advanced HIV receiving ARV drugs in selected countries.

Source: UNAIDS, 2008

compared with 2004. However, access to ARVs remains low in many countries, where often fewer than a third of those with advanced HIV have access. Moreover, few countries are likely to meet the 2015 targets as set out in the MDGs. A case study of HIV/AIDS in Uganda is presented in Box 6.3.

Box 6.3

HIV and AIDS in Uganda: a success story?

The occurrence of an HIV/AIDS epidemic in Uganda was first identified during the 1980s, with the first case of 'slim disease' being reported in 1982. Since then, the disease has gone on to claim the lives of over a million people, and it has had a significant impact on the life expectancy of the population, the agricultural work-force, and the health and education sectors. Another lasting legacy of the epidemic has been over a million orphaned children (Avert, 2010b). However, in the fight against HIV/AIDS, Uganda is repeatedly cited as a success story. During the 1980s the Ugandan government was quick to recognize the devastating health impacts of the disease and its potential to cripple social and economic development in the country. With prevalence rates reaching 29 per cent in urban areas, the government implemented a high-profile HIV/AIDS control programme in 1987, promoting family values and faithfulness with slogans such as 'love faithfully' and 'zero grazing' (Allen and Heald, 2004). Interestingly, the use of condoms did not feature prominently in publicity material from this era, due, allegedly, to President Museveni's own beliefs, although by the mid-1990s the ABC (abstain, be faithful, use a condom) approach was being widely promoted. Another key element of the government's approach to tackling HIV/AIDS was its decentralized and cross-sectoral approach to HIV/AIDS planning, and particularly the empowerment of the NGO community. The AIDS support organization TASO (http://www.tasouganda.org) started as a community social support group providing family assistance and counselling before going on to become the largest indigenous HIV/AIDS NGO in Africa.

During the 1990s, and what would later be termed the 'second phase' of Uganda's HIV epidemic, HIV prevalence rates began to fall dramatically from a peak of 15 per cent in 1991 to 5 per cent in 2001. This decline in prevalence was directly attributed to a successful awareness raising and prevention campaign that effectively changed the sexual behaviour of much of the population. Cohen (2003) reports that between 1988 and 1995 fewer Ugandans were having sex at younger ages, levels of monogamy increased significantly, and condom use rose steeply among unmarried, sexually active men and women. Campaigns that have included ABC education, free condoms, testing, counselling and more recently the provision of free antiretroviral drugs continued to be implemented throughout the 2000s with the aid of bilateral and multilateral financial assistance. Since 2000, HIV prevalence rates have stabilized at around 6–7 per cent, and Uganda's relative success in combating HIV/AIDS has been celebrated widely in the discourse of development.

A recent report from the Ugandan government, however, suggests that infection rates have begun to increase in recent years (over 100,000 new infections occurred in 2008, with fewer than 70,000 AIDS-related deaths), and that there has been some reversal in the uptake and practice of preventative sexual health among the general population (Government of Uganda, 2010). The reasons for this apparent reversal are yet to be fully understood, but some have suggested an element of 'AIDS fatigue' and complacency among both the population and, to some extent, donors (Mullan, 2008). The fact that Uganda has been so successful in its HIV/AIDS care, prevention and treatment has meant that other concerns now take precedence among the population. Another view is that the epidemiological data on the epidemic have been open to manipulation and misinterpretation, and that high rates of infection and AIDS-related deaths have been under-reported. In an article for *The Lancet* medical journal in 2002, Justin Parkhurst (2002: 80) argues:

> Today the country can be praised for the fact that the adult HIV-1 prevalence rate is most likely much lower than 10 per cent, and continues to decline, but still there remains no evidence that the nation ever saw a 30 per cent fall in prevalence rates. That misinterpretations such as this one have endured could be attributable to various pressures that might exist, particularly in low-income and middle-income countries, which allow success stories of this kind to go unchallenged. Specifically, the notion of donor fatigue . . . combined with an overall reduction in development funds available to Africa . . . can produce political pressure to present an image of success to maintain funds.

Mullan (2008) similarly cites field evidence that suggests a 12 per cent rather than 6 per cent prevalence rate, and goes on to argue that to ignore the reality of the situation and dilute the HIV/AIDS message would be a 'travesty'. Clearly, the immediate and future challenge for Uganda is to build upon its past success and guard against complacency.

6.5.2 Tuberculosis

Until the late 1990s, the incidence of tuberculosis (commonly referred to as TB) was declining, yet in 2005 the World Health Organization declared it an emergency within Africa, stating that the epidemic had reached 'unprecedented proportions' (WHO, 2005c). In that same year, over 2.5 million new cases of TB and over half a million deaths were recorded. TB is a highly infectious disease of the lungs, transmitted from one person to another via airborne droplets, and in its active form causes chronic cough, fever, weight loss and death in approximately half of untreated cases (Dye *et al.*, 2006). Infection does not necessarily lead to disease, since it can remain in a latent form for many years, but active tuberculosis can develop rapidly in those whose immune systems have been weakened, for instance through HIV. Indeed, this relationship with HIV has accounted for the resurgence of TB in recent years: 79 per cent of all TB cases in Africa during 2007 were also HIV positive (WHO, 2009b).

The challenges for controlling the disease include increasing the rate of detection, particularly among young children – the most vulnerable group, yet the most difficult to diagnose. Since 1995, the WHO's DOTS (Directly Observed Treatment, Short-course) Strategy has been widely implemented throughout Africa (WHO, 2006b). The five components of DOTS are:

- *Political commitment with increased and sustained financing* – the creation of national and international partnerships, national legislation, and securing adequate funding from a variety of sources.
- *Case detection through quality-assured bacteriology* – investing in a laboratory network which can improve bacteriological diagnosis.
- *Standardized treatment with supervision and patient support* – establishing consistent, standardized and accessible treatment for all TB cases, and ensuring patients take their drugs to reduce the likelihood of drug resistance.
- *An effective drug supply and monitoring system* – the sustained provision of free, quality-assured TB drugs.
- *Monitoring and evaluation system and impact measurement* – establishing reliable and accurate monitoring and reporting systems.

The approach has achieved some success throughout Africa in terms of the treatment of TB, although it has reportedly been less successful in detecting new cases of the disease (Dye *et al.*, 2006; Obermeyer *et al.*, 2008). Moreover, there has been some concern that DOTS has struggled to make an impact on the disease due to TB–HIV co-infection. Corbett *et al.* (2006) cite the case of the DOTS Strategy implemented within South Africa's gold-mining industry, where, despite rigid monitoring and treatment regimes, cases of TB increased dramatically in line with HIV infection throughout the 1990s. The Stop TB Strategy, launched by the WHO in 2006, takes the DOTS approach further, and has set ambitious objectives aligned to MDG 6. These include achieving universal access to diagnosis and treatment, reducing the socio-economic burden of TB, protecting the poor and vulnerable from TB, and supporting the development of new tools to combat the disease (WHO, 2006b).

6.5.3 Malaria

Malaria is a parasitic infection transmitted to humans through the bites of infected female *Anopheles* mosquitoes, which breed close to stagnant or slow-moving water. Once in the bloodstream, the malaria parasite can cause a high fever and serious damage to the nervous system, liver and kidneys. Of the four types of malaria affecting humans, *Plasmodium falciparum* is responsible for most cases and morbidity in Africa. It is more common during the rainy season, when farmers are busy producing the household's food. Mosquito larvae can be killed by draining swamps and spraying pools with insecticides, such as DDT. Sprays may also be used inside dwellings, and insecticide-impregnated mosquito nets over beds provide important protection. However, mosquitoes are becoming resistant to certain chemicals and anti-malarial drugs (e.g. chloroquine), and the incidence of the disease has actually increased in recent years. Mosquitoes also transmit dengue fever and yellow fever.

It is estimated that there are over 300 million acute cases of malaria across the world each year, resulting in over a million deaths, some 90 per cent of which are in Africa, particularly among young children (RBMP, 2008). Of the thirty-five countries that account for some 98 per cent of malaria deaths globally, thirty are located in sub-Saharan Africa. Whereas the incidence of malaria is negligible in North African countries, many sub-Saharan countries have serious problems with the disease. In the West African countries of Niger and Mali, an estimated 469 and 454 deaths per year per 100,000 of the population, respectively, were due to malaria in the period between 2000 and 2005. By contrast, in Togo, where 54 per cent of children under the age of five were sleeping under insecticide-treated bednets, malaria deaths were relatively low, at 47 per 100,000 of the population (World Bank, 2008e). It does seem that richer countries have less malaria, and poorer countries more. Malaria is both a disease of poverty and a cause of poverty. It has been estimated that it is responsible for an economic growth penalty of up to 1.3 per cent annually in some African countries, leading to a decline in per capita GDP in many sub-Saharan African countries (RBMP, 2008).

While malaria has received significant publicity in recent years, some 800,000 African children under the age of five still die from the disease each year. Many of the newer drugs, which are more effective in preventing and combating the disease, are unfortunately much more expensive than the older (and now less effective) drugs, which in some cases has meant that drugs are less widely available among poor communities.

There have been a number of important initiatives in the last decade to bring malaria under control, particularly in sub-Saharan Africa. A major boost to the coordinated control and wider publicity of malaria was the establishment in 1998 of the Roll Back Malaria Partnership (RBMP) by the WHO, the United Nations Children's Fund (UNICEF), the United Nations Development Programme (UNDP) and the World Bank. The RBMP is working towards meeting the malaria target of MDG 6 by reducing all preventable deaths to near zero by 2015, and aims to 'enable sustained delivery and use of the most effective prevention and treatment for those affected most by malaria by promoting increased investment in health systems and incorporation of malaria control into all relevant multi-sector activities' (RBMP, 2008). In April 2000, the meeting of African heads of state in Nigeria signed the Abuja Declaration to tackle malaria, since when some twenty African states have reduced or abolished taxes on insecticide-treated nets to increase their affordability. Over half of the malaria-endemic African states have also established strategic plans to achieve the goals set by RBMP and at Abuja.

Africa's best hope of eradicating malaria lies in the development of a relatively cheap and widely accessible vaccine. A number of organizations are engaged in research and advocacy concerning such a vaccine. For example, the African Malaria Network Trust (AMANET) was established as the African Malaria Vaccine Testing Network (AMVTN) in 1995, primarily to plan and conduct vaccine trials. Similarly, the European Malaria Vaccine Initiative (EMVI) was established in 1998, funded mainly by the European Commission and the national governments of the Netherlands, Norway, Ireland, Sweden and Denmark. EMVI aims to accelerate the development of experimental malaria vaccines in Europe and developing countries (EMVI, 2008).The Bill and Melinda Gates Foundation has played an important role in funding the PATH Malaria Vaccine (MVI) initiative, set up in 1999 to research into possible vaccines and ensure their availability and accessibility in poor countries (MVI, 2008). The Malaria Vaccine Technology Road Map was launched on 4 December 2006 at the Global Vaccine Research Forum in Bangkok, and is funded jointly by the Gates Foundation, the PATH MVI and the Wellcome Trust. It calls for joint action among the world's leading international health organizations to accelerate the development and licensing of an effective malaria vaccine by 2025. This would have a protective efficacy of more than 80 per cent against the disease and would provide protection for over four years. PATH MVI and the Wellcome Trust are already engaged in various clinical trials in UK and in The Gambia.

Many deaths in Africa are caused by the combination of malaria and other health problems, such as diarrhoea, measles, HIV/AIDS and tuberculosis. Pregnant women who have both HIV/AIDS and malaria are at very high risk of developing anaemia and malarial infection of the placenta. This can lead to babies having low birth weights and higher rates of infant mortality. As a WHO report has shown, adult men and non-pregnant women with HIV/AIDS may have a greater chance of contracting severe malaria, especially those with advanced immuno-suppression. WHO (2004c) argues that prevention and treatment programmes of the two diseases should be better coordinated so that they mutually reinforce each other.

6.5.4 Yellow fever

Yellow fever is also transmitted by mosquitoes, but this time the *Aedes* species. Some thirty-three African countries and over 500 million people are at risk. In the past, large epidemics have led to fatality rates as high as 85 per cent, but at present there are some 200,000 estimated cases of yellow fever annually, with about 30,000 reported deaths (although under-reporting is common). The number of cases has risen in recent years, even though an effective vaccine that provides at least ten years' protection has been available for sixty years. The symptoms of the disease are initially fever, headache, loss of appetite and jaundice (hence *yellow* fever), and later internal haemorrhaging, kidney failure and coma. Yellow fever is now the only disease which requires an international certificate of vaccination to travel into and out of infected areas.

6.5.5 Other infectious diseases

Measles is a highly infectious disease that in 2002 caused the death of over a third of a million children and adolescents in Africa. Transmission occurs via water droplets in the air, and its relatively mild symptoms of fever and cough can lead to potentially fatal complications that include pneumonia, diarrhoea and encephalitis. In recent years, however, the disease has declined as a result of comprehensive immunization programmes supported by the Measles Initiative and other organizations. As well as being relatively cheap ($1 per person), the measles vaccine is highly effective: over 95 per cent of children who receive it develop immunity for life. In 2008, the WHO reported that deaths from measles across Africa had fallen 89 per cent between 2000 and 2007 – to an estimated 45,000. As of 2007, outbreaks of the disease were limited mainly to Ethiopia, Nigeria, Niger and the Democratic Republic of the Congo (WHO, 2008e).

It is worth mentioning that some diseases that were once widespread throughout Africa have now been brought under control due to sustained improvements in reporting, monitoring and treatment. Smallpox once claimed the lives of thousands of people each year, until it was eradicated by a global vaccination programme during the 1970s. More recently, there has been similar success in efforts to eradicate poliomyelitis, a disease that used to affect thousands of children under the age of five, causing paralysis. Through the efforts of the Global Polio Eradication Initiative, launched in 1988, the number of cases has declined dramatically: in Africa in 2009, polio remained endemic only in Nigeria, where 321 cases were recorded (GPEI, 2009).

6.6 Other health issues

Although attention has tended to focus on the impact and burden of communicable infectious diseases in Africa, closer inspection of Table 6.1 reveals a significant number of non-communicable causes of death. Some diseases which in the past have been associated only with the developed world are becoming significant public health problems across Africa as a result of behavioural change linked to development, globalization and urbanization. Among these, the rise of cardiovascular disease, including coronary heart disease, cerebrovascular disease and hypertension, is of particular concern. Data for 2002 suggest that 1 million people died from these diseases in Africa, with Egypt (210,000), Nigeria (202,000) and Ethiopia (101,000) accounting for the most cases. World Health Organization estimates (WHO, 2011) suggest that this figure increased to around 1.25 million in 2008 (11.5 per cent of all deaths). The causes of cardiovascular diseases are diverse, but they are usually associated with an unhealthy lifestyle characterized by the consumption of fatty and salty foods, smoking tobacco, and a lack of adequate exercise. Countries affected are usually those that have made progress along the so-called 'epidemiological transition' (Omran, 1971), in which improvements in public health have led to increased life expectancies and changing patterns of consumption. In Africa, however, there is significant spatial diversity in the extent to which places have undergone this transition, although urban and peri-urban areas are particularly affected (Mensah, 2008).

The pattern for cancer, the second leading cause of illness due to non-communicable disease, is similar. Cancer caused the deaths of half a million people in Africa in 2002, with the highest rates again reported in Egypt (39,000), Nigeria (79,000) and Ethiopia (39,000). Growth of the disease is linked to lifestyle factors, particularly the use of tobacco, but it also reflects HIV prevalence.

Dealing with cardiovascular diseases and cancer alongside infectious diseases has been described as a 'double burden' for Africa (Mbewu and Mbanya, 2006), and it constitutes a major emerging challenge for the continent's healthcare systems.

Similarly, personal injuries are an under-reported yet major cause of death and poor health throughout Africa. An estimated 1 million people died through injury, either intentional or unintentional, in 2008, with road traffic accidents accounting for approximately 29 per cent of the total (see Figure 6.11). Growth in the number of vehicles, combined with a poor-quality road network and the lack of road safety laws, has made road traffic injuries a major public health concern. Africa has the highest rate of road traffic fatalities in the world, with an average rate of 39.3 per 100,000 population (although the total number of recorded fatalities is less than in other regions) (WHO, 2009c). In the majority of cases, pedestrians (particularly young children) are more vulnerable than motorists, on account of having to share the road with often unsafe and/or unlicensed vehicles. Crashes

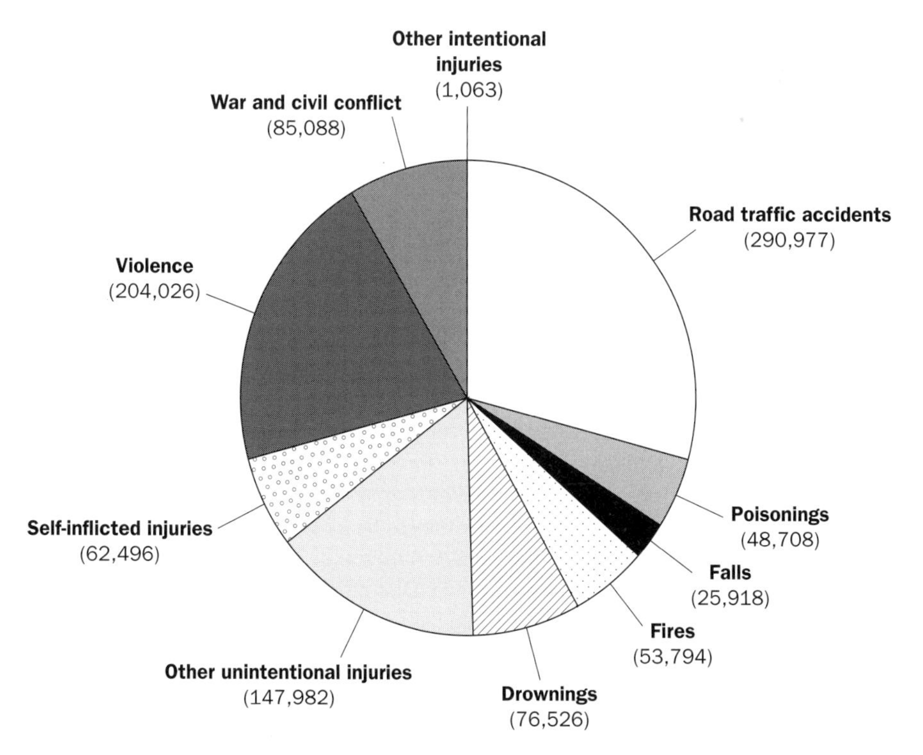

Figure 6.11 Deaths through injury in Africa, 2008.

Source: WHO, 2008a

involving poor-quality, unregulated minibuses are so common in Nigeria that the vehicles are known locally as *danfo* – which translates as 'flying coffins'.

It is worth noting that war and violence also account for almost one-third of all deaths through injury (see Figure 6.11). As we shall see in Chapter 7, conflict has been endemic to Africa virtually every year since the Second World War, and Africa as a region has the highest rate of deaths and injuries as a result of warfare in the world (Bowman *et al.*, 2006). Of particular concern is the trend throughout West and central Africa towards mobilizing child soldiers, many of whom will subsequently suffer a lifetime of poor physical and mental health (Pearn, 2003).

Finally, there is increasing recognition of mental illness as a key health issue throughout Africa. Persistent poverty, livelihood insecurity, ill health and the wider impacts of HIV/AIDS on families and communities have all been regarded as contributing to the rise of common mental disorders, such as depression and anxiety (Baingana *et al.*, 2006; WHO, 2006a). Moreover, armed conflict has resulted in significant numbers of people, particularly women and children, living with post-traumatic stress disorder. In addition to a range of other mental disorders, including epilepsy and schizophrenia, Baingana *et al.* (2006) suggest that alcohol abuse, among both adults and children, is emerging as a major health issue. Similarly, in East Africa, the use of the psychoactive plant khat has become increasingly common (see Box 6.4). Addressing the stigmatization and subsequent marginalization of those suffering with mental disorders remains one of the first priorities for mental health services across the continent.

Box 6.4

Khat-chewing in the Horn of Africa: a mental and social health issue

Khat is the common name for the small evergreen shrub *Catha edulis*, which is believed to be native to either Ethiopia or Yemen, but is now cultivated extensively as a cash crop throughout East Africa. The significance of the plant lies in its stimulant properties: the young leaves contain relatively high concentrations of cathinone, an amphetamine-like chemical which, if consumed in sufficient quantities, can generate feelings of heightened concentration, excitement and a state of euphoria. Traditionally, khat has been cultivated in Ethiopia and Yemen, where it has been used as a medicinal plant and stimulant probably since the twelfth century, particularly among Muslim communities (Al-Hebshi and Skaug, 2005). However, the cultivation and use of khat have expanded significantly throughout the Horn of Africa in recent years as coffee prices have declined. In Ethiopia and Somalia, in particular, it has now become part of the social life of all segments of the population (Feyissa and Aune, 2003). Khat production has also been fuelled by a buoyant export market: estimates suggest that Ethiopia earned around $209 million in 2010 from its export of khat to such places as the UK, the USA, Canada, Djibouti and Somalia (Tradearabia, 2010). While the increase in popularity of khat has undoubtedly produced economic benefits for those involved in its production, many have started to question its wider impact in terms of its effects on the health of individuals and society.

Khat-chewing typically occurs among groups of males who will consume the leaves usually over a four-hour period (see Plate 6.5). Although experiences vary

Plate 6.5 A man chewing khat in western Ethiopia (Alan Dixon).

depending on the quality and strength of the khat, it is normal for participants to experience a period of heightened stimulation followed by the onset of depression. The reported effects of long-term use, however, include depression, insomnia, gastro-intestinal problems, dental problems, paranoia and, in some cases, psychosis (Houghton, 2004; Al-Hebshi and Skaug, 2005). There are, therefore, clear physical and mental health issues for those who are psychologically addicted to the drug, and these are compounded by the fact that frequent users may struggle to sustain employment or maintain normal family relationships. Consequently, it is common for frequent users to 'drop out' of society, leading to them becoming entrenched in poverty and marginalized by their communities. Increasingly, even those who use khat on an infrequent basis are facing discrimination. For example, Feyissa and Aune (2003) report that non-khat-using labourers in Ethiopia are paid significantly more than those who admit to using the drug.

For Ethiopia, khat represents a significant health challenge for the future. It is a lucrative crop, so ever more farmers are converting their farmland to its production, and export earnings from the drug are a critical (and increasing) source of revenue for the Ethiopian government. At the same time, however, ever more people, particularly the young, see khat as a means of temporarily escaping poverty and underdevelopment. This is engendering a moral panic in some sections of Ethiopian society, yet the authorities seem to be doing little to address the problem.

6.7 Health systems

Much of this chapter has drawn attention to the diverse range of health issues facing people in Africa today, and while some of the determinants of these issues are societal, behavioural and environmental in nature, it is impossible to ignore the fact that they are also products of weak national health systems. Indeed, the African Union's *Africa Health Strategy* (African Union, 2007: 4), argues that the high burden of disease in Africa continues because:

- Health systems are too weak and services too under-resourced to support targeted reduction in disease and achieve universal access.
- Health interventions do not match the scale of the problem.
- People are not sufficiently empowered to improve their health, nor adequately involved, while cultural factors play a role in health-seeking behaviour.
- The benefits of health services do not equitably reach those with the greatest disease burden.
- There is widespread poverty, marginalization and displacement on the continent.
- There is insufficient action on the inter-sectoral factors impacting on health.
- Environmental factors and degradation are not sufficiently addressed.

The strategy goes on to argue that, 'A vicious cycle remains in which poverty and its determinants drive up the burden of disease, while ill health contributes to poverty. Investment in health could therefore contribute to economic development' (African Union, 2007: 4).

Clearly, although there is widespread recognition of the linkages between other 'development issues' and health, there is a fundamental need to address Africa's health systems; to understand why, after decades of investment and reform, they continue to be weak and under-resourced; and to identify ways in which health systems can be strengthened for the future.

In 2008, representatives from national governments and international organizations, including the WHO, the United Nations Programme on HIV/AIDS, the World Bank and the African Development Bank, met to discuss a strategic approach to achieving the health-related Millennium Development Goals. The meeting resulted in the adoption of the Ouagadougou Declaration on Primary Health Care and Health Systems in Africa, which was signed by all African states (WHO, 2008f). The declaration acknowledged that the challenges facing health systems are diverse and range from a lack of financing, infrastructure, medicines and human resources to poor information systems and corruption

(Kirigia and Barry, 2008). These factors mean that many people throughout Africa do not have access to health services, either because the government cannot afford them, and hence the distribution of clinics and medicines is limited, or because, where services are provided, they are too expensive for the majority of people. Consequently, the strengthening of health systems so that they are able to provide essential, equitable and good-quality health services was one of the key priorities of the framework for implementation of the Ouagadougou Declaration (see Box 6.5).

Box 6.5

The future of Africa's health systems? The Ouagadougou Declaration on Primary Health Care

In April 2008, over 600 participants and country representatives from around Africa met in Ouagadougou, Burkina Faso, to attend the International Conference on Primary Health Care and Health Systems in Africa. The conference, organized by UNICEF, UNFPA, UNAIDS, the African Development Bank and the World Bank, identified the need for 'accelerated action by African governments, partners and communities to improve health'. A declaration, signed by forty-six of Africa's health ministers, specifically acknowledges the interrelationships between wider development objectives and health, and the need to accelerate progress towards the Millennium Development Goals. While the declaration itself calls for action from Africa's governments and the international community on such issues as primary healthcare development and the need for financial support, a subsequent 'framework for implementation' was produced which outlines nine specific priorities for African countries:

1. *Leadership and governance for health.* Countries are encouraged to develop comprehensive national health policies that integrate a primary healthcare approach throughout. These should particularly strengthen health systems to achieve the Millennium Development Goals, particularly those that relate to HIV/AIDS, tuberculosis, malaria, child health, maternal health, trauma and the emerging burden of chronic diseases.
2. *Health service delivery.* The delivery of health services should be improved so that they are affordable and effective for those who need them. This includes investing in infrastructure, engaging with communities, and integrating different health services to improve efficiency and equity of access.
3. *Human resources for health.* Countries should improve the quality, numbers and distribution of their healthcare workforce. The need for better financing for health workers is also identified, as this will increase staff motivation and retention. Resources should also be targeted at mid-level health workers who can deliver preventive, curative and rehabilitative healthcare.
4. *Health financing.* In addition to developing a comprehensive health financing policy, the framework suggests that at least 15 per cent of a country's national budget should be allocated to healthcare. It goes on to emphasize the importance of appropriate accounting, planning, monitoring and evaluation mechanisms at national and local levels.

5 *Health information systems.* Each country should establish a national health information system (NHIS) which can monitor health status and health services at all levels. Establishing systems for gathering statistics on population change, births and deaths is considered essential.
6 *Health technologies.* Countries should increase the provision of and access to medicines, vaccines, biological equipment and medical technology. This includes promoting dispensary and prescribing practices, ensuring the safety and quality of medicines, and monitoring drug resistance.
7 *Community ownership and participation.* Countries should allow communities a say in how their health services are delivered and facilitate a more 'bottom-up' integration of information so that health services are more demand-driven. Community-based health organizations should also be strengthened, as should links between these and external NGOs.
8 *Partnerships for health development.* Due to the large number of health initiatives (e.g. Roll Back Malaria and Stop TB), there are concerns about the fragmentation (and replication) of health services which erode the state provision. Partnerships for health therefore need to be strengthened and should address inter-sectoral collaboration, public–private–NGO relationships, and South–South cooperation within Africa.
9 *Research for health.* Citing the statistic that only 10 per cent of global health research funding is allocated to solving the health problems of 90 per cent of the population, the framework proposes that countries should allocate more funding towards research within ministries of health, and establish networks through which research findings might be disseminated. Participation in South–South and 'equitable' North–South partnerships for technology transfer is also encouraged.

Source: WHO (2008g)

One of the main challenges facing the development of health systems is the need for a coordinated, cross-sectoral approach to healthcare. So-called 'vertical' health programmes, which target a specific issue, such as malaria or sanitation, have been popular among donors and NGOs. While many of these have been successful, there are concerns that they represent lost opportunities to deliver other health benefits to target communities and have diverted human and financial resources away from other programmes (Chen and Hanvoravongchai, 2005; WHO, 2006a). Adopting a sectoral approach, which supports the 'horizontal' integration of health programmes, creates an environment where various stakeholders work together within a framework of shared policies and strategies, in which health goals are aligned and mutually reinforcing. One of the emerging challenges here, however, is how to ensure alignment and integration with private health services, which have prospered in the absence of adequate state-subsidized healthcare. Public–private partnerships – in which private health providers manage specific services that continue to be regulated by the state – are increasingly seen as a means of building on the strengths of both sectors, with partnerships already established in Ghana and South Africa (WEF, 2008; IFC, 2007). The drive towards horizontal integration also has implications for the international donor community, which, despite contributing over 16 per cent of healthcare spending in Africa in 2003, continues to place restrictions and conditions on the activities and programmes it supports. However, while improved coordination between donors and a sectoral approach to funding should be encouraged, donors must be confident that their funds are being used in an effective and efficient manner. The responsibility therefore lies with national governments committing to long-term, strategic health system planning, and establishing accountable and transparent systems of finance allocation.

Another target of health system reform is to address the acute shortage of skilled health professionals. For the African countries highlighted in Table 6.6, only Zambia and Algeria have more than one health worker per 1000 population, while Malawi has only one physician per 50,000 people. These shortages are a result of years of under-investment in training and recruitment, poor working conditions, and, in recent years, the migration of health workers to countries with higher salaries and better employment conditions. For example, between 1993 and 2002, a total of 1734 health workers migrated from Kenya to Saudi Arabia, the UK, the USA and Zambia, among other countries (WHO, 2006c; Awases *et al.*, 2004). Measures to reverse this trend start with better budgetary support for health training and employee salaries, but other approaches have involved training nurses to take on some of the responsibilities of doctors, and devolving some healthcare to local community members – with the latter strategy particularly effective in HIV/AIDS treatment and care.

Acquiring reliable data on the health of the population is of fundamental importance to health planning, yet most African countries do not have adequate systems in place to record morbidity and mortality, let alone the more positive outcomes of treatment and care. Without this information, it is difficult to make informed decisions on where and which health services are needed and how these are changing over time, which interventions are effective, or indeed which priorities should be relayed to the donor community for development assistance (Ashraf, 2005). The lack of information technology and skilled

Table 6.6 Density of the health workforce per 1000 population in selected countries, 2004

	Algeria	*Zambia*	*Lesotho*	*Sierra Leone*	*Malawi*	*Ethiopia*	*Africa*
Physicians	1.131	0.116	0.049	0.033	0.022	0.027	0.217
Nurses and midwives	2.231	2.015	0.623	0.356	0.589	0.22	1.172
Dentists and technicians	0.306	0.045	0.009	0.001	–	0.001	0.035
Pharmacists and technicians	0.203	0.095	0.034	0.066	–	0.019	0.063
Environmental and public health workers	0.081	0.094	0.031	0.026	0.002	0.019	0.049
Laboratory technicians	0.283	0.13	0.081	–	0.004	0.038	0.057
Other health workers	0.163	0.305	0.013	–	0.057	0.104	0.173
Community health workers	0.034	–	–	0.237	–	0.264	0.449
Administrative and support staff	1.947	0.99	0.01	0.001	–	–	0.411
Total	6.378	3.79	0.85	0.72	0.674	0.692	2.626
Total numbers	199,407	41,429	1532	3721	8309	48,972	–

Source: WHO, 2006c

personnel, unreliable electricity sources, and a lack of consistency in monitoring methods have all led to poor health information systems. While the development of well-resourced, comprehensive health information systems incorporating functioning civil registration and medical certification practices is the long-term goal for governments, in the short term the Sample Vital Registration (SVR) system is currently regarded as a practical alternative. This involves community-based monitoring of morbidity and mortality in a representative sample of sites within a country, and includes the technique of verbal autopsy, whereby a questionnaire is administered to family members of a deceased person to identify the cause of death (Setel *et al.*, 2005).

SVR and verbal autopsy initiatives are parts of a wider attempt to encourage and support community participation in health service provision. This may involve the provision of palliative care for the sick, which has now become commonplace as a result of the HIV/AIDS burden, or it may involve community members playing a role in monitoring and treatment, as has occurred in the DOTS approach to TB treatment. In recent years, there has been a significant drive towards the provision of reproductive healthcare and family planning at the community level, through the training of community health workers. In Tanzania, for example, various local NGOs are supporting community-based reproductive health by training voluntary workers, who are remunerated for their efforts with bicycles and umbrellas (Janowitz *et al.*, 2000); and in Uganda, community health workers have successfully piloted a scheme to provide contraceptive injections to local women (Stanback *et al.*, 2007). Community-based healthcare also recognizes the importance of traditional medicine, and particularly the role of traditional healers within Africa culture (Iyiani *et al.*, in press). The *Africa Health Strategy* outlines the need to integrate traditional medicine into national health systems, albeit with an appreciation of its strengths and weaknesses (African Union, 2007), and many regard traditional healers as key members of the community who can be utilized in the promotion and provision of more mainstream health services (Giarelli and Jacobs 2003; Alem *et al.*, 2008).

6.8 Conclusion

This chapter has drawn attention to a wide range of health issues that have resulted in a huge burden of ill health and low life expectancies across the continent. HIV/AIDS continues to be the most significant health issue, accounting for over 1.7 million deaths per year and exacting a huge socio-economic impact on those people and their families who are living and coping with the disease. Indeed, it constitutes one of Africa's biggest development challenges as it continues to drain human capital and economic productivity. However, there are encouraging signs that the epidemic is stabilizing in some countries as ARVs become more widely available.

Although the HIV/AIDS epidemic has tended to dominate the development agenda in recent years, this chapter has also shown how poor environmental health, particularly the lack of clean water and sanitation, continues to be a major cause of disease, sickness and death. Over 885,000 people die each year through diarrhoeal diseases, the vast majority of which could be prevented by investments in public health infrastructure and education.

A recurrent theme of the chapter has been how the poor and marginalized are more vulnerable to disease and ill health, which subsequently compounds poverty and marginalization in a vicious cycle. Women, in particular, occupy a subordinate position to men in most societies and as such are less able to make informed decisions about their lifestyle and health. They also have less access to health services. Consequently, improving maternal and reproductive health raises the challenge of addressing these underlying gender inequalities in addition to the provision of dedicated health services for women.

While the health of Africa's people is inextricably linked with embedded poverty, gender inequalities, the persistence of conflict, and the environment itself, there is growing recognition of the need to reform Africa's health services. Acquiring funding and investment for health services is a clear priority, but in the absence of domestic income there is a reliance on donor funding, which may come with conditions attached. Pursuing an all-embracing, primary healthcare strategy which seeks multi-sectoral integration, multi-stakeholder (including community) participation and public–private partnerships is regarded by the African Union as the way forward.

Summary

1. Africa continues to be crippled by a huge burden of disease and ill health.
2. Women and children are disproportionately affected by ill health.
3. Cardiovascular disease and diabetes are becoming more prevalent as Africa becomes more 'developed'.
4. HIV/AIDS continues to exact a huge burden of ill health on the population, although levels are stabilizing in many countries due to more widespread availability of ARVs and increasingly effective prevention programmes.
5. Health system reform demands improvements in the supply and distribution of health services, and the creation of favourable employment conditions for healthcare workers.

Discussion questions

1. Despite their vulnerability to HIV/AIDS and a range of maternal health issues, women continue to live longer than men. How might this be explained?
2. How will Africa's health challenges in 2030 differ from those in 2011?
3. Thinking about the images of Africa portrayed in the media, do the data in Table 6.1 surprise you in any way?
4. How did the impact of and response to the 2009–2010 swine flu epidemic differ between Africa and Europe?

Further reading

Jamison, D.T., Feachem, R.G., Makgoba, M.W., Bos, E.R., Baingana, F.K., Hofman, K.J. and Rogo, K.O. (2006) *Disease and Mortality in Sub-Saharan Africa*, Washington, DC: World Bank.

Useful websites

Partnership for Maternal, Newborn and Child Health: http://www.who.int/pmnch/en/. Supports the global health community in working towards achieving MDGs 4 and 5.

RBMP (Roll Back Malaria Partnership): http://www.rbm.who.int/. The global framework to implement coordinated action against malaria. Contains numerous publications on action against malaria, key epidemiological trends and country assessments.

UNAIDS (Joint United Nations Programme on HIV/AIDS): http://www.unaids.org. UNAIDS is a UN programme that coordinates international efforts to respond to HIV/AIDS. The website holds country reports on the HIV/AIDS situation, statistics on prevalence rates, and epidemiological studies.

World Health Organization: http://www.who.int. The WHO is the directing and coordinating authority for health within the United Nations system. The website holds information on all the major health issues affecting Africa and the world, and the various initiatives that are dealing with these.

7 Conflict and post-conflict

7.1 Introduction

Africa has been ravaged by war and conflict in recent years. Between 2000 and 2010, armed conflict affected twenty-four African countries and killed millions of military personnel and civilians. Looking beyond the tragedy of battle-related deaths, however, the implications of conflict for development are far reaching, and as Asha-Rose Migiro, Deputy Secretary-General of the United Nations, stated in 2010, 'the costs of war in Africa have cancelled out the potential impact of 15 years of development aid' (UN, 2010b: 1). Conflict has resulted in the collapse of economic systems, state infrastructure and effective governance, and more people die from disease and starvation due to the loss of essential services than from the direct impacts of military action. Those who manage to survive through conflict situations face numerous development challenges, ranging from displacement and livelihood insecurity to being orphaned or infected with HIV. There are, therefore, clear and demonstrable linkages between the widespread poverty and underdevelopment that exists throughout Africa and the persistence of armed conflict.

The relationship between poverty and conflict is, however, complex. Among the diverse (and often contested) causes of conflict in Africa, which include colonial legacies, ethnic divisions and economic inequality, is poverty and underdevelopment itself. Indeed, as is discussed in this chapter, poverty has been regarded as a key driver of conflict as well as a ubiquitous consequence, and breaking this mutually reinforcing relationship remains a key challenge in periods of both peace and conflict. The first half of this chapter examines the nature of conflict in Africa, its geographical scope and complex interrelated causes, and, critically, how it has affected the population and hindered development efforts. The second half discusses the ways in which development has approached conflict in terms of instituting mechanisms for resolution and prevention, and planning for post-conflict reconstruction.

7.2 The scope of conflict in Africa

7.2.1 What do we mean by 'conflict'?

It is perhaps useful to start by considering what precisely is meant by 'conflict', although it is difficult to provide a universal definition. As Unwin (2002) points out, 'conflict' is a term that is used synonymously with 'war', which usually implies armed conflict, driven by a struggle for power, which results in a specific number of deaths per year. The Uppsala Conflict Data Project (UCDP) defines an armed conflict as: 'a contested incompatibility that concerns government or territory or both where the use of armed force between two parties, of which at least one is the government of a state, results in at least 25 battle-related deaths in a single calendar year' (Wallensteen and Sollenberg, 2001: 643). Similarly, 'major armed conflict', which is also classed as 'war', constitutes at least 1000 battle-related deaths per year (Gleditsch *et al.*, 2002). The UCDP also uses a typology of conflicts, which includes:

- inter-state armed conflict, which occurs between two or more states;
- extra-state armed conflict, which occurs between a state and a non-state group outside its own territory;
- internationalized internal armed conflict, which occurs between the government of a state and internal opposition groups with intervention from other states;
- internal armed conflict, which occurs between the government of a state and internal opposition groups (Gleditsch *et al.*, 2002: 619).

Clearly, these definitions consider conflict occurring at different geographic scales, for primarily political reasons, but it has been argued that conflict can exist in a much broader sense: it can involve exclusively civilians or military personnel, and it can be characterized by lingering violence that does not produce battle-deaths. Brzoska (2007) suggests that economically motivated conflict between non-state actors and violence against unarmed civilians are becoming increasingly common, and arguably more significant than armed conflict involving military personnel.

7.2.2 The geography of armed conflict

As highlighted in Table 7.1, forty-two African countries have experienced minor or major armed conflict since 1950, and in many cases this conflict can be attributed to long and protracted civil wars. For example, between 1950 and 2008, Angola and Ethiopia experienced armed conflict in one form or another over a period of forty-six and forty-five years, respectively, with a combined estimated total casualty count in excess of 2 million people (Luckham *et al.*, 2001). It should be pointed out, however, that estimating the number of deaths from such conflicts is extremely difficult, not least because of the problems of differentiating between direct battlefield casualties and civilians who die as a result of associated famine or the collapse of infrastructure.

The data presented in Table 7.1 disguises the fact that conflict has often been unevenly distributed within and between countries. This was particularly evident in Sudan, where the long-running civil war was confined to the southern half of the country, and in the Democratic Republic of the Congo (DRC), where the internationalized internal conflict continues to be localized around the eastern border with the Great Lakes region (see Box 7.1). In contrast, the conflict in Mozambique and Somalia during the 1980s and 1990s enveloped the whole of these countries, and similarly the ongoing political violence in Zimbabwe is nationwide in nature. Obviously, the geographical extent of the conflict influences the magnitude of direct and indirect causalities, although, as Luckham *et al.* (2001) point out, larger impacts are felt when conflict occurs in core regions that are centres for governance and economic productivity, rather than in peripheral areas.

Although it can be said that there are very few parts of the continent that have remained untouched by armed conflict, an analysis of the long-term trends reveals that conflict may have peaked in the late 1990s (see Figure 7.1). While this is undoubtedly encouraging for Africa's development prospects, the challenge is to ensure that those countries in a post-conflict state do not fall back into conflict, since, as is discussed in the following section, past conflict tends to be a good predictor of future conflict.

Table 7.1 Minor and major armed conflicts in Africa, 1950–2008

Country	*1950–present*	*Total number of years in which conflict occurred*
Algeria	1954–1962, 1991–2008	24 years
Angola	1961–2007	46 years
Burkina Faso	1985, 1987	2 years
Burundi	1965, 1991–1992, 1994–2006, 2008–present	16 years
Cameroon	1957–1961, 1984, 1996	6 years
Chad	1966–1972, 1976–1984, 1986–1987, 1989–1994, 1997–2002, 2004 – 2008	29 years
Central African Republic	2002, 2006	
Comoros	1989, 1987	2 years
Congo, Republic	1993–1994, 1997–1999, 2002	6 years
Côte d'Ivoire	2002–2004	3 years
Democratic Republic of the Congo (Zaire)	1960–1962, 1964–1967, 1977–1978, 1996–2001, 2006–2008	17 years
Djibouti	1991–1994, 1999, 2008	6 years
Egypt	1951–1952, 1956, 1967, 1969–1970, 1973, 1993–1998	12 years
Equatorial Guinea	1979	1 years
Eritrea	1997–2000, 2003	5 years
Ethiopia	1960, 1964–2008	45 years
Gabon	1964	1 year
Gambia	1981	1 year
Ghana	1966, 1981, 1983	3 years
Guinea	2000–2001	2 years
Guinea-Bissau	1963–1973, 1998–1999	13 years
Kenya	1952–1956, 1982	6 years
Lesotho	1998	1 year
Liberia	1980, 1989–2003	15 years
Madagascar	1971	1 year
Mali	1990, 1994, 2007–2008	4 years
Mauritania	1957–1958, 1975–1978	6 years
Morocco	1953–1958, 1971, 1975–1989	22 years
Mozambique	1964–1974, 1977–1992	27 years
Niger	1991–1992, 1994–1997, 2007–2008	8 years
Nigeria	1966–1970, 2004	6 years
Rwanda	1990–2002	13 years
Senegal	1990–1993, 1995, 1997–1998, 2000, 2001, 2003	10 years
Sierra Leone	1991–2000	10 years
Somalia	1978, 1982–1996, 2001–2002, 2006–2008	21 years
South Africa	1966–1988	23 years
Sudan	1963–1972, 1976, 1983–2008	37 years
Tanzania	1978	1 year
Togo	1986, 1991	2 years
Tunisia	1953–1956, 1961, 1980	6 years
Uganda	1971–1972, 1974, 1978–1992, 1994–2007	32 years
Zimbabwe	1967–1968, 1973–1979	9 years

Source: SIPRI, 2010

Box 7.1

Conflict in the Great Lakes region

The conflict that spread throughout the Great Lakes region during the 1990s and 2000s was also known as 'Africa's World War' because it eventually involved the armed forces of eight different countries and cost the lives of over 5 million people. The conflict provides a good illustration of how conflict can be driven and reproduced by a range of complex interrelated factors, such as poverty, inequality, ethnicity, resource endowment, colonial history and geography.

The Rwandan Civil War and genocide of 1994 is usually cited as the origin of the modern-day conflict in the Great Lakes region, although this itself was a product of colonial and post-colonial policies that resulted in ethnic inequalities. The overthrow of the Hutu Rwandan government by the Tutsi-led Rwanda Patriotic Front (RPF) in April 1994 resulted in the dispersal of thousands of Tutsi and Hutu refugees in neighbouring Tanzania, Burundi and eastern parts of the Democratic Republic of the Congo (DRC; then Zaire). Subsequently, within the DRC, the fleeing Hutu perpetrators of the genocide (the *interahamwe* militia groups) were able to capitalize on the poor security situation in the country and reorganize themselves to lead incursions into Rwanda, taking revenge on the post-genocide RPF government and the Tutsi population. The spill-over of the conflict into the DRC was soon consolidated when Hutu militia also began targeting the country's ethnic Tutsi population, the Banyamulenge, with support from Zairian (DRC) government forces and President Mobutu Sese Seko.

Increasing violence and persecution of the Banyamulenge throughout 1996 ultimately led to an exchange of fire between Rwandan and Zairian troops, and the emergence of the Alliance of Democratic Forces for the Liberation of Zaire (AFDL), which formed from several opposition groups and was headed by Laurent-Désiré Kabila. The AFDL received substantial support from both Rwandan and Ugandan forces, and between October and November 1996 they engaged in the systematic destruction of the large Hutu refugee camps along the Zaire–Rwanda border, forcing many Hutus to flee back into Rwanda. Six months later, Kabila was installed as President of the renamed DRC, having achieved a victory over government forces across the country. This first period of conflict would become known as the First Congo War. The Second Congo War began just a few months later, and this would ultimately cost the lives of over 5 million people.

Once in power, Kabila sought to centralize control over the country and suppress minority groups seeking democratic reform and decentralization. Meanwhile, it became increasingly clear that the Rwandan and Ugandan troops remaining in the eastern part of the country, purportedly for security reasons, were actually there to exploit the region's natural resources and mineral deposits. Kabila's reaction to their reluctance to leave was to expel all Rwandan and Ugandan officials from the DRC. In response, the Rwandan and Ugandan governments backed Banyamulenge rebels, creating the Rally for Congolese Democracy (RCD), which consolidated territorial gains in the eastern provinces. At the same time, the Ugandan government helped establish the Movement for the Liberation of Congo (MLC), consolidating its exclusive influence in the north-eastern region. In 1998, however, Angolan, Zimbabwean and Namibian troops entered the conflict in support of Kabila's forces (although

Angola used the opportunity to attack UNITA rebels in its own province of Cabinda). Despite all the parties signing the Lusaka Ceasefire Agreement in 1999, sporadic conflict continued across the country, and notably around the city of Kisangani, where Rwandan and Ugandan troops engaged in open combat for the first time. In 2001, Laurent-Désiré Kabila died following an assassination attempt and was replaced as President by his twenty-nine-year-old son, Joseph Kabila. After several high-profile meetings with the Uganda and Rwandan governments, both agreed to pull back their forces, and in 2002 peace deals were signed, leading to the withdrawal of foreign troops from the DRC. A transitional government was established in 2003 and elections held in 2006, which saw Joseph Kabila returned to power in a fragile peace.

Although the DRC has not erupted into full-scale civil war, many of the underlying causes of tension remain, and conflict has persisted in the eastern region of the country, in both North and South Kivu provinces. Hutu militia groups retain a presence in these areas and, according to Rwanda, continue to threaten ethnic Tutsis along the border. Most recently, Rwanda has been accused of covertly supporting Laurent Nkunda, the leader of a rebel Tutsi group that has been fighting both DRC government forces and Hutu militia groups. Nkunda has been heavily criticized for his human rights abuses, including the recruitment of child soldiers, indiscriminate use of violence against civilians, and rape. The humanitarian crisis that engulfed eastern DRC between 2008 and 2009 was widely attributed to the actions of Nkunda's forces. He was arrested in 2009 as part of a deal between the DRC and the Rwandan government, which allowed Rwandan forces into Kivu to pursue *interahamwe* militia in return for Rwanda's cooperation in Nkunda's capture.

The conflict has had a significant impact on development throughout the region. Thousands of people have been displaced from their land and livelihoods, children have been denied education and have been forced to fight, women have been victims of sex crimes, and HIV/AIDS infection rates have increased significantly. Despite an official ceasefire and the best efforts of various peacekeeping initiatives, the situation in the DRC remains fragile as various groups continue to seek control of mineral resources and exploit ethnic tensions along the eastern border.

Sources: Ewald *et al.*, 2004; Lun, 2006; Turner, 2007

7.3 The causes of conflict

It is extremely difficult to establish the exact causes of Africa's many complex, interrelated conflicts. Much research has sought to identify risk factors – the specific underlying social, cultural, economic, political and environmental conditions which predispose a country to conflict. The presence of certain conditions, however, does not necessarily guarantee that conflict will occur. In fact, conflict is usually triggered by sudden change or a specific event. A good example is the Rwandan genocide in which a reported 800,000 people were killed between April and July 1994. The cause of the conflict has been attributed to the underlying inequalities between Hutu and Tutsi ethnic groups which had existed and been reproduced in Rwandan society for generations. Indeed, the roots of the conflict can be traced back to the ethnic segregation and political manipulation of the Belgian colonial government during the 1920s. However, it was the shooting down of President Habyarimana's

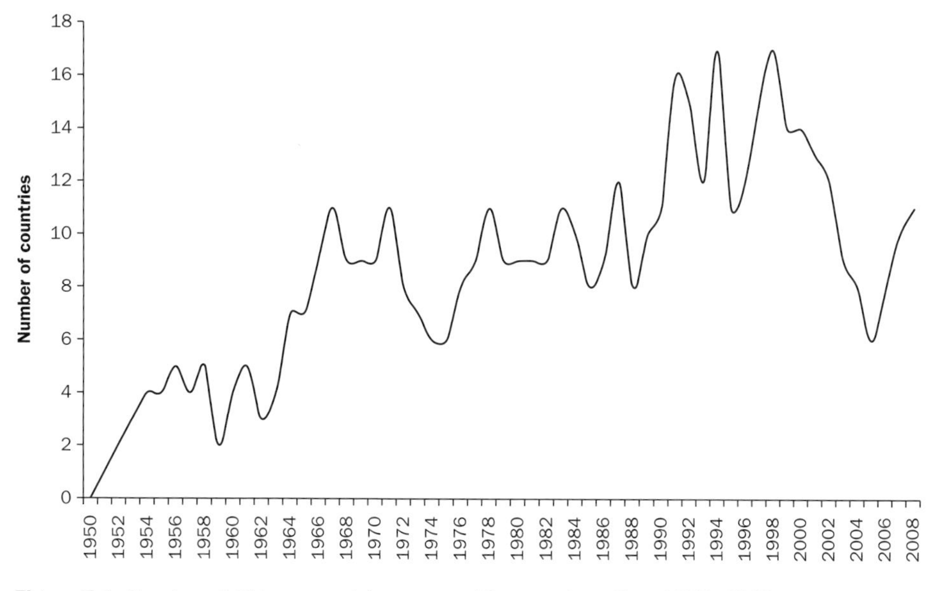

Figure 7.1 Number of African countries engaged in armed conflict, 1950–2008.

Source: SIPRI, 2010

plane on 6 April 1994 that triggered the mass killings and, some would argue, the subsequent years of conflict throughout the Great Lakes region (see Box 7.1).

Compounding the causal relationship between conflict and specific risks and triggers is the fact that these can change over time. As Luckham *et al.* (2001) point out, the factors that sustain a conflict are not necessarily those that caused it. Conflict itself can lead to political, economic and social transformations, which in turn create the conditions for further conflict. In this way, conflict can be seen as reproducing itself, and in some cases it becomes embedded in society and effectively is a 'way of life'. Luckham *et al.* (2001) also suggest that twentieth-century conflict in Africa can be viewed as occurring in several historical stages, during which some causes and drivers of conflicts have become more significant than others. First, conflict occurring as part of a process of liberation against colonial powers was largely endemic throughout Africa during the 1950s, 1960s and 1970s (e.g. the Mau Mau Rebellion in Kenya during the 1950s, and the Mozambique War of Independence, 1964–1974). Second, the first post-colonial conflicts were motivated by the need to redefine national identities and colonial boundaries – as, for example, occurred in the Nigerian Civil War between 1967 and 1970. Third, conflict was fuelled by rivalries during the Cold War, when states and rebel groups aligned themselves politically and economically with either the US or the USSR. An example here is the Angolan Civil War, fought between 1975 and 1991, which was famously characterized by US and South African support for UNITA and FNLA forces, while the opposing MPLA was supported by the Soviet Union and Cuba. Fourth, many African conflicts within the 1970s and 1980s can be considered 'reform' wars, characterized by insurgencies against post-colonial governments that had failed to deliver on their promises of development and prosperity (e.g. the activities of the National Resistance Army in Uganda between 1979 and 1986). Finally, according to Luckham *et al.* (2001), the conflict of the post-Cold War era is typified by fragile states and the opportunistic privatization of war, as has been evident in the DRC since 1997.

7.3.1 Poverty, inequality and underdevelopment

The persistence of poverty, inequality and underdevelopment is probably the most cited explanation for conflict in Africa. The commonly held view is that those who live in poverty and experience inequality have little to lose by joining a rebellion or supporting a conflict that promises an improvement in quality of life (Collier and Hoeffler, 2004). This has often been offered as one explanation for the child-soldier phenomenon, common in the recent conflicts in Sierra Leone, Liberia and Côte d'Ivoire (Achvarina *et al.*, 2009; Maconachie and Binns, 2007b). However, it is arguably the underlying form of governance and the lack of development and investment in infrastructure that create the conditions of poverty that may ultimately lead to conflict. Although economic inequality within society – that is, wealth existing alongside poverty – is also regarded as an underlying risk factor, it is important to view this in the wider context of the social and political processes that have created it (Cramer, 2003). Moreover, as Ostby (2008) points out, it is the inequality that manifests between specific groups (rather than individuals) that causes resentment and grievances that are likely to evolve into conflict. Clearly, the conflict that occurred during South Africa's apartheid era emanated from economic inequality between groups, which could be traced to complex historical, social and political processes and transformations. While the uprisings and conflict that occurred in North Africa during 2011 did not occur as a result of inequality among the populations of those countries, they have been attributed to a lack of development in terms of human rights and democratic freedom.

7.3.2 Ethnicity and religion

As has been well documented elsewhere, the Rwandan genocide of 1994 placed ethnic politics firmly at the forefront of attempts to understand and predict conflict in Africa. Although there is some debate as to how influential a country's ethnic composition and differences are as causes of conflict, several studies have proposed a heightened risk where one ethnic group is dominant within society (e.g. Collier and Hoeffler, 2004). It has also been suggested that, while ethnicity is often proposed as a cause of conflict, ethnic tensions tend to reflect underlying social and economic inequalities. In the case of Rwanda, ethnic tensions date back to the German and Belgian colonial governments that arbitrarily segregated Hutu from Tutsi, and installed the latter in positions of power. Seeking to quash growing demands for independence from the Tutsi elite during the 1950s, the colonial government then transferred power and land from Tutsi to Hutu. Subsequent conflict, while fundamentally driven by unequal power relations, increasingly became associated with ethnicity, to the point where the systematic eradication of Tutsis became the goal of various Hutu militia groups in 1994. Similarly, in neighbouring Burundi, it has been argued that ethnic tensions were manipulated by both Hutu and Tutsi political elites to further their own interests during the civil war of 1993 (Ngaruko and Nkurunziza, 2005). More recently, ethnic tensions have once more been cited (particularly by the world's media) as the driver of conflict in the Darfur region of Sudan (see Box 7.2).

Box 7.2

The conflict in Darfur: ethnic tensions or climate change?

The Darfur region of Sudan came to the attention of the world's media in 2003, when a violent armed conflict suddenly erupted between indigenous nomadic pastoralist Arabs and farmers of the Fur, Masalit and Zaghawa ethnic groups. Despite a relatively peaceful coexistence of these groups in the past, during which farmers allowing pastoralists access to land and water, sporadic outbreaks of violence attributed to ethnic tensions over land and resources first began to occur during the 1980s. In 2003, however, tensions rapidly escalated into a regional conflict that to date has claimed the lives of over 2.4 million people and has led to an ongoing humanitarian crisis in the region.

Shortly after the outbreak of violence in 2003, and particularly as the conflict began to escalate, it became clear that a process 'ethnic cleansing' (some would argue genocide) was being perpetrated throughout Darfur by government forces and state-sponsored local Arab militia, known as *janjaweed*. Indeed, since the late 1990s, the Arab government of Sudan had been replacing local administrators in Darfur with its own political figures, with the aim of supporting Arab pastoralist groups over farmers (de Waal, 2005). The need to retaliate against this oppression was the trigger for what essentially became framed by the development community as yet another of Africa's longstanding ethnic conflicts.

In 2007, however, the UN published its *Sudan: Post-Conflict Environmental Assessment* report, which challenged the view that Darfur's conflict was rooted in ethnic divisions. Instead, the report blamed the conflict firmly on land degradation, desertification and an 'unprecedented' scale of climate change (UNEP, 2007: 60). It suggested that rainfall throughout the region has declined significantly and this has caused drought, water shortages and the degradation of agricultural land. This, combined with a growing population and poor environmental governance, has purportedly led to the breakdown of established social systems in Darfur and the outbreak of conflict due to competition for scarce resources. The Darfur conflict was subsequently characterized in the media as the first of many climate-induced conflicts that would inevitably affect Africa in the years to come.

While few would deny that sustaining a livelihood in Darfur has been a challenge, many denounced the environmental degradation argument as being too simplistic, citing evidence that people in the region have adapted to environmental change for centuries without resorting to conflict. Indeed, in a detailed analysis of rainfall trends in the region, Kevane and Gray (2008) report that while there has been some variability in rainfall from one year to the next, there is little evidence of a gradual, long-term decline in rainfall during the thirty years preceding the conflict. Moreover, they note that the significant droughts of 1984 and 1990 did not lead to ethnic conflict. The roots of the conflict, they suggest, are overwhelmingly political, and ultimately rest with 'an elite ruling the country from the capital that has preferred to exclude peripheral populations from genuine participation in political processes and has repeatedly revealed a willingness and ability to use large-scale violence, often against civilian populations, in response to perceived threats from the peripheral regions' (Kevane and Gray, 2008: 10).

Although often cited as a factor increasing the risk of conflict, there is scant evidence of conflict occurring solely as a result of religious differences (Moller, 2006). For example, tensions between Islam, Christianity and indigenous religions that emerged during conflict in Liberia and Sierra Leone did not really represent fundamental religious grievances; rather, religion was used to radicalize existing conflicts. In Sudan, however, although the drivers of the conflict extend beyond religion, there is little doubt that the gradual Islamicization of the state and the introduction of sharia law have fuelled tensions between the north and south.

7.3.3 Natural resources

In recent years, much attention has focused on the role of high-value natural resource endowments in fuelling conflict in Africa, and it has been suggested that a strong statistical correlation exists between a developing county's reliance on natural resources and its susceptibility to civil war (Collier and Hoeffler, 2001; Elbadawi and Sambanis, 2002). Certainly, during the 1990s and 2000s, attempts to control access to such natural resources as diamonds, gold, timber and coltan were blamed for civil wars in the DRC, Liberia, Sierra Leone and Angola (Maconachie and Binns, 2007a, 2007b).

The relationship between natural resources and conflict is, however, complex. In the well-documented case of Sierra Leone, it remains unclear whether the desire to gain access to diamond mines constituted the original cause of the conflict between the state forces and the Revolutionary United Front (RUF) (Richards, 2003). Nonetheless, once the conflict began, there is little doubt that control of these areas became a major objective (Maconachie and Binns, 2007c; see Plate 7.1). According to UNEP (2009), natural resources are linked to conflict in three ways. First, the economic inequality that results from unequal access to the benefits of natural resources can lead to grievances and trigger the outbreak of conflict, as was arguably the case in Sierra Leone. Second, where conflict exists for other reasons, high-value natural resources may become strategic targets for military forces who see their

Plate 7.1 Diamond miners shovelling gravel in eastern Sierra Leone (Tony Binns).

control as a means of increasing revenue and fuelling further conflict. This has been the situation in eastern DRC, where the forces of several countries sought to gain control over diamond mines and coltan-rich areas, despite fighting what was initially an ethnic-based war. Third, because of the wealth generated by natural-resource exploitation, there is often little incentive to halt the conflict and concede territory and access rights. Natural resources, therefore, can become a barrier to peace-building.

In examining these relationships further, several preconditions for natural-resource-fuelled conflict emerge. It has been argued that the failure of states to manage and distribute the benefits of natural resources in a fair manner can precipitate violent conflict. Corruption, land appropriation, tolerance of illegal exploitation or simply the lack of enforceable laws can create the conditions for grievance among the population, especially those who are living in acute poverty. Removing control of natural resources from the state can therefore be seen as a way of addressing inequality. Many authors, however, have suggested that greed rather than a desire for redistributive justice is a major driver of seeking natural-resource control (Collier and Hoeffler, 2004; Maconachie and Binns, 2007a). Furthermore, it has been argued that, rather than seeing weak, fragile states as a causal factor behind resource conflicts, the opposite is true: the presence of high-value natural resources (and a dependency on them) actually erodes state functioning because resource-centred governance tends to neglect other infrastructural considerations (Fearon and Laitin, 2003; Collier, 2007). This has arguably been the case in Nigeria, where most of the country's economic activity centres on oil production, and where concerns are increasingly being raised about the fragile nature of the state and its inability to govern a diverse, multi-ethnic population (Maier, 2001). Indeed, it is hardly surprising that many have written of Africa's 'resource curse'.

7.3.4 Geographical factors

In addition to the presence and distribution of specific natural resources, a range of other geographical factors has been proposed as contributing to the risk of conflict in Africa (ADB, 2009). The presence of distinct geographic boundaries, including rivers, mountains and forests, can provide geographical isolation for insurgency groups, as was the case for the anti-government Eritrean People's Liberation Front (EPLF) and the Tigrayan People's Liberation Front (TPLF) forces operating from the northern Ethiopian Highlands during the civil war in the 1970s and 1980s. The size of a country and the distribution of population within it can also be influential. It has been argued that large populations located in peripheral areas, away from the capital, may be ignored by government, thereby fuelling resentment and conflict (Collier and Hoeffler, 2004). This may well have been a contributing factor towards civil conflict in sub-Saharan Africa's two largest countries, DRC and Sudan, where central government has struggled to be effective in distant regions that offer safe havens for rebel forces. On the other hand, many of Africa's smaller countries, such as Rwanda, Burundi, Liberia and Sierra Leone, have been engaged in civil conflict, too. In the case of Rwanda, population pressure and tension over land availability have been cited as two of many factors driving the ethnic conflict in 1994 (Magnarella, 2005).

Many have argued that the roots of modern conflict in Africa lie in the arbitrary geographical boundaries that were imposed by colonial governments during the late nineteenth century. One simply has to look at the present-day borders between Ethiopia and Somalia, Kenya and Tanzania, and Senegal and The Gambia to realize that these were drawn with little consideration of culturally and ethnically contiguous areas. In the case of Ethiopia and Somalia, conflict in the late 1970s was driven by Somali claims over the disputed, ethnically Somali, Ogaden territory, which had been allocated to Ethiopia by colonial powers in the 1950s. Conversely, Sudan was effectively managed as two distinct regions during the colonial era, yet after independence and unification tensions emerged between the less developed, ethnically diverse south and the more developed, mainly Arab north.

7.3.5 Globalization, interconnectedness and the spill-over effect

It has been argued that an increasing interconnectedness of the world system, which has included transport, telecommunications and markets, in combination with the collapse of Cold War rivalries during the 1990s, has provided an environment in which conflict can easily flourish. Luckham *et al.* (2001) suggest that the immediate reduction in funding and arms supplies after the Cold War led to the instability of many states, and subsequently both state and rebel forces looked to the growing number of private suppliers of arms to provide the resources to assert control over territory. Meanwhile, as access to global markets improved, new sources of revenue, such as natural resources, have been exploited. In this context, a further consequence of the seemingly arbitrary nature of Africa's borders is the way in which conflict has spilled over from one country to the next. In the case of the conflict in the DRC, where no fewer than nine state forces were fighting at any one time during the 1990s, the conflict has become defined more by the interconnections between natural resource markets, arms suppliers and the mobility of different ethnic groups, rather than the state versus rebel groups. According to Gleditsch (2007), conflict is twice as likely in a country where one of its neighbours is at war.

7.3.6 Conflict triggers

While the factors discussed above heighten the *risk* of conflict, it is usually specific political, social, economic or environmental events that *trigger* conflict. Military coups, for example, have triggered conflict in Burundi, the DRC and Angola, although 'bloodless' coups – such as those in Benin, Burkina Faso, The Gambia and Ghana – have also frequently occurred (Meredith, 2005; see Plate 7.2). In recent years, there have been several examples of

Plate 7.2 Arch 22, Banjul, The Gambia to commemorate the military coup of 22 July 1994 (Tony Binns).

government elections that have precipitated civil conflict and unrest. In Ethiopia (2005), Kenya (2007), Zimbabwe (2008), Nigeria (2011) and Côte d'Ivoire (2011), violence erupted in response to voter intimidation and alleged electoral manipulation.

An emerging concern is the relationship between climate change and conflict. According to the IPCC (2007), temperature increases and water shortages in Africa will have a dramatic effect on agricultural land and food security, with between 75 million and 250 million people affected by the 2020s. Environmental degradation, food insecurity and water scarcity will put pressure on people's livelihoods, create competition for land and resources, and ultimately result in violent conflict (Brown *et al.*, 2007). Indeed, in 2007, UN Secretary-General Ban Ki-moon claimed that the impact of climate change was a significant cause of the conflict in Sudan's Darfur region (see Box 7.2). However, some argue that the links between climate change and conflict are not clear cut, and that analyses tend to ignore complex socio-economic and political determinants, as well as the impressive adaptive capacity of Africa's people (Hendrix and Glaser, 2007; Abraham, 2007).

7.4 The impacts of conflict

In 2005, the UN's *Human Development Report* suggested a 'strong association between low human development and violent conflict', with nine out of the ten lowest-ranked countries (which were all African) having experienced violent conflict since 1990 (UNDP, 2005: 154). The implication here is that conflict has a serious impact on those variables used to calculate the Human Development Index – GDP, life expectancy and education – although, as is explored in this section, the impacts of conflict are significantly more diverse and wide-ranging. The impact of conflict on people and their livelihoods can also be long lasting, extending well beyond the period of conflict itself. Since few African countries have escaped conflict, understanding the interrelated impacts and legacies of conflict has become a fundamental component of most development strategies, particularly those involved in post-conflict reconstruction and peace-building. A critical consideration in understanding the impacts of conflict, however, is that the consequences of conflict are not always negative. While this is undoubtedly subjective – depending upon which 'side' one is on and the underlying causes of the conflict – periods of conflict are fundamentally characterized by a dynamic restructuring of power relationships from the macro-economic to the household levels. Conflict can create new forms of inequality and poverty, but it can also lead to the development of new opportunities for people, which can be consolidated in the post-conflict period (see Section 7.6).

Before discussing the specific impacts of conflict in detail, it is worth considering the issues surrounding the reporting of these impacts. Here, the reliability of data becomes a key problem. As Luckham *et al.* (2001) point out, government records may be physically destroyed during the conflict, making it difficult to establish a benchmark from which demographic, economic and social change can be measured. Data may also exist only at an aggregated, nationwide level, which can mask the effects of localized conflict. Perhaps more importantly, it is often difficult to establish the impacts of conflict relative to what would have occurred in its absence, particularly given the already impoverished status of many African countries that face a range of other development challenges. Statistical information, therefore, is at best unreliable; and most would argue that, for these reasons, the effects of conflict tend to be greatly underestimated.

7.4.1 Casualties

A distinction should be made between 'battle deaths' – military personnel and civilians killed during conflict – and 'total war deaths', which include the former as well as deaths from disease, starvation and crime (Lacina and Gleditsch, 2005), although the latter are especially difficult to estimate. The African Development Bank (ADB, 2008) suggests that 1.6 million battle deaths occurred in Africa between 1960 and 2005, some 24 per cent of the global total, yet these may constitute only a fraction of total war deaths. For instance, in Angola, there were an estimated 1.5 million total war deaths between 1975 and 2002, of which 165,475 were recorded battle deaths (Lacina and Gleditsch, 2005). Similarly, war in the DRC between 1998 and 2007 cost the lives of over 5.4 million people (2.1 million after the formal end of the war in 2002), yet fewer than 10 per cent of these were battle deaths (Coghlan *et al.*, 2008). The gross disparities between battle deaths and total war deaths are even more stark in the case of Ethiopia, where the 1976–1991 conflict resulted in just 16,000 battle-related deaths but perhaps 2 million total war deaths, mainly due to famine (Lacina and Gleditsch, 2005).

7.4.2 Macro-economy and government

The poorest, least developed of Africa's countries, including Niger, Sierra Leone, Somalia and the DRC, have all been associated with recent conflict (Francis, 2008), yet it is difficult to determine whether their impoverished status is a consequence or a cause. The reality is probably both: underdevelopment increases the risk of conflict, while conflict exacerbates economic decline and underdevelopment. Oxfam's *Africa's Missing Billions* report suggests that conflict in twenty-three African countries between 1990 and 2007 cost around $300 billion in total, with conflict shrinking a country's economy (GDP) by an average of 15 per cent per annum (Oxfam, 2007). Again, the economic costs incurred as a result of conflict are difficult to quantify, but they can be divided into direct and indirect costs. The former are those costs that involve expenditure to support or manage the effects of conflict. Military spending, for example, will increase significantly during periods of conflict, and effectively divert money away from other essential services or infrastructure (see Plate 7.3). In 1999, a year into its conflict with Ethiopia, Eritrea poured a staggering 34.4 per cent of its GDP into the military (in 2003, several years after the ceasefire, military expenditure remained at 20.9 per cent) (SIPRI, 2010). This is illustrated in Figure 7.2, which also shows the level of military spending as a proportion of GDP in several other conflict-stricken countries. The increase in military spending in Burundi between 1993 and 2005 can be linked to the country's civil war in that period. Similarly, in Sudan, spending drops after the end of the north–south conflict in 2001, while the subsequent rise in spending during 2004 can be linked to the conflict in Darfur. Other direct costs include medical and repatriation expenses for the injured, the costs of civilian displacement (see Section 7.4.6), and the cost of rebuilding essential infrastructure, although this usually occurs during the post-conflict period (see Section 7.6).

The diversion of funds to meet the direct costs of conflict inevitably results in a range of indirect economic costs, which can be seen in terms of opportunities lost through the lack of investment in essential peacetime infrastructure, services and economic development activities. The deterioration of infrastructure, through either physical destruction or lack of investment, together with ineffective governance, creates the conditions of economic decline. Under these circumstances, tax revenues decrease as economic activity shifts towards an informal economy characterized by short-term opportunistic behaviour and segmented markets, as occurred throughout Somalia following the collapse of the state

Plate 7.3 Display of armaments in Johannesburg, South Africa (Etienne Nel).

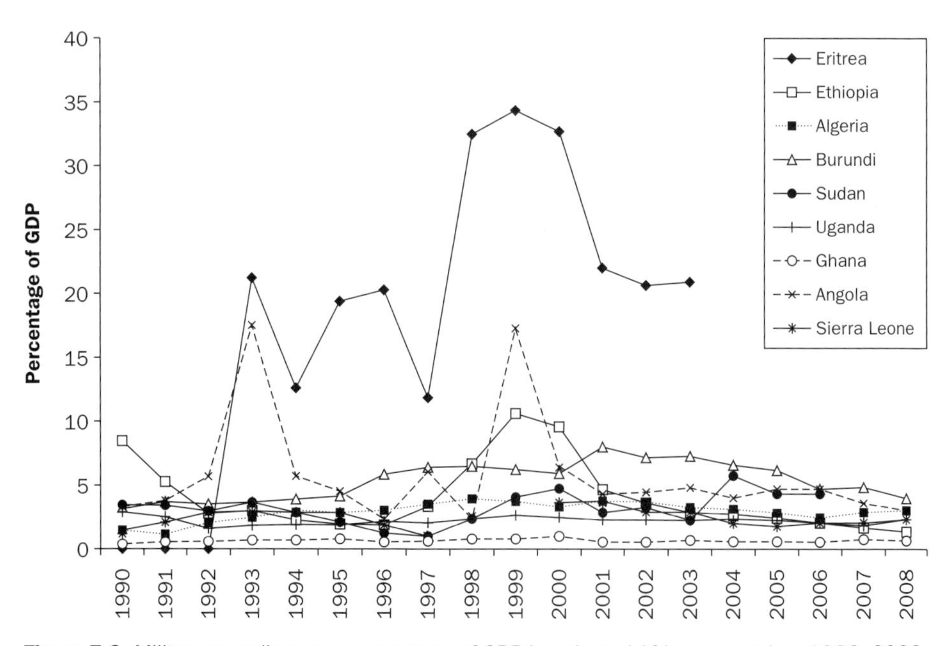

Figure 7.2 Military spending as a percentage of GDP in selected African countries, 1990–2008.

Source: World Bank, 2009b

in 1991 (Luckham *et al.*, 2001). There may also be a significant 'brain drain', as those with skills in such sectors as health and education flee the country. Bilateral and multilateral aid and funding for development projects and NGOs are also likely to be suspended as donors become concerned about operational issues, or money being diverted into military spending. Similarly, foreign direct investment by private firms may be scaled down or halted, while the occurrence of conflict will significantly reduce the likelihood of future foreign investment due to the perceived risks of insecurity.

The overall effects of this reduced investment and economic decline are multi-sectoral in nature: health, education, development, energy, roads, water, environment and social services are all likely to be affected, albeit it in different ways (see Table 7.2).

7.4.3 Tourism

Tourism is an essential source of foreign exchange and it contributes a significant proportion of GDP in such countries as Kenya, Tanzania and South Africa. However, even brief periods of conflict or civil unrest can affect tourist receipts as visitors are sensitive to the security situation. The violence that followed the Kenyan presidential elections in 2007 resulted in a halving of tourist revenue during 2008, with serious implications for employment in the sector (see Box 8.6, below; BBC, 2008c). Similarly, tourism in Zimbabwe has declined significantly since the beginning of the 2000s as a result of violence linked to the controversial land reform programme. The physical destruction of tourism infrastructure, such as buildings and wildlife, means that tourism takes a long time to recover during the post-conflict period.

7.4.4 Agriculture and rural livelihoods

In Sierra Leone, agriculture accounts for approximately 50 per cent of GDP, with more than two-thirds of the population involved in the agricultural sector. This situation is echoed throughout Africa's less developed countries, and hence it is easy to see why conflict in rural areas can have a catastrophic effect on a country's development through the disruption of rural livelihoods, as indeed was the case in Sierra Leone (Maconachie and Binns, 2007c). Land, homes and agricultural resources, such as livestock and farming equipment, may be destroyed or rendered inaccessible, leading to the displacement of

Table 7.2 Summary of conflict impacts on different sectors

	Direct impacts	*Indirect impacts*
Agriculture	Destruction and contamination of land Destruction or theft of livestock and equipment	Displacement of people Difficult rural livelihoods
Industry	Destruction and looting of factories and equipment Disruption of supply and marketing channels	Reduced economic activity Unemployment Perceived investment opportunities too risky
Education	Destruction of schools Shortage of teachers	Brain drain Legacy of poor education
Health	Destruction of hospitals and health infrastructure Looting of equipment	Disease and ill health Increased morbidity and mortality Public health neglected
Tourism	Destruction of infrastructure (buildings, roads, wildlife)	Declining tourist receipts

people, and, in the case of Somalia, a national shortage of economically significant livestock resources (Ahmed and Green, 1999). Even if land is still accessible and agricultural activities can continue, normal agricultural production may be disrupted by the destruction and collapse of infrastructure, such as water supply, road networks and extension services, as well as the lack of finance and marketing opportunities. The presence of armed groups may also place increasing pressure on farming: Luckham *et al.* (2001) cite the case of RENAMO insurgents in Mozambique, where soldiers forced local people to give up their food and water supplies. A shortage of agricultural labour may also occur as a result of the forced conscription of men and children, as occurred in Sierra Leone.

Despite the severe disruption to the functioning of many rural livelihoods, however, evidence suggests that people adapt to the new scenarios created by conflict. Faced with a decline in the able-bodied male workforce, women are likely to take on new roles. Forms of agriculture that require less maintenance, and are less vulnerable to destruction, may also be developed. Luckham *et al.*(2001) cite the example of a shift to cassava cultivation during the civil war in Sierra Leone. Cassava is less labour intensive, can be harvested when required, and cannot be looted as easily as other crops.

7.4.5 Environmental impacts

The impacts of conflict on the environment can also be considered in terms of their direct and indirect impacts, and in the wider context of the institutional and governance vacuum created by conflict. Direct impacts include the physical destruction or contamination of land, resources and wildlife, usually caused by explosives or their debris. However, the indirect impacts are arguably even more significant.

Indirect impacts on the environment occur largely through the actions and coping strategies of civilians: in the absence of infrastructure, governance and livelihood opportunities, people are often forced to exploit natural resources. Deforestation, for example, can increase in response to local fuel shortages and the need for construction materials, but illegal timber production and marketing can also form a lucrative economic activity for both civilians and the military. The profit from 'conflict timber' consequently acts as an incentive for further conflict (see Section 7.3.3). In the absence of markets and 'normal' agricultural resources, wildlife poaching tends to proliferate and 'bush meat' becomes an important source of food. The recent increase in bush-meat trading throughout central Africa, and particularly in the DRC, is believed to have developed as a result of livelihood insecurity and food shortages caused by persistent conflict. Alarmingly for conservationists, this trade has included endangered species, such as mountain gorillas.

These pressures tend to be highest where there are displaced people, particularly where refugees are concentrated. For example, it has been estimated that around 3750 hectares of forest were lost within three weeks of the arrival of refugees from Rwanda in the South Kivu region of the DRC in 1994 (UNEP, 2005). Similarly, in Karago, western Tanzania, where refugees from Burundi have been hosted since 1999, there is widespread evidence of serious soil erosion and land degradation caused by the destruction of vegetation. The demands placed on water resources by refugees can also lead to water shortages, and a deterioration of water quality through human-waste contamination. However, these effects may be limited to the short term, and indeed, as Black (1998) points out, there is evidence that refugees adapt to their new environment in ways that do not necessarily lead to long-term environmental degradation.

7.4.6 Displacement and refugees

Faced with the destruction of land, resources and livelihoods, many people in conflict zones are forced to migrate in search of security. Those leaving their country of origin assume the status of a refugee, while those who remain are classified as internally displaced persons (IDPs). In both cases, displacement has a range of social, economic and cultural consequences for those on the move and for those already residing in host areas. Not only have displaced people lost their livelihood assets, such as land and resources, but they may have lost their kinship and social connections. Displaced people subsequently become 'outsiders' in the areas where they settle, to the extent that this may create tensions within host communities. Such tensions have recently emerged in South Africa in response to the influx of Zimbabwean refugees.

Recent data from the UNHCR (2010) suggest that in 2009 there were over 6.4 million IDPs throughout Africa, and over 2.3 million refugees. From the data highlighted in Table 7.3, it is clear that conflict in Somalia, the DRC and Sudan has had a dramatic effect on the displacement of people: over 2 million people have been displaced *within* the DRC alone, while just under half a million of its population have fled to neighbouring countries. Many of those displaced will inevitably be drawn towards relief camps, where there is a chance that their basic needs will be met, until the security situation improves and repatriation can occur. Yet, the high concentration of people in relief camps creates other challenges. Food and water tend to be limited, and hence malnutrition and disease are endemic. For example, in the relief camps established around Goma in the aftermath of the Rwandan genocide in 1994, an estimated 50,000 people died as a result of cholera and dysentery linked to poor sanitation (Goma Epidemiology Group, 1995). The mixing of various diasporas within these camps also led to further ethnic violence.

7.4.7 Women

The consequences of conflict differ greatly between men and women, in a manner which usually reflects the wider existing gender divisions in society. Since men are more likely

Table 7.3 Refugees and internally displaced people in selected African countries, 2009

	Internally displaced people (assisted by UNHCR)	*Refugees (originating within country)*	*Refugees (arrivals)*
Kenya	399,000	9620	358,928
Chad	170,531	55,014	314,393
Sudan	1,034,140	368,195	152,375
DRC	2,052,677	455,852	185,809
Uganda	446,300	7554	127,345
Ethiopia	–	62,889	121,886
Tanzania	–	1204	118,731
Congo, Republic	–	20,544	111,411
Cameroon	–	14,766	99,957
Egypt	–	6990	94,406
Somalia	1,550,000	678,309	1815
Côte d'Ivoire	519,140	23,153	24,604
Angola	–	141,021	14,734
Eritrea	–	209,168	4751

Source: UNHCR, 2010

to be directly involved in fighting, and account for the majority of deaths, it is women and children who are most affected by the legacies of conflict. For example, the collapse of health services will have a greater impact on women, who are dependent upon reproductive and maternal healthcare. One effect of the disproportionate loss of males during conflict is an increase in the work burden of women, who are forced to extend their role in economic production, in addition to continuing their household and childcare activities. The death of an estimated 20 per cent of Rwanda's male population during the genocide left 60 per cent of women widowed and created a disproportionate number of female-headed households within the country. Although some would argue that there are some benefits to this in terms of women's empowerment, in reality it is often difficult for women to access land and credit, particularly if inheritance laws favour men (Luckham *et al.*, 2001).

Although recent research suggests that the gender-based violence during conflict is influenced by its prevalence in pre-conflict periods, there is little doubt that sexual and domestic violence towards women increases significantly during conflict periods. This can be explained as resulting from the 'breakdown' of society, but also by the fact that sexual violence has been used as a strategic weapon to spread fear and destabilize communities (El Jack, 2003). For example, in recent conflicts in Sudan, Uganda and particularly Rwanda (during the 1994 genocide), rape was used extensively as an 'ethnic cleansing' tool in an attempt to exercise power over specific ethnic groups. The effects of sexual violence are long term: victims of rape may find themselves ostracized within their communities and unable to marry, thereby compromising economic security. But sexual violence can also affect women's health in numerous ways, ranging from amateur abortion attempts to HIV infection and long-term psychological issues.

7.4.8 Children

In terms of vulnerability, children are arguably in a similar situation to women. Indeed, many children born into Africa's long-running conflict hotspots in West Africa, the Great Lakes region and the Horn of Africa have known little else. In addition to experiencing poor health and nutrition, displacement and a lack of education opportunities (and perhaps because of this), children are also being recruited into military forces. Recent estimates suggest that there are around 100,000 active child soldiers in Africa (Vautravers, 2009). While children have always been involved in fighting during periods of conflict throughout history, several explanations have been proposed for their increasing participation in African conflicts in recent years. The long duration of many of these conflicts has led to a decline in the availability of adult male combatants, and hence forces have turned their attention to children as potential recruits. Children are often psychologically manipulated by armed groups who emphasize the moral justifications of their cause; and, as occurred during the conflicts in Sierra Leone and Liberia, drugs such as marijuana and cocaine may be administered to instil fearlessness and a willingness to fight. In many cases, however, children have been forcibly abducted. Alternatively, there is evidence that some children join armed forces simply because they perceive this to be the only route out of a life of poverty (which itself has been created partly by long-term conflict). Others, meanwhile, are motivated by a desire to avenge the deaths of friends and relatives. Finally, it has been argued that the widespread availability of cheap and light firearms, which can be operated easily by children, has been a major driver (Vautravers, 2009).

Soldiering has a long-term effect on the lives of those children involved. In addition to sustaining physical injury, many are engaged in violent conflict during their formative learning years, and hence violent behaviour can become an embedded, 'normal' way of life. Indeed, this has been offered as one explanation for the way in which conflict tends to re-emerge within society. A key challenge for the post-conflict situation, therefore, is how to reintegrate physically and psychologically scarred child soldiers into society (Albertyn *et al.*, 2003).

7.5 Conflict resolution

As was pointed out above, many of Africa's conflicts have become protracted because there has been little incentive for armed groups who are profiting from it to stop fighting. This is particularly the case where access has been gained to lucrative natural resources, or where a return to pre-war inequalities is economically or ideologically unattractive for participants. Nonetheless, as is highlighted in Table 7.1, conflicts *do* end, usually due to a combination of several factors. First, decisive military action by one group may comprehensively defeat the opposition, as occurred in the Angolan government's victory over UNITA forces in 2002. Second, the intervention of external forces in a peacekeeping role may swing the balance of military action or lead to the demobilization of armed groups. Throughout the 1990s, ECOMOG (the Economic Community of West African States Monitoring Group) was active as a peacekeeping force in Liberia, Guinea-Bissau and Sierra Leone (see Box 7.3), although the effectiveness of external peacekeeping interventions has been mixed. While the end of the civil war in Sierra Leone in 2000 is widely attributed to the intervention of British troops who restored relative security to the country, UN peacekeeping operations in Rwanda (1994) and Somalia (1992–1995) failed to prevent an escalation of conflict.

Box 7.3

Regional peacekeeping: the role of ECOMOG in Liberia

ECOWAS, the Economic Community of West African States, was formed in 1975 with the aim of forging closer economic links and monetary union between participating countries, of which there were fifteen in 2010. Since its formation, ECOWAS has been dominated by Nigeria, which has the largest population and economy of the constituent countries. Among its founding principles was the need to maintain peace and security throughout the region, and in 1981 it enshrined in its constitution a protocol relating to mutual assistance on defence, legitimizing a collective response to any conflict that threatened the security of member states. Subsequently, in 1990, a regional peacekeeping force known as ECOMOG (Economic Community of West African States Monitoring Group) was deployed in the ongoing Liberian Civil War, with the aim of imposing a ceasefire between Samuel Doe's government forces and Charles Taylor's National Patriotic Front of Liberia, and establishing the conditions for an interim government. Around 3500 troops were initially deployed, with this figure rising to around 10,000 just before the force was withdrawn in 1998. However, despite ECOMOG's mission being considered a success in terms of restoring peace and security to Liberia, controversy has surrounded the nature of its intervention.

Very soon after arriving in Liberia and consolidating their position in Monrovia, ECOMOG forces were drawn into the conflict in a way that was perceived by many to go beyond their peacekeeping mandate. The rebel leader Charles Taylor regarded the Nigerian-led ECOMOG forces as invaders, and attacks on the peacekeepers themselves precipitated a shift from humanitarian peacekeeping to offensive military action. Consequently, ECOMOG was perceived to have aligned itself with the Doe government in Monrovia, and then, following Doe's death, as one of various factions vying for state control. The situation was compounded by political divisions between ECOWAS's francophone and anglophone states: Burkina Faso and Côte d'Ivoire opposed troop deployment and were critical of anglophone Nigeria's political ties with Doe's government (Burkina Faso and Côte d'Ivoire themselves were sympathetic towards Taylor). In order to placate these concerns, leadership of the ECOMOG forces was handed to a Ghanaian general, Arnold Quinoo, but this merely led to further problems of command, given that over 70 per cent of the troops were Nigerian. At various stages in the conflict, ECOMOG forces were accused of assisting all of the increasing number of rebel groups in one way or another, and being inconsistent in their approach to military action. The legitimacy of ECOMOG peacekeepers was eroded further by reports of widespread corruption and looting, earning them the alternative acronym of 'Every Car Or Movable Object Gone' among the local population.

In theory, a regional peacekeeping intervention such as ECOMOG's should have been considerably more effective than a typical UN-led peacekeeping operation, not least because the West African force possessed a more detailed and nuanced understanding of the culture, environment and politics of the region. Indeed, a lack of 'local knowledge' is often cited as a major problem for multinational UN peacekeeping operations. However, it was arguably the political closeness of all the states involved that led to confusion and disagreements over the precise role of ECOMOG troops. Furthermore, fear of the escalation and spill-over of the conflict into other parts of the region had a significant influence on the extent to which each state supported the intervention, particularly when the impartiality of ECOMOG was questioned. This lack of support – in terms of finance, resources or troops – effectively diminished the opportunities for clear and decisive action, and undoubtedly resulted in a more prolonged and bloody conflict.

Sources: Brown, 1999; Tuck, 2000; Olonisakin, 2008

Finally, with or without military victory, conflicts may be resolved through diplomacy and brokered negotiations culminating in a peace deal, the crux of which is usually some sort of power-sharing arrangement. Peace agreements and power-sharing initiatives, however, can be fragile, and at worst can lead to the renewal of hostilities. At their most cynical, power-sharing agreements can involve what Omeje (2008) regards as 'elite co-optation' – that is, the incorporation of influential individuals from opposition groups into positions of power – to bring conflicts to a close and effectively weaken the legitimacy of opponents. In Zimbabwe, for example, the merger between Robert Mugabe's ZANU party and Joshua Nkomo's opposition ZAPU party in 1987 following civil war between the two groups secured Mugabe's one-party control of the state and effectively silenced all opposition movements. More recently, the appointment of opposition Movement for Democratic Change (MDC) leader Morgan Tsvangirai to Prime Minister in 2009 can be viewed in similar terms. Power-sharing arrangements can also be problematic in that they effectively reward those involved in violent conflict, while at the same time marginalizing the more moderate voices in society (ADB, 2009). Such polarization in representation may lead to the emergence of new inequalities, and new opposition movements, hence providing a breeding ground for future conflict. Power sharing may also be unsustainable where parties control specific resources or territories, and are reluctant to give up their claims and dominance in those areas (this is a major issue hindering conflict resolution in the DRC). Disarmament is another key issue: parties are likely to retain their arms and military wings in anticipation of the failure of the conflict-resolution process.

7.6 The post-conflict period

Perhaps in response to the peak in the number of Africa's armed conflicts during the 1990s, much attention among development academics, practitioners and international organizations has focused on the so-called 'post-conflict period' – a transitional period between a state of conflict and peace, during which significant challenges emerge for peace, reconstruction, rehabilitation and development. Although individual circumstances dictate the nature and duration, and indeed the formal designation, of post-conflict periods, the period tends to be characterized by intense social, political and economic change. While ceasefire agreements and power sharing may bring conflict to an end, they do not signal an instant return to normality or a state of peace and security. Indeed, to the majority of people caught up in violent conflicts, there is little difference between the conflict and the immediate post-conflict periods, since very little changes in terms of state functioning, access to resources and livelihood opportunities. Post-conflict periods may also see a continuation of violence as small pockets of armed conflict in isolated areas, or they may be relatively short lived and simply provide an intermission between conflicts, as occurred in the Great Lakes region during the 1990s. Either way, the post-conflict period gives rise to a range of complex, interrelated development challenges, which to some extent are influenced by the original causes of conflict as well as the way in which the conflict ends (Green, 2001). Clearly, post-conflict rehabilitation and development are likely to be faster and easier where conflict has been short in duration, and where the impacts have been kept to a minimum. But it has also been argued that where the conflict itself is perceived to have been a liberation struggle, or initiated by external actors, then the rehabilitation will be easier than if the causes are more deeply entrenched in groups within society (Green, 2001; see Plate 7.4).

An important consideration during the post-conflict period, particularly for those involved in planning and development, is that a return to the 'normality' of pre-conflict is not necessarily desirable. For example, the inequalities that led to conflict may have been entrenched in pre-conflict government structures and societal institutions, and hence rebuilding these during the post-conflict period may simply reproduce the conflict at some point in the future. The post-conflict period therefore presents valuable opportunities for change which often go beyond socio-political reforms. The damaged economic sector can be restructured, enabling inefficient or subsidized industries or agriculture to be replaced with more profitable alternatives. There is also the opportunity for pre-conflict gender inequalities to be redressed, particularly given the redefining of women's roles in economic production and social support during the conflict period itself. This has occurred in post-conflict Liberia, where President Ellen Johnson-Sirleaf, the first woman to be elected head of an African country, has spearheaded a programme of gender mainstreaming that aims to increase sensitivity to women's issues in all government departments.

Until fairly recently, post-conflict reconstruction and development within many African countries have largely been undertaken in an *ad hoc* manner, involving numerous international and regional actors. In 2005, however, the New Partnership for Africa's Development (NEPAD) programme of the African Union launched its vision for post-conflict

Plate 7.4 Conflict memorial in Koidu, eastern Sierra Leone (Tony Binns).

reconstruction in the continent, providing a comprehensive framework identifying common issues, phases, dimensions and processes that should be addressed in pursuit of sustainable peace, security and development (NEPAD, 2005). While highlighting the common needs of post-conflict reconstruction, the strategy also emphasizes the importance of recognizing the individual circumstances of each country's conflict – that is, the often unique and complex causes and consequences of conflict that should explicitly inform the reconstruction process. Box 7.4 summarizes the NEPAD approach, which characterizes post-conflict reconstruction in terms of three phases (emergency, transition and development), in which five dimensions of reconstruction should be addressed.

Box 7.4

The NEPAD strategy for post-conflict reconstruction

Dimensions	*Emergency phase*	*Transition phase*	*Development phase*
Security	Establish a safe and secure environment	Develop legitimate and stable security institutions	Consolidate local capacity
Political transition, governance and participation	Determine the governance structures, foundations for participation, and processes for political transition	Promote legitimate political institutions and participatory processes	Consolidate political institutions and participatory processes

Dimensions	*Emergency phase*	*Transition phase*	*Development phase*
Socio-economic development	Provide for emergency humanitarian needs	Establish foundations, structures, and processes for development	Institutionalize long-term developmental programme
Human rights, justice and reconciliation	Develop mechanisms for addressing past and ongoing grievances	Build the legal system and processes for reconciliation and monitoring human rights	Established and functional legal system based on accepted international norms
Coordination and management	Develop consultative and coordination mechanism for internal and external actors	Develop technical bodies to facilitate programme development	Develop internal sustainable processes and capacity for coordination

Security

Security is of paramount importance. It facilitates emergency operations and reduces the likelihood of further outbreaks of violence. Security activities may involve the protection of returning refugees and IDPs, institutions and infrastructure. They may also focus on demobilization activities, the control of small arms and repatriation, while security sector reform is a key component of reconstruction activities.

Political transition, governance and participation

This dimension centres on the development of political and administrative institutions, through a process that maximizes the participation of a range of stakeholders. Activities include strengthening democratic civil society and government administrative organizations, facilitating the participation of ex-combatants in the political process, and developing mechanisms for conflict prevention and resolution.

Socio-economic development

This dimension reflects the need to rehabilitate or reconstruct basic social and economic services, and essentially rebuild livelihood opportunities for the population. Activities include the provision of humanitarian assistance (during the emergency phase), facilitating repatriation, building infrastructure such as roads, airports and telecommunications, rehabilitating markets and the financial sector, and restoring industry and agriculture.

Human rights, justice and reconciliation

In the post-conflict environment, there is a need to rebuild trust, ensure human rights are upheld, and promote reconciliation in the interests of building a sustainable peace. Activities may include judicial reform, establishing human rights monitoring systems, and instituting a truth and reconciliation process.

Coordination, management and resource mobilization

Building the capacity for coordination, management and resource mobilization is fundamental to facilitating the activities in each of the other dimensions. Reconstruction and rehabilitation require planning, cooperation and the coordination of other development strategies.

Source: NEPAD (2005)

Common to all conflict is an emergency period that immediately follows the cessation of hostilities. During this phase, a major goal will be establishing security, possibly through the deployment of African Union or UN peacekeeping forces. These should then provide an operational environment suitable for an emergency response in the form of humanitarian assistance from international NGOs, such as UNHCR (United Nations High Commission for Refugees) and ICRC (International Committee of the Red Cross). Following the emergency phase, the transition phase is characterized by the appointment of an interim government which oversees reconstruction until elections are held and a legitimate government is formed. There should also be a transition away from emergency relief to a wider process of reconstruction and recovery, characterized by the rebuilding of infrastructure, economy, security and social services, such as education and health. Finally, in the development phase, security operations are scaled down, and the responsibilities of external agencies are likely to be handed over to internal actors and government. However, external aid may continue for some time in support of sectoral development programmes, political reform and economic development. Although this phase may last up to ten years, it is likely that the repercussions of conflict will have to be dealt with for many years thereafter.

However, the NEPAD strategy has been criticized. For instance, Murithi (2006) highlights the absence of any reference to gender mainstreaming, which is an essential component of post-conflict reconstruction. Similarly, it is suggested that the framework does not adequately address the role that civil society groups can play in the immediate aftermath of conflict, and how they can work with the various other actors. Finally, it has been criticized for not explicitly addressing the role of culture in post-conflict reconstruction, again in terms of how indigenous systems of conflict prevention and resolution can be developed and integrated within civil society to avoid the outbreak of violent conflict. As Murithi points out, however, the key challenge lies in popularizing and institutionalizing such a comprehensive strategy among African governments and the various external actors involved in order to prevent *ad hoc*, uncoordinated responses in the future.

7.6.1 Disarmament, demobilization and reintegration

As can be seen in Box 7.4, there are many challenges for post-conflict reconstruction throughout all of these three periods in terms of security, governance and participation, socio-economic development, human rights, reconciliation and coordination. Within NEPAD's 'security' dimension, one of the key challenges for African countries in a post-conflict situation has been how to disarm, demobilize and reintegrate (DDR) rebel forces back into society and thereby reduce the risk of future conflict. Since 1989, DDR initiatives managed by the UN have been an integral part of post-conflict resolution in such countries as Angola, Liberia, Sierra Leone and Burundi. The DDR process usually begins with disarmament, characterized by the collection and destruction of weapons, a process which is often publicized widely to symbolize the end of the conflict. For example, in July 2007, various African heads of state attended a 'flame of peace' ceremony in Côte d'Ivoire, during which President Laurent Gbagbo and Prime Minister Guillaume Soro each set fire to a pile of weapons to signify the launch of the post-conflict disarmament process.

However, disarmament programmes have not always been successful. Despite numerous efforts by the Nigerian government to disarm ethnic groups in the Niger Delta, using the incentives of cash and employment, initiatives have failed to reduce the number of arms and levels of violent conflict (ADB, 2009). Meanwhile, in Liberia, the first attempts at disarmament in 2003 led to sporadic outbreaks of violence when large numbers of ex-combatants were turned away from under-resourced demobilization camps that were unable to cope with the demand (Daboah *et al.*, 2010).

Once disarmament has taken place, demobilization attempts to discharge combatants from their armed forces. This usually involves the processing of ex-combatants within demobilization camps, where they receive a discharge package. This has varied between countries, but common elements of these packages include finance, land allocation, job training and healthcare (with an emphasis on HIV/AIDS awareness). In Uganda during the 1990s, the school fees of veterans' children were paid for one year as part of the discharge package (Kingma, 1997). Clearly, the emphasis here is on providing a social and economic incentive for ex-combatants to return to civilian life.

However, it is perhaps the reintegration of ex-combatants within society that poses the greatest challenge for the whole DDR process (see Plate 7.5). As Solomon (2008) points out, there is the overarching question of 'reintegration into what?', since most ex-combatants will have been socialized into a vastly different way of life and hence cannot simply be deposited within a community and expected to revert to their pre-conflict livelihood system. First, as occurred in Liberia, there may be problems gaining access to land as a result of land expropriation by displaced peoples during the conflict. Second, ex-combatants may struggle to gain acceptance and forgiveness from their communities due to the erosion of trust during conflict. Resentment may be compounded if ex-combatants receive financial assistance through reintegration packages while local communities do not. Finally, many ex-combatants will suffer from mental health problems which render them unable to reintegrate fully into community life. In recognition of this, reintegration programmes in both Liberia and Sierra Leone have included psycho-trauma counselling and human rights training, alongside the provision of education and employment opportunities. While the DDR programmes in these two countries have been largely successful, it is widely acknowledged that a sustainable peace depends upon a long-term political and financial commitment to providing employment and sustainable livelihood opportunities for ex-combatants (Daboah *et al.*, 2010; Solomon, 2008; Hill *et al.*, 2008).

The challenges of DDR are much more acute when children have been directly involved in conflict. Children experience conflict in different ways to adult combatants: they are more likely to suffer from psycho-social trauma and aggression disorders, and, of course, they will not have gained a complete education (Albertyn *et al.*, 2003). Yet, because demobilization activities have largely adopted a policy of 'one man, one weapon' – an ex-combatant must give up a weapon in order to qualify for a discharge package – children have been excluded from many DDR programmes (Gislesen, 2006). In recent years, however, UNICEF and Save the Children have established separate DDR programmes for children in Angola, Sierra Leone, Sudan, the DRC and Rwanda. As well as providing

Plate 7.5 Ex-combatant training to be a blacksmith in Panguma, Sierra Leone (Tony Binns).

counselling, healthcare and education, children's DDR has focused on facilitating family reunification, since many children have been forcibly displaced as a result of conflict or, for various psycho-social reasons, are initially reluctant to return home.

7.6.2 Justice and reconciliation

A central element of building peace in the post-conflict period is bringing to justice those who have perpetrated war crimes, while simultaneously engaging in a wider process of reconciliation among all those involved in conflict. As Ginty and Williams (2009) suggest, the underlying rationale here is that it is important to ascertain the 'facts' of a conflict, in particular who is responsible for the killing and/or human rights abuses. In doing so, victims and perpetrators can reflect on their experiences and begin the process of rebuilding trust and functional relationships – important elements of most peace agreements and the path towards democratization. Non-judicial Truth and Reconciliation Commissions (TRCs) have increasingly formed part of the post-conflict reconstruction process throughout Africa in recent years, most famously in South Africa, where the TRC sought to investigate and highlight the injustices and abuses that occurred during the apartheid era. South Africa's TRC pioneered the idea of conditional amnesty, which allowed participants to gain amnesty from prosecution in exchange for the truth about their experiences and crimes (although amnesty was refused in some cases where the testimony of witnesses did not fulfil specific criteria). More recently, TRCs have been established in Sierra Leone (2002), Liberia (2005) and the DRC (2004), while demands for a TRC process were made in Zimbabwe throughout 2009.

While TRCs constitute a form of 'restorative justice', in that they focus on rebuilding the relationship between victims and perpetrators of crime within society, most post-conflict African countries have also engaged in a process of 'retributive justice' by prosecuting perpetrators in courts of law. For many, criminal investigations and war crimes tribunals represent the most effective way of reinforcing the rule of law, removing the perpetrators of crime from power, and deterring war crimes and human rights abuses in the future. While many war crimes tribunals throughout Africa have been held at the domestic level, there has been a shift in recent years towards embedding prosecutions within a framework of international law. In 1994, the International Criminal Tribunal for Rwanda (ICTR) was established to bring to justice the planners and leaders of genocide, crimes against humanity and other war crimes. Among those individuals standing trial was the former prime minister of Rwanda, Jean Kambanda, who pleaded guilty to genocide and was subsequently sentenced to life imprisonment.

However, despite significant progress in bringing the perpetrators of genocide to justice, and the completion of thirty-three cases, the ICTR has been heavily criticized for its long duration: as of 2010, the trials of twenty-three people were still in progress, with a further three awaiting trial. Meanwhile, thousands of lower-ranking individuals accused of complicity in the genocide have been brought to trial more conventionally through the Rwandan national court system, although the sheer number of cases and detainees have placed immense pressure on the domestic judicial and prison systems, which had all but collapsed during the 1994 conflict. In response to these pressures, the Rwandan government announced in 1999 that it would speed up the tribunal process by referring the accused to *gacaca* courts, a traditional community-based system for conflict resolution (see Box 7.5).

Box 7.5

Indigenous justice and reconciliation: *gacaca* courts in Rwanda

Despite trials of individuals commencing in 1996, Rwanda's judicial and prison systems were ill prepared to cope with the sheer number of prosecutions against suspected perpetrators of crimes during the genocide, and by the end of the 1990s an estimated 120,000 suspects remained in prison awaiting trial. At the time, it was estimated that completing these cases through the conventional courts would take more than 100 years. To manage the backlog of cases, in 2001 the Rwandan government reinstituted a traditional community approach to justice known as *gacaca* courts (in Kinyarwanda, the national language, *gacaca* means 'grass'; in the past, these hearings were conducted in the open air).

Although initially devised to speed up the number of prosecutions, many saw the potential of *gacaca* courts to act as mechanisms for establishing truth and reconciliation, which could help rebuild trust within communities, and in Rwandan society more generally. While the more serious genocide crimes were dealt with by the ICTR and the Rwandan national courts, those suspected of murder, assault and damage to property were diverted to the *gacaca* courts. Suspected perpetrators of violence could be given the opportunity to explain their actions, often in front of their victims, and seek forgiveness as a means of 'moving on'. The courts retain the power to impose sentences, including community work or imprisonment, although those who openly confess to their crimes are given a reduced sentence. The judges, known as *inyangamugayo*, which translates as 'those who detest dishonesty', are elected from within the community and can include women and young people. Following periods of training for *inyangamugayo* and the piloting of *gacaca* courts in specific areas, the system was introduced nationwide in 2005, with the expectation of having completed all the trials by June 2009. With the process nearing completion at the time of writing in 2010, over 12,000 *gacaca* courts have been engaged in over a million cases, resulting in the conviction of over 800,000 people.

Although the *gacaca* courts are generally considered to have had a positive impact on Rwandan society and the rebuilding of trust within communities, they have been criticized on several grounds. First, many have perceived the whole process as a form of 'victors' justice', in which Paul Kagame's Tutsi-led government sought retribution for the crimes committed by Hutu during the genocide. Crimes committed by the Rwanda Patriotic Front during the genocide, and subsequently during the spill-over conflict in the DRC 1996, are not considered within the remit of the *gacaca* courts. There has also been concern that the large number of convictions (approximately 14 per cent of Rwanda's total population in 1994) reflects political manipulation in the process, because judges are under pressure to convict as many people as possible. Furthermore, Longman (2009) suggests that in some cases the government encouraged *gacaca* courts to bring charges against those who were openly critical of the regime. Other critics have highlighted the lack of legal representation for the accused during the trials, which has led to convictions on the basis of unreliable witness testimonies. Finally, it has been suggested that the lengthy duration of the *gacaca* system has seen participation in it decline in recent years, as communities become weary and cynical of the process.

Despite these issues, the legacy that *gacaca* leaves behind is one of a pioneering innovation in community-based retributive and restorative justice for post-conflict periods. The central lesson to be drawn, however, is the need to avoid political interference in the process, and to ensure fairness and equity on all sides in terms of accountability for crimes committed.

Sources: Kanuma, 2003; Malan, 2008; Longman, 2009

The approach of Sierra Leone has been slightly different to that of Rwanda in that it commenced both restorative and retributive approaches to justice and reconciliation following the end of the 1991–2001 civil war. A TRC was established to document the causes, nature and extent of human rights violations and war crimes, and particularly the role of national and international organizations in the conflict. The TRC also broke new ground in that it collaborated with UNICEF in establishing suitable mechanisms through which children could be involved in the process of reconciliation and reintegration (May, 2006). Simultaneously, a Special Court for Sierra Leone (SCSL) was established in 2002 to prosecute the key perpetrators of war crimes and crimes against humanity. Most notably, the ex-president of Liberia, Charles Taylor, was indicted in 2003, and later transferred for trial to the International Criminal Court (ICC) in the Hague following concerns over regional security had his trial occurred in Sierra Leone. The ICC itself, established in 2002, has also been extensively involved in investigations of crimes against humanity, genocide and war crimes in the DRC, Uganda, the Central African Republic, Kenya and Sudan. In 2008, much media attention focused on its issuing of an arrest warrant for Sudan's President Omar al-Bashir relating to his responsibility for crimes against humanity and war crimes in Darfur. However, this was highly controversial within Africa, and was officially opposed by the African Union.

7.6.3 Preventing future conflict

There is a general consensus that the key to preventing future conflict in Africa lies in addressing and reducing the underlying causes, as outlined earlier in this chapter. From an economic perspective, the African Development Bank and the World Bank have emphasized the importance of facilitating economic growth throughout the continent, which in the long term will increase people's income, raise educational levels and promote livelihood diversification to the extent that conflict will become a less attractive option (Michailof *et al.*, 2002; ADB, 2009). Furthermore, the ADB regards the sustainable and equitable management of natural resources as central to both economic growth and conflict prevention; ensuring that revenues from natural-resource exploitation benefit the country as a whole rather than the minority and can reduce regional inequalities and the risk of conflict. However, there is a wider consensus that the elimination of poverty – in its economic, social or cultural sense – constitutes the way forward in building human security and preventing conflict in Africa (Salih, 2008).

Strengthening democratic institutions and instituting good governance at both the local and the national level is also widely regarded as a means of reducing the risk of conflict. Multi-party democracies, characterized by open and fair elections, independent judiciaries and central banks, transparency in economic activities, and a commitment to decentralization and public participation, have been regarded as a prerequisite to peace, security and economic growth (Michailof *et al.*, 2002; ADB, 2009). However, many have questioned the extent to which 'Western-style' democracy has been, and indeed might be, successful in Africa, given the diversity of ethnic groups and the continuing significance of tribal leadership and traditional institutions (Gebrewold, 2008). It has been suggested that the incorporation and promotion of traditional African values and philosophies into systems of democratic governance might embed a better framework for conflict prevention and resolution in society (Omeje, 2008). One example is the Swahili concept of *ubuntu*, which centres on group cooperation, respect, forgiveness, tolerance and generosity. This concept informed South Africa's widely respected Truth and Reconciliation Commission.

While recognizing what needs to be done is relatively easy, the implementation of economic, social and political reform to prevent conflict is much more challenging, particularly for those countries mired in a post-conflict period. It is therefore unsurprising that since the late 1990s, when conflict in Africa peaked, a significant amount of external aid and assistance has been linked to conflict-prevention initiatives and post-conflict development strategies. Within the discourse of development, poverty reduction strategies have sought to emphasize the links between human development, on the one hand, and security and conflict prevention, on the other. Similarly, the concepts of 'nation-building', 'good-governance', 'democratization' and indeed 'post-conflict reconstruction' itself began to enter the mainstream development discourse, and subsequently bilateral and multilateral aid interventions. For example, in 2001, the UK's Department for International Development established its 'Africa Conflict Prevention Pool', an initiative which aimed to support conflict-prevention capacity within Africa, post-conflict reconstruction and security sector reform. Specific interventions have included the provision of civil society peace initiatives in Kenya following the post-election violence in 2007, funding for peace talks between the Lord's Resistance Army (LRA) and the Ugandan government, and training assistance for security forces in Sierra Leone (DfID, 2008c). The World Bank, meanwhile, launched its State and Peace-Building Fund in 2008, with a budget of over $100 million, funding such projects as land sector reform in Liberia and support for gender-based NGOs in Côte d'Ivoire (World Bank, 2009b).

7.7 Conclusion

This chapter has drawn attention to the complex relationship between armed conflict in Africa and poverty, inequality and development. While one should be cautious about generalizing, given the diversity of experiences throughout Africa, it is clear that those countries affected by conflict are more likely to have inadequate health, education and economic infrastructure; they are more likely to experience environmental degradation and resource exploitation; and they are significantly more at risk of future conflict than those countries that have remained largely peaceful. Although it has been suggested that poverty and underdevelopment themselves fuel armed conflict (the so called poverty–conflict nexus), this has not always been the case. Many countries, for example Malawi and Zambia, have remained poor but have not descended into armed conflict. Clearly, although our current understanding of the nature of African conflicts helps us to predict, plan for and institute measures to prevent conflict, history has shown that conflict is often a result of a volatile, unpredictable mix of specific events and circumstances.

Possibly more than at any time in Africa's history, the continent and the rest of the world now appear better equipped to minimize the likelihood of further conflicts. The peak in the number of armed conflicts during the 1990s appears to have placed conflict firmly on the development agenda, while the Rwandan genocide of 1994, the US intervention in Somalia in 1995 and the Darfur conflict have provided sobering lessons for the international community in terms of what should be done, and what can be done.

Summary

1. Most African countries have been involved in some form of armed conflict during the last fifty years, with the number of conflicts peaking in the 1990s.
2. The causes of conflict in Africa are complex, diverse and interrelated. They have included competition for natural resources, ethnic tensions, colonial boundary legacies and persistent poverty.
3. Although conflict can have a direct impact on populations in terms of war casualties, the effects of poverty, underdevelopment and insecurity caused by conflict usually have longer-term impacts on mortality.
4. Conflict can have a major impact on the agricultural sector and lead to chronic food insecurity.
5. Many of Africa's refugee crises have been caused by conflict.
6. There is a marked gender dimension to conflict: women and children suffer greater long-term consequences than men.
7. UN peacekeepers have had mixed results in preventing and resolving conflict throughout the continent.
8. The post-conflict period poses new challenges for development in terms of establishing security, demobilizing armed groups, rebuilding institutions and infrastructure, and removing the original causes of conflict.

Discussion questions

1. Why might the impacts of conflict be difficult to assess?
2. What challenges face the establishment of security during the post-conflict period?
3. In what way is the boundary between the conflict and post-conflict periods often blurred?
4. How has globalization influenced armed conflict in Africa?
5. In what ways have Africa's conflicts affected rural livelihoods?

Further reading

African Development Bank (2009) *African Development Report 2008/2009: Conflict Resolution, Peace and Reconstruction in Africa*, Oxford: Oxford University Press.

Collier, P. and Hoeffler, A. (2004) 'Greed and grievance in civil war', *Oxford Economic Papers*, 56 (4): 563–595.

Francis, D. (2008) *Peace and Conflict in Africa*, London: Zed Books.

Ginty, R.M. and Williams, A. (2009) *Conflict and Development*, Abingdon: Routledge.

Luckham, R., Ahmed, I., Muggah, R. and White, S. (2001) *Conflict and Poverty in Sub-Saharan Africa: An Assessment of the Issues and Evidence*, IDS Working Paper 128, Brighton: Institute of Development Studies.

Turner, T. (2007) *The Congo Wars: Conflict, Myth and Reality*, London: Zed Books.

Useful websites

Centre for Conflict Resolution (South Africa): http://www.ccr.org.za/. The website of this research centre provides a wide range of current information and analysis of the scope of conflict, peacekeeping activities and post-conflict reconstruction across the continent.

NEPAD (the New Partnership for Africa's Development): http://www.nepad.org/. NEPAD is a programme of the African Union which aims to promote sustainable growth and development throughout Africa. The need for peace and security is implicit in many of NEPAD's strategies and development initiatives.

SIPRI (the Stockholm Peace Research Institute): http://www.sipri.org/. SIPRI is an independent international institute dedicated to research into conflict, armaments, arms control and disarmament. The website contains research reports, publications and databases with statistics on conflict and security trends and military expenditure.

UNHCR (United Nations High Commission for Refugees): http://www.unhcr.org. This UN agency leads and coordinates international action to protect refugees worldwide. The website has numerous reports, publications and updates outlining the state of Africa's conflicts and their implications for IDPs and refugees throughout Africa.

8 African economies

8.1 Introduction

It is a sad fact that many African countries were, in relative terms, economically worse off at the beginning of the twenty-first century than they had been at independence in the 1960s. A lamentable record of poor governance, manifested in political instability, endemic corruption and civil strife, together with environmental catastrophes, such as drought, has had a disastrous effect on economic and social progress. For the majority of ordinary Africans, 'development', in the shape of a recognizable improvement in quality of life, has simply not occurred. In fact, some writers have described the 1980s in particular as the 'lost decade' as far as development in Africa is concerned (Onimode, 1992: 1). In many sub-Saharan African countries, life expectancy is still less than 50 years (compared with 79 years in the UK), infant mortality is over 100 per 1000 live births (UK is 5 per 1000) and the adult literacy rate is below 50 per cent (UK is 99 per cent) (UNDP, 2007). At a time of increasing globalization in so many different ways, Africa generally lags well behind every other region of the world. For example, in the area of telecommunications, it is a staggering fact that in the mid-1990s, before the massive advances in mobile-phone technology, there were more fixed land-line telephones in New York City than in the whole of sub-Saharan Africa (APIC, 1996).

Considering the position of Africa in the global economy, although the continent in the late 1990s and early 2000s had the world's fastest-growing economy (Equatorial Guinea), it also included many of the world's worst-performing economies (Collier, 1998). In the 1990s, some economists referred to the progressive 'marginalization' of the continent, with its steadily falling share of world exports (reflecting a deterioration in producer prices) and the declining proportion of direct private investment into developing countries that was destined for Africa. As Collier observed at the time, 'the only international economic sphere in which Africa has remained non-marginal is aid' (Collier, 1995: 541). In fact, for a number of African countries, aid from a variety of donor agencies, together with remittances from migrants, represents their main form of participation in the world economy. In 2004, Africa received over four times the amount of net official ODA (overseas development assistance) transfers per head of population than were received in Asia, representing 37 per cent of total world ODA transfers. The Democratic Republic of the Congo (DRC) received 11 per cent of all ODA going to Africa in 2004, amounting to $2.8 billion, a figure which represents 39 per cent of the country's total GDP of $7.1 billion. Other major ODA recipients in 2004 were Tanzania, Ethiopia and Mozambique (each of which received $1.5 billion) (World Bank, 2007b).

Since the 'wind of change' brought independence to much of Africa during the 1960s, some commentators would argue that Africa has steadily lost the niche in the world economy that it had under colonialism (Agnew and Grant, 1997). Considering levels of foreign direct investment (FDI), a useful indicator of links with the global economy,

African countries in the 1990s were among the lowest in the world. In fact, a large number of African countries received FDI amounting to less than 0.5 per cent of the value of their GNP. Given the generally low levels of GNP in Africa, this means that the real value of FDI is exceedingly small. World Bank statistics showed that the three West African states of Burkina Faso, Niger and Togo received no FDI whatsoever in 1996 (World Bank, 1998: 230). As Collier observed around the same time, 'Africa is currently attracting only those investments which cannot be located elsewhere, such as mineral extraction or production for the domestic market. The major internationally footloose investments are simply by-passing Africa as a location' (Collier, 1998: 22).

Since 2000, there has been a significant improvement in FDI inflows to Africa, from $9 billion in 2000 to $88 billion in 2008. But inflows fell by 36 per cent in 2009 due to the economic crisis and, as the NEPAD–OECD Africa Investment Initiative points out, 'FDI in Africa, with a share of about 5 per cent of global flows, remains small compared to flows to and among industrialised and major emerging countries' (NEPAD–OECD, 2010: 1).

8.2 The size of African economies

This chapter will focus on the nature of African economies, their key features and the factors that have shaped them and may shape them in the future. Table 8.1 shows some key economic statistics for the countries of Africa. Using Gross Domestic Product (GDP) as a measure of the market value of all the output produced by a nation in one year, it is interesting to note that South Africa's GDP ($239.5 billion) was more than twice that of the country with the next highest GDP in Africa, Algeria ($102.3 billion). South Africa's GDP in fact represented 38 per cent of the total GDP for sub-Saharan African (SSA) countries, which is indicative of its economic position within the continent (see Box 8.1).

Table 8.1 Africa: key economic indicators

Country	*Population (millions) 2005*	*GDP ($ billion) 2005*	*GDP per capita ($) 2005*	*GDP per capita (annual growth rate %) 1990–1999*	*GDP per capita (annual growth rate %) 2000–2005*	*Percentage of population living on less than $1 a day 1990–2005*
Algeria	32.9	102.3	3112	–0.4	2.9	<2
Angola	15.9	32.8	2058	–1.7	6.3	–
Benin	8.4	4.3	508	1.1	1.0	30.9
Botswana	1.8	10.3	5846	3.7	5.6	28.0
Burkina Faso	13.2	5.2	391	0.9	1.8	27.2
Burundi	7.5	0.8	106	–2.8	–1.1	54.6
Cameroon	16.3	16.9	1034	–2.1	1.8	17.1
Cape Verde	0.5	1.0	1940	2.8	3.0	–
Central African Republic	4.0	1.4	339	–1.1	–1.6	66.6
Chad	9.7	5.5	561	–0.8	8.9	–
Comoros	0.6	0.4	645	–0.6	0.3	–
Congo, Republic	4.0	5.1	1273	–2.4	1.7	–
DRC	57.5	7.1	123	–8.2	–0.5	–
Djibouti	0.8	0.7	894	–4.3	0.2	–
Egypt	74.0	89.4	1207	2.4	2.0	3.1
Equatorial Guinea	0.5	3.2	6416	17.4	21.3	–

Table 8.1 Continued

Country	*Population (millions) 2005*	*GDP ($ billion) 2005*	*GDP per capita ($) 2005*	*GDP per capita (annual growth rate %) 1990–1999*	*GDP per capita (annual growth rate %) 2000–2005*	*Percentage of population living on less than $1 a day 1990–2005*
Eritrea	4.4	1.0	220	6.4	–3.3	–
Ethiopia	71.3	11.2	157	–0.1	3.6	23.0
Gabon	1.4	9.6	5821	–0.5	–0.6	–
Gambia	1.5	0.5	304	–0.4	1.2	59.3
Ghana	22.1	10.7	485	1.6	2.6	44.8
Guinea-Bissau	1.6	0.3	190	–1.0	–1.8	–
Guinea	9.4	3.3	350	1.1	0.7	–
Ivory Coast	18.2	16.3	900	–0.3	–2.5	14.8
Kenya	34.3	18.7	547	–0.6	0.9	22.8
Lesotho	1.8	1.5	808	2.7	2.6	36.4
Liberia	3.3	5.3	135	–3.2	–0.7	–
Libya	5.9	38.8	6621	–	0.9	–
Madagascar	18.6	5.0	271	–1.3	0.1	61.0
Malawi	12.9	2.1	161	2.0	0.2	20.8
Mali	13.5	5.3	392	0.9	2.8	36.1
Mauritania	3.1	1.9	603	0.0	0.6	25.9
Mauritius	1.2	6.3	12,715	4.2	3.2	–
Morocco	30.2	51.6	1711	1.1	2.1	<2
Mozambique	19.8	6.6	335	2.4	5.3	36.2
Namibia	2.0	6.1	3016	0.8	2.8	34.9
Niger	14.0	3.4	244	–1.4	–0.2	60.6
Nigeria	131.5	99.0	752	0.4	3.1	70.8
Rwanda	9.0	2.2	238	1.2	2.3	60.3
São Tomé and Príncipe	0.2	0.1	451	0.5	–	–
Senegal	11.7	8.2	707	0.0	1.9	17.0
Seychelles	0.1	7.2	8,209	3.3	–1.4	–
Sierra Leone	5.5	1.2	216	–5.2	7.9	57.0
Somalia	8.2	–	–	–	–	–
South Africa	46.9	239.5	5109	–0.8	2.4	10.7
Sudan	36.2	27.5	760	2.1	4.2	–
Swaziland	1.1	2.7	2414	0.6	0.6	47.7
Tanzania	38.3	12.1	316	0.2	4.2	57.8
Togo	6.1	2.2	358	–0.4	–1.1	–
Tunisia	10.0	28.7	2860	3.3	3.5	<2
Uganda	28.8	8.7	303	3.5	2.1	–
Zambia	11.7	7.3	623	–2.1	2.7	63.8
Zimbabwe	13.0	3.4	259	0.6	–6.4	56.1
NORTH AFRICA	152.9	313.4	–	1.4	2.2	–
SUB-SAHARAN AFRICA	743.7	629.5	–	–0.6	1.9	–
ALL AFRICA	896.6	942.9	–	0.0	1.9	–

Sources: World Bank, 2008a; UNFPA, 2007; UNDP, 2007

Box 8.1

South Africa in Africa

South Africa is the 'economic powerhouse' of Africa. The South African economy towers above that of all other African countries, with a GDP in 2005 of $239.5 billion, more than twice that of the next-largest economy, Algeria ($102.3 billion). In fact, South Africa's economy represents 38 per cent of the total GDP for sub-Saharan Africa and over 25 per cent of the entire continent (see Tables 8.2 and 8.3).

In addition to its economic strength, South Africa has a relatively well-developed infrastructure, with good road, rail and air networks. For example, the country has more than one-quarter of Africa's railway track and one-fifth of its paved roads (Binns, 1998). South African Airways (SAA) is the largest Africa-based airline, which, together with a number of other large international airlines, is a member of the worldwide Star Alliance network. In January 1997, South Africa was ranked sixteenth in the world in terms of level of connection to the internet, just behind Denmark, but ahead of Austria (Ahwireng-Obeng and McGowan, 1998). The country is also rich in mineral resources, producing some 54 per cent of the world's platinum, 33 per cent of its chromium and 27 per cent of its gold in 1993 (Africa Institute of South Africa, 1995).

Table 8.3 shows comparative statistics for eight selected African countries, including South Africa. Examining a range of non-economic variables, South Africa also appears to be much better off than most other African countries, although with an estimated HIV/AIDS adult (ages 15 to 49) prevalence rate of 18.8 per cent in 2005, life expectancy at birth declined from 64 years in the mid-1990s to 51 years in 2005. With an infant mortality rate of 55 per 1000 live births, South Africa ranks among the top twelve African countries, while its 17.6 per cent adult illiteracy rate is one of the lowest on the continent (UNDP, 2007). In 2005, South Africa was ranked 121 out of 177 in the Human Development Index (HDI), with an HDI of 0.674. Whereas several North African countries (Algeria, Egypt, Libya and Tunisia) have higher HDI values than South Africa, from sub-Saharan Africa only Gabon (119) is ranked higher (UNDP, 2007).

But such national-level statistics can conceal considerable social and spatial variation within the boundaries of a single country, especially in the case of South Africa, which of course has been a highly unequal society for decades (Durning, 1990). For instance, the HDI value of the white population may well be twice that of the black population. It has been suggested that before the end of apartheid in 1994, the white minority had an HDI similar to that of Spain (0.878), while the vastly more numerous black population had an HDI just above that of the Republic of the Congo (0.462) (UNDP, 1994). Adedeji, writing in 1996, commented, '5 per cent of the population own 88 per cent of the country's wealth; four large corporations control 81 per cent of share capital, and 50,000 White farmers own 85 per cent of all agricultural land' (Adedeji, 1996: 15). And Turok asserted, 'The apartheid legacy in the economic field is that a minority of a White minority established a kind of "skyscraper economy" which benefited relatively few and marginalised the great majority' (Turok, 1993: 8). So, while South Africa seems to be relatively favoured in economic and social terms when compared with other sub-Saharan African countries, it will probably be several generations before apartheid's legacy of unacceptable inequality is finally overcome.

Table 8.2 South Africa and other significant African economies

Country	*Population (millions) 2005*	*Land area (,000 sq km) 2005*	*Population density (people per sq km) 2005*	*GDP ($ billion) 2005*	*GDP per capita (purchasing power parity, $) 2005*	*HDI rank (out of 177) 2005*
Algeria	32.9	2382	13.8	102.3	7062	104
Congo (DRC)	57.5	2267	25.4	7.1	714	168
Côte d'Ivoire	18.2	318	57.2	16.3	1648	166
Egypt	74.0	995	74.4	89.4	4337	112
Morocco	30.2	446	67.7	51.6	4555	126
Nigeria	141.4	911	144.3	99.0	1128	158
South Africa	47.9	1221	38.6	239.5	11,110	121
Tunisia	10.0	155	64.5	28.7	8371	91
Sub-Saharan Africa	743.7	23,628	31.5	629.5	–	–
UK	60.2	242	249	2,001.9	33,238	16
USA	299.8	9159	32.7	12,416.5	41,890	12

Sources: World Bank, 2008a; UNFPA, 2007; UNDP, 2007

Table 8.3 Comparative statistics for selected African countries

Country	*Land area (,000 sq km)*	*Population (millions) 2005*	*Average annual population growth rate (%) 1975–2005*	*Life expectancy at birth 2005*	*Infant mortality (per 1000 live births) 2005*	*GDP per capita ($) 2005*	*HDI rank (out of 177) 2005*
South Africa	1221	47.9	2.1	51	55	5109	121
Botswana	567	1.8	2.7	48	87	5846	124
Egypt	995	74.0	2.1	71	28	1207	112
Ghana	228	22.5	2.6	59	68	485	135
Kenya	569	35.6	3.2	52	79	547	148
Mali	1220	11.6	2.5	53	120	392	173
Nigeria	911	141.4	2.8	47	100	752	158
Zimbabwe	387	13.1	2.5	41	81	259	151
UK	242	60.2	0.2	79	5	36,509	16
USA	9159	299.8	1.0	78	6	41,890	12

Source: UNDP, 2007

Many white South Africans freely admit that South Africa is a 'Third World' country. This view was echoed in 1990 by Pik Botha, Foreign Minister in F.W. de Klerk's National Party government, at a conference in Pretoria on the theme 'Southern Africa towards the Year 2000'. Botha (1990: 216) commented, 'we are still evolving from an agricultural and mining economy to a manufacturing economy . . . South Africa ought to be classified as a developing country and not a developed country . . . and we ought to qualify for international advantages granted to developing countries'. Despite such sentiments, many white South Africans frequently display a strong reluctance to accept the fact that they are 'Africans', a term which was formerly reserved for both South Africa's and the continent's black population, rather than *all* inhabitants of Africa. There has been a longstanding dualistic tendency among South African politicians, academics and others, such that in one sense they display a somewhat myopic and parochial view in focusing their attention entirely on South Africa, while in another they have tended to look over and beyond the rest of the African continent and towards their traditional links with Europe and North America. For many South Africans (and by no means just the white population), civilization and development go no further than the Limpopo River along the country's northern border with Zimbabwe.

The enduring links with Europe and North America are still reflected in South Africa's trading network. In 2006, South Africa's main trading partners were Germany, the USA, Japan, the UK and China, followed by Saudi Arabia, France, the Netherlands, Italy and Iran (Editors Inc., 2006). Manufacturing (e.g. assembled automobiles and components) now accounts for over 60 per cent of exports while minerals (e.g. platinum, gold and coal) have declined to about 30 per cent. Imports are dominated by petroleum products, machinery, raw materials, chemicals and consumer goods. Europe still overwhelmingly dominates South Africa's trading relationships, with the EU accounting for over 40 per cent of its imports and exports, as well as 70 per cent of foreign direct investment (SouthAfrica.info, 2008). However, the importance of South Africa's trading activities in the rest of Africa should not be underestimated. As early as 1995, the South African government had already established twenty-two trade missions in African countries and it was recognized that sub-Saharan Africa represented a potentially enormous market for South African businesses. Trade with countries in the Southern African Development Community (SADC) has increased in recent years and now accounts for 16 per cent of South Africa's total trade (SouthAfrica.info, 2008).

South Africa and the southern Africa region

The statistics presented in Table 8.4 indicate the overwhelmingly dominant position which South Africa occupied among the fourteen SADC countries in 2005. Apart from the Seychelles and Botswana, South Africa has the highest GDP per capita ($5109) and a total national GDP of $239.5 billion, which represents over 70 per cent of the combined GDP of the SADC countries.

The Southern African Development Coordination Conference (SADCC) originated among the so-called 'frontline states' of Angola, Botswana, Mozambique and Tanzania, which in the 1970s were collectively concerned about the Rhodesian and South-West African conflicts. This attempt at regional cooperation was formalized with the Lusaka Declaration of April 1980, when nine southern African states came

Table 8.4 The fourteen SADC countries

Country	*Population (millions) 2005*	*Land area (sq km)*	*Population density (people per sq km) 2005*	*GDP ($ billion) 2005*	*Share of SADC GDP (%) 2005*	*GDP per capita ($) 2005*	*GDP average annual growth (%) 1990–2005*
Angola	16.1	1247	12.9	32.8	9.7	2058	1.5
Botswana	1.8	567	3.2	10.3	3.0	5846	4.8
Congo (DRC)	58.7	2267	25.9	7.1	2.1	123	–5.2
Lesotho	2.0	30	67	1.5	0.4	808	2.3
Malawi	13.2	94	140.4	2.1	0.6	161	1.0
Mauritius	1.2	2	600	6.3	1.9	5059	3.8
Mozambique	20.5	784	26.1	6.6	1.9	335	4.3
Namibia	2.0	823	2.4	6.1	1.8	3016	1.4
Seychelles	0.1	0.45	222	0.7	0.2	8209	1.5
South Africa	47.9	1221	39.2	239.5	70.8	5109	0.6
Swaziland	1.1	17	64.7	2.7	0.8	2414	0.2
Tanzania	38.5	884	43.6	12.1	3.6	316	1.7
Zambia	11.5	743	15.5	7.3	2.2	623	–0.3
Zimbabwe	13.1	387	33.9	3.4	1.0	259	–2.1
Total	227.7	9066.5	92.6 (average)	338.5	100	2452.6 (average)	1.11 (average)

Source: UNDP, 2007

together in the SADCC: Angola, Botswana, Lesotho, Malawi, Mozambique, Swaziland, Tanzania, Zambia and Zimbabwe. Namibia joined the group after gaining independence in 1990.

The principal aims of the SADCC were to strengthen links, promote regional cooperation and integration and reduce economic dependence on South Africa. Within the continent as a whole, which has had a somewhat patchy record with regard to regional groupings, it has been suggested that the SADCC was 'one of Africa's few successful attempts at regional cooperation' (Morna, 1990: 49). As a tangible result of the SADCC's endeavours, air links and satellite telecommunications between member countries have been improved, and there has been considerable success in obtaining aid from international bodies, such as the World Bank, the European Union and, most notably, the Scandinavian countries. But, on the negative side, relatively little progress has been made in reducing external dependence and in particular the region's dependence on South Africa.

In the early 1990s, with the end of the Cold War and the emergence of democratic rule in South Africa, the emphasis among SADCC member states gradually switched from specific projects to the coordination of sectoral plans and programmes. In October 1993, the Treaty of Windhoek formalized the replacement of the SADCC with the SADC, and the ten countries were joined by a further two – South Africa in 1994 and Mauritius in 1995. In 1996, President Mandela was elected chair of the SADC for three years. Then, in September 1997, at its Blantyre summit meeting, the SADC somewhat surprisingly admitted a further two countries – the Democratic Republic of the Congo (formerly Zaire) and the Seychelles – thus enlarging the community to its current fourteen member states. South Africa strongly supported admitting the DRC, not least because of the considerable business interest of the South African state-owned utility Eskom in the DRC's hydro-electric power resources. In fact, even before it was admitted to the SADC, hydro-electric power from the south of the DRC was being used in several SADC states.

Between 1990 and 2005, economic growth in the SADC as a whole averaged 1.11 per cent per annum (see Table 8.4), although there was considerable variation in average annual growth rates between the relative 'high-fliers', such as Botswana and Mozambique (4.8 per cent and 4.3 per cent, respectively), and the less successful economies of Zambia, Zimbabwe and the DRC, all of which had negative annual economic growth rates (0.3 per cent, –2.1 per cent and –5.2 per cent, respectively).

Despite optimism about the SADC's considerable potential both in the region and throughout the whole of Africa, the organization has suffered criticism and scepticism from various quarters. For instance, Sidaway (1998: 559) suggests that 'The core business of the SADC lies in the putting together of bids for development programmes and in soliciting finance for these. In practice, all SADC-mediated development programmes function on the basis of external aid.' In 1995 alone, various donors pledged over $4 billion of aid and development funding to the SADC.

However, the community has held together, and there is widespread recognition of its important role in promoting regional integration, developing intra-regional trade and infrastructure (transport, telecommunications, water, power), achieving economies of scale in manufacturing production and controlling disease and ethnic conflict. The SADC has already brokered agreements on shared water resources, drug trafficking and food security, and it hopes to set up a customs union by 2012 as part of a strategy to stimulate industrial development and boost trade and investment in the region. Such a union would put the SADC in the top twenty regional economies of the world, with a combined GDP of an estimated $300 billion in 2012 (Editors Inc., 2006).

South Africa's agenda for the future development of the SADC undoubtedly recognizes the opportunity for exploiting resources and markets in the region, and this is seen as a potential threat by the other member states. As Johann Wingard, vice-president of the South African Chamber for Agricultural Development in Africa, commented, 'Our economic border is no longer the Limpopo River – there are unbelievable opportunities out there' (quoted in Pickard-Cambridge, 1998: 13). The other SADC members can be very suspicious of South Africa, on account of its massive economy, its strong world profile, and not least because the SADCC was originally established to counter South Africa's apartheid regime. As one South African diplomat has commented, 'There is a sense in which SADC is joining South Africa, not that South Africa is joining SADC' (quoted in Sidaway, 1998: 568).

The overwhelming dominance of South Africa within the SADC has been likened to neo-imperialism and hegemony, and could well lead to pressure for some form of import restriction from neighbouring countries if trade with South Africa all flows in one direction. Furthermore, with much concern about unemployment levels in South Africa, it is likely that opportunities for migrants from neighbouring countries might decline. A further concern relates to a damaging 'brain drain' among educated professionals from neighbouring countries as they are attracted to higher salaries in South Africa.

The question must also be raised about South Africa's ability to divert much-needed resources into a range of international initiatives when there are so many complex and expensive priorities to be dealt with at home. Foreign Minister Alfred Nzo, in his address to the UN General Assembly in June 1994, reminded the international community of his country's domestic priorities:

> uppermost in our minds, however, are the responsibilities which our new Government of National Unity has towards the people of South Africa. Our primary goal is to strive to create a better life for all our people . . . [Consequently,] South Africa will have extremely limited resources for anything which falls outside the Reconstruction and Development Programme.
>
> (Nzo, 1994: 12–13)

South African businesses have certainly recognized the potential for penetration into other parts of Africa since 1994. Their aggressive expansion is exemplified in the operations of three very different companies, chosen from a long list. The Anglo American Corporation has established mining operations in Angola, Botswana, Ghana, Guinea, Ivory Coast, Mali, Namibia, Senegal, Tanzania, Zambia and Zimbabwe; SA Breweries is involved in beer and soft-drink production in Botswana, Lesotho, Mozambique, Swaziland, Tanzania and Zambia; and the South African hotel chain Protea has set up hotels in Botswana, Egypt, Kenya, Malawi, Mauritius, Swaziland, Uganda, Zambia and Zimbabwe (Ahwireng-Obeng and McGowan, 1998).

South Africa obviously has an important role to play in both the region and the continent during the twenty-first century. It has been suggested that economic, political and demographic power in Africa rests with just four countries: Egypt in

North Africa, Nigeria in West Africa, Kenya in East Africa and South Africa in southern Africa (Davies, 1996). As former US Under-Secretary for International Trade Jeffrey E. Garten (1996: 8) has commented, 'South Africa is more than an inspiring story. It is a nation that represents the last best hope of a continent.'

Table 8.5 shows the five African countries with the largest GDPs, and it can be seen that the South African economy dominates the entire continent, with other countries having significantly smaller economies. Table 8.6 lists the five countries with the largest per capita GDPs, and shows that the Indian Ocean island state of Mauritius is well ahead of its closest rivals, with a per capita GDP of $12,715 in 2005. Interestingly, all five of these countries have relatively small populations, ranging from Libya, with 5.9 million people, to the Seychelles, with only 100,000. The per capita GDP figures for the countries with the largest populations are significantly lower than those in the top five, apart from South Africa, which, with a population of 46.9 million in 2005, had a per capita GDP of $5,109 (see Table 8.7). Meanwhile, Nigeria, Africa's most populous country, with 131.5 million people, had a per capita GDP of only $752 in 2005, while the DRC, with 57.5 million people, had a per capita GDP of only $123 – the second lowest in the world, according to the UNDP (2007), after the central African state of Burundi ($106).

It is interesting to compare economic growth rates over a relatively long period of time. Annual growth rates of GDP, as measured on a per capita basis, vary considerably between African states for the two periods 1990–1999 and 2000–2005. For both periods, the annual per capita growth rates for the North African countries were higher than for SSA, but between 2000 and 2005 the rate of growth accelerated in the latter from –0.6 to 1.9 per

Table 8.5 African countries with the largest Gross Domestic Product

Rank	*Country*	*GDP ($ billion) 2005*	*GDP as % of total Africa GDP 2005*
1	South Africa	239.5	25.4
2	Algeria	102.3	10.8
3	Nigeria	99.0	10.5
4	Egypt	89.4	9.5
5	Morocco	51.6	5.5
	Total Africa	942.9	100

Source: UNDP, 2007

Table 8.6 African countries with the largest GDP per capita

Rank	*Country*	*GDP per capita ($) 2005*	*Population (millions) 2005*
1	Mauritius	12,715	1.2
2	Seychelles	8209	0.1
3	Libya	6621	5.9
4	Equatorial Guinea	6416	0.5
5	Botswana	5846	1.8

Source: UNDP, 2007

cent per year. Table 8.8 shows the top five countries in terms of annual increase in per capita GDP for 1990–1999 and 2000–2005. For both periods, Equatorial Guinea had the highest growth rates of 17.4 per cent and 21.3 per cent, respectively. The only other country which was in the top five for both periods was Botswana, with growth rates of 3.7 per cent and 5.6 per cent, respectively.

At the other end of the economic spectrum, Table 8.9 shows the five countries with the lowest annual per capita GDP growth rates in the same two periods. In the first period, the

Table 8.7 GDP per capita for Africa's most populated countries

Country	*Population (millions) 2005*	*GDP per capita ($) 2005*
Nigeria	131.5	752
Egypt	74.0	1207
Ethiopia	71.3	157
DRC	57.5	123
South Africa	46.9	5109

Source: UNDP, 2007

Table 8.8 African countries with highest annual growth in per capita GDP, 1990–1999 and 2000–2005

Country	*GDP per capita annual growth (%) 1990–1999*	*Country*	*GDP per capita annual growth (%) 2000–2005*
Equatorial Guinea	17.4	**Equatorial Guinea**	21.3
Eritrea	6.4	**Chad**	8.9
Mauritius	4.2	**Sierra Leone**	7.9
Botswana	3.7	**Angola**	6.3
Uganda	3.5	**Botswana**	5.6
NORTH AFRICA	1.4	**NORTH AFRICA**	2.2
SUB-SAHARAN AFRICA	–0.6	**SUB-SAHARAN AFRICA**	1.9
ALL AFRICA	0.0	**ALL AFRICA**	1.9

Source: UNDP, 2007

Table 8.9 African countries with lowest annual growth in per capita GDP, 1990–1999 and 2000–2005

Country	*GDP per capita annual growth (%) 1990–1999*	*Country*	*GDP per capita annual growth (%) 2000–2005*
DRC	–8.2	**Zimbabwe**	–6.4
Sierra Leone	–5.2	**Eritrea**	–3.3
Djibouti	–4.3	**Ivory Coast**	–2.5
Liberia	–3.2	**Guinea-Bissau**	–1.8
Burundi	–2.8	**Central African Republic**	–1.6
NORTH AFRICA	1.4	**NORTH AFRICA**	2.2
SUB-SAHARAN AFRICA	–0.6	**SUB-SAHARAN AFRICA**	1.9
ALL AFRICA	0.0	**ALL AFRICA**	1.9

Source: UNDP, 2007

DRC had an annual growth rate of –8.2 per cent, while in the latter period Zimbabwe recorded a growth rate of –6.4 per cent. No country featured among the worst five performers during both periods.

The figures shown in these tables generally reflect specific events in the respective countries: for example, political instability and/or civil war in the DRC, Liberia, Ivory Coast, Sierra Leone and Zimbabwe. Those countries which performed particularly well during either or both of the two periods also generally have a 'story behind the statistics', such as oil exploration in Equatorial Guinea and Chad, mineral wealth in Botswana and Angola, the development of a manufacturing sector in Mauritius, and post-conflict reconstruction in Sierra Leone after the end of the civil war in 2001 (see Box 8.2).

Box 8.2

Growth without development in Equatorial Guinea

For a number of years, the small central African state of Equatorial Guinea has had the distinction of boasting the most rapid economic growth rate of all African countries. In 2007, its real GDP growth was estimated to be 9.8 per cent, yet the country was still ranked well down the list of countries (127th out of 177) in the UNDP's *Human Development Report* (UNDP, 2007) – seven places below its ranking the previous year. Among the 600,000 people who live in Equatorial Guinea, life expectancy at birth was only 50.4 years in 2005, and 57 per cent of households did not have access to an improved water source. The country's maternal mortality rate in 2005 was 680 per 100,000 live births, while the under-five mortality rate in that year was the highest among UNDP-classified 'medium human development' countries, at 205 deaths for every 1000 live births. Malaria is the main cause of death among children, but there is a low level of vaccination against a range of childhood diseases, such as measles, diphtheria and whooping cough. As the African Development Bank comments, 'Despite sustained growth and a considerable influx of revenues, the social situation in Equatorial Guinea is way behind all expectations' (ADB, 2008: 297). Although detailed human development statistics for the country are difficult to obtain, it seems that 60 per cent of people in urban areas and 70 per cent in rural areas were living in extreme poverty on less than $1 a day between 1994 and 2001, and a household survey conducted in 2006 concluded that 76.8 per cent of the population was 'poor' (ADB, 2008).

Equatorial Guinea, with a total area of only 28,051 sq km, consists of five inhabited islands and a portion of the mainland. The country gained its independence from Spain in 1968 and, following a coup in 1979, President Teodoro Obiang Nguema Mbasogo has ruled with almost total control. Supposedly democratic presidential elections in 1999 and 2004 are widely regarded as having been seriously flawed. Large oil reserves were discovered offshore in 1995, and oil production increased from 81,000 to 363,000 barrels a day between 1998 and 2007. This enabled the country to become the third-largest African petroleum exporter after Nigeria and Angola. There are three major oilfields (Zafiro, Ceiba and Alba), with significant international investment from a number of petroleum companies from the USA (e.g. Exxon Mobil, Chevron, Triton Energy), Australia, Malaysia, South

Africa and Switzerland. The rights to explore in the country's newest oilfield were granted to the Chinese National Offshore Oil Company, with two new oil rigs completed in 2009. Large natural gas reserves have been discovered off Bioko Island, where the capital city, Malabo, is located. In 2001, the government created GEPetrol, a national oil company, and Sonagas a natural gas company.

The former mainstays of the economy – cocoa, coffee and timber – are now relatively insignificant when compared with oil production and a growing construction industry. The government is attempting to liberalize and diversify the economy, placing more emphasis on agriculture, tourism and fishing and an open investment regime that aims to attract greater private sector foreign investment. In 2007, the USA was the largest bilateral foreign investor in Equatorial Guinea with over $12 billion of investments, while significant trade was undertaken with the European Union, focusing on petroleum and timber to Europe and a range of manufactured goods from the EU to Equatorial Guinea.

However, a lack of transparency and widespread corruption make it difficult for foreign companies to conduct business in Equatorial Guinea, and in the World Bank's assessment of the business environment, the country was ranked 165th in the world, with an average time to set up a business of 136 days, compared with just 20 days in Cameroon (ADB, 2008). In recent years, there has been considerable investment in telecommunications, power and water supply, roads, airports and ports, as well as schools and healthcare, yet the country still has a long way to go, since just a decade ago these services were in an extremely poor state, even by African standards. It remains to be seen whether at least some of the massive wealth being generated by oil and gas exports will be channelled into improving the quality of life for the bulk of Equatorial Guinea's population.

A report published by Human Rights Watch in 2009 was highly critical of the Equatorial Guinea regime, arguing that President Obiang has used the oil boom to entrench and enrich himself at the expense of the country's people (HRW, 2009). In launching the report, the Business and Human Rights Director of Human Rights Watch commented, 'Here is a country where people should have the per capita wealth of Spain or Italy, but instead they live in poverty worse than in Afghanistan or Chad . . . This is a testament to the government's corruption, mismanagement and callousness toward its own people' (HRW, 2009: 2). There have been several high-profile scandals: for example, when President Obiang bought two mansions in Washington at a cost of $3.8 million in 2004; and when his eldest son spent $35 million on property in California in 2006. Meanwhile, press freedom in Equatorial Guinea is strictly controlled, the political opposition has been constantly harassed, and in 2008 President Obiang won 99 out of 100 seats in the parliamentary elections. Human Rights Watch has called on him and his government to ensure complete public disclosure of how they manage the country's oil wealth, and has urged foreign governments, notably the USA and Spain, to exert pressure on the government to improve its human rights record (HRW, 2009).

8.3 The shape of African economies

Table 8.10 shows the contributions of the three main economic sectors – agriculture, industry and services – to the GDP of African countries, together with the percentage of employment in each sector. Such a breakdown gives a good insight into the structure of each economy. Table 8.11 shows the sectoral breakdown regarding employment for 1997 and 2007. If we look at the relative significance of the different sectors in the various African economies, in terms of both contribution to GDP and employment, we can see a fundamental difference between the North African countries and those of SSA. In the latter, the agricultural sector is often still the dominant sector in terms of employment, while employment in industry and services is less significant. In North African countries, the proportion of employment in agriculture is about half of the SSA figure, while employment in industry and services is about twice the average rate of SSA countries (Table 8.10). Agriculture in the North African countries contributes only between 2 per cent (Libya) and 13 per cent (Egypt) to GDP, whereas in SSA it contributes over 50 per cent of GDP in such countries as the Central African Republic, Guinea-Bissau, Liberia and Somalia. In some SSA countries, agriculture employs over 80 per cent of the working population: for example, Angola, Burundi, Ethiopia, Malawi, Niger and Rwanda. Employment in agriculture is less than 5 per cent in Djibouti and Eritrea, both small countries with important port facilities and large service sector employment. The other SSA country with under 5 per cent employed in agriculture is Nigeria, which has well-developed service and industry sectors. In those countries where industry contributes a large proportion of the GDP, this is generally due to the development of the oil industry: for example, Algeria and Libya in North Africa, and Angola, the Republic of the Congo and, most notably, Equatorial Guinea in SSA. As we saw earlier (Table 8.8), Equatorial Guinea has maintained the highest economic growth rates in the continent over the last fifteen years (Box 8.2). In the North African countries, industrial employment averaged 22 per cent in the period 1996–2005, whereas it averaged only 11 per cent in SSA, with countries such as Burundi, Ethiopia, Malawi, Rwanda and Tanzania having less than 5 per cent employed in the industrial sector. In North Africa, employment in the service sector averaged 44 per cent (1996–2005), whereas in SSA the average was only 27 per cent. But this average figure was affected by considerable service sector employment in small countries such as Djibouti, Eritrea and the Seychelles, and also in two of the largest economies, Nigeria and South Africa, which each had a service sector employing over 60 per cent of the working population.

Table 8.11 compares employment in the three main sectors in 1997 and 2007. In both North Africa and SSA, the proportion of the working population employed in agriculture declined during the period, but the figures are still considerably higher than in developed economies and the EU, where in 2007 only 3.9 per cent were employed in agriculture. Employment in industry shows only a small increase in both North Africa and SSA, with the former now approaching the level of industrial employment in developed economies and the EU, but SSA still lagging well behind. Across the continent, employment in the service sector increased during the period, with a higher growth rate in SSA, but with both SSA and North Africa still having significantly lower service sector employment levels than developed economies and the EU. However, both Nigeria and South Africa have service sector employment which is at a similar level to that found in more developed economies.

A characteristic feature of many African economies is the dominance of a single or small number of commodities – what might be termed 'one-product economies'. As Table 8.12 shows for a five-year period in the 1990s, in addition to the dominance of oil in several

Table 8.10 Contributions of agriculture, industry and services to GDP and employment

Country	*Agriculture value added (% of GDP) 2007*	*Industry value added (% of GDP) 2007*	*Services value added (% of GDP) 2007*	*Employment in agriculture (%) 1996–2005*	*Employment in industry (%) 1996–2005*	*Employment in services (%) 1996–2005*
Algeria	8	61	30	21	26	53
Angola	10	68	22	85	5	10
Benin	32	13	54	–	–	–
Botswana	2	51	47	26	24	50
Burkina Faso	33	22	44	90	5	5
Burundi	35	20	45	93	3	4
Cameroon	19	29	52	61	9	23
Cape Verde	9	17	74	–	–	–
Central African Republic	56	16	28	–	–	–
Chad	23	44	32	80	10	10
Comoros	40	5	55	80	10	10
Congo, Republic	5	60	35	–	–	–
Congo (DRC)	42	28	29	–	–	–
Djibouti	17	23	60	2	8	80
Egypt	13	36	51	30	20	50
Equatorial Guinea	3	92	5	–	–	–
Eritrea	18	24	58	4	19	77
Ethiopia	46	13	40	92	3	5
Gabon	5	60	35	60	15	25
Gambia	33	9	58	75	19	6
Ghana	36	25	38	55	14	31
Guinea-Bissau	62	12	26	82	8	10
Guinea	17	45	38	76	10	14
Ivory Coast	23	26	51	68	12	20
Kenya	23	19	58	19	20	61
Lesotho	15	45	40	60	17	23
Liberia	66	16	18	70	8	22
Libya	2	81	17	17	23	59
Madagascar	27	15	58	78	7	15
Malawi	34	20	45	90	3	7
Mali	37	24	39	80	8	12
Mauritania	13	47	41	50	10	40
Mauritius	5	25	70	14	36	50
Morocco	12	29	59	44	20	36
Mozambique	28	27	45	81	6	13
Namibia	11	31	58	32	12	56
Niger	40	18	42	90	6	4
Nigeria	33	39	28	3	22	75
Rwanda	36	14	50	90	3	7
São Tomé and Príncipe	17	13	70	–	–	–
Senegal	15	22	63	77	10	13

Seychelles	3	25	72	10	19	71
Sierra Leone	44	24	32	–	–	–
Somalia	65	10	25	71	10	19
South Africa	3	31	66	10	25	65
Sudan	32	28	41	80	7	13
Swaziland	12	45	43	–	–	–
Tanzania	45	17	37	82	3	15
Togo	43	23	34	65	5	30
Tunisia	11	27	62	55	23	22
Uganda	29	18	53	69	9	22
Zambia	22	38	40	70	7	23
Zimbabwe	19	24	57	66	10	24
NORTH AFRICA	9	47	44	33	22	44
SUB-SAHARAN AFRICA	27	28	45	62	11	27
ALL AFRICA	25	30	45	59	12	29

Sources: World Bank, 2008a, 2009c; UNDP, 2007

Table 8.11 Employment in agriculture, industry and services, 1997 and 2007

	Employment in agriculture (% of total employment) 1997	*Employment in agriculture (% of total employment) 2007*	*Employment in industry (% of total employment) 1997*	*Employment in industry (% of total employment) 2007*	*Employment in services (% of total employment) 1997*	*Employment in services (% of total employment) 2007*
North Africa	35.4	32.8	19.9	20.6	44.7	46.6
Sub-Saharan Africa	72.1	64.7	8.5	9.6	19.4	25.7
Developed economies and EU	6.1	3.9	28.3	24.5	65.6	71.5

Source: ILO, 2008

economies, copper production represented 50 per cent or more of Zambia's exports, while tobacco was dominant in Malawi, coffee in Burundi, Ethiopia and Uganda, and cocoa in São Tomé and Príncipe. Malawi's dependence on tobacco has increased in recent years, providing 53 per cent of the country's export revenues in 2007 and 70 per cent in 2008. In such a poor country, with over 50 per cent of the population living below the poverty line and with few mineral resources, this heavy dependence on tobacco is problematic, not least because of falling world tobacco prices and increasing international pressure to reduce the production and use of tobacco. One-product economies are particularly vulnerable to fluctuations in world commodity prices and, where a single commodity overwhelmingly dominates exports, as with tobacco in Malawi, such price fluctuations can have serious economic and social implications for the country and its people. A move towards diversifying production and exports could help to reduce the impact of such price fluctuations, but in a country such as Malawi, with its limited natural resources and

Table 8.12 African countries dependent on a single primary commodity for export earnings (annual average of exports, in $, 1992–1997)

	50% or more of export earnings	*20–49% of export earnings*	*10–19% of export earnings*
Crude petroleum	Angola, Republic of the Congo, Gabon, Nigeria	Cameroon, Equatorial Guinea	Algeria
Natural gas		Algeria	
Iron ore		Mauritania	
Copper	Zambia		DRC
Gold		Ghana, South Africa	Mali, Zimbabwe
Timber (African hardwood)		Equatorial Guinea	Central African Republic, Gabon, Ghana, Swaziland
Cotton		Benin, Chad, Mali, Sudan	Burkina Faso
Tobacco	Malawi	Zimbabwe	
Arabica coffee	Burundi, Ethiopia	Rwanda	
Robusta coffee	Uganda		Cameroon
Cocoa	São Tomé and Príncipe	Ivory Coast, Ghana	Cameroon
Tea			Kenya, Rwanda
Sugar		Mauritius	Swaziland

Source: Cashin *et al.*, 1999

inadequate basis for industrial development, it is difficult to identify the most appropriate strategy, other than perhaps an expansion of the service sector. As Table 8.10 shows, the contribution of the service sector to Malawi's GDP in 2007 had already reached 45 per cent, overtaking both the agricultural (34 per cent) and industrial sectors (20 per cent).

8.3.1. Agriculture

As we have seen, the contribution of the agricultural sector to the GDP of individual national economies varies across the continent, ranging from less than 3 per cent in Botswana, Equatorial Guinea, Libya, the Seychelles and South Africa to over 60 per cent in Guinea-Bissau, Liberia and Somalia. But even in countries where agriculture contributes about 30–40 per cent to GDP, the agricultural sector employs the bulk of the population, as in Burkina Faso (90 per cent), Burundi (93 per cent), Ethiopia (92 per cent), Malawi (90 per cent), Niger (90 per cent) and Rwanda (90 per cent). The agricultural sector grew in productivity in the period 2002–2006, though 2004 was a difficult year for West Africa, as Table 8.13 shows.

The agricultural sector in most African countries can be broadly divided into 'traditional' and 'modern' sectors (see Chapter 4). In countries where a large proportion of employment is in agriculture, and particularly in SSA, it is likely that the bulk of these people are employed in the traditional sector, characterized by generally small family farms

Table 8.13 Growth rate of the agricultural sector, 2002–2006

	2002	*2003*	*2004*	*2005*	*2006*
North Africa	2.2	8.1	–2.5	0.5	6.9
West Africa	1.8	–1.8	–18.7	4.1	4.6
Central Africa	4.4	5.0	3.9	4.8	4.9
East Africa	–5.9	1.4	8.1	2.3	3.5
Southern Africa	9.8	0.3	1.6	2.3	0.8
Total Africa	2.3	3.3	–3.9	3.4	5.0

Source: World Bank, 2007b.

with relatively low-level technology and using mainly family labour to grow predominantly food crops for household consumption and perhaps some market sales when harvests are good. The key staple crops are grains such as maize, millet, sorghum, rice and wheat, with the main root crops including yams, cassava and sweet potato. In addition, many families usually grow various green vegetables, as well as fruit crops such as banana, paw-paw, avocado and a range of citrus fruits. Many family farmers also grow varying amounts of 'cash crops', such as groundnuts, coffee, cocoa, tea and sugar cane.

In contrast to family farming, the 'modern' sub-sector is characterized by generally larger farms, greater use of technology, including irrigation and fertilizers, with paid workers growing crops mainly for domestic or overseas markets: for example, wheat, cocoa, coffee, sugar cane, fruits and, increasingly in recent years, vegetables and flowers destined for international markets (see Box 10.4, below)

In addition to these two sub-sectors, the livestock sector is important in several African countries, such as Chad, Mali, Mauritania and Niger, where it provides employment and a range of animal products, including meat, milk and hides (see Chapter 4). The livestock sector can also be divided into traditional pastoral activities, involving the herding of animals often over long distances in arid and semi-arid areas, and the modern livestock sector that typically involves less mobile ranching, similar to that practised in parts of Europe and North America, with improved pastures, sometimes irrigation and controlled feeding.

Fishing is also often included as part of the broad 'agricultural sector'; and, as with farming and pastoralism, the fishing sub-sector includes both traditional and modern activities (see Chapter 4). It can be further subdivided into marine (sea) fishing and inland fishing in lakes and rivers. While there are significant cultural groups that specialize in fishing, (e.g. the Bozo of Mali and the Fanti of Ghana), a large proportion of Africa's population engage in fishing to some extent, and the contribution of fish to household protein intake should not be overlooked (see Chapter 6).

It is vital for both national economies and human development that the recent growth and investment in the agricultural sector should continue into the future. With some 64.7 per cent of employed people (over 190 million) in SSA working in the agricultural sector, and some 229 million extremely poor people living in rural areas, African governments simply must give priority to this sector (ILO, 2008; see Box 8.3).

Box 8.3

Supporting the agricultural sector: a vital strategy for economic growth and poverty alleviation

In the last two decades, the agricultural sector has made considerable progress in African countries, and particularly in some of the poorer countries of sub-Saharan Africa. Agricultural growth can play a key role in reducing poverty and hunger, and thus helping to achieve Millennium Development Goal 1 (eradicating extreme poverty and hunger), as well as contributing to the achievement of Goals 4 (reducing child mortality) and 5 (improving maternal health). In addition to providing food, the agricultural sector provides employment and a social safety net during difficult times. It also contributes to exports, creates a demand for inputs from other sectors, and can act as an important catalyst for local economic development and off-farm activities (ILO, 2008). In 2002, the New Partnership for Africa's Development and the African Union set up the Comprehensive Africa Agriculture Development Programme (CAADP) to focus on agriculture-based development to end hunger and reduce poverty and food insecurity, and to increase opportunities in the export market.

Productivity growth in the agricultural sector increased from 1.65 per cent per year in the period 1984–1993 to 1.83 per cent in the period 1994–2003, with growth in both domestic food production and agricultural exports. By 2005, the target agricultural growth rate of 6 per cent set by the CAADP had been achieved in thirteen countries: Angola, Burkina Faso, Republic of the Congo, Eritrea, Ethiopia, The Gambia, Guinea-Bissau, Mali, Mozambique, Namibia, Nigeria, Senegal and Sudan (Badiane, 2008). It is recognized that higher agricultural growth generally leads to lower poverty rates, since farmers, who generally comprise the poorest section of the population, are able to earn higher incomes, which in turn leads to higher incomes more generally in the rural sector. A report from the International Food Policy Research Institute has shown that poverty and hunger were significantly reduced between 1993 and 2006 in certain SSA countries where steady growth has occurred, such as Ghana (from 52 per cent to 28 per cent) and Uganda (56 per cent to 31 per cent) (IFPRI, 2008).

However, public investment in agriculture fell from 6.4 per cent in 1980 to 4.5 per cent in 2002. In light of this, African leaders meeting in Maputo in 2003, agreed to work towards allocating 10 per cent of their national budgets to agriculture. (Apparently, both India and China spent about 10 per cent on agriculture during the 'Green Revolution', which played such an important role in increasing food production and reducing hunger.) However, by 2007, only seven African countries had achieved the 10 per cent investment target: Burkina Faso, Cape Verde, Chad, Ethiopia, Mali, Malawi and Niger (IFPRI, 2008). Writing in 2008, at a time when the prices of basic foodstuffs were rising alarmingly in international markets, Badiane (2008: 10) commented,

> African countries are in the middle of the strongest economic recovery in the past 40 years . . . If the economic and agricultural sector performance of the past decade is to be sustained and broadened to accelerate growth and reduce poverty, African countries and their partners need to focus on

boosting the supply response to the rise in international food prices during the next two or three years.

In an upbeat assessment of the situation and the future potential of African agriculture, Badiane urged African leaders to achieve the 10 per cent budget allocation to the sector as soon as possible.

The significance of the agricultural sector was also recognized by the World Bank in its *World Development Report 2008* (World Bank, 2008e), which, for the first time in twenty-five years, was devoted to agriculture. Meanwhile, in the same year, the African Development Bank announced an increase of $1 billion (to $4.8 billion) for its agricultural portfolio, while bilateral agencies in the USA (USAID), UK (DfID) and Sweden (SIDA) pledged further support to strengthening Africa's agricultural sector (IFPRI, 2008). With such a strong worldwide focus on Africa's agricultural sector, it must be hoped that its productivity will continue to improve and the benefits will be distributed widely and fairly.

8.3.2 Industry

A review of the scale of industrial production in Africa reveals the significance of four countries – South Africa, Egypt, Algeria and Nigeria. South Africa, with its industrial production valued at $44,053 million in 2005, representing 31.7 per cent of national GDP, dominates the continent and accounts for about a third of total production in sub-Saharan Africa. However, considering the continent as a whole, the North African countries together account for almost a half of the value of industrial production, with this region dominated by Egypt ($37,781 million, 35.6 per cent of GDP) and Algeria ($37,114 million, 52.9 per cent of GDP). The only other country in sub-Saharan Africa with a significant level of industrial production in 2005 was Nigeria, with $29,849 million, representing 50.6 per cent of GDP (ADB, 2008; World Bank, 2007c).

In the past, Africa's industrial sector has been characterized by extractive industries, notably the mining of minerals such as copper, gold, platinum and diamonds and, increasingly, drilling for oil and natural gas. Minerals were typically exported in a raw state, with little processing being done in Africa. The industrial sector's contribution to GDP has increased in recent years as a result of high world prices for oil and gas and certain mined minerals, though the recession of 2008–2009 had a negative impact on this. In 2006, the industrial sector as a whole contributed 41.5 per cent of Africa's GDP, a significant increase over the figure of 36.5 per cent for 2000–2005. However, the manufacturing sector experienced a relative decline from 12.8 per cent in 2000–2005 to 10.9 per cent of GDP (ECA, 2008; see Plate 8.1).

8.3.3 Extractive industry

African industry is still overwhelmingly dominated by mining and crude oil in terms of production and exports. For example, in Equatorial Guinea, one of Africa's smallest countries in terms of both size and population, industrial production accounted for 91.2 per cent of GDP in the period 2000–2006. This largely reflects the discovery and development of large oil and gas reserves since the mid-1990s, such that the country is now sub-Saharan

Plate 8.1 Petrochemical complex, Sasolburg, Free State Province, South Africa (Etienne Nel).

Africa's third-largest oil producer (after Nigeria and Angola). In 1995, the multinational petroleum company Exxon Mobil discovered the large Zafiro Field, with estimated reserves of 400 million barrels, and in 1999 Triton Energy, a US independent, discovered the La Ceiba Field with an expected production of approximately 133,000 tonnes a day. Daily oil production in Equatorial Guinea rose between 1998 and 2001 from around 10,800 to around 28,000 tonnes. Between 2000 and 2005, this country had the distinction of achieving the continent's most rapid rate of economic growth, with a GDP annual increase of over 24 per cent (World Bank, 2008a). However, there is very little industry in Equatorial Guinea and, despite government efforts to attract foreign investment, it has been largely unsuccessful in creating an atmosphere conducive to investor interest. With an output of 15.2 million tonnes in 2009, Equatorial Guinea was Africa's seventh-largest oil producer, though the bulk of the country's population seems to have gained little from the oil revenues, and poverty remains widespread (see Box 8.2).

At the end of 2009, it was estimated that Africa had proven oil reserves of 16.9 billion tonnes – 9.6 per cent of the world's total reserves. Oil production in 2009 amounted to 459.3 million tonnes, representing 12 per cent of the world total. Among African countries, Libya and Nigeria had the largest estimated reserves in 2009, with 5.8 and 5.0 billion tonnes, respectively. Angola was in third place, with 1.8 billion tonnes (BP, 2010; Table 8.14). These three countries and Algeria account for over 70 per cent of the continent's oil production (with Nigeria the top producer, followed by Angola, Algeria and Libya). Other oil-producing countries are Cameroon, Chad, Congo, Côte d'Ivoire, the DRC, Egypt, Equatorial Guinea, Gabon and Tunisia (BP, 2010; Mbendi, 2009).

In 2009, it was reported that considerable oil reserves had been discovered in the Lake Albert Basin in western Uganda, the first oil discovery in East Africa. With Ugandan reserves estimated to be as great as 1.5 billion barrels, this is comparable to the oil reserves of the Republic of the Congo or Equatorial Guinea (1 metric tonne of crude oil is equivalent to 7.33 barrels; see Table 8.14). The UK-based oil company Tullow Oil has signed contracts with the Ugandan government and is investing over £380 million in exploration.

Table 8.14 Proven oil reserves and oil production from African countries, 2009

Country	*Proven oil reserves (billion tonnes) 2009*	*Oil production (million tonnes) 2009*
Algeria	1.5	77.6
Angola	1.8	87.4
Chad	0.1	6.2
Congo, Republic	0.3	14.1
Egypt	0.6	35.3
Equatorial Guinea	0.2	15.2
Gabon	0.5	11.4
Libya	5.8	77.1
Nigeria	5.0	99.1
Sudan	0.9	24.1
Tunisia	0.1	4.1
Other Africa	0.1	7.6
Total Africa	16.9	459.3

Source: BP, 2010

There is concern, however, about the impact on the environment in the Murchison Falls National Park and carbon-dioxide emissions associated with the flaring of gas. President Museveni is keen for the country and its people to reap benefits from the oil, though parallels are frequently drawn with Nigeria, where, after more than fifty years of oil exploration, there has been considerable environmental pollution in the Niger Delta region and widespread poverty persists (*Guardian*, 2010b).

The mining of minerals has been an important activity in some African countries in both the formal and the informal economies (see Plates 8.2 and 8.3). Africa supplies about 46 per cent of the world's diamonds (USGS, 2009). Botswana's high per capita GDP ($5,846), relative to other African countries, is largely due to the country being one of the world's leading producers of gem-quality diamonds since the early 1980s, accounting for about 70 per cent of export earnings. Since independence from UK in 1966, four highly mechanized deep open-pit diamond mines have opened, the first at Orapa in 1972, followed by Letlhakane (1977), Jwaneng (1982) and Damtshaa (2002). The Jwaneng mine has the distinction of being the single richest diamond mine in the world. Two mining companies dominate diamond production, with the largest, Debswana, owned equally by the Botswanan government and the De Beers Company of South Africa, while the smaller Bamangwato Concessions Limited also has substantial government involvement. All diamonds are sorted and valued by the Botswana Diamond Valuing Company, a subsidiary of the Debswana Diamond Company. Botswana is a key participant in the Kimberley Process, an association of diamond-producing and -importing countries, commercial diamond firms and NGOs that has implemented a certification scheme for international trade in rough diamonds, designed to prevent trade in so-called 'blood' or 'conflict' diamonds (Maconachie and Binns, 2007b).

Diamond mining in Sierra Leone is very different from that in Botswana, yet in the 1970s diamonds provided 70 per cent of the country's export revenue (Binns, 1982b). The main difference is that while the Botswanan mining is highly mechanized in well-guarded, deep open mines, diamonds in Sierra Leone are found mainly on the banks and in the beds of rivers that have progressively eroded the diamondiferous kimberlite dykes. During the 1960s and 1970s, an Alluvial Diamond Mining Scheme was in place, and through this local people could obtain licences to undertake mining legally. There was also a government-

Plate 8.2 Coal mine in Utrecht, KwaZulu-Natal Province, South Africa (Etienne Nel).

Plate 8.3 Illegal artisanal coal mining, South Africa (Tony Binns).

run Diamond Buying Office, though a large element of the mining and marketing of the country's diamonds was, and remains, under the control of Lebanese and other entrepreneurs. While many people in Sierra Leone are permanently engaged in diamond mining, the majority of miners are active only in the dry season, having worked on their farms during the rainy season. As a result, the mining and farming economies are closely linked, with farmers selling large quantities of their produce to miners (Binns and Maconachie, 2005).

During Sierra Leone's civil war (1991–2002), diamond mining became anarchic, and the challenge in the post-conflict period is to bring the situation under control so that the national and local economies can share the benefits of this valuable resource. The relatively easy access to and 'lootability' of diamond reserves were key factors fuelling the civil war, and gave rise to the concept of blood diamonds. Unlike Botswana, where the diamond reserves are tightly controlled, in Sierra Leone the lootability and smuggling of an accessible resource with a high value-to-weight ratio have led to both political instability and the loss of valuable revenue.

The civil war in Sierra Leone and resource conflicts elsewhere in Africa have prompted researchers to investigate the so-called 'resource curse syndrome'. However, it remains unclear whether an abundance of natural resources is a blessing or a curse for political and socio-economic development (Ross, 1999; Rosser, 2006; Maconachie and Binns, 2007b; see Chapter 7).

Gold has been a key feature of the South African economy for well over a century. The country has the world's largest reserves of the metal, but production fell steadily to 272,100 kg in 2006 (from a peak of 713,447 kg in 1975) and employment in gold mining contracted from 564,452 in 1987 to only 159,984. Understandably, with such a decline in both employment and production, this contraction in the industry has had serious ramifications for the economy, communities and individual households (see Box 8.4). Africa also supplies an estimated 62 per cent of the world's platinum and palladium, with over 95 per cent of both coming from South Africa (USGS, 2009).

Box 8.4

Changing fortunes for South Africa's gold-mining industry

For over 140 years, the strength of Africa's largest economy, South Africa, has been based on an extraordinary dependence on the mining industry, and particularly gold. The country's unique geological resources have made it the world's most important supplier and store of a vast range of strategic minerals. It was the discovery of gold in the Johannesburg area – known as the Witwatersrand – in 1886 that ultimately drove the country's industrial revolution, transforming the economy from being overwhelmingly based on agriculture to one in which the mining, energy and industrial sectors are predominant. Over the last thirty years, however, the traditional dominance of the mining sector has been severely affected by a series of dramatic changes that have altered the traditional dependence on gold mining, in particular. Down-scaling of the gold industry, massive redundancies (over half of the workforce) and the severe blow dealt to 'downstream' industries and workers' livelihoods have had a devastating impact on mining towns.

Although South Africa has the world's largest known gold reserves, production and employment in gold mining have been on a downward trend for some time, with the 2006 production of 272,100 kg representing the lowest total since 1955 (see Table 8.15 below). Between 1975 and 1999, gold production declined by 37.7 per cent, and between 1999 and 2006 there was a further decline of over 38 per cent. Between 1987 and 1999, the number of gold-mining jobs fell by a massive 57.9 per cent. There was a particularly large decline in employment (90,072 jobs) in just two years, 1997 and 1998, representing a 26 per cent fall, while in the same two-year

Table 8.15 South Africa: gold production and employment, 1975–2006

Year	*Production (kg)*	*Employment*
2006	272,100	159,984
2005	294,700	160,634
2004	337,200	179,964
2003	373,200	198,465
2002	398,500	199,378
2001	395,000	201,673
2000	430,800	216,982
1999	444,427	237,732
1998	464,217	255,855
1997	503,878	345,927
1996	498,324	345,902
1995	523,808	386,407
1994	579,290	397,474
1993	619,318	390,890
1992	613,737	414,000
1991	599,194	451,177
1990	602,997	483,737
1989	601,524	522,217
1988	612,367	536,368
1987	601,774	564,452
1986	638,047	553,656
1985	670,753	515,913
1984	680,678	513,653
1983	674,944	499,795
1982	663,695	485,786
1981	655,745	490,445
1980	672,416	473,769
1979	703,273	457,792
1978	704,576	441,282
1977	699,887	432,252
1976	713,390	408,116
1975	713,447	388,159

Source: Statistics South Africa, 1975–1999, 2008

period gold production fell by 39,661 tonnes (7.9 per cent). Between 1987 (the peak year for employment in gold mining) and 2006, there was a fall of 72 per cent in numbers employed in the industry. The country's share of world production dropped from 66.6 per cent in 1981 to 32.8 per cent in 1993, and in 2007 China overtook South Africa as the world's largest gold producer.

The effects of these changes have been selectively felt in the country's seven major goldfields, with the two worst-affected areas being Free State and Klerksdorp in North West Province. These two areas alone have lost a staggering 100,000 and 30,000 mining jobs, respectively. In certain centres, such as Stilfontein and Welkom, which were developed as new towns during the boom in gold mining, the radical decline in production has been economically and socially catastrophic. Unemployment rates, particularly among low-skilled black labourers, have increased from almost zero to 65 per cent in many areas over the last three decades. In the absence of any significant alternative employment opportunities, few people have been able to find work, with the exception of some marginal, informal sector activity. The effective absence of a state welfare system in South Africa, most notably medium- and long-term unemployment benefit, has aggravated the situation (Binns and Nel, 2001; Nel and Binns, 2002).

Throughout the country, the Mineworkers Union has responded with understandable dismay to the closures. In the case of the Free State, where the country's greatest losses have occurred, the Mineworkers Development Agency has initiated programmes to retrain redundant miners in new skills, and has assisted with the establishment of small business operations, such as poultry farming. Probably the most comprehensive responses to the declining fortunes of gold mining have come from the various local governments in the areas, which either directly or in collaboration with the private and public sectors have sought to restructure their local economies by creating new forms of employment. For example, in the mid-1990s, Klerksdorp embarked upon a joint initiative to market the area to potential investors and attract new industries, notably textiles. Retraining programmes for redundant miners have been initiated in collaboration with a local tertiary-level educational institution, and support has been given to emerging small businesses, including the manufacture of pottery, wrought iron and leather products. One of the most innovative responses in the Stilfontein area involved selling old miners' houses, effectively at cost, to retired people from the large cities who wished to live in a small, quiet and less expensive country town.

In the Free State city of Welkom, in collaboration with neighbouring towns and the private sector, the Free State Goldfields Development Centre was set up in the early 1990s to attract new investment and encourage local economic restructuring and job creation. This has provided much encouragement for new economic activities, including tourism promotion, support for agricultural activities and agro-industry, as well as an impressive gold-jewellery production factory, linked with a jewellery training school, with the latter established in collaboration with a mining company (Binns and Nel, 2001; Nel and Binns, 2002).

Zambia is a country where mining has long been a significant element in the economy, and economic fortunes have depended largely on fluctuations in the value of minerals on the world market. The main copper-producing area is in and around the towns of Ndola, Kitwe and Chingola in the Copperbelt region, some 350 km north of the capital Lusaka, and close to the border with the DRC. In 1969, Zambia produced 720,000 tonnes of copper, making it the world's fourth-largest producer. From the mid-1970s, however, declining copper prices and falling investment in the mining sector meant that production had fallen to only 256,884 tonnes by 2000. Privatization of state assets, including the mines and manufacturing, in the 1990s led to significant job losses, which, in the case of the primary Copperbelt economic centre of Ndola, amounted to the loss of approximately 75 per cent of manufacturing plants and some 9000 jobs in manufacturing by 2000. Unemployment in Ndola rose from 13.5 per cent in 1990 to 33.2 per cent in 2000 (CSO, 2007). In neighbouring Kitwe, some 317 companies closed between 1992 and 2002, meaning 3499 people lost their jobs. Unemployment rose to 45 per cent and poverty to 75 per cent in the early 2000s (CSO, 2007). More recently, between 2000 and 2006, there was an increase in the copper price from $0.82 per pound to $3.15 per pound, although there is still considerable unemployment in the Copperbelt towns, and many people are engaged in various informal sector activities and urban agriculture. In addition to its important copper production, Zambia produces coal and uranium, and it is the world's largest producer of cobalt, supplying about 20 per cent of global demand.

Mauritania, in West Africa, is also heavily dependent on mining, having made its first exports of iron ore to Europe in 1963. In 2007, iron ore still accounted for 50 per cent of the country's exports, although a fall in world demand has subsequently led to cutbacks in production. The country is keen to diversify its economy, which is dominated by fishing and mining, while almost 70 per cent of the nation's food is imported.

In terms of the structure of African economies, some countries are dangerously dependent on mineral fuel exports. For example, in Algeria, Equatorial Guinea, Libya and Nigeria, mineral fuels account for over 90 per cent of national export earnings (USGS, 2009).

8.3.4 Manufacturing industry

Africa has always had a strong indigenous tradition of manufacturing, although non-traditional industrial performance since the early 1960s has generally been disappointing (see Plates 8.4 and 8.5). This performance can be measured by Manufacturing Value Added (MVA), which is the difference between the gross output of the manufacturing sector and the sum of physical purchases and service inputs before any provision is made for depreciation. The whole of Africa's share of world MVA rose fractionally from 0.7 to 0.8 per cent during the 1960s, but had fallen by the late 1980s to 0.3 per cent. In 1989, for the countries south of the Sahara (including South Africa), manufacturing contributed only 10 per cent to the region's GDP, compared with over 30 per cent for industrial market economies and 24 per cent for other developing countries.

In 1989, five countries accounted for almost 60 per cent of sub-Saharan MVA. These were, in order of importance, Nigeria, Zimbabwe, Kenya, Côte d'Ivoire and Ethiopia.

Plate 8.4 Tannery in old city of Fes, Morocco (Tony Binns).

Plate 8.5 Village blacksmith, northern Nigeria (Tony Binns).

While the last four of these had a small positive growth in MVA between 1980 and 1989, ranging from 0.3 per cent in Côte d'Ivoire to 3.5 per cent in Kenya and Zimbabwe, Nigeria experienced a negative (–5.0 per cent) growth rate (Binns, 1994). However, more recent MVA statistics suggest that further changes have occurred, such that in 2005 annual MVA growth was 8.2 per cent in Nigeria, 5.01 per cent in Kenya, 5.0 per cent in Ethiopia and 4.85 per cent in Côte d'Ivoire. Perhaps unsurprisingly, given the political turmoil in that country, Zimbabwe experienced a negative MVA growth of –17.0 per cent (ECA, 2008).

Factors preventing the further development of manufacturing industry include the low per capita income level and size of population, which affect the size and purchasing power of the internal market, the absence of supportive infrastructures in the form of transport and reliable power supplies, together with the nature of the existing industrial base and the countries' poor resource endowment. In addition, African countries have suffered from a heavy debt burden, adverse terms of trade and the poor performance of their agricultural sectors, which have affected consumer demand, the level of farm inputs and supplies of raw materials.

South Africa is the most diversified industrialized economy, whereas in other African countries there is a focus on traditional industries, such as food processing and textiles. However, in Tunisia, there has been steady development of the electrical and electronics industries, while textiles and clothing have experienced a decline.

There are some notable exceptions to the poor record in the post-independence development of manufacturing industry. For example, the Indian Ocean island state of Mauritius has successfully built up its manufacturing sector since gaining independence from the UK in 1968. Faced with high unemployment, the government created an Export Processing Zone (EPZ) in 1971 and has developed a reputable textile industry, with some 275 garment factories in 1999. Legislation gives investors in EPZ enterprises tax relief, duty exemption on most imports, unlimited repatriation of capital and profits and cheap electricity. Mauritius is viewed as both a successful trade liberalizer and an economy with extensive government intervention in the labour market, in terms of enforcing minimum wage levels, monitoring working conditions and elaborate wage-settlement procedures and conciliation mechanisms in relation to disputes with trade unions. Per capita exports of manufactures from Mauritius rose in the 1990s from $200 to more than $900, while in 1999 monthly wages averaged $340, seven times the level of Ghana, for example. With a per capita GDP of $4404 in 2005, compared with an average of $572 for sub-Saharan Africa, Söderbom and Teal (2003: 3) suggest, 'If other African countries could grow as fast as Mauritius, poverty could effectively be eliminated in a single generation.'

In a study of a sample of manufacturing firms in five SSA countries (Ghana, Kenya, Nigeria, South Africa and Tanzania), Söderbom and Teal (2003) found that the average number of employees was highest in Nigeria (246) and lowest in Tanzania (78), with South Africa having 206, Kenya 110 and Ghana 83. Labour productivity and capital intensity per employee were highest in South Africa and lowest in Ghana. Whereas over half of the South African firms were exporters, only 8 per cent of Nigerian firms exported. Firms with more than 100 employees were generally exporters, whereas the smaller firms were not. Söderbom and Teal examined what is needed for manufacturing firms in other African countries to be as successful as those in Mauritius. A key factor is undoubtedly the level of efficiency. More efficient firms are generally more successful in entering the export market, and what they learn from this process can then help them to enhance their technical efficiency. In the labour-intensive production of garments, where Africa has a potential cost advantage, Söderbom and Teal suggest that large firms should use more labour-intensive technology. African manufacturing enterprises and governments might also try to emulate the example of Mauritius, where success has been linked to such factors as favourable tax treatment for exports, macro-economic stability, good levels of technical expertise to raise efficiency, trade preferences and a good knowledge of export markets.

Developing export markets and attracting foreign investment are important factors in the growth of manufacturing enterprises. Business surveys conducted in a number of African countries have revealed four major concerns in attracting foreign investment: financing, corruption, infrastructure and inflation. A good local investment climate is important, so that the costs of certain services which are important to manufacturing are at a reasonable level: for example, land and buildings, telecommunications, water, electricity and transport. Firms often face considerable uncertainty and risk in investing in enterprise development in African countries, and the high risk often makes them adopt a more conservative product mix with a lower expected rate of profit (Bigsten and Söderbom, 2006).

Among almost 2000 firms surveyed in Nigeria in 2007–2008, over 60 per cent identified an unreliable electricity supply as their main problem, with an average of twenty-seven power outages in a typical month. A similar enterprise survey in Mozambique found that 20 per cent of firms identified access to finance as a major obstacle to their operations, while in Kenya over 20 per cent of firms mentioned regulations and tax as their biggest problem (Enterprise Surveys, 2008). Infrastructure is a serious problem for the development of business enterprises in many African countries, and it is common for firms to buy infrastructure services or provide them themselves: for example, constructing road access, installing electricity generators and ensuring a reliable water supply. In a survey of Nigerian businesses, almost a quarter identified corruption as a major issue, saying that they had to pay bribes in order to get things done, and well over 50 per cent of firms felt they were expected to make gifts in order to gain permits for construction (Enterprise Surveys, 2008).

8.3.5 Service sector

An employment survey undertaken in Kenya in 1979 revealed that about 20 per cent of Nairobi's labour force was employed in manufacturing, 12 per cent in construction, 30 per cent in commerce and finance, 8 per cent in transport, 16 per cent in administration and social services, and 10 per cent in domestic service (Binns, 1994). Over thirty years later, this pattern remains typical of large African cities. The importance of administrative and service employment, including domestic service, is a common feature of many of them. Administration, commerce and finance are particularly important in most capital cities, notably Yaounde in Cameroon, Nouakchott in Mauritania and Kigali in Rwanda. The service sector accounted for 65 per cent of GDP in South Africa in the period 2000–2006 (see Plate 8.6), and was considerably higher in countries where port services were particularly significant: for example, Djibouti (79.9 per cent) and Cape Verde (74.3 per cent) (ECA, 2008). Djibouti's economy is based on service activities connected with its strategic location on the Red Sea, and it is both a transit port for the region and an international trans-shipment and refuelling centre.

Cape Verde, with a population of approximately 500,000 and an area of only 4000 sq km, has few natural resources and imports about 90 per cent of its food. The country has a mainly service-oriented economy, which includes ship repairing and an international airport on the island of Sal. The service industry sector accounts for about three-quarters of the country's GDP, and there has been a steady growth in tourism since 2000 (see Box 8.5). Tourism has also developed significantly in many other African countries over the last two decades. South Africa and Egypt dominate the continent in terms of international tourism receipts, with Morocco and Kenya in third and fourth positions (see Plate 8.7). But tourism can be adversely affected by political events, such as the 1994 coup in The Gambia and violent clashes between different ethnic groups following the disputed presidential elections in Kenya in December 2007 (see Box 8.6). In both cases, there was a sharp decline in tourist arrivals and receipts, and it took some time for these to recover after peace was restored.

Plate 8.6 Gateway regional shopping complex, Durban, South Africa (Etienne Nel).

Plate 8.7 Montecasino hotel complex, Johannesburg, South Africa (Tony Binns).

Box 8.5

Cape Verde: development in an island economy

A Portuguese colony until 1975, Cape Verde is one of Africa's smallest countries, with an area of just 4033 sq km (slightly larger than the US state of Rhode Island or the UK county of Suffolk). Consisting of an archipelago of ten volcanic islands, nine of which are inhabited, Cape Verde is situated some 600 km off the West African coast, opposite Senegal and Mauritania. The population in 2008 was just under 500,000, growing at an annual rate of 1.4 per cent. With a per capita GDP in 2005 of $1940, an average life expectancy of 71 years and an under-five mortality rate of 35, Cape Verde is, in both economic and human development, better off than many other SSA countries. It was ranked 102nd out of 177 countries in terms of Human Development Index in 2005, putting it in the top half of the 'medium human development' countries. It is also on target to achieve most of the Millennium Development Goals by 2015 (UNDP, 2007). In terms of economic growth, Cape Verde had an average annual per capita GDP growth rate of 2.8 per cent in the 1990s and a growth rate of 3.0 per cent between 2000 and 2005 (World Bank, 2008a). The African Development Bank estimated that real GDP grew by 7.6 per cent in 2008, while inflation fell to under 3 per cent (ADB, 2008).

Cape Verde suffers from relatively poor natural resources, with low rainfall, serious water shortages, a rugged terrain and wind erosion. Agriculture, livestock and fishing account for only 9 per cent of GDP, and in 2006 the national unemployment rate was 18.3 per cent (ADB, 2008). Money remitted by the large (approximately

500,000) Cape Verdean diaspora amounted to $139 million in 2007. The country's economy is dominated by service industries, accounting for 74 per cent of GDP in the same year. Tourism, in particular, is growing rapidly and has considerable potential. The tourist industry expanded by 12.7 per cent between 2000 and 2003, and then by 15.6 per cent between 2004 and 2007. By 2015, it is anticipated that there will be 1 million tourists annually (World Bank, 2008a). But severe water and power shortages will hinder tourism development unless they are addressed, so expensive desalination plants are likely to be needed. Furthermore, about 90 per cent of food must be imported. Tourism currently has limited linkages with local communities as it is generally located in large resort centres.

Cape Verde is located at a strategic crossroads of mid-Atlantic air and sea lanes. There is an international airport on the island of Sal, built in 1939 and upgraded in the 1970s, after which it was used as a refuelling stop by South African Airways en route to Europe and USA (over-flying and landing rights were denied to the apartheid regime by most other African countries). In 2004, the airport served over a million passengers on tourist and scheduled flights. The country has another international airport near Praia, the capital, and a third was opened in December 2007 in Boa Vista. The ports at Mindelo and Praia have been upgraded, with important ship-repair facilities, fishing fleets and fish-processing plants. Since the European Union lifted its ban on fish imports from Cape Verde in 2005, there is considerable potential for further development of the fishing industry, notably lobster and tuna. There are plans to transform Cape Verde into a shipping-support platform and regional air-transportation hub (ADB, 2008).

The country has a stable democratic system, and since 1993 the government has encouraged privatization and foreign investment, which in the last decade has involved the large-scale building of tourist hotels and associated infrastructure. Some 109 investment projects have been attracted to the country, amounting to total capital of $416 million. Most of the FDI has come from the UK, Italy and Spain (Canary Islands), and 90 per cent of it has gone into tourism development. More recently, investors from China and the United Arab Emirates have become involved, too.

Over 70 per cent of imports and exports are with the European Union, which has granted Cape Verde special partnership status. The country became the 152nd member of the World Trade Organization in December 2007, following changes to import tariffs. The previous year, Cape Verde had received net overseas development assistance of $138 million, while in 2007 external aid amounted to just over 10 per cent of GDP. By September 2008, the World Bank had approved some twenty-three development projects totalling over $270 million. These related to road construction, HIV/AIDS and poverty reduction. With serious shortages of energy and large oil imports, Cape Verde is developing wind farms and is even looking into building a nuclear power plant. The key challenge is to create a sustainable balance between economic development and use of the country's natural resource base and infrastructure in light of the rapidly growing tourist industry.

Box 8.6

Tourism in Kenya

Kenya is one of Africa's main destinations for international tourists, as Table 8.16 shows. Tourism accounts for about 10 per cent of the country's GDP, making it the third-largest contributor to the economy (after agriculture and manufacturing) and the third-largest foreign exchange earner (after tea and horticulture). In 2007, the tourism sector maintained an upward growth momentum, with over 1.8 million international tourist arrivals, compared with 1.6 million in 2006 (see Table 8.17). Tourism receipts also increased – from $749 million in 2006 to $872 million in 2007, almost double the value of tourist receipts the country had earned in 1990 ($443 million).

Statistics for 2006 indicate that the largest tourist groups were from two countries: the UK, with over 1 million hotel bed-nights, followed by Germany (930,000 bed-nights). Other significant groups came from Italy, France, the rest of Europe and the USA (GOK, 2007a). The pattern is broadly similar to the 1980s, when Germany, the UK, the USA and Switzerland dominated tourist arrivals.

Table 8.16 International tourism receipts and tourist arrivals in selected countries (with large or fast-growing tourist industries)

Country	*International tourism receipts ($ million) 2004*	*International tourist arrivals (,000s) 2004*	*Average annual growth in tourist arrivals (%) 2000–2004*
Angola	66	194	39.7
Armenia	86	263	55.5
Australia	15,191	4,774	3.4
Cambodia	603	1,055	22.7
China	25,739	41,761	7.5
Costa Rica	1,358	1,453	7.5
Egypt	6,125	7,795	11.1
France	40,841	75,100	–0.3
Kenya	486	1,199	6.7
Mexico	10,796	20,600	0.0
Mongolia	185	301	21.7
Morocco	3,924	5,477	6.4
New Zealand	4,790	2,348	7.1
South Africa	6,282	5,998	3.2
Thailand	10,034	11,737	5.2
Uganda	266	512	27.6
Ukraine	2,560	15,629	24.9
United Kingdom	28,221	27,800	3.5
USA	74,547	46,100	–2.6
Yemen	214	274	39.2

Source: World Tourism Organization, 2006

Table 8.17 Kenya: international tourist arrivals and tourism earnings

	1990	*1995*	*2000*	*2003*	*2004*	*2005*	*2006*	*2007*
International tourist arrivals (,000s)	814	896	899	927	1199	1479	1600	1816
International tourism receipts ($ million)	443	486	283	347	486	579	749	872

Sources: GOK, 2007a, 2008; World Tourism Organization, 2009

Tourism development in Kenya has been a post-war phenomenon. In the latter part of the colonial period, an increasing number of wealthy tourists arrived, mainly by ship, to visit the newly established national parks. Annual tourist numbers increased in the 1950s from around 20,000 to 40,000, but it was after independence in 1963 that tourists were attracted in much larger numbers, arriving mainly by air. The rapid growth of Kenyan tourism was due to such factors as:

- better air links and airports, such as the opening of Moi International Airport in Mombasa in 1979;
- increased marketing by European tour operators;
- promotional efforts of the Kenya Tourist Development Corporation (KTDC); and
- better internal transport, enabling the development of popular two-centre beach and safari holidays.

The full exploitation of tourism's potential has been an obsession of successive Kenyan governments, and the considerable investment in tourist infrastructure, hotels, game parks and transport has to be balanced against tourist receipts. Most of Kenya's tourist facilities are located in the southern half of the country, in three main locations:

- along the Indian Ocean coastline around Malindi and Mombasa;
- in Nairobi and surrounding areas; and
- in the up-country game parks, such as Amboseli, Maasai Mara and Tsavo (see Plate 8.8).

The game parks cover about 10 per cent of Kenya's land surface, and in 2006 there were some 2.3 million visitors to the country's national parks and reserves (GOK, 2007a).

In July 2007, the Kenyan government launched the ambitious Kenya Vision 2030, a plan for national development that 'aims at making Kenya a newly industrializing middle income country providing high quality life for all its citizens by the year 2030' (GOK, 2007b: 2). This plans views tourism as one of six key sectors for development, with Kenya becoming one of the ten most popular long-haul destinations in the world, 'offering a high-end, diverse and distinctive visitor experience that few of her competitors can offer' (GOK, 2007b: 5). Three specific goals have been identified for the tourist sector by 2012:

Plate 8.8 Amboseli National Park and Mount Kilimanjaro, Kenya/Tanzania (Tony Binns).

- to quadruple tourism's contribution to the GDP to over $1 billion;
- to raise international visitor numbers to 3 million and to increase spending per visitor to $850; and
- to increase the number of hotel beds to approximately 65,000, with more emphasis on high-quality service.

Vision 2030 also identifies a number of 'flagship tourist projects' to be achieved by 2012:

- The development of three resort cities: one on the northern part of the coast, one in the south and a third in Isiolo, in the centre of the country, to the north of Meru.
- To bring more tourists to the game parks by marketing the lesser-visited parks and upgrading their facilities.
- To launch a 'premium parks initiative' to cater for the higher end of the tourist market in places such as Maasai Mara and Nakuru.
- A 'niche products initiative' to identify four key sites in western Kenya to provide 3000 beds in high-cost accommodation for cultural tourism, eco-tourism and water sports.
- To promote cultural tourism in Kenyan homes by certificating 1000 units of home-stay accommodation.
- To develop a 'business visitors' initiative, with five new international hotels built in Nairobi, Mombasa, Kisumu and possibly Isiolo (GOK, 2007b: 6).

In 1988, it was estimated that tourism accounted for 9.0 per cent of total Kenyan employment, representing 110,600 employees. However, wages of hotel staff were lower than in other sectors of the economy, with the possible exceptions of agriculture and domestic service. Tourism-related employment included soft-drink and beer brewing, handicraft production, hotel entertainment, car rental, photographic studios, local restaurants and many others. During the 1980s, Kenya was successful in developing local food production for the tourist sector, and the importation of food declined sharply – from 77 per cent in 1984 to only 14 per cent in 1988 (GOK, 2007a).

Approximately 78 per cent of the major coastal hotels, 67 per cent of Nairobi hotels and 66 per cent of lodges in national parks and game reserves have some foreign investment. It could be argued that Kenya's 'open-door' policy towards foreign investment in tourism, combined with a policy favouring the employment of local people, has enabled the country to establish a large number of good-quality hotels and has helped Kenyans to gain sufficient specialist knowledge to manage those hotels efficiently. Many of Kenya's top businessmen and politicians are involved in the tourist industry and have shares in the multinational firms owning the hotels. These people tend to promote tourism through government aid, even when a venture is not in the best interest of the national economy. Private enterprise gains most revenue from tourism, while the Kenyan state subsidizes such enterprise through infrastructural facilities and conservation of natural attractions, such as the game parks.

The effect of tourism on wildlife is another important consideration. It is in the interests of the tourist industry to protect wildlife, and the Kenyan government uses this argument as its main justification for maintaining the national parks and game reserves. However, farmers and herdsmen living near the parks are increasingly unhappy about damage to their crops and loss of their animals due to wild animals. Although such damage is, in principle, financially compensated by the government, the compensation is often inadequate and irregular. With such a rapidly growing population, pressure on land is considerable, and there is often a fundamental conflict of interests between the domestic farming and livestock economy and preservation of game parks for tourists.

In Kenya and neighbouring Tanzania, large areas of the savanna grasslands were historically home to pastoralists, such as the Maasai, and concerns have been expressed about the displacement of such ethnic groups in favour of game park tourism. Wild game and groups such as the Maasai have coexisted for centuries, and the Maasai have never destroyed large numbers of wild animals, yet now they are being forced out of their homelands. As Mowforth and Munt (2009: 255) comment, 'As the Maasai have been excluded, so the tourists have been allowed access.' George Monbiot, who has written frequently on this topic, comments, 'Several times I was told by [tourism] conservation officials that the Maasai had to be kept out because the tourists did not want to see them there' (quoted in Mowforth and Munt, 2009: 256). Some critics have also questioned the economic benefits of allocating large areas of Kenya's land to wildlife-based tourism.

Future planning must aim to balance the needs of indigenous people with those of the ever-growing numbers of tourists. The effects of tourism on local society and culture vary geographically, but they are perhaps greatest in coastal Kenya, where most tourists take their holidays, where the Swahili culture prevails and where the moral ethics of Islam pervade local society. Careful land-use planning will be needed in the future, with tourism development zones designated after the evaluation of competing demands and possible environmental and social effects. During the 1970s and 1980s, much of the development of beach-based tourist facilities happened without any planning controls, resulting in uncontrolled ribbon development and severe pressure on facilities, local society and the environment. Sea pollution from beach hotels and pressure on natural resources, such as drinking water, land and marine ecology, are other environmental effects of the rapidly growing tourist industry.

Heavy reliance on foreign capital and expertise has biased Kenya's tourism sector in favour of the luxury market, much to the advantage of the foreign entrepreneurs who control the industry. Recent government policy, however, has tried to popularize domestic tourism and increase the 'Kenyanization' of the industry with more local ownership and management. But this process of indigenization must proceed cautiously, since a hurried replacement of expatriates could damage the country's favourable reputation as a tourism destination.

Probably the biggest threat to Kenya's tourist industry is political and economic instability, both in Kenya and overseas. The industry endured two particularly difficult years in 2007 and 2008, and a continuing worldwide recession might well impact on tourist numbers. Certainly, the oil crisis of 1972–1974, the recession of the early 1990s and the terrorist bombings in August 1998 and November 2002, blamed on al-Qaeda, all had significant repercussions on the international tourism industry. Political instability has also affected Kenyan tourism in the past, as in the mid-1970s and early 1980s, when instability in neighbouring Uganda resulted in a significant fall in tourist numbers and hotel occupancy in 1977 and 1982. More recently, the disputed presidential elections of 27 December 2007 led to widespread rioting in the country, with serious clashes between members of different ethnic groups, some 1500 deaths and over 600,000 people losing their homes. Large parts of the country were affected, most notably the Rift Valley and western Kenya, where conflict occurred mainly, albeit not exclusively, between the Kikuyu and Luo communities. Many overseas governments advised their citizens not to travel to Kenya, and in March 2008 it was estimated that the post-election violence had already resulted in a 44 per cent decline in receipts from international tourists. In late January 2008, usually a peak time for international visitors, Peter Mbogua, sales and marketing manager of Serena Hotels, had admitted, 'At one of our lodges in Tsavo, next week I have two days where I don't have a single client. That lodge would normally be enjoying an occupancy rate of about 85–90%' (Oyuke, 2008). Meanwhile, Rufus Mwachiru, chairman of the Kenya Association of Tour Operators, commented that about 20,000 of his members had lost their jobs, and the tourism industry was expected to lose about $84 million per month during the first quarter of 2008 (Oyuke, 2008).

The effects of this serious instability are likely to have long-lasting implications for Kenya's tourism industry, and it could take many years to rebuild visitor confidence and achieve the tourist numbers and financial receipts of 2007. On 27 December 2008, a year after the disputed elections, thousands of Kenyans were still awaiting resettlement and continued to live in temporary camps.

The Kenyan tourist industry is therefore facing an uncertain future, with political and economic difficulties at home and overseas. The government sees great potential in the development of the industry, but success will ultimately depend on fostering stable conditions in the country, and on the nature and flexibility of contractual relationships between Kenyan-based firms and those in the industrialized countries from which most of Kenya's tourists originate.

A major concern about international tourism in relatively poor countries, such as Kenya, is the extent to which local people, and particularly the poor, benefit financially and in other ways. There have been recent attempts to apply Value Chain Analysis (VCA) to tourism activities by disaggregating the range of activities, inputs, outputs and flows, and assessing what proportion of the profit goes to poor people. About 60 per cent of holiday costs are paid in developed countries, for air travel and to cover the administrative costs of travel agents and tour operators. However, whereas international tour companies and airlines receive the largest shares of tourist expenditure, Ashley and Mitchell (2007) suggest that a significant proportion (typically between 25 per cent and 50 per cent) of tourist spending can reach the poor from expenditure in local restaurants, shopping, local transport and excursions.

In countries where there are generally high rates of unemployment and no state benefit system, tourism provides essential, though often poorly paid, jobs for hotel workers and others, from which they can acquire skills that might subsequently be used elsewhere in the economy. Local farmers can also benefit from tourism where the food supply chain is geared towards sourcing local (rather than imported) produce to feed tourists in hotels and restaurants. The agricultural supply chain, which provides such items as fruit and vegetables for the tourist industry, can sustain many more poor households than jobs in hotels and restaurants. An important indirect benefit of tourism that can have a wide impact on local people is the improvement of infrastructure, such as roads, telecommunications and domestic air services. For example, a study undertaken in Tanzania found that the impact of upgrading rural roads in connection with tourism development transformed 'the distributional impacts of tourism from an urban bias to one where rural households experience twice the level of welfare gains compared with urban households' (Mitchell and Ashley, 2007: 3).

But there are pressures associated with a growing international tourist industry that are more social and cultural in character. Tourists are primarily interested in seeing 'traditional' Africa, the primitive and exotic, and many have outdated, stereotypical views of Africa and Africans. Cultural performances and traditional handicrafts, emphasizing curious and primitive aspects of the traditional culture, have become vehicles for boosting the tourist industry (see Plate 8.9). There is a danger here of so-called 'zooification', which involves turning indigenous people into one of the 'sights' of the tourist's trip. Mowforth and Munt (2009: 265) suggest that the process of zooifying local people 'leads inevitably to a position of powerlessness for them as well as a complete loss of human dignity'. With the clash of cultures, the display of wealth, leisure and different moral standards, tourists can provoke and demoralize a local population. For example, young Gambians or Kenyans living near tourist resorts may try to imitate visitors' clothes and behaviour. Also, because of greater wealth and education, the tourist is usually in a dominating position over the local resident. This can generate resentment among local people and may lead to open hostility. Young people may become involved in semi-legal and illegal activities as 'beach boys', petty traders and beggars. Money changing, drug selling, prostitution and theft are other such activities.

8.4 Conclusion

While South Africa's economy dominates the continent and is reasonably diverse, a large number of Africa's national economies are small and fragile, being heavily dependent on a small number of products and exports, as they were in the colonial period. There is an urgent need to strengthen and diversify these economies in order to reduce their vulnerability to global factors and price fluctuations. The Zambian economy, for example, has

Plate 8.9 Tourist market at Sidi Bou Said, Tunisia (Tony Binns).

often suffered from downward movements in the price of copper, while political disturbances in Kenya associated with the elections in 2007 had serious effects on the tourism industry (see Box 8.6). The Kenyan economy also suffered badly in April 2010, when exports of flowers and vegetables to Europe worth $1.3 million a day, were severely affected by air-traffic restrictions as a result of ash in the atmosphere from an Icelandic volcano (*Guardian*, 2010c).

Attracting both domestic and foreign investment is vital if the agricultural sector, which is still dominant in many countries, is to become more productive, while such investment is also equally important in developing the industrial sector and upgrading infrastructure associated with, for example, a growing tourist industry. But conditions have to be attractive for such investment, and it is important that the host countries and their people gain some benefit from the process. Too often in the past, the awarding of contracts has been far from transparent and the only people who have benefited have been the political and business elites.

Summary

1. Some economists have referred to the progressive marginalization of Africa in economic terms, with its falling share of world exports and a declining proportion of direct private investment.
2. Many African economies are small by world standards. The South African economy is more than twice as large as the next largest, Algeria.
3. The economies of African countries are frequently dominated by a small number of commodities, which can lead to vulnerability due to price fluctuations or changes in trading arrangements.
4. The economic structure of the North African countries is different from that of sub-Saharan countries, with a considerably higher proportion of employment in industry and services in North Africa and a correspondingly smaller proportion engaged in agriculture.

5 African countries need to diversify their economies and attract higher rates of both domestic and overseas investment for agriculture, industry and service sectors.

Discussion questions

1. What have been the main legacies of the colonial period in terms of the present-day character of African economies?
2. Select an African country whose economy is heavily dependent on a single commodity. Investigate the price fluctuations over the last two decades and suggest how these have impacted on the national economy?
3. In comparing the economies of different African countries, to what extent might South Africa provide a model that could be copied elsewhere?
4. Why is the economy of Equatorial Guinea so strong, yet its people still so poor?
5. Examine the economy of one of Africa's island states and evaluate possible strategies for its future development.

Further reading

Collier, P. (1995) 'The marginalisation of Africa', *International Labour Review*, 134 (4/5): 541–557.

Collier, P. (1998) *Living down the Past: How Europe Can Help Africa Grow*, Studies in Trade and Development No. 2, London: Institute of Economic Affairs.

Mitchell, J. and Ashley, C. (2007) *Can Tourism Offer Pro-poor Pathways to Prosperity?*, London: ODI.

Onimode, B. (1992) *A Future for Africa*, London: Earthscan.

Rosser, A. (2006) *The Political Economy of the Resource Curse: A Literature Survey*, IDS Working Paper 268, Brighton: Institute of Development Studies.

Useful websites

Africa Development Indicators (World Bank): http://data.worldbank.org/data-catalog/africa-development-indicators. A detailed collection of data on Africa, containing over 1600 indicators covering fifty-three African countries over the period 1961–present.

African Economic Outlook: http://www.africaneconomicoutlook.org/en/. An invaluable online resource with data and statistics, economic news, in-depth focus on particular economic sectors and country profiles.

World Tourism Organization: http://www.unwto.org/index.php. A specialist agency of the UN, with news about global tourism, facts and figures, and links to tourism bodies in individual member states.

9 Developing Africa

> We can say without fear of equivocation that in the Third World, over the last fifty years, no single ideal has been as obsessive, in the thinking of African governments, as the goal of development . . . [N]o government dares leave the issue of development out of its political language and rhetoric.
>
> (Ukaga and Afoaku, 2005: 8)

9.1 Introduction

In light of Africa's all too apparent socio-economic backlogs and persistent poverty (see Chapter 2), relative to other continents, strategies and policies designed to promote 'development' have been hallmarks of the last few decades (Seck and Busari, 2009). It is apparent that while Africa has been performing relatively better economically in the first years of the twenty-first century than it did in the twentieth, growth is occurring from a very low base and only limited success has been achieved in raising overall welfare levels. In addition, there are very real concerns that the Millennium Development Goals are unlikely to be achieved within the set time-frames in the continent.

As will be outlined below, 'development' is a value-laden term that has proven difficult to define categorically (Power, 2003; Moss, 2007). Despite this, efforts by international agencies, governments, non-governmental organizations and community groups, and to some degree the private sector, have variously sought to advance national economic growth, support industrialization, address poverty, ensure self-sufficiency and raise overall living conditions. The nature of such interventions cannot be divorced from both prevailing international conceptions of what constitutes development and internal political and social realities.

Some of the most striking features of what has come to be referred to as the 'development experiment' are, first, the degree to which development theory and practice have evolved from a narrow focus on Western concepts of modernization through to more situationally specific endeavours at self-reliance (Potter *et al.*, 2008). Allied with this are internally initiated endeavours at local, community and regional development levels, in the quest to attain African solutions and self-reliant development. Third, and associated with the experimentation with different approaches, have been the rather limited results that have been achieved at the aggregate level through applied interventions, which is both a cause for concern and motivation to identify more appropriate interventions.

After providing a brief overview of what development is understood to be, and why it has been regarded as important to pursue, this chapter presents a historical overview of evolving development theory and practice, with specific reference to Africa. The last section of the chapter examines a range of core issues that are impacting on Africa's current development prospects, including poverty relief interventions, local and regional development initiatives, the Millennium Development Goals, the interrelated issues of trade, aid and debt, and the current nature of Africa's economic performance.

9.2 Defining 'development'

The concept of 'development' is an elusive one to define, especially as its meaning has evolved over time (Moss, 2007). Aside from the obvious use of the term in education and psychology, from an economic and human well-being perspective it has been in common currency for some 150 years. Development has traditionally been associated with belief in the value of industrialization, the achievement of higher economic returns, and frequently faith in the assumed benefits of 'modernization', whereby colonies and later independent states were encouraged to emulate the economic trajectories followed by Western states (Pieterse, 2001). However, limited success in attaining widespread improvements in the human condition through such purely economic-based conceptions of development led to a questioning of what was meant by the term and how to achieve it. In the 1970s, Dudley Seers pointed out that, in order to achieve development, issues of unemployment, inequality and poverty needed to be addressed, over and above previous efforts to raise economic growth, which had clearly not led to wealth distribution (in Potter *et al.*, 2008). In practice, this led to a greater focus on addressing the 'basic needs' of people from the 1970s, in contrast to simple efforts to achieve economic growth. By the 1980s, themes of social progress and human welfare started to filter into discussions about development, leading to the later construction of the internationally applicable Human Development Index, which provides an aggregate measure for comparing the development status of all countries in the world based on life expectancy (a surrogate for health and well-being), education and income (Willis, 2005). In the late 1990s, Amartya Sen extended the understanding of what development is through his 'development as freedom' thesis, which views dimensions of political and social freedom as critical to the effective attainment of true improvements in well-being (Sen, 1999).

Over and above these conventional understandings of what development is, more critical authors have questioned the Eurocentric bias of definitions of the term. From the 1970s, neo-Marxist scholars argued that development often leads to a situation of 'dependency' on Western powers. In recent years, there have been appeals to try to identify 'alternative' forms of development that are not locked into Western conceptualizations but rather grounded in human-centred, locally relevant strategies (Hettne, 1995). Work by Robert Chambers and Walter Stöhr, arguing for development to be seen as rooted in local action, resulted in the 'development from below' thesis (Willis, 2005). More recently, anti- and post-development theorists, such as Escobar, have questioned the whole basis of development and sought locally based alternatives, grounded in local practices, as opposed to what they regard as Western impositions (Schuurman, 1993; Power, 2003). We should also bear in mind that Africa has not accepted Western interventions unquestioningly. Indeed, Africa has a rich history of defining and seeking development on African terms and based on traditional values of social support. Efforts to promote African socialism and pan-Africanism, and regional self-reliance through the Lagos Plan of Action are cases in point (Ukaga and Afoaku, 2005; Moss, 2007; see also Box 10.2, below).

Hence, the term 'development' is an ideologically and politically laden concept, as indeed are the different strategies identified to 'achieve' development in practice. Clearly, development involves change and 'improvement' in economic and social conditions, which cannot be divorced from local political conditions, the need for changes in human welfare and associated freedoms to fulfil human aspirations. Within this conceptualization, however, care needs to be taken not to 'impose' external preconceptions on other societies; rather, local values and objectives must be respected. As Moss (2007: 2–3) argues,

> there is no agreement on what exactly 'development' means . . . At its broadest, the development question asked here is, 'how can the standards of living be improved in Africa?' At the same time, it must be recognised that these questions are not merely technical. Development is ultimately not about bricks and budget systems, but about social change.

9.3 Africa's need for development

As was discussed in Chapter 2, Africa clearly lags behind most other parts of the world in a range of key social and economic indicators, which both calls for and justifies either locally initiated and/or externally driven endeavours to try to reduce poverty, raise incomes and improve overall levels of well-being. Table 9.1 indicates how Africa performs in socio-economic terms, relative to global averages. From the outset, it should be pointed out that averaged figures for the whole continent of Africa mask the relative successes achieved in such countries as Botswana, Mauritius and Tunisia, and the far poorer performances of such countries as Niger and Malawi.

As Table 9.1 indicates, Africa is lagging behind the rest of the world in a range of key indicators that include not only economic scores but measures of human welfare, such as nutrition and healthcare. This has prompted national and international concern and involvement since the 1950s by bodies such as the UNDP (United Nations Development Programme) and the aid programmes of Western nations. It is estimated that a total of $568 billion (in today's terms) has been spent in the 'development' of Africa since the 1950s, yet sadly the continent still lags significantly behind other world regions, and minimal economic growth occurred in many parts of Africa in the late twentieth century (Easterly, 2007). As Moss (2007: 87) argues, if one examines the last forty years of development progress, it is apparent that 'economic growth was barely able to keep up with population growth, and in nearly half of the continent, the average person was actually poorer in 2000 than they were in 1970'.

Various reasons can be put forward to explain the poor development success of Africa and what Barrett *et al.* (2008: 1) refer to as 'persistent poverty'. While some of this is certainly due to local mismanagement and the weak local economic and environmental base, one cannot ignore the reality that Africa struggles to compete on an unequal playing field in a hostile global trading environment. The static prices of commodities and the persistence of agricultural subsidies in the North discriminate against producers in the

Table 9.1 Key development indicators

Key indicator	*Global average*	*Africa's average*
Infant mortality rate (per 1000)	46	74
Life expectancy (years)	69	55
Percentage population living on under $2 per day in 2005	48	65
Gross National Income per person parity in $ in 2008	10,090	2660
Human Development Index	0.753	0.514
Calories per person per day in 2005	2768	2098

Sources: FAO, 2009; PRB, 2009; UNDP, 2009

South. This takes place despite the rhetoric of free trade, emphasizing that the global economic system is dominated by the North. On top of this, much of the foreign aid from, and development policy espoused by, foreign powers has been politically motivated. The aim has been either to advance national interests or, especially during the Cold War, to try to buy the support of often dictatorial regimes. Rather more critically in this regard, Patrick Bond (2007), in his evocatively titled book *Looting Africa*, argues that Africa's key problems are exploitative debt, financial dominance by the North, misdirected aid, unfair trade, distorted investment, capital flight, a 'brain drain' to the North and the negative role played by ruling elites. Concerns over the validity of the historic interventions and guidance offered by the World Bank and the International Monetary Fund (IMF), and often the negative role played by multinational corporations in terms of payment of low wages, asset stripping and the repatriation of profits, also need to be factored into these debates (Boaduo, 2008).

According to Paul Collier (2007: 180), several 'development traps' hinder the current and future development of the poorest countries in the world, especially those in Africa:

- The Conflict Trap, including civil wars and coups, which collectively conspire to ensure the persistence of economic stagnation and the dependence on the production of primary products.
- The Natural Resource Trap – associated with the 'resource curse' – is when there is an over-reliance on an often narrow range of export commodities to the detriment of the development of the rest of the economy.
- Being landlocked with poor neighbours.
- Bad governance and associated ills of poor state guidance and planning which characterize many 'failing states'.

A range of additional considerations can be added to this list, including: polarization within societies, political tensions, the reality that many countries are resource poor and often lack adequate land to ensure food security, climate extremes, shocks (both economic and environmental), the prevalence of disease (especially malaria and HIV/AIDS), high levels of population growth and low levels of human and social capital development, weak government and financial institutions, corruption, political instability, low savings rates and low levels of productivity growth (Mistry, 2005; Ndulu *et al.*, 2007). To this list must be added weak levels of democracy, and the 'weak state', hyper-inflation, the debt crisis and externally imposed structural adjustment that saw the weakening of currencies, the loss of economic independence and retarded development in almost all African countries (Moss, 2007). More recently, concerns have also been raised about the increasing 'digital divide' related to retarded progress in information and communication technology development, which is increasing the gap between Africa and the rest of the world in terms of knowledge and information sharing (Mutula, 2008). Finally, the issue of potential climate change cannot be ignored, since Africa, and southern Africa in particular, seems to be facing the most severe risk of declining food production (Addison and Tarp, 2010; de Janvry and Sadoulet, 2010).

In concluding this section, it is perhaps appropriate to reflect on the reality that the challenge of achieving development is not purely located within the African context. At a broader level, as Easterly (2007: 329) notes, the keys to 'growth' seem harder than ever to identify and, 'in the new millennium, a remarkably broad group of academics and policymakers seem to agree that, after all, maybe we don't know how to achieve development, although they are reluctant to say so, exactly'.

9.4 Development theory and practice over time: reflections from twentieth-century Africa

Rather soberingly, Colin Leys (2005: 109) refers to 'the rise and fall of development theory' in his reflection of the past half-century of development theory and practice. However, it would be incorrect to argue that all endeavours in the 'development experiment' have failed. Rather, we need to identify which elements in each of the successive development theories and practices have enjoyed some degree of success in order to maximize future development prospects.

A wide range of development theories and associated strategies have been tried, as is evident in Table 9.2. It should be noted that the various development theories and practices detailed in the table are not mutually exclusive; nor do they necessarily replace each other in direct sequence. Rather, Table 9.2 indicates when they emerged and the era in which they were particularly in vogue. Further, they can exist in parallel in single countries (i.e. national growth strategies and basic needs support), while in other cases strategies have not been adopted by all countries (e.g. self-reliance).

9.4.1 Development in the late-colonial era

In the late colonial era, the European powers tended to treat their African colonies as adjuncts of their national states (see Chapter 1). The colonies served as sources of raw materials for European industry, suppliers of food, markets for European produce and, in times of war, sources of troops (Barratt Brown, 1995). The economies of the colonies were often transformed from pre-colonial self-sufficiency to mono-economies supplying a single product for European markets: for example, copper in Northern Rhodesia (now Zambia), and groundnuts and cotton from Senegal and Mali. While it is difficult to generalize, according to Vance's Mercantile Model (1970, cited in Potter *et al.*, 2008), colonialism imposed urban and transport systems that were developed essentially as conduits to supply raw materials from the colony to the imperial power and to redistribute products from the West. The dominance of port cities in Africa (which often became the colonial capitals) and the development of transport routes not necessarily to where the bulk of the population lived but rather to the resource nodes – the mines, the areas of European farming and the European towns – created skewed economies and, according to such theorists as Andre Frank and Paul Baran (cited in Willis, 2005), 'dependency'. In this line of thinking, colonial economies lost their historical self-sufficiency and their economies were rearticulated to ensure their absolute dependence on external control, markets and capital. Since the independence era (generally the 1950s and 1960s), this dependence has tended to persist through economic linkages established in the colonial era with the European powers (Willis, 2005).

In the late colonial era, development was conceptualized as focusing on the creation of the infrastructure to promote growth – ports, railways and so on – and the associated investment in large mining and agricultural undertakings. The building of key ports, such as Mombasa, Beira and Accra, and the expansion of the East African and southern African rail networks, linked to core mining areas such as the Witwatersrand in South Africa and the Copperbelt in central Africa, are examples of these processes. In terms of agriculture, Africa was often viewed as a 'giant farm' that could supply industrial crops, such as cotton, and food products, such as groundnuts, to the European markets. Development here involved the expropriation of land to provide farmland for European farmers, such as tobacco farms in Southern Rhodesia (now Zimbabwe), and coffee and tea estates in Kenya and Mozambique. Alternatively, it involved the design of mega-schemes that were run

Table 9.2 Development theory and practice over time

Time-frame	*Development theory*	*Applied practice*	*Examples*
Late colonial	Classical economics	Western-style project-based development	Tanganyika groundnut scheme Gezira irrigation scheme in Sudan
From the 1950s	Modernization/diffusionist thinking	Stages of growth/national development strategies	Lake Volta in Ghana Lake Kariba in central Africa
From the 1960s	Dependency theory	Pursuit of self-reliance/Afro-socialism	Ujamaa in Tanzania State socialism in Zambia and Ethiopia
From the 1970s	Basic needs approach	Pursuit of infrastructural and capacity-building interventions	Continent-wide
From the 1980s (following the debt crisis)	Neo-liberalism/monetarism	Pursuit of structural adjustment programmes/lending	Ghana and Kenya
From the 1980s	Human development/Alternative development	Eco-development/bottom-up development	Self-reliance strategies in Tanzania
From the 1990s	Anti- and post-development	Search for alternatives, new conceptualization of development anchored in social movements and localism	

Sources: Adapted from Barratt Brown, 1995; Pieterse, 2001; Potter *et al.*, 2008

essentially as corporate undertakings to supply single products, such as the cotton scheme at Gezira on the Nile in Sudan, the Office du Niger on the Niger Delta that was designed to produce cotton and rice (see Box 2.3, above), and the East African groundnut scheme in Tanganyika (now Tanzania) (see Box 9.1; Best and de Blij, 1977; Willis, 2005; Potter *et al.*, 2008). In most cases, these externally imposed schemes paid little regard to environmental and social issues, including poor soil quality, the risk of water-borne diseases, the aridity of many areas, social constraints on the employment of women in Muslim areas and traditional farming practices. While some of these mega-schemes failed (such as the one in Tanganyika), others (such as Gezira and the Office du Niger) continue to this day. However, even the latter are often only shadows of what was originally anticipated (Deiemar, 2004).

Box 9.1

The East African Groundnut Scheme

One of the most dramatic development interventions undertaken in the late colonial era in Africa is variously known as the 'Tanganyika Groundnut Scheme' and the 'East African Groundnut Scheme' (Wood, 1950, in SJSU, n.d.; Rizzo, 2006). This particular intervention had all the trappings of a fervent belief in modernization – the notion that vast schemes would transform developing regions and that, in this case, Britain could source its domestic food requirements from vast agricultural undertakings in its colonies. Unfortunately, for the designers of the scheme, the project suffered from a significant lack of adequate scientific research, and the rather naive belief that foreign ideas and technology could simply be imposed in a different physical and socio-economic context. Sadly, the scheme now stands not as a development success story, but rather as an 'anti-model' of what can go wrong when development interventions are inappropriate and inadequately planned.

The genesis of the scheme lay in the fact that post-war Britain was suffering chronic food shortages and needed to find alternative sources of supply. In this instance, shortage of cooking oil prompted a plan to transform hundreds of thousands of acres of land in East Africa into a giant groundnut plantation. Running in parallel with the necessity of meeting Britain's needs was a genuine desire to bring about meaningful development in Africa and encourage the modernization of the region (Best and de Blij, 1977). Buoyed by recent successes in the war, British leaders believed in the power of scientific inventions and interventions, and in the ability of strong leadership to drive change.

But the scheme showed all the hallmarks of ill-conceived planning that was largely driven from London in a 'top-down' manner with little contact with on-the-ground realities and constraints. The planners had insufficient knowledge of the local climate and soils, misjudged the negative socio-economic impacts of the scheme, and underestimated the logistical challenges of trying to develop a vast scheme thousands of miles from Britain (Rizzo, 2006).

The scheme had its origins in the thinking of Frank Samuel, who in 1946 headed the United Africa Company, a subsidiary of Unilever. Samuel came up with the plan of producing vegetable oil from a vast plantation in the British Protectorate of Tanganyika (now Tanzania) and neighbouring protectorates (Wood, 1950, in SJSU, n.d.; Birchall, n.d.). The British Labour government welcomed the idea and sent a research team to Africa to investigate the potential of such a scheme. Led by John Wakefield, the Director of Agriculture in Tanganyika, the team reported favourably, arguing that the apparent barrenness and low productivity of the chosen area was due only to traditional local farming practices and that Western technology could rectify the problems. But the team's soil and rainfall studies were inadequate and contributed to the eventual failure of the scheme, while the local Wagogo people were not consulted (Binns, 1994). As a result of the favourable report, the Minister of Food in Britain, John Strachey, authorized £25 million to be spent on the cultivation of the first 150,000 acres (607 sq km) of scrubland over six years. The full scheme envisaged the clearance of 2.4 million acres in Tanganyika, Kenya and Northern Rhodesia (now Zambia) (Wood, 1950, in SJSU, n.d.). The goal was bold and ambitious – that 'the south would be transformed into a thriving agricultural–industrial region within a matter of years, and thousands of Africans would be brought into the modern sector of the economy' (Best and de Blij, 1977: 425). The Overseas Food Corporation, headquartered in London, was established to oversee the project, while an ex-military man, Major General Desmond Harrison, was placed in charge of operations in Tanganyika (Birchall, n.d.).

The site of the first phase of the scheme was Kongwa. This proved to be a major challenge in itself. There was a nearby railway to the closest port of Dar es Salaam, but the line had been irreparably damaged in floods, requiring all of the equipment for the scheme to be brought up a dirt track once it had been liberated from some major congestion at the port (Birchall, n.d.). The first objective of the scheme was to clear thousands of acres of bush to provide land for the first plantings. However, the labourers had to contend with wild animals and frequent bee attacks, while the soil (a mixture of clay and sand) became hard and abrasive when dry (Rizzo, 2006), such that it ruined most of the first sixteen tractors within months of their arrival; even bulldozer blades were blunted. In desperation, hybridized tractors based on old Sherman tanks were drafted in, and the bush was cleared by dragging anchor chains across the land. That still left the problem of what to do with tree roots, though. Moreover, there were issues in terms of food supply and accommodation for the vast staff of 1000 Europeans and 60,000 Africans (Wood, 1950, in SJSU, n.d.).

Other challenges included a flash flood in the first year, a shortage of equipment, delayed decision-making because everything had to be agreed by London, and the workers' antipathy to Harrison's military-style discipline (Birchall, n.d.). Unsurprisingly, labour unrest, machinery failure and mismanagement soon became serious problems (Best and de Blij, 1977). The first crop was disappointing, and the baking sun and lack of rain in the second year dashed hopes of a significantly improved harvest. Out of the original target of 150,000 acres, only some 40,000 were cleared in the first year. In the second year, only 2000 tonnes of groundnuts were harvested from 1000 tonnes of seeds planted (Wood, 1950, in SJSU, n.d.). An attempt was made to grow sunflowers in 1949, but drought destroyed the crop. In January 1951, the British government finally cancelled the project, which had cost some £49 million, had generated virtually nothing for Britain, and had left Kongwa as an unusable dust-bowl (Birchall, n.d.).

In summary, the project suffered from: inadequate research, particularly the failure to undertake proper soil and rainfall surveys or to field-test equipment; its top-down and over-centralized nature; the unrealistic faith placed in Western technology; and the failure to consult with local people (Binns, 1994; Rizzo, 2006). Negative impacts on the area included the long-term discouragement of future investment and food-price inflation as a result of the wages earned by some, which left many others hungry. Meanwhile, labour competition with sisal estates led to local labour shortages in that industry. In short, 'the ill-fated East African Groundnut scheme illustrates the problems of attempting large-scale commercial agriculture in a tropical environment without adequate pilot testing and proper understanding of the region's human and environmental potentials and limitations' (Best and de Blij, 1977: 425).

9.4.2 The start of the development age: independence and development through modernization, from the 1950s

The era of development is generally associated with the post-Second World War era of independence. President Truman's famous inaugural speech in 1949, in which he declared that the 'underdeveloped' nations of the world needed to benefit from the technological and scientific progress of the West, is often taken as the starting point of the 'development era' (Willis, 2005). Within this context, the American model of self-reliant development along market lines was held up as a blueprint, and development was conceived as simply being a process of 'catching up' with the West. In this era, development was seen as a process of 'modernization': that is, adopting the technology and values of the West and transforming society such that sustained growth could be achieved following 'take-off' to a phase of high growth. Growth was regarded as 'diffusing' from the West to the core areas within the former colonies, from where, in turn, it would diffuse to peripheral areas, and thereby raise overall levels of development. Economic definitions of development prevailed, and development was seen as synonymous with the attainment of high levels of economic growth (Barratt Brown, 1995; Potter *et al.*, 2008).

Authors such as W.W. Rostow (1960) developed idealized models of how traditional societies could evolve through a phase of rapid growth, referred to as the 'take-off', before eventually reaching a level of high mass consumption (see Figure 9.1). In the same era, John Friedmann identified the 'core–periphery' framework, in which he argued that growth could 'diffuse' from a series of growth-points (i.e. the 'core' areas of developing countries) to their peripheries and thus raise overall levels of development (Barratt Brown, 1995; Potter *et al.*, 2008). This thinking encouraged newly independent governments to plan large, multifaceted development schemes, such as Lake Volta in Ghana, which sought to link dam construction with the generation of electricity, the development of heavy industry, irrigated farming and fish farming (see Box 9.2; Best and de Blij, 1977). Such schemes were conceptualized as the key ingredients required to achieve the desired 'take-off'. In parallel, many national governments formulated national development strategies in which the ideas of Rostow and Friedmann clearly underlay conceptions of how to promote development (Ukaga and Afoaku, 2005; see Plate 9.1).

Despite the initial optimism that characterized the pursuit of development from the 1950s, expectations were seldom fulfilled, and schemes such as Lake Volta were hindered by limited spin-offs and the unanticipated impacts of water-borne diseases. The limited results generally led to a questioning of the nature of the strategies applied and the viability of the American model that had been adopted without question. The United Nations' reflections on the outcomes of the 'first development decade' suggested that the results were modest, at best. It is therefore not surprising that development theorists and national states started to question the efficacy of development and began to search for alternatives (Barratt Brown, 1995; Potter *et al.*, 2008).

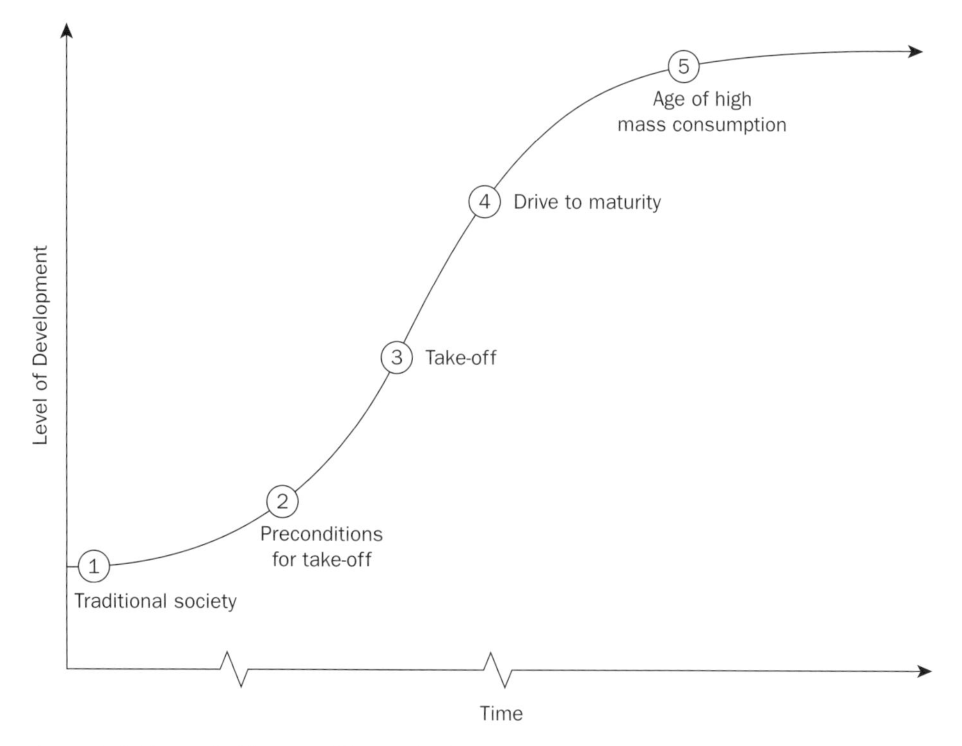

Figure 9.1 Rostow's 'Stages of Economic Growth'.

Source: Redrawn from Binns, 1994

Plate 9.1 Barrage Lalla Takerkoust, near Marrakech, Morocco (Tony Binns).

Box 9.2

The Volta River Project

The early independence era in Africa was characterized by the pursuit of large, show-piece mega-projects. Spurred on by belief in the potential offered by pursuit of 'modernization' planning, and encouraged by President Truman's speech and commitment to address underdevelopment, many newly independent countries in Africa actively sought and implemented what were essentially large, 'top-down' development schemes designed to accelerate modernization. The outcomes of these endeavours were mixed, partly because of their failure to engage actively with those people most affected by the imposed changes.

One of the most impressive modernization schemes of the early independence era was the Volta River Project on the Volta River in Ghana (formerly the British colony of Gold Coast), which in 1957 was the first sub-Saharan country to gain independence (Best and de Blij, 1977). This scheme was promoted in the 1960s as a symbolic representation of the country's new-found independence and economic potential (Ghana Web, n.d.). It undoubtedly brought many benefits to the country; however, as with so many other projects of this nature and size, it also had unintended, undesirable consequences.

One of the key physical features of Ghana is the Volta River, which dominates the physical environment of the eastern half of the country (see Figure 9.2). Albert Kitson, a geologist, first proposed damming the river in 1915, but the sheer cost of the endeavour meant the idea was shelved for over thirty years. Then, in 1949, the Gold Coast's colonial government commissioned a report on damming the Volta for power-generation purposes and independently started planning the new port of Tema at the mouth of the river, just to the east of Accra, the capital city (Ghana Web, n.d.; Fobil and Attuquayefio, 2003). However, with an estimated cost of some £230 million, the project was again delayed. Finally, after independence, it was decided that the dam could be economically justified on account of the need to generate power for the smelting of bauxite at Tema, sourced from mines at nearby Kpong. Following lengthy negotiations with the British and American governments and various aluminium-producing corporations, the newly independent government of Kwame Nkrumah secured loans from the International Bank of Reconstruction and Development, the United States Agency for International Development and private British and American banks (Best and de Blij, 1977; Fobil and Attuquayefio, 2003).

Construction of Akosombo Dam on the Volta River began in 1962, creating Lake Volta, one of the largest artificial lakes in the world, which covers some 8500 sq km and has a 5500-km shoreline (Gyau-Boakye, 2001; Fobil and Attuquayefio, 2003). In 1961, the Volta River Authority was set up to oversee the development of the project and to manage the operation. The scheme was initially designed around four key components. First, the dam and the associated hydro-electric power station, which was designed to generate 912 megawatts (some 98 per cent of total national supply); second, the 145,000-ton-capacity aluminium smelter at Tema; third, an electricity supply network to most of Ghana and even to neighbouring Togo and Benin; and, fourth, the modern deep-water port at Tema and associated road and rail links. Secondary foci included, water supply to Accra and Tema, inland water-borne transport, inland fishing and irrigation of the Accra plains (Best and de Blij,

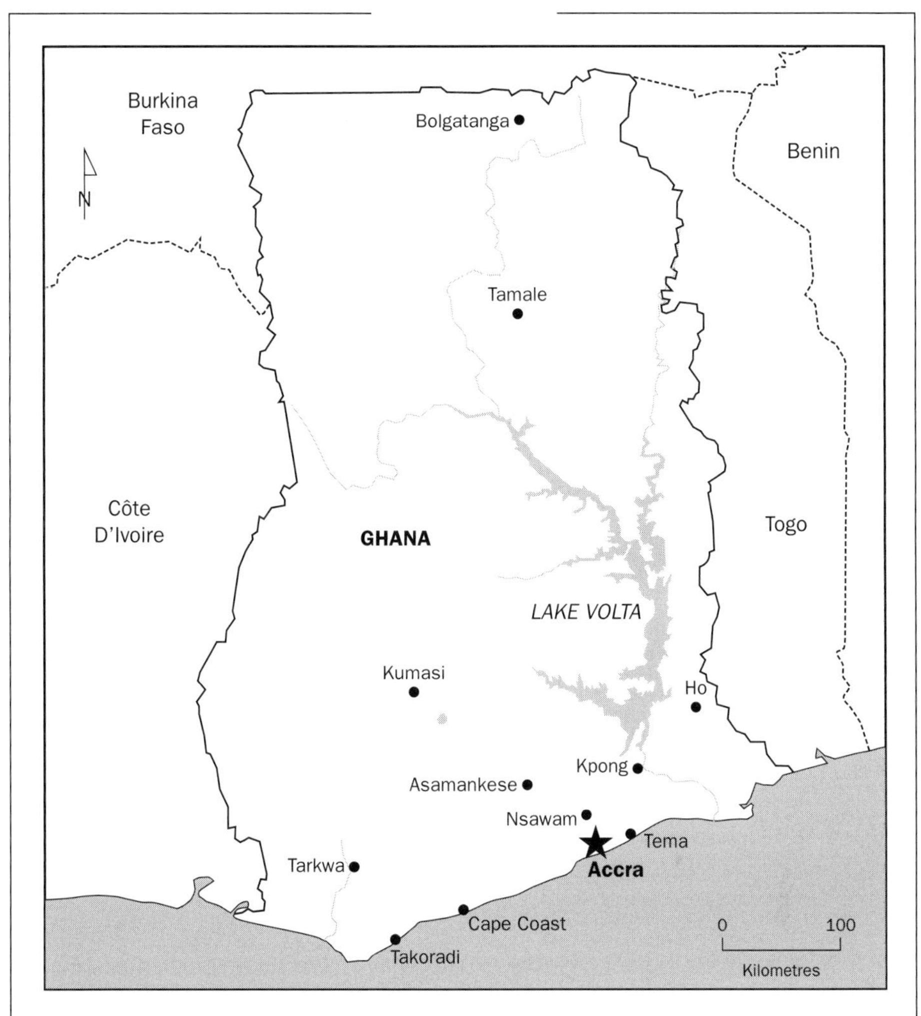

Figure 9.2 Lake Volta in Ghana.

1977; Gyau-Boakye, 2001). The construction required the resettlement of 80,000 people to fifty-two new villages near the lake (Best and de Blij, 1977).

Completion of the project in the late-1960s has provided significant long-term benefits for the country. These include: major industrial development in Tema; the associated generation of foreign-exchange earnings and employment; a reliable supply of electricity; lake transport; lake tourism; farming along the lake's shoreline; and fishing (Gyau-Boakye, 2001). Unfortunately, as has so often been the case with large, 'top-down' mega-projects, there have also been significant, unanticipated negative outcomes. The incidence of schistosomiasis (bilharzia) in the population near the lake has increased from 2 per cent to 32 per cent, while cases of malaria, and initially river blindness, have risen steeply, too (Gyau-Boakye, 2001). The enforced migration of 80,000 people has led to poverty for many and the associated spread of HIV/AIDS (Sauve *et al.*, 2002). At the social level, many of these people have found it difficult to cope with their resettlement and the loss of their lands and ancestral sites, and they have resented having to change their livelihood strategies to either fishing or farming. Resettlement also often increased population numbers to unsustainable levels in the recipient areas, leading to land degradation and out-migration. By the early 1970s, some 20,000 people had already left the resettlement areas (Fobil and Attuquayefio, 2003). In response to many of these social challenges, the Volta River Authority actively promotes educational programmes in the area and it has introduced new economic activities, such as shrimp farming (Fobil and Attuquayefio, 2003).

In addition to the social and health considerations, the sheer size of the lake has caused a range of physical problems. These include: seismic activity; siltation in the dam and estuary; flooding and scouring downstream; negative effects on estuarine life in the delta; micro-climatic changes; and an increase of aquatic weed on the lake itself, which has negatively impacted on transport and fishing (Gyau-Boakye, 2001).

Over the long term, the scheme has also experienced significant operational challenges. In 2007, the aluminium smelter at Tema had to be shut down due to power shortages caused by low water levels on the lake. However, major expansion is now taking place. Plans to develop a second refinery, new power-generating facilities and the opening of new bauxite mines were announced in 2008 (Kpodo, 2008).

While the Volta River Project undoubtedly helped Ghana to proclaim its independence and attain a degree of 'modernization', this clearly came at considerable cost, which has impacted most on those directly affected by the construction of the dam. 'Top-down' planning, often in pursuit of Western-style projects, has often not led to the anticipated 'trickle down' of benefits to everyone. This fact surely justifies the parallel or alternative pursuit of 'bottom-up' projects grounded in communities to try to maximize local benefits and ensure greater empowerment.

9.4.3 Alternative conceptions of development: dependency theory and African self-reliance, from the 1960s

From the 1960s, the dependency theorists, such as Frank and Baran (see above), began to question the relevance of the Western model of development, which they saw as creating and enforcing a dependent form of development based on the West (see Figure 9.3, which shows how transport links were structured in the colonial era). Instead, they argued that the states in the developing world needed to 'de-link' from their historical dependence on the West and identify situationally appropriate development strategies, which the theorists referred to as the pursuit of 'self-reliant' development (Barratt Brown, 1995; Willis, 2005; Potter *et al.*, 2008).

In practice in Africa, there was a clear overlap between the questioning of the Western model and an increasing interest shown in socialist models that were in vogue at the time in eastern Europe and China. While few African countries, with the possible exception of Ethiopia, directly pursued eastern-style socialism, many attempted to throw off what were perceived to be the 'shackles' of colonial oppression through efforts to promote unique forms of African development, such as 'Afro-socialism' and the active courting of links with the USSR, China and the Eastern Bloc. These strategies had their roots in pan-Africanism, as espoused by key African leaders such as Kwame Nkrumah from Ghana, and the belief in African self-reliance as advocated by such leaders as Kenneth Kaunda in Zambia and Julius Nyerere in Tanzania (Davidson, 1994; Ayeni, 1997; see Box 10.2,

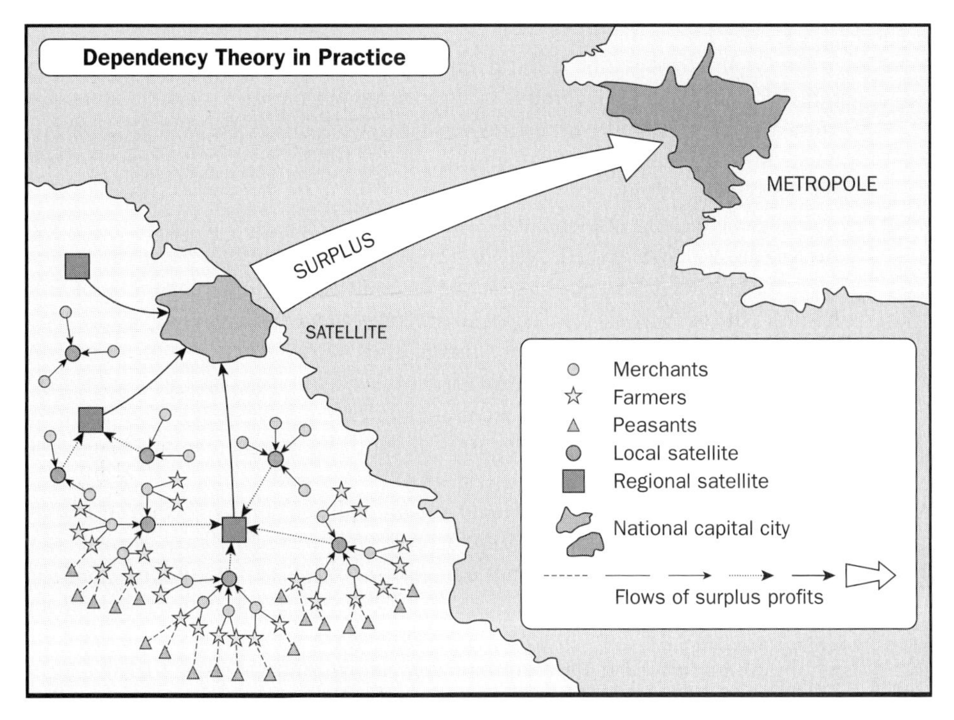

Figure 9.3 Dependency theory in practice.

Source: Adapted from Potter *et al.*, 2008

below). In countries such as Ghana, Zambia, Ethiopia and Tanzania, this era was characterized by bold attempts to implement nationally appropriate strategies that often had a strong rural focus and an appeal to 'self-reliance' principles. The most extreme case was Tanzania, which actively sought to sever links with the external world and pursued an ambitious policy of rural development through a villagization development scheme known as '*Ujamaa*' (Davidson, 1994; Ayeni, 1997; see Box 9.3). Although motivated by efforts to raise rural living standards, interventions proved costly and often disrupted traditional life when people were required to live in central villages distant from their farmlands. Parallel efforts to ensure industrial self-sufficiency were hindered by a lack of local skills, resources and funds. As a result, and despite the enthusiasm with which these programmes were pursued, true economic independence from historical and economic ties with the West proved difficult to attain (Ayeni, 1997).

Box 9.3

Ujamaa, villagization and rural development in Tanzania

Challenges posed by trying to achieve development through modernization approaches in Africa encouraged the popularity of 'dependency theory', which argued for the shedding of colonial bonds and the pursuit of self-reliance. The parallel emergence of a new group of leaders in Africa, such as Kenneth Kaunda and Julius Nyerere, who advocated policies of humanism and 'African socialism', led to the implementation of rural development schemes and the creation of new rural service villages anchored in basic socialist principles. Probably the most well known of these, albeit perhaps the least successful, was the villagization policy pursued in newly independent Tanzania in the 1960s and 1970s (Kjekshus, 1977; Barratt Brown, 1995).

This scheme was implemented just after Tanzania gained independence from Britain in 1961, when well over 90 per cent of the country's population lived in scattered rural settlements. Agriculture's contribution to the nation's GDP was falling and most farmers were engaged in growing food for family subsistence, using traditional low-level technology. Inequalities were increasing in terms of the distribution of income and the provision of basic services, such as healthcare and education, and the country was becoming increasingly dependent on foreign loans and grants. It was hoped that an aggressive rural development policy would help to address these issues (Binns, 1994).

Less than six months after the end of colonial rule, Julius Nyerere, Tanzania's president and leader of the ruling party, TANU (Tanganyika African National Union), published a pamphlet entitled *Ujamaa: The Basis of African Socialism*, in which he urged a return to 'traditional values', according to which everyone had a right to be respected, an obligation to work and a duty to ensure the welfare of the whole community and the common ownership of basic goods (Binns, 1994). *Ujamaa* is a Swahili word that translates as 'familyhood' and relates to the concepts outlined by Nyerere in his pamphlet (Best and de Blij, 1977: 426). Later, in 1962, in his inaugural address as President, Nyerere introduced the idea of villagization, emphasizing the importance of developing the agricultural sector, which, he argued, could be achieved only if farmers lived in villages. He said,

> unless we do [this,] . . . we shall not be able to use tractors; we shall not be able to provide schools for our children; we shall not be able to build hospitals, or have clean drinking water, it will be quite impossible to start small village industries, and instead we shall have to go on depending on the towns for all our requirements; and even if we had a plentiful supply of electric power we should never be able to connect it up to each isolated homestead.
>
> (Nyerere, quoted in Binns, 1994: 110)

Critical to this vision was the perceived need for farming to be done communally, and for people to live close together in villages where services and marketing support could be concentrated (Best and de Blij, 1977).

Five years later, the Arusha Declaration of February 1967 was a landmark in Tanzanian, and indeed African, political history. Nyerere announced the nationalization of banks, trading organizations and the largest multinational corporations operating in the country. He also called for a halt to the accumulation of private wealth by party and government leaders (Hyden, 1980) and committed the country to the pursuit of self-reliance (Coulson, 1982). In essence, the Arusha Declaration outlined the key elements of socialist development, stressing the use of local ideas and resources. This was reinforced in September 1968 with Nyerere's paper on 'Socialism and Rural Development', whose Swahili title was '*Ujamaa Vijijinies*', which literally translates as 'Socialism in the Villages'. In this paper, Nyerere again rejected rural capitalism and turned the *ujamaa* of his 1962 paper into a national policy, such that rural workers had the responsibility to establish and/or encourage *ujamaa* villages, building on traditional mutual aid in the extended family of Tanzanian society (Binns, 1994).

The main aims of the new rural development strategy were:

1. The establishment of self-governing village communities to improve living standards by providing social infrastructure, such as healthcare and education, while also providing facilities for the marketing of crops and livestock.
2. Better use of rural labour, taking advantage of economies of scale to increase communal production (although the emphasis on communal production was later reduced).
3. The dissemination of new values and the avoidance of exploitation. Village councils would oversee land reform, allocating land among private cultivators.
4. The mobilization of people for national defence by using the villages as paramilitary organizations (Hyden, 1980; Coulsen, 1982; Kahama *et al.*, 1986; Binns, 1994).

Government and other organizations would supposedly help to explain the underlying principles of *ujamaa* villages, promote good leadership among farmers, and plan village sites, food cultivation and service provision. Village development was encouraged through a series of 'operations', but these campaigns by regional party and government administrations were often carried out hurriedly, with insufficient planning, little consultation with the people and limited understanding of existing farming and pastoral systems. Nyerere took a personal interest in the formation of the *ujamaa* villages, for example by initiating 'Operation Dodoma', a government-planned programme to move all the people in that region into villages. Consequently, the number of new villages in Dodoma increased from 75 in 1970 to 246 in 1971 (Binns, 1994).

On 6 November 1973, Nyerere announced that all Tanzanians would have to live in villages by the end of 1976, and at the same time some fundamental policy changes were made. Emphasis on communal production, for example, was dropped in favour of block farming, designed to promote economies of scale and village production planning, but not requiring a wholehearted commitment to the principles of *ujamaa*. From this time, the new villages were called 'development villages', and the idea was that they would serve as multi-purpose cooperatives, taking on all crop marketing and credit functions. This major thrust of compulsory villagization between 1973 and 1976 was probably the largest resettlement effort in the history of Africa (Hyden, 1980). Although most people did not have to move long distances, they often had to agree to abandon their former residence and land. People were instructed to move to the nearest new village or trading centre, and political efforts were made to create *ujamaa* villages with 250 or more families out of the enlarged units. Initially, people were persuaded to move into the new villages through the promise of such services as water, schools and dispensaries. However, sometimes these, and food supply, proved to be inadequate. Villages were at first concentrated in the poorer regions of Tanzania, and by September 1974 there were 5000 with 2.5 million residents, representing 20 per cent of the country's total population.

In that year, the government decided to abandon its strategy of persuasion in favour of forced relocation. Between May and December 1974, Operation Sogeza (moving) implemented this new policy, and by 1975 a large majority of the rural population was resident in the new villages. In 1976, 3 million people were living in 7000 *ujamaa* villages (Best and de Blij, 1977). Thereafter, a series of bad harvests and food shortages encouraged the government to increase agricultural production targets, and peasant farmers were each required to plant a minimum of three acres of food crops and one acre of cash cops. By the late 1970s, however, many village-based initiatives had collapsed and the future looked unpromising (Binns, 1994).

The main problems encountered by the villagization programme were:

- For the majority of people in most villages, their private farms still remained their primary focus of interest, and there was some conflict as to how much labour and attention should be given to communal activities. The first five years of *ujamaa* did not have any significant impact on peasant agricultural production, and the productivity of the land did not improve. Yields per hectare in communal farming areas were well below those of private farming and there was often confusion over the distribution of income from the communal farm. Communal work actually ceased in some villages by the early 1970s. The bigger the village, the more difficult it was for members to achieve a common sense of purpose. Petty-capitalist farmers often did best, as they were more willing to take risks and modernize their farming methods. Poorer peasants often chose these more successful farmers as their village leaders.
- In many *ujamaa* villages, the absence of a reliable system of financial control caused problems, perhaps due to the low level of education of leaders and villagers. Embezzlement of funds was common and administrative and technical staff often failed to serve the villages adequately and loyally.
- Impressive progress was achieved in the provision of schools, dispensaries, water supplies and other rural infrastructure, and marked improvements in healthcare and adult literacy were recorded. However, schools and dispensaries often lacked even basic supplies and electrical power and water supply were frequently inadequate. Also, in spite of better rural service provision, young people were still attracted to education and work in the towns, a move that was often supported by parents because it could yield additional family income.
- The principles of *ujamaa* and communalization frequently conflicted with well-established networks of social institutions, such as women's groups and cattle-owner associations. Furthermore, the social effects of villagization must have been traumatic at times, particularly where force was used to resettle people. There is still much debate about whether villagization reduced or increased inequality within and between villages and households.
- Where tractors and other machinery were introduced, there were problems of maintenance and lack of fuel, particularly after the 1973 oil crisis.
- The environmental knowledge that peasants had prior to villagization was often invalid in the new settlements, where soil conditions and other factors of production were different. There was concern in some areas that nucleation (concentrated settlements) could lead to erosion as the carrying capacity was exceeded.
- In many cases, resettled people continued to farm their now-distant ancestral lands at great cost in terms of time and physical energy. Simultaneously, significant declines in export crop production were noted, namely in cotton, coffee, sisal, tea and tobacco.

- The scheme has been criticized for focusing on the siting of villages, normally on a road to enhance external access, while paying little regard to local mobility patterns, access to the best farmland and local traditions (Kjekshus, 1977; Hayden, 1980; Kahama *et al.*, 1986; Binns 1994; Mapolu, n.d.).

Following food shortages and near famine in 1974, which led to the need to import 880,000 tonnes of maize, wheat and rice, and was attributed to the upheaval in rural areas, cooperatives were completely abolished in 1976 (Best and de Blij, 1977; Kahama *et al.*, 1986), after which the buying of crops and the provision of farm inputs became very unreliable.

In the decade after the Arusha Declaration, the majority of peasants had been subjected to numerous government directives and orders without witnessing much economic development. In 1974, Nyerere admitted that the country had exhausted its financial resources (*Time*, 27 January 1975). The villagization programme generated massive resentment among peasant farmers. Most writers agree that it was a failure, and the Tanzanian government has since reversed many of its earlier policies. It is noteworthy that underplaying the potential role of industrial development in favour of rural development negatively affected internal industrial supply levels (Kahama *et al.*, 1986).

As a result, Tanzania's bold efforts to ensure rural service delivery and national self-sufficiency fell well short of expectations. While access to certain social services clearly improved, the failings of the intervention are indicative of the inherent problems of 'top-down' schemes that do not involve local people in the decision-making process and where politics rather than rationality drives policy. The scheme also reveals the challenges of pursuing the concept of 'self-reliance' to its logical conclusion, as proposed by the dependency theorists.

At a broader level, the pursuit of African self-reliance and associated adherence to the principle of pan-Africanism culminated in the Lagos Plan of Action of 1980 – a bold attempt to lay the basis for a continent in which internal linkages and trade were to be encouraged. However, mistrust, the production of like products (which restricted inter-African trade opportunities) and continuing dependence on Western trade and technology, hampered such bold thinking (Ukaga and Afoaku, 2005; Janneh, 2006).

9.4.4 The basic needs approach, from the 1970s

From the 1970s, it became apparent that 'take-off' had not occurred in much of Africa, so there was a need to rethink what development involved and how to achieve it. In parallel, the International Labour Organization and research undertaken by Keith Hart into the informal sector in Kenya, revealed the key role played by non-formal economic activities in the livelihoods of a significant percentage of the African population (Barratt Brown, 1995). In turn, this led the donor agencies to adopt a new approach to funding and support of development that prioritized the 'basic needs' of Africa's people. Concrete action shifted from the development of mega-projects and national development strategies to direct support of the poorest communities through the extension of primary healthcare, the sinking of wells and education support (see Plate 9.2). However, the scale of the development backlog in Africa rather limited the effectiveness of these well-intentioned interventions (Barratt Brown, 1995; Potter *et al.*, 2008).

Plate 9.2 Water hand-pump in Eastern Cape Province, South Africa (Etienne Nel).

9.4.5 Structural adjustment and neo-liberalism, from the 1980s

Efforts at national development in Africa hit an almost insurmountable barrier in the early 1980s, when most African countries found themselves unable to repay the debts they had accumulated in the 1970s, following the oil-price hike of that decade and global recession. In order to avoid a collapse of the Western banking system, starting in Mexico in 1982, the IMF and the World Bank introduced a system that allowed a rescheduling of the debt repayments in return for acceptance of a package of 'structural adjustment' measures, which obligated recipient countries formally to accept neo-liberal economic prescriptions, including the withdrawal of the state from much of the economy, rationalization of the civil service, support for the private sector, market liberalization, encouragement of investment and currency revaluation (Barratt Brown, 1995; Bond, 2007). In turn, new conditional loans could be secured. During the course of the 1980s, most African countries were obligated to accept these structural adjustment packages (SAPs). While the World Bank claimed some success, extreme hardship was experienced in Africa, incomes and economic growth rates fell, unemployment rose and social service expenditure declined. Consequently, the 1980s came to be known as the 'lost decade' for much of Africa. With few exceptions, significant market-led growth did not occur (Barratt Brown, 1995; Bond, 2007). While SAPs and neo-liberalism were not development strategies as such, they led to the rationalization of previous forms of state intervention, the required pursuit of neo-liberal orthodoxy, and the belief that freeing up the markets would draw in trade and investment.

9.4.6 Alternative development, from the 1980s

Persistent poverty, the limited success of the SAPs and the questioning of both Western and socialist models led to the search for alternative, situationally appropriate theories of development from the 1980s onwards. Encouraged by the work of the Dag Hammarskjold Foundation and thinking about 'alternative development' (Hettne, 1995), theorists questioned the beliefs in linear, Western-based development models, which essentially imposed development from the top down on societies that were either ill prepared or ill informed about the need or nature of such interventions.

Alternative development thinking spawned, and was related to, a wide diversity of approaches that emerged in this era, including recognition of the territorial bases for development, and the need to encourage eco-development, to work with communities at their level and to draw upon their indigenous technical knowledge. Perhaps most significant of all was the recognition of the role and place of 'bottom-up' development anchored in the strengths of the host community (Hettne, 1995; Potter *et al.*, 2008). The latter seeks to encourage community-based development and self-reliance through encouraging local people to take the lead and tap into situationally specific resources to improve overall levels of well-being. Numerous case studies from Africa were detailed by such authors as Gooneratne and Mbilinyi (1992) and illustrated the value of community self-reliance, particularly in the face of the harsh realities imposed by SAPs.

9.4.7 Post- and anti-development and the pursuit of local development, from the 1990s

By the 1990s, development policy and theory had come full circle in Africa. Belief in the ability of modernization to enable African countries to emulate the West had been shattered, and a 'development impasse' had arisen (Andreasson, 2005). The debt crisis of the 1980s and SAPs had reinforced perceptions of the subservient role that most African countries play in the global economy, while indigenous efforts, such as Afro-socialism, had proved incapable of meeting the needs of growing populations. At a theoretical level, more critical authors, such as Escobar, Pieterse and Schuurman (see Potter *et al.*, 2008) essentially rejected the Western conceptualization of development (i.e. anti-development thinking) and urged planners to seek more situationally relevant forms and approaches to promote improved well-being. Some authors have also argued for the need to move beyond current understandings of development (i.e. post-development) and to look for approaches that are community based and empowering. While not yet offering a real alternative, these theorists have encouraged the broader academic and development community to seek creative alternatives beyond modernization, many of which relate to exploring 'local' development potential and capacity. This approach has come to the forefront in an era when the decentralization of state activities is facilitating greater forms of self-reliance among both communities and local governments (Andreasson, 2005; Willis, 2005).

9.5 Development practice in the twenty-first century

In the twenty-first century, development practice has become more eclectic than in previous decades. While many countries still strive to promote national development strategies, such interventions are now tempered by the legacy of SAPs, the requirement of accession to the World Trade Organization to reduce state intervention, principles of decentralization and the acknowledged role of community groups, non-governmental organizations and the private sector in development. In addition, within Africa there are new drives to pursue African-based development options, anchored on historic principles of pan-Africanism and self-reliance, while, at the global level, poverty reduction strategy plans (PRSPs) and the Millennium Development Goals (MDGs) have set the parameters through which global support is applied and assessed. This section of the chapter explores the practical outcomes of these changes, which include the enhanced role of local and community development, the new role played by regional integration in African development, the roles of trade, aid and debt, and the significance of PRSPs and the MDGs.

9.5.1 Local and community-based development

Two broad themes characterize local development in Africa. The first is the formal empowerment of local governments through processes of decentralization to engage more directly with local development challenges. The second relates to processes that have always been in existence, namely the actions of local community groups who wish to improve living and economic conditions in their locality. In the view of Gooneratne and Obudho (1997), given the scale of the African economic crisis, communities have to pursue local development options. They also recognize the key role that NGOs can play in supporting local initiatives, often in the absence of state support. A particular concern for a range of authors is the need for governments to allow for, and facilitate, 'local self-reliance' among communities (Gooneratne and Mbilinyi, 1992; Taylor and Mackenzie, 1992).

In terms of the theme of local government decentralization, driven partially by SAPs and partially by the widespread pursuit of democratic engagement, there has been a dramatic shift in most countries from direct central government control over local governments to acknowledgement and facilitation of local control and decision-making. The introduction of decentralization policies in countries as diverse as Ghana and Zambia has characterized this new era, and the World Bank has noted how widespread this process has been (Binns *et al.*, 2005; Gooneratne and Obudho, 1997; Hope, 2008). However, on-the-ground evidence in countries with well-established track records of decentralization over more than a decade indicates that deep-rooted operational challenges impede progress (Binns and Nel, 2002; Nel and Binns, 2003b; Egziabher and Helmsing, 2005). These include a lack of skilled staff in localities, limited funds, and what Stockmayer (1999) has termed the consequential 'decentralization of poverty', which has shifted responsibility from the central state to the local level.

A variation of local development is the more focused approach of local economic development (LED), which has attracted considerable attention across Africa. While many countries, such as Swaziland and Zambia, have expressed interest in the LED approach and have introduced some policy support, applied practice across Africa remains limited (Gooneratne and Obudho, 1997; Egziabher and Helmsing, 2005; Hampwaye, 2008). The one exception is South Africa, which is regarded as a leader in the policy and practice of LED (Rodriguez-Pose and Tijmstra, 2005). LED is now a legal requirement of local governments in that country and, interestingly, local governments have been challenged to implement policies of 'developmental local government', which implies being conscious of encouraging development-related outcomes from all of their actions (Nel 1999; Nel and Binns, 2003a; Nel and Rogerson, 2005). Experience varies widely in the country – from that of small, impoverished rural municipalities, which are able to support only limited community projects in such activities as community tourism and farming, to the large metropoles, which are pursuing globally competitive marketing and investment strategies (Rogerson, 1997). Johannesburg and Cape Town have positioned themselves on the world stage as 'global cities', replete with modern airports, sports stadiums, convention centres, business and tourism support programmes and various forms of informal sector support (Nel and Rogerson, 2005; see Box 5.6, above, and Box 9.4). However, evidence suggests that outcomes, though often impressive, seldom devolve down to those communities that are most in need in the big cities. In smaller urban centres, the lack of progress is often more evident where municipalities are unable to effect change, commonly through a lack of resources and staff. A net result of the limitations experienced by developmental local government in both large and small centres has been widespread civil sector protests against local government's restricted delivery in recent years (Nel *et al.*, 2009). At a broader level in Africa, there is growing recognition of the role that local governments can play in development processes, as indicated by the recent establishment of the Municipal Development Partnership for Eastern and Southern Africa. Based in Harare, Zimbabwe, this organization is promoting research and collaborative exchanges between local governments. It also works with the Africa Local Government Action Forum, which has a primary focus on the promotion of LED (MDP, 2006).

Box 9.4

Local economic development in Durban, South Africa

The twenty-first century has been characterized by the devolution of planning and developmental responsibilities to local governments and local agencies in many parts of the world, including in numerous African countries. One of the key aspects of such devolution involves enhanced local economic planning and development interventions generically referred to as 'local economic development' (LED), which enjoys widespread recognition around the world (Pike *et al.*, 2006). This term is generally associated with local government actions, referred to as 'urban entrepreneurialism', and/or the actions of local community groups, with or without the support of non-governmental organizations and local business organizations. The focus of such actions tends to be on responding to local economic crises, and/or responding to new growth opportunities, encouraging employment creation and business development and responding to social development needs (Blakely, 1989; Pike *et al.*, 2006).

Though common practice for decades in Europe and North America, it is only relatively recently that LED has been practised to any significant degree in Africa. The reasons for this are twofold: first, the traditionally centralized nature of state control in Africa which allowed little scope for local action; and, second, the severity of economic crises that have overlapped with the devolution of powers to local levels in recent years. Nevertheless, as Egziabher and Helmsing (2005) note, LED is starting to feature in countries as diverse as Ethiopia, Kenya, Uganda, Zambia, Swaziland and South Africa (Nel and Rogerson, 2005; Hampwaye, 2008).

Within this recent movement into LED, South Africa has come to be regarded as something of a front-runner, not only within Africa, but in the Global South more generally (Rodriguez-Pose and Tijmstra, 2005). South Africa's prominence in this regard is partly due to its tradition of having large and, by African standards, well-resourced cities, with relatively independent and strong local governments that have the ability and means to influence local development processes. Furthermore, after the fall of apartheid in 1994, the new, democratically elected government instituted a series of measures designed to empower communities and ensure that local development was actively pursued by all local authorities. For instance, the 1996 National Constitution mandated local governments to engage in socio-economic development in the areas under their jurisdiction, and the 1998 Local Government White Paper introduced the concept of 'developmental local government' (Binns and Nel, 2002; Nel and Binns, 2003a). This prompted local governments to consider the developmental implications of all of their activities. Subsequent LED policy papers, and the requirement that LED must form part of regular integrated development planning and promote pro-poor development, have made LED a hallmark of South African local government activities (Nel and Rogerson, 2005).

The diverse city of Durban, with a population of 3.2 million people, lies on the east coast of South Africa. It hosts the busiest port in Africa, is the premier domestic tourism destination, with a string of five-star hotels on its 'Golden Mile', has one of the world's biggest convention centres and the biggest single-phase shopping centre in the southern hemisphere, has two oil refineries and one of the biggest Toyota car manufacturing plants in the world, and is the country's second-biggest concentration of industrial activity (see Plate 9.3). In addition, though, it is a city of extreme poverty, with an unemployment rate exceeding 30 per cent and hundreds of thousands of people living close to the poverty line (Nel *et al.*, 2005; Robinson, 2008). The LED imperative therefore requires the city authorities both to address chronic poverty and simultaneously to promote Durban as a key, secondary node in the global economy.

These ideas were encapsulated in the 'Long-Term Development Framework' adopted by the city in 2002, which committed Durban to enhancing the overall quality of life of its citizens through a threefold strategy of meeting basic needs, strengthening the economy and building skills and technology (Durban Unicity, 2001). To help achieve these goals, the city established an Economic Development Unit, actively promoted partnerships with business through a series of planning forums, and included LED as part of its long-term integrated development planning process.

Plate 9.3 Durban CBD, South Africa (Etienne Nel).

The realities of Durban's development needs and opportunities are such that both pro-poor and pro-market strategies need to be pursued in parallel. Regrettably, it would appear that the only significant success has been achieved with the latter, despite the urgent need to address chronic problems of unemployment and poverty. In line with other major cities in the world, partnership formation has proved critical, not only in the regular hosting of joint planning forums in the city, but in the initiation of mega-projects. These include the joint city–private sector development of the 'edge city' at Umhlanga Ridge, to the north of Durban, and the associated shopping centre development, redevelopment of the beach front, two new casino complexes and the opening of a water-world complex. Key city projects include upgrading the transport infrastructure, building the convention centre and waterfront development (Nel *et al.*, 2005). These have undoubtedly enhanced the city's standing as a key tourism, business, retail and industrial node, but there is little evidence that the poor have shared in these benefits.

In terms of pro-poor interventions, in the 1990s the city attempted to promote township economic development with almost no success ('townships' in South Africa are low-income areas occupied almost exclusively by black people). This was due to a lack of resources, a failure to involve the beneficiaries in decision-making, and the persistence of poverty. More recently, development teams have targeted the poorest areas, such as Cato Manor and Inanda, where they have attempted to encourage entrepreneurial training, small business support and market development. In terms of project development, support for a Business Development Unit and the noteworthy Warwick Triangle represent concerted efforts to provide skills training and facilities to the poorest entrepreneurs. Warwick Triangle, a key transport node through which hundreds of thousands of commuters pass daily, is also home to thousands of informal sector traders. The council has provided trading facilities, toilets, medical- and child-care in this area, and has won the recognition of the World Bank as a leading example of small business support (Nel *et al.*, 2005; Economic Development Unit, n.d.; see Plate 9.4).

Plate 9.4 Warwick Junction market, Durban, South Africa (Etienne Nel).

Over and above such direct interventions, the city's Economic Development Unit promotes investment in the city, coordinates economic data collection, supports tourism development and provides general business advisory support. The city is also known for the nature of its procurement policies, which, wherever possible, give preferential consideration to tenders from low-income and disadvantaged communities (Economic Development Unit, n.d.).

These activities and policy commitments show how one of Africa's leading cities has committed itself to the active pursuit of LED. The range of projects adopted is impressive, as are the engagement with the private sector and efforts to promote pro-poor development. There is no doubt that Durban's role as a key industrial, tourism, business and retail node has been significantly enhanced. Unfortunately, for the underprivileged majority in the city, there is minimal evidence that a significant reduction in poverty has been achieved. It seems likely that this will remain a key failing of LED in Durban and other African cities, where integrating the poor into the formal economy has proved to be particularly challenging.

At the community level, recognition in the 1970s that development along the lines proposed by the diffusionist and modernization approaches was not succeeding in addressing mass poverty in Africa galvanized a reinterpretation of development interventions and approaches. This included acknowledgement that 'basic needs' intervention was probably a more appropriate line to follow. There was also finally recognition that the informal or 'second' economy is often the largest part of the economy in many areas (Pacione, 2001; Hope, 2008). In the 1980s, the scale of the debt crisis and negative economic growth in the poorest communities forced many rural communities back into subsistence and frequent reliance on barter and parallel economic systems. Despite numerous constraints, NGOs have been most active in this area, and governments have attempted various low-level support measures, such as providing market facilities and extension support to farmers and entrepreneurs (see Plate 9.5). In parallel, a significant volume of literature on this dimension of local development has emerged. Prominent in this regard were the research and policy proposals of Gooneratne and Mbilinyi (1992), based at the United Nations Centre for Regional Development's regional office in Nairobi, Kenya. In addition, Baker (1990) and Baker and Pederson (1992), as well as other researchers based at the Africa Studies Centre in Uppsala, Sweden, discussed the realities of rural and urban livelihoods, and local economic adaptation in order to survive. While the overall economic situation has improved in many parts of Africa since the 1980s, self-reliance initiatives at the local/community and family levels remain absolutely critical for survival for the majority of the poor. As outlined by Taylor and Mackenzie (1992) and Egziabher and Helmsing (2005), initiatives such as the production of charcoal, basic metalwork and handicrafts are often critical in ensuring economic survival (see Box 9.5). Recent writings on rural and urban livelihood strategies indicate just how important it is for communities to have multiple livelihood strategies and sources of income. While often poorly understood and difficult to support directly, this form of local development is an essential survival mechanism for a significant number of Africa's residents. Future research and policy support in these areas will be critical for the long-term well-being of these people (Egziabher and Helmsing, 2005).

Plate 9.5 An NGO extension agent in Ethiopia delivers spades and energy-efficient stoves to a rural community (Alan Dixon).

Box 9.5

Community beekeeping in Bondolfi, Zimbabwe

Persistent poverty in Africa, the loss of developmental momentum and the economic collapse and stagnation associated with the debt crisis and structural adjustment of the 1980s obliged thousands of community groups to turn inwards – to traditional or new forms of community self-reliance – in order to make ends meet. In many cases, the 'lost decade' of the 1980s forced many communities out of the money economy and into reliance on barter systems and various forms of subsistence (Taylor and Mackenzie, 1992). Extensive research undertaken by the United Nations Centre for Regional Development in Africa has indicated just how widespread community or local self-reliance has become in supporting the livelihoods of millions (Gooneratne and Mbilinyi, 1992). Community self-reliance is seen as having a key role to play, because it is an effective local response to marginalization, it tends to rely on indigenous skills, it is often ecologically appropriate and it is cost effective. Within this context, there is clearly a role for non-governmental organizations and faith-based groups to play in terms of the supply of information, encouragement and support (Burkey, 1993; Scoones and Chibudu, 1996). A key asset of self-reliance strategies is the fact that it draws on indigenous skills and often centuries-old practices of farming, resource management, traditional manufacturing and trade.

Zimbabwe has, sadly, experienced significant political and economic turmoil for several decades. Within this context, hundreds of thousands of people have been forced out of the formal economy, have lost access to social services and have become increasingly self-reliant. It is often easier to promote collective community action within traditional rural village settings rather than the densely settled cities. The community of Bondolfi lies 40 km south of the central Zimbabwean town of Masvingo. In the broader area there are some 10,000 people, and the key assets in the community are a few stores, a church, a mission and a school (Illgner *et al.*, 1998). While most people rely on the cultivation of traditional crops, such as maize, the severity of the economic crisis experienced in the mid-1990s prompted local community members, in collaboration with mission staff, to assess the community's alternative assets and skills. Traditionally, honey has been collected from natural hives for use in domestic beer brewing and rituals. It was recognised that, given the high price that honey can command, there was scope to expand the volume collected. Fortuitously, around this time, the mission was visited by a volunteer worker who had gained experience in Kenya in making simple, cost-effective hives out of natural materials. These hives generated higher yields than traditional honey-collection methods from wild hives (Nel *et al.*, 2000).

As a consequence of this, in 1995 members of the community formed the Bondolfi Beekeepers Association. The seventy members of the association work collectively in the manufacture of hives and basic safety equipment and the collection and processing of the honey, while the mission has provided basic training and, critically, a market outlet, as it transports the honey to local retailers in Masvingo (Illgner *et al.*, 1998). Under optimum conditions, and with access to an average of ten hives per member, income from honey sales can be nearly 150 per cent higher than from traditional crop farming. It is important to note that as beekeeping is a part-time occupation, members can still produce their usual field crops, so the beekeeping has become a key element in a multi-livelihood strategy (Nel *et al.*, 2000).

The success of this scheme clearly indicates that community-based development, which draws upon local skills and available resources, has the ability to enhance community incomes significantly. Simultaneously, one should not underestimate the key role played by an NGO, in this case the local mission, in helping to address key skill and logistical shortfalls.

9.5.2 Regional development: the search for collective self-reliance

A distinctive feature of regional development practice and policy in Africa is widespread adherence to the principles of cross-border linkages within the continent for purposes of collective self-reliant development, facilitating trade, social, cultural and economic exchange, peacekeeping and the overall promotion of pan-Africanism. International connectivity has had a relatively long history in the continent. The world's oldest customs union is the Southern African Customs Union, formed in 1910, which now includes South Africa, Botswana, Namibia, Lesotho and Swaziland (Kyambalesa and Hougnikpo, 2006).

Makinda and Okumu (2008) argue that the proliferation of regional groupings in Africa came about because of the perceived need for both collective security and development in the post-Cold War era, and in response to regional conflicts within the continent. According to Adejombi and Olukoshi (in Cambria Press, 2008), slow economic progress and increasing marginalization of the continent at the global level have also given impetus to new regional development strategies. The small size of national economies and the logic of establishing a collective voice through supranational arrangements (Griffiths, 1995) has created what Bell (1987: 108) argues is a 'powerful case' for regional cooperation.

The post-independence era in Africa was characterized by clear commitment in many African countries to the principle of pan-Africanism and the determination of a unique and collaborative vision for the continent (Makinda and Okumu, 2008). This was initially advocated by first-generation independence leaders, such as Lumumba, Kenyatta, Nyerere, Kaunda and Nkrumah. The last of these argued for the concept in his book *Africa Must Unite* (Griffiths, 1995; see Box 10.2, below). These concepts crystallized in 1980 with the Lagos Plan of Action, which laid the basis for seeking greater degrees of self-reliance through supranational arrangements. Earlier, in 1963, the Organization of African Unity had been formed as a loose political union among most of Africa's states. This has since evolved into the African Union, established in 2001, which has now set up a pan-African parliament and is seeking to form an African economic community (Janneh, 2006).

In addition to continent-wide initiatives, Africa boasts a significant range of customs and nascent economic unions that have variously assisted with such issues as the provision of unified telecommunications networks in southern Africa and the formation of regional peacekeeping forces in West Africa (see Box 7.3, above). As Ayeni (1997: 53) notes, when these organizations succeed, they 'have serious repercussions for processes of regional development all over the continent'. The key unions are: the Southern African Customs Union (SACU); the East African Community (EAC), first established in 1967 and revived in 2000, leading to the establishment of a customs union in 2005 (Kyambalesa and Hougnikpo, 2006); the Economic Community of Central African States (ECCAS), which was established in 1985 to promote regional economic cooperation, free trade, a customs union and eventually a common market in central Africa; and the Economic Community of West African States (ECOWAS), established in 1975, which is seeking collective 'self-sufficiency' through the development of an economic and monetary union and a trading bloc (Konadu-Agyemang and Panford, 2006). In southern Africa, there is the Southern African Development Community (SADC; see Box 8.1, above), which was formed in 1992 to replace an earlier political union in the region. This organization strives to promote socio-economic, political and security cooperation. Currently, twelve countries have formed a free trade area and progress has been made in a range of joint infrastructure, trade and healthcare projects (Kyambalesa and Hougnikpo, 2006). Another union is the Common Market for Eastern and Southern Africa (COMESA), formed in 1994 to replace the Preferential Trade Area, which has established a free trade area between its nine member countries. The next stage is to expand free trade arrangements with the EAC and SADC, which, if realized, will create a free trade zone encompassing nearly half of all the countries in Africa, with a combined GDP of some $624 billion (Kyambalesa and Hougnikpo, 2006; BBC, 2008b).

At an even grander level are the proposed activities of the African Economic Community (AEC), which seeks to utilize the above-mentioned regional groupings as 'pillars' for its proposed activities. The AEC was founded in 1991 through the Abuja Treaty and has helped promote the development of the regional trading blocs (Niang, 2006; Kyambalesa and Hougnikpo, 2006). It is supported by all of Africa, with the exception of the Arab Maghreb Union. Future goals are a continent-wide customs union by 2019, and an African common market by 2023. The future establishment of an Africa Free Trade Zone (AFTZ) will be another key step in the attainment of the goals of the AEC (Janneh, 2006).

At a political level, the establishment of the Organization of African Unity (1963) and its replacement with the African Union (AU) in 2001 have been the primary continental forms of regional political collaboration and cooperation. Headquartered in Addis Ababa, the AU seeks to promote socio-economic and political integration, to achieve consensus and common positions on important issues, and to establish peace and security (Konadu-Agyemang and Panford 2006; Makinda and Okumu, 2008). A pan-African parliament includes representatives from every country on the continent, with the exception of Morocco (due to ongoing disagreement over the future of Western Sahara). Mechanisms to promote peacekeeping, democracy and development have also been put in place (Mohamoud, 2007). In 2007, a Union Government for Africa was mooted.

A parallel and associated international initiative that has been formally adopted by the AU is the New Partnership for Africa's Development (NEPAD), established in 2001 through the merger of South Africa, Algeria and Nigeria's Millennium Partnership Plan and Senegal's OMEGA Plan (Konadu-Agyemang and Panford, 2006; see Box 10.3, below). NEPAD seeks to put in place continent-wide mechanisms to eradicate poverty, promote sustainable growth and development, integrate Africa into the world economy and accelerate the empowerment of women (Niang, 2006; Mohamoud 2007). Partnerships have since been developed with many of the world's key financial bodies, and programmes focusing on agriculture, science, e-schools, infrastructure and building continental institutions have been established. In practice, though, slow progress, the lack of civil society participation, perceptions that NEPAD is working too closely with 'Washington Consensus' organizations, and the dominance of South Africa in the organization are concerns for several member states (Makinda and Okumu, 2008).

UNCTAD (2009) argues that while substantial progress has been made to establish regional institutions, intra-African trade, investment and people's mobility have not increased significantly in recent years, meaning Africa still has some of the most fragmented markets in the world.

9.6 Recent theoretical and applied responses to persistent poverty

9.6.1 Asset- and livelihood-based approaches

Persistent poverty in the South, and particularly in Africa, has prompted a reassessment of development interventions and ways in which to respond to poverty and encourage livelihood diversification. One recent approach to understand and respond to poverty is the 'asset-based approach', which examines the stock of productive, financial, physical, natural, social and human assets controlled by households and individuals, and determines their position in society and their ability to respond to shocks. Strengthening assets can build resilience to shocks and provide the poor with security, be it derived from the land, their families or their skills, should they lose access to income or perhaps experience unanticipated external disruption to their well-being (Barrett *et al.*, 2008).

Related to this has been the identification and development of the sustainable livelihoods framework, which identifies the capabilities, assets and activities that contribute to the livelihoods of the poor (see Figure 4.2, above). This approach provides a mechanism to identify appropriate entry points for interventions and allows better sequencing of interventions to support the poor. At the core of the framework is the identification of the 'asset pentagon', comprising the various 'capitals' to which people might have access: natural, human, financial, social and physical (Potter *et al.*, 2008). The approach recognizes that survival in the South often requires reliance on multi-livelihood strategies, and it has influenced recent interventions and approaches to development pursued by development and donor organizations, such as UNDP and CARE (Potter *et al.*, 2008).

9.6.2 Pro-poor growth and poverty reduction strategy papers

In the early years of the twenty-first century, international development efforts and associated global support have tended to revolve around two core themes – poverty reduction strategy papers (PRSPs) and the Millennium Development Goals, which collectively seek to achieve the recently identified goal of 'pro-poor growth' (World Bank, 2002b). This has a conceptual link with Pieterse's idea of 'reflexive development', which refers to the way in which mainstream development agencies have softened their economic interpretations and policies to allow for more people-centred development in response to both the shortcomings of traditional development interventions and the criticism of those who have argued for alternative forms of development (Pieterse, 1998). Pro-poor growth ultimately seeks to promote policies that simultaneously achieve growth and ensure the participation of poor people themselves (Ravallion, 2004). It advocates the need to promote empowerment and investment in poor people and to encourage their participation in the economy, while encouraging a sound investment climate to promote aggregate growth (Bigsten and Shimeles, 2004).

While pro-poor development ultimately underlies much of the work of NGOs, social movements and civil societies, its objectives are increasingly filtering into mainstream practice. One of the most significant influences can be seen in the structure and focus of the development plans or PRSPs that developing countries are now required to formulate. PRSPs can, in some ways, be regarded as successors to SAPs. They are strategic documents drafted by developing countries around which the World Bank, the IMF and other donors coordinate their development efforts (Potter *et al.*, 2008). For the poorest countries, producing PRSPs is obligatory in order to achieve debt relief and conditional lending. These documents focus on both macro-economic and structural considerations for the management of economies and associated social sector reform, with specific emphasis on development outcomes that benefit the poor. While host countries assume ownership of the process, critics argue that most strategies continue to rely on external prescriptions of reducing state involvement, engaging with globalization and issues of trade that are not always favourable to the host countries and require such countries to take responsibility for their own poverty reduction (Potter *et al.*, 2008).

A World Bank/IMF review undertaken in thirty African countries in 2002 suggested that levels of popular participation in the planning process and increased access due to the decentralization of state administration were valuable attributes of the PRSP process. However, the study also recognized that inadequate support was being given to agriculture, and that many countries lacked the requisite skills and planning capacities to implement change adequately (World Bank, 2002b).

9.6.3 Making Poverty History

At a broader level, anti-poverty campaigns became a rallying call among social groups, governments and academics in the North during the early twenty-first century. The 'Make Poverty History' campaign (Make Poverty History, 2010), the efforts of pop-culture figures, such as Bono, and theories about the 'end of poverty' (Sachs, 2005) have all featured prominently in the media, as have appeals to reduce debt and address the scourge of poverty. The degree to which this translates into securing significant improvements in global, and particularly African, well-being remains to be seen.

9.7 The Millennium Development Goals

In 2000, the United Nations General Assembly formally adopted the Millennium Development Goals (MDGs) as a concerted global initiative to coordinate donor efforts and national interventions to improve the welfare and living conditions of the world's poor (UN, 2008b, 2010c). The MDGs are not a 'development intervention' in the classic economic sense, but are aligned rather more with conceptions of 'human development' and 'basic needs'. Mistry (2005: 675) goes so far as to argue that 'the MDGs are, in fact, poverty reduction goals that have surprisingly little to do with fostering development'.

There are eight defined goals, with associated targets and indicators designed to give focus to efforts to reduce extreme poverty and hunger, promote equality, combat disease and ensure environmental sustainability through a global partnership for development (see Table 9.3). The authors of the MDGs anticipated that they could be attained by 2015.

While the MDGs are clearly well intentioned, and aid agencies and national governments have aligned their development interventions with them, they are not a development strategy, as such. Instead, they should be seen as a long-overdue effort to raise global living standards, and in so doing provide a base on which more sustainable endeavours can be built.

In terms of attainment of the MDGs, while it appears that many countries in South America and Asia have already achieved, or are on track to achieve, most of them (Clemens *et al.*, 2007), the same cannot be said about Africa. Globally, while the United Nations estimates that the number of people living in extreme poverty fell from 1.8 billion in 1990 to 1.4 billion in 2005 (UN, 2010d), Potter *et al.* (2008) argue that results are 'patchy' and, of all the continents, that Africa is least likely to attain the goals. For instance, in 2008, 40 per cent of Africa's population survived on less than $1 per day, compared with an average of 19.2 per cent of the population in all developing countries (ODI, 2010). In 2004,

Table 9.3 The Millennium Development Goals

Goal	*TARGETS*
1 To eradicate extreme hunger and poverty	• To halve the proportion of people living on less than $1/day • To halve the proportion of people who suffer from hunger
2 To achieve universal primary education	• To ensure all boys and girls complete primary schooling
3 To promote gender equality and empower women	• To eliminate gender disparities in primary and secondary education
4 To reduce child mortality	• To reduce by two-thirds the mortality rate of under-fives
5 To improve maternal health	• To reduce the maternal mortality ratio by 75 per cent
6 To combat HIV/AIDS, malaria and other diseases	• To halt and begin to reverse the spread of HIV/AIDS • To halt and begin to reverse the incidence of malaria and the spread of other diseases
7 To ensure environmental sustainability	• To integrate principles of sustainable development into country policies and to reduce the loss of environmental resources • To halve the proportion of people without access to safe drinking water • To improve the lives of 100 million slum dwellers
8 To develop a global partnership for development	• To promote open trade and finance • To address the trade and debt relief needs of the poorest countries • To address the special needs of landlocked and small-island developing states • To make debt sustainable • To develop youth employment • To provide affordable drug access • To make available new technologies, especially information and communications

Source: UN, 2010a

Africa's poverty rate was 41 per cent, which shows just how slight the improvement has been (Clemens, *et al.*, 2007). That said, it is important to note that there are clear 'front-runners', such as Benin, Mali and Ethiopia, while conditions in some twelve African countries appear to have worsened (ODI, 2010). In 2007, it was estimated that forty-two of the forty-seven countries considered in Africa were 'off-track' to meet the goals (Clemens *et al.*, 2007).

In September 2010, the United Nations appealed for an 'extra push' in terms of aid, trade and debt relief to help meet the MDGs (UN, 2010d). This appeal noted that aid flows were at an all-time high of $120 billion in 2009, but that this was still $20 billion less than had been pledged by the G8 countries in 2005, while the funding gap for Africa was $16 billion. Aside from the scale of the development backlog that needs to be addressed, various authors have questioned the institutional capacity in Africa to address this, as well as the realism of the goals in terms of the scale of financial commitment required from donor countries, which has not been forthcoming at the desired rate (Clemens, *et al.*, 2007; Easterly, 2009). It is important to note that Africa's seemingly poor performance and assumed inevitable 'failure' to meet the MDGs do not adequately acknowledge that the continent started from a weaker base than all other continents and that questions over the reliability of the scores can be raised statistically (Easterly, 2009). Mistry (2005) further contends that poverty reduction needs to go hand in hand with greater efforts at development if 'true' progress is to be attained.

Despite these concerns and debates about whether Africa might attain the MDGs by 2015, they remain key objectives. If they are achieved in the current decade or later, they will surely lay a foundation for sustainable growth and development.

9.8 Aid, trade and debt

The three interlinked topics of aid, trade and debt are undoubtedly some of the most controversial issues impacting on the current development prospects of Africa. All three ultimately exist because of the key role which globalization now plays in our contemporary world, be it in terms of economic interdependence between countries, the desire to access necessary raw materials and goods, and/or the ability to respond to perceived development or humanitarian shortfalls in developing countries. From a development perspective, the supply of aid has been one of the key ways in which to foster strategic development interventions, while encouraging trade is regarded as a mechanism through which to promote the development of the broader economy. By contrast, debt has often inhibited key investment and has negatively impacted upon growth prospects. Therefore, these three issues, and the ways in which they are playing themselves out in the twenty-first century, provide a critical context that can either encourage or stifle broader development interventions.

9.8.1 Aid

Aid refers to the transfer of resources, usually from the North to the South. Historically, the supply of aid had links with modernization theory, when international support was seen as a way to help developing countries address lags in development and achieve 'take-off'. Over the decades, it has assumed various guises. These include government-to-government and government-to-NGO/community transfers, and private or NGO support to groups in the South (Willis, 2005). Such support can take various forms, including financial grants or loans, direct assistance with government operations, food aid, military support, technical advice, and the supply of emergency relief and skills. Aid can also be classified as bilateral (transfers directly from one country to another) or multilateral (when an international institution, such as the World Bank, which represents numerous donors, initiates the support) (Moss, 2007). Official state aid to the South, called ODA (Official Development Assistance), currently averages 0.41 per cent of Gross National Income from the OECD states, despite calls from international agencies since the 1970s that a figure of 0.7 per cent is more appropriate (Potter *et al.*, 2008). Nevertheless, the value of aid received in Africa increased from $13 billion in 2001 to $24 billion in 2004 (Moss, 2007). This is equivalent to over 5 per cent of the GDP of Africa itself, and if South Africa and Nigeria are excluded from the equation, the figure rises to over 10 per cent.

While aid is undoubtedly of critical importance, for example in times of famine, there are times when it has been criticized for generating dependence and for being siphoned off by elite groups (Omotoye *et al.*, 2007). In addition, politically tied aid is highly controversial, with some of the key recipients, both now and historically, being countries aligned with the political interests of the donor countries (Wade, 2004; Omotola and Saliu, 2009). Equally contentious is military assistance. Critics of aid have argued that it can become a 'bottomless pit', but Sachs (2005) has argued that an aid-financed 'big push' is now needed if countries are to break out of the 'poverty trap'.

In the case of World Bank assistance, the focus of aid support has evolved according to the bank's changing agenda – from support for infrastructure projects in the 1960s, to targeting management in the 1970s, to structural reform in the 1980s, and more recently to focusing on issues of governance (Moss, 2007). While aid can be critical, it can also lead to external interests eclipsing local ones in terms of what can be funded and supported (Omotoye *et al.*, 2007). For example, despite the reality that nearly two-thirds of Africa's population depend on near-subsistence agriculture for their livelihoods, support for this sector has fallen by two-thirds, while more money is going into support for social services (Wade, 2004). Further, Mistry (2005) argues that while linking aid to achieving the MDGs is laudable, long-term challenges are being created by reducing support to other sectors, particularly productive activities.

9.8.2 Trade

Trade is an equally contested concept politically. Many of the trading links dominating Africa over the last 200 years have been put in place to meet the needs of other countries, initially for the supply of slaves, then for Africa's resources, primarily minerals, and more recently for markets for their products (Potter *et al.*, 2008). While trade has long been regarded as a key mechanism to promote economic development, particularly in light of the export-led success of South East Asian countries, the ability to penetrate global markets has been more difficult to attain in the case of Africa (Moss, 2007). This state of affairs has been caused by deteriorating prices received for Africa's produce, which is dominated by the export of raw materials, poor competitiveness, trade barriers and poor domestic performance in the manufacturing sector (Moss, 2007). At the same time, World Trade Organization (WTO) requirements that African countries should lower trade barriers have extended market opportunities for foreign firms in Africa, and no doubt have extended the product range at a lower cost to consumers (Willis, 2005). However, Africa's fledgling industries have often foundered in the face of cheap foreign products, especially from China, and countries have had little to export apart from their raw materials (Wade, 2004). In the case of the latter, EU and North American local-producer food subsidies discriminate against African producers. In 1999, it was estimated that if the West truly opened its markets to developing countries, the latter's economies would benefit by $700 billion, while in 2001 a UN report estimated that poor countries lose a total of $2 billion each day because of unjust trade rules (Omotoye *et al.*, 2007). Africa's poor trading performance is exemplified by the fact that its share of global trade was 6 per cent in 1980 but had fallen to a mere 2 per cent by 2002.

While African countries have traditionally struggled to make effective objections to the nature of the current global trading regime, in recent years opposition from key developing countries (particularly Brazil and India) at global trade discussions (such as those held in Geneva in 2006) has clearly impacted on the ability of the WTO to achieve global consensus (Wade, 2004; Potter *et al.*, 2008). The Africa Progress Panel (2008) has argued that Africa needs a workable international trade policy and support to create jobs and take advantage of global opportunities. In addition, fair and ethical trade might benefit communities in the South by allowing them to sell such products as coffee and tea directly to Western consumers who wish to make ethical choices about what they purchase (Potter *et al.*, 2008; see Box 4.6, above).

9.8.3 Debt

African debt is equally controversial. While it may be argued that African countries only have themselves to blame for getting into debt, one should not ignore the context in which the debt crisis arose. In the 1970s, there were economic crises in the West associated with the devaluation of the US dollar and oil-price hikes. This, coupled with injudicious commercial loans to the South, low or negative economic growth globally, limited trade and investment, and currency devaluation in Africa, made debt repayment impossible for many countries in the 1980s (Willis, 2005). In 1970, Africa's debt stood at $7 billion. However, between 1980 and 1999, Africa's debt tripled from 28 per cent to 72 per cent of GDP, equivalent to $216 billion (Moss, 2007; Potter *et al.*, 2008).

Subsequent structural adjustment policies and rescheduled debt repayment obliged many African countries to cut their social service expenditure, and several of them became net exporters of currency. In the case of Nigeria, debt service payments were equivalent to between 20 and 30 per cent of total exports between 1998 and 2000 (Omotola and Saliu, 2009). It is therefore understandable that the 1980s is regarded as the 'lost decade' in terms of socio-economic development in Africa (Barratt Brown, 1995). It is estimated that between 1970 and 2002, Africa received $540 billion in loans and paid back $550 billion, yet in 2002 the debt still stood at $293 billion (Omotola and Saliu, 2009). More recent debt write-offs for the poorest countries are to be welcomed, as are global campaigns on this issue, such as Jubilee 2000 (Potter *et al.*, 2008), but the debt burden remains for many countries. A total of thirty-eight countries in the world, of which thirty-two are in Africa, have been identified as HIPCs (highly indebted poor countries). In 2005, in return for meeting externally imposed criteria for good governance and making commitments to address poverty, eighteen of them were granted debt relief. Critics argue that while this is clearly beneficial, locking countries into Western-imposed criteria parallels earlier SAPs and, in some senses, maintains the status quo by not addressing the fact that Africa remains on the lowest rung of the global economic ladder (Omotola and Saliu, 2009). More positively, those HIPCs that have benefited from debt relief have started to increase investment in education and healthcare. For the future, Omotola and Saliu (2009) argue that debt relief and effective aid are inextricably linked, requiring the joint development of applied solutions to Africa's development challenges.

9.9 Current economic growth trends in Africa

Despite the general failure of development interventions to raise overall levels of human welfare in Africa, there are signs that real growth is starting to occur in the formal sector and among the associated middle class. For much of the period from 1995 to 2005, economic growth rates in Africa averaged 5 per cent per annum or more. It is also noteworthy that five of the ten fastest-growing economies in the world in the 2000s were in Africa, and in general terms the continent has been one of the best-performing areas in the world (Allen, 2010). Significantly, the global recession of 2008–2009 appeared to have less severe effects in Africa than in other parts of the world, and the rate of recovery has been more rapid there than elsewhere. Even though growth rates fell to 2.5 per cent in 2009 (*OECD Observer*, 2010), the continent may soon see a return to a growth rate of 5 per cent per annum (Devarajan, 2010). Economists believe this rapid growth and quick recovery are attributable to the pursuit of sound economic policies, managing inflation, avoiding protection, reducing fiscal deficits, and macro-economic stabilization. The rapid recovery is also attributed to the resumption of demand by emerging countries, notably China (Kasekende *et al.*, 2010; *OECD Observer*, 2010).

While all of this is positive, such growth needs to be seen in context, as these figures generally represent percentage increases from an exceptionally low base. In addition, resource- or oil-rich countries have generally fared better, skewing continental statistics (*OECD Observer*, 2008), and it would be fair to argue that the prime beneficiaries of such growth have been corporations and the emerging middle class. Whether such benefits are passed on to the lower classes in any meaningful way remains to be seen. Also of concern are the fact that investment levels are still only half of those in Asia, and the likelihood that future development prospects will be hindered by continuing massive infrastructural and human-resource deficits (Devarajan, 2010). Furthermore, rising food prices and associated food shortages led to riots in 2010 in such countries as Mozambique, while significant buy-outs of Africa's productive land by Asian and Middle Eastern countries are taking place in anticipation of further global shortages. Recently, South Korea was blocked from proceeding with a deal that would have seen the sale of 40 per cent of Madagascar's arable land (*OECD Observer*, 2008; Draper, 2010; see Chapter 4).

So, while recent growth is gratifying, the question remains: who benefits? Africa today presents a mixed balance sheet. On one level, it boasts some of the world's fastest-growing economies, yet, on another level, it is unlikely to achieve any of the Millennium Development Goals. Consequently, ensuring that growth is translated into improved livelihoods for deprived majorities is clearly a key challenge for the future.

9.10 Conclusion and future prospects

As the preceding discussion has illustrated, the experience of 'development' in Africa has met with mixed results. A series of interventions inspired by varying theoretical persuasions has been attempted but, sadly, it is difficult to argue that significant improvements in human welfare have taken place since the 1950s. It is apparent that the dreams of emulating the West, which characterized the immediate post-independence era, were scarcely realized, while experiments with self-reliance and local development achieved only limited success. Current efforts to achieve poverty relief, attain the MDGs and promote development through regional integration will hopefully, over time, match the current success being recorded in the formal sectors of many African countries and lead to redistribution and improved welfare in general. Principles of 'bottom-up' and local development will be increasingly important, but, as Samli (2008) argues, such approaches must be grounded in effective entrepreneurship if tangible progress is to be made on the ground.

It is inevitable that future development in the continent will be shaped by new global players, notably China and, to a lesser degree, India, Brazil and South Africa (see Box 10.5, below). China's and India's relentless quest for minerals and raw materials to feed their high-growth economies, and South Africa's search for new markets for its industrial products, food produce and communications technologies, has led to significant new investment in the continent (Cropley, 2010). While the establishment of new mines and factories may well perpetuate the problems of isolated nodal development, the extension of mobile-phone networks in countries lacking land-line connections is proving to be a valuable stimulus for isolated communities by allowing them to establish market links. China's economic assistance to Africa doubled in 2006, and it also undertook to train 15,000 professionals (Akukwe, 2007). In parallel, Chinese trade with the continent increased tenfold between 2000 and 2009, meaning it ultimately surpassed the USA's trade with Africa. Major investment in the oil industry, mining, industry and infrastructure is the hallmark of the new Chinese presence, which is signalling a potential realignment of historic economic links between Africa and the North (Cropley, 2009). However, ever-increasing demands for oil and mineral resources from the likes of China and Brazil will inevitably shape future development and could perpetuate the dependent relationships of the 1970s.

Naturally, two key questions for the near future are whether all (or indeed any) of the MDGs will be met and whether mass poverty will be significantly reduced in Africa. The challenges are very real, and while impressive economic growth is taking place in many African countries, indicators suggest that the needs of the poor majority are being met only very slowly (ODI, 2010). If this continues, it will simply perpetuate Africa's relatively poor performance in global terms.

It is interesting to reflect that the countries of the North have recently experienced a period of economic recession and restructuring, marked by enhanced levels of state intervention that are contrary to traditional monetarist prescriptions that have been preached to Africa for decades. Within the context of the now-questionable Western model, it is perhaps appropriate for Africa to start to look inwards for its own solutions. In *The Challenge for Africa*, Nobel Peace Prize-winner Wangari Maathai (2009: 6) states:

> As I write, the world is in a financial crisis, caused in part by lack of oversight and deregulation in the industrialized world . . . For decades, Africa has been urged to emulate this financial system . . . While this structure has enriched the West, practising it without caution has only impoverished Africa. The current crisis offers Africa a useful lesson . . . rather than following blindly the prescriptions of others, Africans need to think and act for themselves.

Summary

1. Understanding of what 'development' is and how to apply it has evolved over time.
2. Development and political considerations are closely linked.
3. There has been a recent shift to local and self-reliant development.
4. Aid, trade and debt all exercise significant influences over the capacity and potential of countries to develop.

Discussion questions

1. What is 'development' and how has understanding of the concept evolved over time?
2. How effective have been the different development interventions?
3. What roles do trade, aid and debt play in assisting or hindering development?
4. What might be the focus of future development strategies in Africa?

Further reading

Bond, P. (2007) *Looting Africa: The Economics of Exploitation*, London: Zed Books.

Niang, A. (2006) *Towards a Viable and Credible Development in Africa*, Raleigh, NC: Ivy House.

Potter, R.B., Binns, T., Elliot, J.A. and Smith, D. (2008) *Geographies of Development* (3rd edn), Harlow: Pearson.

Power, M. (2003) *Rethinking Development Geographies*, London: Routledge.

Useful websites

Make Poverty History: http://www.makepovertyhistory.org/takeaction/. Contains information about several key global anti-poverty campaigns.

United Nations Millennium Development Goals: http://www.un.org/millenniumgoals/. Details progress towards achievement of the MDGs.

United Nations University – Worldwide Institute of Development Economics Research (UN-WIDER): http://www.wider.unu.edu/. Contains research reports and findings on a range of developmental activities.

World Bank: http://www.worldbank.org. Contains a wide range of statistical and published data pertaining to development activities.

10 What future for Africa?

10.1 Crisis and marginalization

It is unfortunately the case that, for many Africans, there has often been very little evidence of an improvement in their livelihoods in recent decades. Although some countries have had periods of significant economic growth, others have experienced stagnation, or even deterioration, in their national economic situation and in the living standards of the majority of their people. Why has this happened? What has gone wrong? Why have development successes been so selective? And what should be the key priorities for the future?

There has been an ongoing debate about such questions, with a wide range of viewpoints expressed. There are those who still blame the legacies of colonialism, others who focus on the effects of the mid-1970s oil crisis and the 2007–2010 global financial crisis, others who stress such problems as rapid population growth, the HIV/AIDS pandemic, political and economic mismanagement, the machinations of transnational companies, the world trading system and the imposition of structural adjustment programmes, and still others who believe that environmental factors are the root cause of Africa's predicament. In reality, a detailed examination of the experiences of African countries would reveal that there is frequently a variable matrix of factors that have affected African states and their people, and it is often difficult to single out any particular factor.

Media reports on Africa frequently make reference to a state of 'crisis'. In fact, the discourse about an 'African crisis' is multifaceted and has developed over a long period of time. Following drought and famine in the Sahel and Ethiopia in the late 1970s and early 1980s, Lloyd Timberlake, for example, commented in 1985, 'Africa's plight is unique. The rest of the world is moving "forward" by most of the normally accepted indicators of progress. Africa is moving backwards . . . The continent's living standards have been declining steadily since the 1970s. Its ability to feed itself has been deteriorating since the late 1960s' (Timberlake, 1985: 7). Almost a decade after Timberlake was writing, an Oxfam report entitled *Africa – Make or Break: Action for Recovery*, warned, 'Africa is balanced on a knife edge. Without recovery, more than 300 million people . . . will be living in poverty by the end of the decade; infant mortality rates, already the highest in the world, will continue to rise; and vulnerability to hunger will increase' (Oxfam, 1993: iii). More recently, at the start of the new millennium, Kofi Annan, Secretary-General of the United Nations, in delivering his Commonwealth Lecture, commented, 'The truth is, Africa is suffering from multiple crises – ecological, economic, social and political' (Annan, 2000: 2). In 2009, Africa's difficulties were being exacerbated by the worst global recession since the 1930s (see Box 10.1).

Box 10.1

Africa and the global recession

Although the global recession of 2008 and 2009 had its origins in North America and Europe, most commentators would agree that the world's poorest countries have suffered considerably from the effects of the most serious economic downturn since the Second World War. In the fourth quarter of 2008, there was a 5 per cent decline in world GDP, and the World Bank predicted that world trade volumes would fall sharply in 2009 for the first time in twenty-seven years (World Bank, 2009a). Since Africa has the dubious distinction of being the world's poorest continent, the global recession has compounded many of the problems which African countries were already facing.

Furthermore, while the world's richer countries have been overwhelmingly preoccupied with coping with their own difficulties, it might be argued that their concerns for the well-being of the world's poorest people have slipped well down their individual and collective agendas. The situation in 2008 and 2009 contrasted markedly with that of 2005, for example, when Africa was the focus of Tony Blair's Africa Commission and the G8 talks in Gleneagles, Scotland. With a great deal of belt-tightening and expenditure cuts, many of the world's richer countries are, perhaps inevitably, feeling less inclined to help less fortunate countries sort out their difficulties. This situation was confirmed by World Bank economist Shanta Devarajan, who commented in January 2009,

> Africa, initially spared the effects of banking failures, is now facing declining capital flows, slowing remittances, stagnating foreign aid and falling commodity prices and export revenues. The continent will almost surely experience a deceleration in growth. And, if history is a guide, this deceleration will have an impact on human development.
>
> (World Bank, 2009g: 1)

An analysis of ten developing countries by the UK-based Overseas Development Institute (ODI) found that bank lending conditions had been tightened in Ghana and Zambia. Meanwhile, in Kenya, remittances in January 2009 were down by 27 per cent compared with January 2008, tourism had declined, and the horticultural industry shed 1200 jobs in the first half of 2009 and experienced a 35 per cent fall in flower exports. The ODI also suggested that there are likely to be increases of 233,000 poor households in Uganda (0.8 per cent of the population) and 230,000 in Ghana (1 per cent of the population) (ODI, 2009). There is evidence that a large number of overseas-funded development projects in Africa have been abandoned, too. For example, Arcelor Mittal, the world's largest steel company, was reported to have frozen a $1.5-billion project in Liberia to build rail and port facilities in exchange for iron ore, and a $3.3-billion Chinese project to build cement plants in Nigeria was shelved (*Time*, 2009).

Ahead of the World Economic Forum in Davos, Switzerland, in January 2009, World Bank President Robert B. Zoellick warned that it could be the gloomiest forum in the meeting's thirty-nine-year history and urged delegates, 'Don't leave developing countries out in the cold' (World Bank, 2009e: 1). Zoellick proposed the setting up of a 'vulnerability fund' for developing countries that were suffering from the global recession. The idea was that developed countries should dedicate 0.7 per cent of their economic stimulus packages to the fund. In a subsequent statement, issued on 8 June 2009, the World Bank (2009f: 1) commented, 'An unprecedented global economic crisis demands unprecedented initiatives to restore growth. The World Bank Group is helping with the financial rescue, but believes that we must remain focused on the human rescue for the many millions left behind.'

One of the first countries to benefit from the World Bank's 'fast-track assistance programme', in February 2009, was the Democratic Republic of the Congo (DRC), which received $100 million to ameliorate the effects of a sharp slowdown in growth, which had been further undermined by the unstable security situation in the eastern part of the country. In the mining areas around Katanga (south-east DRC), many small mines have closed, with an estimated 200,000 workers losing their jobs, a situation which is probably impacting upon at least 1 million people. The grant from the International Development Association's Financial Crisis Response Fast-Track Facility was designed to assist with infrastructure development as well as education and health initiatives, such as providing essential goods and paying schoolteachers' salaries (World Bank, 2009g).

According to a report from the UN's Food and Agriculture Organization (FAO), the world economic crisis has had a significant impact on food security. The report argues, 'The increase in food insecurity is not a result of poor crop harvests, but because high domestic food prices, lower incomes and increasing unemployment have reduced access to food by the poor. In other words, any benefits from falling world cereal prices have been more than offset by the global economic downturn' (FAO, 2009b: 2). This is certainly the case in the Zambian Copperbelt region, where the FAO estimated that some 30,000 people were directly employed in the mining sector in 2008, but by June 2009 some 10,000 of these jobs had been lost, with very significant implications for the poverty and nutritional status of households (FAO, 2009b). Many retrenched workers in the Copperbelt are turning to urban and peri-urban farming in order to ensure household food security and to generate income from crop sales (see Box 5.4, above).

The World Bank showed its concern for the impact of rising costs of fuel and basic foodstuffs on poor countries by setting up the Global Food Crisis Response Programme (GFRP) in May 2008. By June 2009, the programme had disbursed $758 million, with much of the money targeted specifically at feeding poor children and other vulnerable groups, purchasing seeds and meeting the additional expenses of food imports. Associated initiatives have included increasing access to agricultural finance. Many African countries have received considerable support through the GFRP, including Ethiopia ($275 million), Tanzania ($220 million) and Kenya ($50 million), with smaller amounts going to such countries as Mozambique ($20 million), Liberia ($10 million) and Madagascar ($10 million) (World Bank, 2009g).

At the meeting of the Group of Eight (G8) leaders in Italy in July 2009, there was considerable debate about overseas development assistance, and particularly aid to Africa. At the 2005 G8 summit, leaders had agreed to increase aid to Africa by some $22 billion in the period 2005–2010. While the UK and USA have actually increased their aid to Africa, and Japan and Canada have given more than they promised, Italy has delivered only 3 per cent of its commitment, and France only 7 per cent (*Time*, 2009). Oxfam is concerned that over $8 billion could be cut from the G8's total five-year proposed budget of $50 billion, and it was predicted that sub-Saharan Africa's overall annual economic growth rate might fall from 5.5 per cent in 2008 to just 1.5 per cent in 2009, considerably less than the 3 per cent population growth rate (*Time*, 2009). In fact, the decline was less severe, with the average rate of economic growth declining to 3.1 per cent in 2009 (African Economic Outlook, 2011: 1).

While the World Bank should be commended for its efforts in raising awareness of the effects of the global recession on the world's poorest countries, it remains to be seen how effective these measures will be in reducing the impact at the grassroots level in Africa. Serious concerns remain about both the repercussions of the recession on Africa's poorest individuals and groups, and the fact that the richer countries are reducing their support for the poorest of the poor.

There seems little doubt that, according to a range of development indicators, Africa currently has the dubious distinction of being the poorest continent in the world. A large proportion of Africa's estimated 1000 million people (PRB, 2009) would constitute what Paul Collier (2007) calls 'the bottom billion'. In Africa's most populous state, Nigeria, an estimated 92.9 per cent of young people between the ages of fifteen and twenty-four are living in poverty on less than $2 per day (World Bank, 2009d). Meanwhile, in Africa's richest country, South Africa, life expectancy has decreased dramatically from sixty-two years in 1990 to fifty years in 2007, due largely to the HIV/AIDS pandemic. By 2015, it is estimated that 32 per cent of all children in the country will have lost one or both parents as a result of the disease (*Mail and Guardian*, 2009). In addition to these grim statistics, millions of Africans are suffering under corrupt regimes in such places as Equatorial Guinea, Guinea, Chad and Zimbabwe, or are living in conflict-ridden states, such as the Democratic Republic of the Congo and Sudan, or in persistently dysfunctional states, such as Somalia.

In economic terms, despite recent signs of growth, Africa is arguably the world's most marginalized continent, with many countries suffering from a huge debt crisis and remaining heavily dependent upon official aid flows for financial survival. As Potter *et al.* (2008: 369) observe, 'Between 1980 and 1999, sub-Saharan Africa's debt more than tripled to around $216 billion and, although considerably less than that of Latin America ($813 billion), the region's debt increased from the equivalent of 28 per cent of its GNP to 72 per cent.'

Just as the nineteenth-century explorers, missionaries and traders were often pessimistic in their comments about Africa, over a century later there is still much negativity about both the present state of Africa and its future prospects. However, in the last two decades, some hope for the continent's future has been generated by the formation of the African Union (see Box 10.2), debate about the 'African renaissance', and the New Partnership for Africa's Development (NEPAD) (see Box 10.3).

Box 10.2

'Africa must unite!': the African Union

The formation of the African Union (AU) in 2001 can be traced back to the pan-Africanist movement and, more recently, the Organization of African Unity (OAU) (African Union, n.d.). Pan-Africanism had its origins in the dehumanization of Africans during the slave trade, the subsequent struggle against colonialism and the liberation struggles of African-Americans during the twentieth century. It sought to express the pride and achievements of Africans and to free them from all forms of oppression (Mathews, 2008).

There have been various attempts in the past to encourage greater cooperation between African countries, and to ensure that Africa has a louder voice in world policy-making. In 1957, President Kwame Nkrumah of Ghana (the first black African state to gain independence) insisted that 'Africa must unite!' Nkrumah was a key figure in the twentieth-century pan-Africanism movement, which sought to unify native Africans and members of the African diaspora into a 'global African community' (see Chapter 9). Other key pan-Africanists were Marcus Garvey (the Jamaican advocate of the 'Back to Africa' movement), African presidents (notably Jomo Kenyatta of Kenya, Julius Nyerere of Tanzania and Leopold Senghor of Senegal) and popular music stars, such as Bob Marley (Jamaica), Youssou N'dour (Senegal) and Fela Kuti (Nigeria). More recently, Thabo Mbeki (South Africa) and Olusegun Obasanjo (Nigeria) have championed pan-African ideals.

The first Pan-African Congress was held in Paris in 1919, and others followed, most notably the fifth, held in October 1945 in Manchester, UK. Nkrumah was one of the organizers of that congress, which was attended by a wide range of delegates, including people from the African diaspora living in Manchester, as well as returned soldiers, West Indians and a number of future African leaders.

After many more years of debate, the OAU was finally established in 1963, initially, at least, as a vehicle for promoting pan-African unity and coordinating the struggle against colonialism. However, a major weakness of the OAU was that its member states were reluctant to delegate their individual or collective powers to the organization, so its role in the international system was rather limited.

It has been suggested that the ideals of pan-Africanism are still very relevant in the twenty-first century. As Mathews (2008: 36) comments,

> Pan-Africanism remains the most effective vehicle addressing the debilitating problems of Africa. Africa cannot be developed using externally imposed economic paradigms, cultures and models. Africa is in need of an African cultural renaissance. An African socio-cultural renaissance is essential, not only for sustainable growth and development, but also to reverse the ongoing marginalization of the continent in the age of globalization. A united Africa will be able to manage the natural resources of the continent in order to become globally competitive, and this can only be in the interests of African people.

While these may be admirable ideals, the practicalities of achieving a 'united Africa' are, to say the least, both complex and daunting.

However, it is fair to say that the inauguration of the African Union AU was an event of great significance in the institutional evolution of the continent. On 9 September 1999, the heads of state and government of the OAU issued a declaration (the Sirte Declaration) calling for the establishment of an African Union, with a view to 'accelerating the process of integration in the continent, to enable it to play its rightful role in the global economy, while addressing multifaceted social, economic and political problems, compounded by certain negative aspects of globalisation' (African Union, n.d.: 5). The slogan 'African solutions to Africa's problems' was frequently heard among the rhetoric. The Constitutive Act of the African Union was adopted in 2000 at the Lomé Summit in Togo, and the AU officially came into force the following year. At the Durban Summit (28 June–10 July 2002), the AU was launched and the first Assembly of the Heads of States was convened. All African countries are members of the AU, with the exception of Morocco, which had withdrawn from the OAU in 1982 over the latter's recognition of the Saharawi Arab Democratic Republic (SADR), following Morocco's invasion of Western Sahara in 1976.

One significant difference between the OAU and the AU is that the former adopted a strong non-interventionist approach to conflict and crimes against humanity within sovereign states, while Article 4(h) of the AU Constitutive Act of 2000 sanctions intervention in the internal affairs of states where such problems exist (Akokpari, 2008). The AU intends to establish an 'African Standby Force' to deal quickly with situations of insecurity and conflict.

At its inauguration, the AU adopted the New Partnership for Africa's Development (NEPAD) as a blueprint framework for development (see Box 10.3), and together, 'the AU and NEPAD generated a new vision to develop Africa' (Akokpari, 2008: 371). The vision of the AU is of an 'Africa integrated, prosperous and peaceful, an Africa driven by its own citizens, a dynamic force in the global arena' (African Union Commission, 2004: 7).

The issue of forming an African political union subsequently received much publicity in February 2009, when the controversial Libyan leader, Muammar Gaddafi, was rather surprisingly elected as chairman of the AU, despite the outspoken concerns of some other African leaders. The idea of having as AU chairman a leader who had been a dictator for forty years, and who had supported rebel movements in Sierra Leone, Liberia and elsewhere, was just too much for some other heads of state. However, undeterred by the opposition, Gaddafi, in his inaugural speech, expressed his wish to move quickly towards establishing political union, stating, 'I shall continue to insist that our sovereign countries work to achieve the United States of Africa' (BBC, 2009: 1). Gaddafi had previously expressed his desire for a single African military force, a single currency and a single passport to enable Africans to move freely across the continent. However, there has been much scepticism about such an objective, since many African leaders are reluctant to give up their national sovereignty. Ugandan President Yoweri Museveni was particularly outspoken in his opposition to political union, arguing that the continent lacks a common language and that regional political federations are more likely to be feasible and beneficial in such areas as East Africa (allAfrica.com, 2009: 1).

The euphoria and optimism generated by the launch of the AU in July 2002 have been tempered in recent years by the fact that it has not yet established cordial relationships with certain partners, institutions and stakeholders. For example, it has had little impact on alleviating the humanitarian catastrophe in the Darfur region of Sudan or the turmoil in Zimbabwe, where President Robert Mugabe's policies have devastated the economy and livelihoods. The AU also lacks the resources to intervene militarily in conflicts.

Moreover, there has been criticism of the membership arrangements of the AU, whereby all African states are automatically incorporated, no matter how corrupt or brutal their leaders. Further criticism has focused on the inadequacy of civil society consultation on a number of issues, not least the formation of the AU itself and the design and implementation of NEPAD (Akokpari, 2008; Akokpari *et al.*, 2008).

Box 10.3

The African Renaissance and the New Partnership for Africa's Development (NEPAD)

In May 1996, Vice-President of South Africa Thabo Mbeki called for an 'African Renaissance'. Addressing the US Corporate Council on Africa the following year, Mbeki said, 'An African renaissance [is] a real possibility, in which the current period of crisis might be seen as a time of opportunity, which the New Africa must seize for its own advantage' (Akosah-Sarpong, 1998: 665). As key elements of the Renaissance, Mbeki spoke of the need for the emancipation of women and the mobilization of youth, grassroots democracy and sustainable economic development, all themes that Nelson Mandela championed when he became South Africa's first democratically elected president in 1994. Further impetus was given to the call for an African Renaissance during US President Bill Clinton's eleven-day visit to six African countries in March 1998. Later, in his presidential inauguration speech of June 1999, Mbeki referred to the twenty-first century as the 'African century', when the quality of life had to be improved for all (Lester *et al.*, 2000). NEPAD came to be viewed as the vehicle through which the African Renaissance would be delivered.

The strategic framework document for NEPAD arose from a mandate given to the five initiating heads of state (Algeria, Egypt, Nigeria, Senegal and South Africa) by the OAU to develop an integrated socio-economic development framework for Africa. The thirty-seventh summit of the OAU in July 2001 adopted the strategic framework document, and in October 2001 NEPAD was formally launched at the African leaders' meeting in Abuja, Nigeria. The key motivation for NEPAD was to address the challenges facing the African continent. It was argued that such issues as escalating poverty levels, underdevelopment and the continued marginalization of Africa needed a radical new intervention, spearheaded by African leaders, to develop a new vision that would guarantee the continent's renewal.

NEPAD has four key objectives:

- To eradicate poverty.
- To place African countries, both individually and collectively, on a path of sustainable growth and development.

- To halt the marginalization of Africa in the globalization process and enhance its full and beneficial integration into the global economy.
- To accelerate the empowerment of women.

The proposal from Africa's leaders asserts that NEPAD

> Is a pledge by African leaders, based on a common vision and a firm and shared conviction, that they have a pressing duty to eradicate poverty and to place their countries, both individually and collectively, on a path of sustainable growth and development and, at the same time, to participate actively in the world economy and body politic.
>
> (NEPAD, 2001: 1)

Furthermore, the proposal suggests that the continued marginalization of Africa, 'from the globalization process and the social exclusion of the vast majority of its peoples, constitutes a serious threat to global stability' (NEPAD, 2001: 1).

The seventy-page NEPAD document makes reference to Africa's historical legacies and the poor living standards that many Africans experience today, suggesting that Africa's impoverishment is due to 'the legacy of colonialism, the Cold War, the workings of the international economic system and the inadequacies of, and shortcomings in, the policies pursued by many countries in the post-independence era' (NEPAD, 2001: 1). Some would argue that structural adjustment policies imposed by the World Bank and the International Monetary Fund are, in effect, a form of 'neo-colonialism'.

NEPAD calls for a 'new global partnership', and the need for Africa to negotiate a new relationship with its development partners in the North. Such a relationship, it is argued, should lead to conflict prevention, debt reduction, increased development assistance to meet the 0.7 per cent of GDP target, progress in education and health, with better access to inexpensive drugs, technical support and private sector investment.

NEPAD has a number of strengths and weaknesses. The emphasis on partnerships, the strong link between politics and development, and NEPAD's concern to reduce poverty and address underdevelopment are 'at the core of a "new" Pan-Africanism in Africa' (Landsberg, 2008: 208). While Mandela and Mbeki emphasized the importance of grassroots democracy in empowering the people as a key factor in achieving the African Renaissance, President Olusegun Obasanjo of Nigeria was keen to establish a Conference on Security, Stability, Development and Cooperation in Africa (CSSDCA). The latter, according to Landsberg (2008: 209), has been 'instrumental in carving out new African "core values" and mechanisms of peace, security and preventive diplomacy'.

A significant weakness of NEPAD, however, is that, rather than formulating a pro-poor development strategy, its approach has been based on a Western neo-liberal economic model with an overemphasis on the role of foreign direct investment, 'which some would argue is what keeps Africa from developing and is therefore part of the problem' (Murithi, 2005: 145). Thus far, NEPAD has not focused on empowering civil society, finding African solutions and strengthening intra-African cooperation, but instead has supported large, 'top-down' infrastructural projects, suggesting that what Africa needs 'is a massive investment programme designed to promote rapid industrialisation and the building of infrastructure' (Ziring *et al.*, 1995: 128). Other concerns include the lack of political will among some governments, and the somewhat awkward relationship between NEPAD and the African Union. Many donors have preferred to deal with NEPAD, while some African states see it as a South African conspiracy, with its headquarters located at Midrand, between Johannesburg and Pretoria. As Landsberg (2008: 220) observes, 'there are serious tensions between the mother institution, the AU, and its stepchild, NEPAD – this troubled relationship needs to be corrected'.

Bunwaree (2008: 238–239) advocates four key measures to strengthen NEPAD's people-centred focus:

1 To integrate NEPAD within the AU.
2 To give civil society a stronger voice in NEPAD and make NEPAD more people-friendly.
3 To establish a code of conduct to make donor programmes and aid more coherent.
4 To collaborate more with the scholarly community to promote research on NEPAD and regionalism.

Hopefully, a closer relationship between the AU and NEPAD will help to promote a more democratic, pro-poor approach to development in Africa. This was championed by both Nelson Mandela and Thabo Mbeki, and was seen as a key motivation for the establishment of NEPAD in 2001.

In an attempt to identify some of the key priorities for future development in African countries, we will review the situation with reference to important issues in four dimensions of the development matrix: social, environmental, economic and political.

10.2 Social

The bottom line for any future development strategies must be to achieve some significant improvement in the quality of life for Africa's people, most notably through better nutrition, health and education. As we saw in Chapter 6, the daily lives of many Africans are characterized by poverty and sickness due to HIV/AIDS and a wide range of other common ailments, such as malaria, bilharzia, river blindness, pneumonia and tuberculosis. A priority for all families is to achieve food security and to ensure that all members of the household are adequately fed. In Burundi, for example, 38.9 per cent of children under the age of five are suffering from malnutrition, while the corresponding figure in Niger is 39.9 per cent (World Bank, 2009c). Healthy diets are dependent on sufficient quantities of the key food staples, supplemented with good amounts of fruit and vegetables. With improvements in food security and nutrition, adults should feel better motivated and able to work more effectively, while children will be more responsive at school and less susceptible to recurrent bouts of illness.

The availability of clean drinking water and improved sanitation also has implications for the health status of populations, and particularly for levels of infant and child mortality, where gastro-intestinal infections are a major factor. In Chad, only 42 per cent of the country's population have access to an improved water source, while only 9 per cent have access to improved sanitation (World Bank, 2008a). In some rural areas and low-quality urban settlements, access to both clean water and improved sanitation can be very poor.

Improvements in basic healthcare are essential, and in many African countries there is a need to increase government spending on health. Whereas the OECD countries generally spend over 15 per cent of total government expenditure on health, Guinea-Bissau spends only 4 per cent, Nigeria 3.5 per cent and Burundi 2.3 per cent (UNDP, 2009). Meanwhile, the Stockholm International Peace Research Institute reported in June 2010 that, while global military spending had fallen by 35 per cent since the end of the Cold War, in sub-Saharan Africa it has risen by almost a third over the same period (SIPRI, 2010).

Governments should also recognize the significance of community-based self-help initiatives, such as urban and peri-urban agriculture, which can make a valuable contribution to food security and the nutritional status of urban dwellers. Such initiatives must be nurtured and more properly incorporated into future planning strategies (see Box 5.3, above).

Education is another key component in ensuring a better quality of life, but it is denied to many young Africans, particularly girls. Primary school enrolment figures for the whole of Africa in 2005 indicated that 99 per cent of boys and 87 per cent of girls were enrolled. But these figures decline rapidly as children get older, such that secondary school enrolment was only 35 per cent for boys and 28 per cent for girls. Education for girls is particularly important, since maternal educational attainment has a strong correlation with family nutrition and child mortality (DfID, 2005, 2007). In many African countries, girls are married at an early age. For instance, in Niger, 62 per cent of females are married before the age of 19, and the median age of first marriage for females is just 15.2 years. Meanwhile, in Mozambique, 58 per cent of females between 15 and 24 have given birth at least once (World Bank, 2009d).

Differential levels of educational provision are reflected in literacy rates. In the Sahelian countries of West Africa, literacy levels are among the lowest in the continent (and indeed the world), with Mali having an adult male literacy rate of 33 per cent, while female literacy is only 16 per cent (World Bank, 2008a). Improvements in basic education are a crucial element in future development strategies in African countries.

There is clearly still a long way to go in ensuring that all Africans have access to basic needs and experience some tangible improvement in their quality of life.

10.3 Environmental

As we saw in Chapter 3, Africans' management of the environment has been widely criticized. The degradation and desertification discourse is ongoing, with human agency often seen as a key factor in reducing the quality of environmental resources. However, while we should not minimize, for example, the impacts of mangrove clearance and in some places serious deforestation, numerous empirical studies from across the continent have shown that local land management strategies are in fact remarkably sustainable. In light of widespread poverty and low levels of technology in Africa's rural areas, there seems to be no shortage of relatively sophisticated examples of anti-soil erosion, irrigation and terracing systems, and the use of organic fertilizer and composting techniques, while pastoralists generally display a sensitive appreciation of livestock management priorities in 'non-equilibrium' environments (Leach and Mearns, 1996; Reij *et al.*, 1996).

It is now generally appreciated that intercropping and bush-fallowing farming systems are in many cases very sensible adaptations to environment and provide sustainable crop returns for farming families (Richards, 1985, 1986). Meanwhile, in the area of pastoral management, while many governments remain obsessed with the colonial objective of achieving well-stocked and fixed ranching systems akin to those in Europe and North America, local pastoralists appreciate that the vagaries of living in a non-equilibrium environment necessitate appropriate levels of stocking, mobility and livelihood diversification (Scoones, 1995b).

The ever-strengthening climate change discourse suggests that those livelihoods that are dependent on farming, pastoral and fishing activities will need to make significant adaptations in the future, as we saw in Chapter 3. But with a long history of drought cycles, interspersed with periods of average or above-average rainfall, rural Africans have shown a good level of adaptation to climate change in the past. Farmers in north-east Nigeria, for example, do not just keep one seed variety of their staple crop, millet, but instead have a variety of seed types, each of which is suited to different topographies and levels of soil moisture. Diversification, as an adaptation to variable environmental conditions, is a longstanding and well-developed feature of many rural African farming and pastoral communities (see Chapter 4). This must be fully appreciated by development planners and effectively incorporated into future development strategies.

10.4 Economic

Despite impressive rates of economic growth in such countries as Botswana, Equatorial Guinea and Mauritius, the quality of life for the bulk of these countries' populations has not always improved – a case of 'growth without development'. As we saw in Chapter 8, the theory that wealth inevitably 'trickles down' to the poorest and remotest areas and their inhabitants invariably does not work in practice. It seems to be much more common for wealthy areas and people to become richer, while impoverished areas become poorer. These poorer areas and their people, who are distant from the centres of political and economic power, are often neglected when development policies are designed and implemented. As Chambers (1983) and others have shown, poverty is often unseen or misperceived, and therefore remains untreated, because of systemic biases in the research methodologies that provide (often unreliable) data that shape development interventions. In a continent where the agricultural sector employs some 60 per cent of the working population (and in Burundi and Ethiopia employs as many as 93 per cent and 92 per cent, respectively), this sector should receive far greater attention in development strategies. Chambers (1983: 3) calls this 'putting poor rural people first' (see Chapters 4 and 9).

The need for governments and others to increase investment in the agricultural sector has recently received a great deal of attention. In an address to the African Union Assembly in Sirte, Libya, on 1 July 2009, the UN Deputy Secretary-General, Asha-Rose Migiro, said,

> Since time immemorial, agriculture has been the cornerstone of development in every region . . . Agricultural investment creates jobs, can make economic growth more durable, can increase food and nutritional security, and can also have a profound impact on social equality, particularly by improving the situation of women, who account for the bulk of smallholder farmers in Africa.
>
> (UN News Service, 2009: 1; see Box 8.3)

During the post-independence period of the 1960s and 1970s, food production in Africa received inadequate investment, research and development, suffered from inadequate infrastructure, and commanded low prices. The share of government expenditure going to agriculture was still well below 10 per cent for many African countries in the first half of the 1980s. This resulted in poor agricultural performance, with adverse consequences for food availability, nutrition, balance of payments and living standards. Migiro suggested that neglect of the agricultural sector has resulted in high food prices that are unaffordable for the poorest people. She noted that in 2009 some 265 million people in sub-Saharan Africa went hungry, an increase of 12 per cent from the previous year (UN News Service, 2009).

Migiro urged all African countries to increase their national spending on agriculture to at least 10 per cent of their national budgets, and suggested that donors working with African farmers also needed to increase their support for the sector. She also argued that an increase in agricultural aid from the 2009 figure of $1–2 billion per year to $8 billion could result in African countries halving extreme poverty by 2015, which is a target of the first UN Millennium Development Goal (UN News Service, 2009: 1).

But agrarian-focused development strategies must ensure full participation, maximize rural linkages and be predominantly smallholder-based. Smallholder farms can earn an adequate living for most households and have the potential to produce marketable surpluses of food, export crops and livestock. Following independence in 1963, the Kenyan government mobilized smallholders as the focus of its agricultural development strategy. Farmers with less than two hectares of land, representing three-quarters of all farms, increased their share of national farm production from 4 per cent in 1965 to 49 per cent in 1985 (Binns, 1994). In devising agricultural and rural development strategies, it is important to recognize the diversity and complexity of African agriculture and to develop country- or region-specific strategies based on natural- and human-resource endowments (see Chapter 9). In so doing, particular attention should be given to improving the food security of the poor.

Another recent initiative is the Alliance for a Green Revolution in Africa (AGRA), established in 2006 by the Rockefeller Foundation and the Bill and Melinda Gates Foundation to improve the productivity and incomes of resource-poor farmers in Africa. AGRA is focusing its attention on improving seed quality and soil fertility, strengthening farmer access to agricultural inputs and markets, ensuring better access to water, improving agricultural education, encouraging governments to support small farmers and, importantly, increasing understanding and sharing the wealth of knowledge that African farmers have (AGRA, 2009a). At the launch of an AGRA policy initiative in October 2009 to empower Africa to develop home-grown agricultural policies, Kofi Annan, who is now chair of the AGRA board, commented, ‘Unlike farmers everywhere else in the world, African farmers, most of whom are women, receive little or no support from their governments. We must change this’ (AGRA, 2009b). With a $15-million grant from the Gates Foundation, the initial focus is on five African countries (Ethiopia, Ghana, Mali, Mozambique and Tanzania). The activities of AGRA should be seen alongside the African Union and NEPAD’s Comprehensive Africa Agriculture Development Programme (CAADP), which aims to eliminate hunger and reduce poverty by increasing government investment in agriculture to a minimum of 10 per cent of national budgets and raising agricultural productivity by at least 6 per cent. The four ‘pillars’ of the CAADP are: sustainable land and water management; market access; food supply and hunger; and agricultural research (CAADP, 2009). Some progress is already being made, as in Malawi, where seed and fertilizer vouchers have boosted productivity to such an extent that the country is now a net exporter of maize. Meanwhile, in Rwanda, similar policies have led to increases in food production of 15 per cent in 2007 and 16 per cent in 2008, thus strengthening national food security (AGRA, 2009b).

As we saw in Chapter 8, many African economies are dominated by a single product and export commodity. In connection with developing the agricultural sector, another aspect of the economic dimension in future development priorities is the diversification of exports. There are already many examples of flower and food exports from African countries (see Box 10.4), but there is considerable potential to expand such production. Research in Kenya in the 1990s revealed that large producers of fruits and vegetables, such as green beans, avocadoes and mangoes, dominated the export trade and controlled such key facilities as airport-based cold storage and air-freight space, while smaller producers experienced difficulties in developing export chains because of their limited access to these facilities. However, some smaller producers were beginning to supply the UK Asian retail market, where quality controls were not as strict as in the high-street supermarkets, and wholesalers seemed more willing to source produce from small producers (Barrett *et al.*, 1999). The European market seems to have an insatiable appetite for the year-round availability of fresh produce, which, with appropriate terms of trade and certain basic facilities at various points in the supply chain, could be sourced from Africa. But in an era of climate change and much debate about the ethics of 'food miles', it seems that such production will be more closely scrutinized in the future, perhaps to the detriment of African farmers.

Box 10.4

Corporate social responsibility and flower exports from South Africa

Since the mid-1990s there has been a major shift in corporate behaviour from a situation where companies were interested only in profit and showed remarkably little concern for alleviating poverty in poor countries. Firms that sell directly to consumers are increasingly vulnerable to criticism, and even activism, in cases where their social and environmental standards are perceived as falling short. Corporate social responsibility (CSR) has evolved into a powerful discourse within the mainstream of corporate activities. The majority of the UK's 100 largest companies now claim to be adherents to the CSR agenda, with many large transnationals employing individuals, or even whole departments, to manage CSR issues.

According to Blowfield and Frynas (2005: 503), CSR is best viewed as an umbrella term encapsulating the following key ideas:

- Companies have responsibilities for their impact on society and the natural environment, sometimes beyond legal compliance and the liability of individuals.
- Companies have a responsibility for the behaviour of others with whom they do business (for example, within supply chains).
- Business needs to manage its relationship with wider society, whether for reasons of commercial viability, or to add value to society.

In South Africa, two transnational companies, the petroleum giant Shell and UK retailer Marks and Spencer (M&S), have become involved with local communities in sourcing wild flowers for the UK market. The 580-hectare Flower Valley farm is located near Gansbaai on the western edge of the Agulhas Plain in Western Cape Province (Ashwell *et al.*, 2006; see Figure 10.1). The farm is located within the Cape Floristic Region (CFR), the smallest and richest of the world's six floral kingdoms, and the only one located exclusively within a single country. The main vegetation type within the biome is known locally as *fynbos* ('fine leaved bush'). Unfortunately,

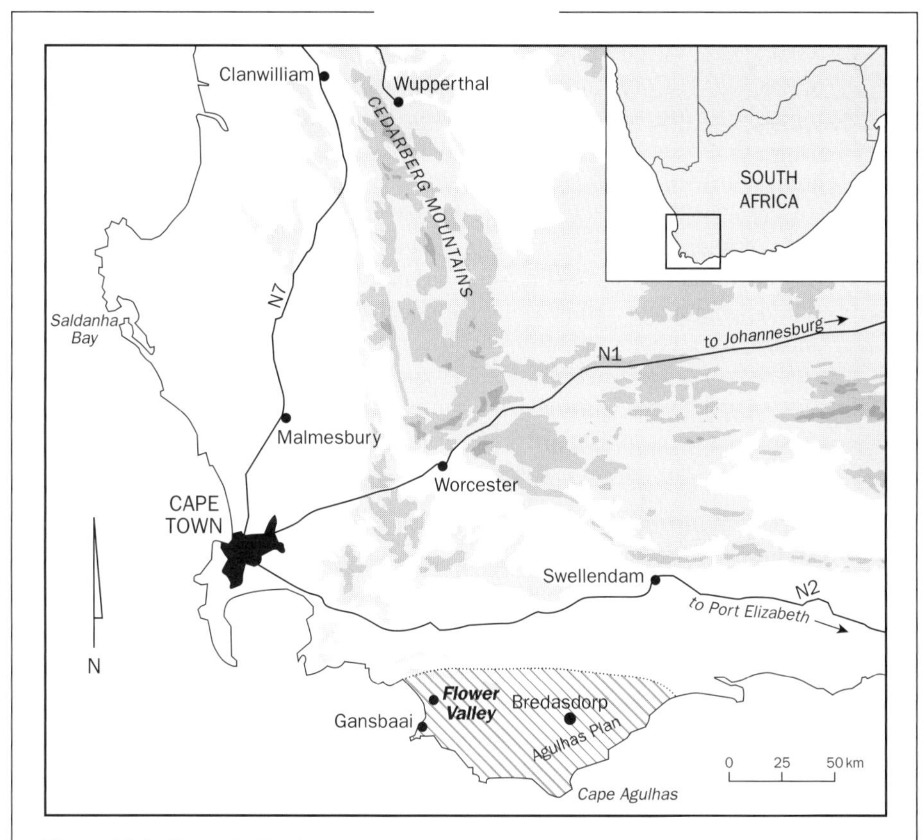

Figure 10.1 Flower Valley in its regional context.

the CFR is threatened by conversion to agricultural land use, poor fire management, alien species infestation and infrastructural development (Ashwell *et al.*, 2006). *Fynbos* flower farming is one of the main components of the agricultural sector in the CFR, with both cultivated and wild flowers harvested for domestic and export markets.

The legacies of apartheid remain on the Agulhas Plain, such that the white minority population still owns the vast majority of land, while unemployment within black communities often exceeds 50 per cent. These unemployment problems have been exacerbated by the arrival of Xhosa people attempting to escape conditions of extreme impoverishment in the former Transkei and Ciskei homelands of the Eastern Cape.

Until the late 1990s, the Flower Valley farm was owned by a commercial farmer who sold locally gathered wild flowers to the Amsterdam flower market. UK-based conservation NGO Fauna & Flora International (FFI) secured funds to purchase the farm in 1999 in order to protect the landscape from possible conversion into a vineyard. The Flower Valley Conservation Trust (FVCT) was set up locally to create a business that linked social investment with biodiversity (FVCT, 2009). Instead of using plantations, stems are picked from plants naturally occurring across the farm's landscape – so-called 'wild harvesting'. With strong support from South African organizations, such as the South African National Botanical Institute (SANBI), and multilateral agencies, such as the Global Environment Facility (GEF), Flower Valley is being used a testing ground for the development of a sustainable harvesting code of practice. The plan is to establish flower picking rates that allow sufficient time for the ecosystem to rejuvenate. The existence of this ethical dimension, defined by the project's conservation and community-development components, enables the product to be directed towards more lucrative niche markets.

Between 1999 and 2004, the project operated with a great deal of enthusiasm and enjoyed high levels of recognition among the international donor community. It was nominated as one of the Shell Group's 'Legacy Projects' to mark the 2002 World Summit on Sustainable Development (WSSD), and a grant of $100,000 was donated for investment in infrastructure on the farm. In 2003, a private company, Fynsa (Pty) Ltd., was established to source and market sustainably harvested wild flowers to overseas and domestic markets, with direct sales to retailers in order to maximize returns at the farm level.

However, the project lacked a sound business strategy, so significant personnel changes were made during 2005. Fortunes improved when the Shell Foundation and M&S joined forces to provide further inputs through the Small Scale Suppliers Programme (Shell Foundation, 2009). Through this programme, the Shell Foundation is investing $1 million (and its expertise) into developing the producer end of M&S supply chains for three flower- and fruit-growing schemes in Africa, including Flower Valley. This intervention has been critical for Flower Valley, as supply-chain management had been a serious problem and had made it difficult to guarantee a consistent supply of quality product. The environmental and social dimensions of the Flower Valley project were also highly compatible with M&S's recently rebranded image. Thus, the relationship came at a good time for the retailer and the farm, which were both at critical junctures in their histories.

The involvement of M&S, which has invested heavily in its 'Behind the Label' programme, has increased the need for Flower Valley to be able to demonstrate its ethical credentials explicitly. During 2006, much effort was put into the development of the sustainable harvesting component of the project, together with associated monitoring systems and a code of practice. In order to assure social standards, Flower Valley has, at the instigation of M&S, become a member of WIETA (the Wine and Agricultural Ethical Trade Association), a Western Cape-based body that provides auditing services concerned with improving working environments in agriculture (WIETA, 2009). WIETA's auditing methodology enjoys an excellent reputation among its stakeholders due to its thoroughness and integrity, and as such it is perceived to be superior to commercially available audits.

Although the project was still in its early stages in 2009, a wide range of benefits had already accrued to the local area. In economic terms, Flower Valley went from strength to strength, with sales approaching £500,000 gross in 2005 and increasing to approximately £700,000 in 2006. By June 2006, 330,000 bouquets had been sold in over 200 M&S stores. Significant progress was also made during 2008, when sales of bouquets in M&S stores, ranging in price from £3.99 to £19.99 (a Mother's Day special), totalled £4.1 million, representing 4 per cent of M&S's flower sales. The retailer has a substantial profit margin of 40–45 per cent.

In an attempt to diversify its markets, in 2009 Fynsa concluded a deal with the large South African supermarket chain Pick'n Pay to supply *fynbos* bouquets to Western Cape stores. It is also looking at possibilities for developing markets in Scandinavia, and through a smaller UK retailer. In 2009, the project employed thirty-two permanent packing-shed staff, and between ten and forty casual workers, depending on demand. It is likely that the Pick'n Pay contract will lead to recruitment of a further ten permanent staff (personal communication, March 2009; see Plate 10.1).

The project has provided clear benefits for the local community, especially through its creation of jobs in an area of high unemployment. In addition to the packing-shed workers, Fynsa has seven suppliers on the Agulhas Plain, including FVCT, which employ eighty to ninety pickers between them. Shell Foundation (2009) projections suggest that 3200 people will eventually benefit from improved livelihoods, better jobs and more stability as the project model is disseminated more widely throughout the region.

Donor support for the project has enabled an early learning centre to be set up for workers' children, and local women to be trained as childcare practitioners. In addition, efforts have been made to overcome the seasonal nature of employment in flower picking by identifying opportunities to develop diversified products, such as handmade paper. On-farm worker housing has also been improved. An assistant supervisor commented that her job was one of the best she could aspire to in the local area, as it was well paid (with double pay on Sundays) and provided three weeks' holiday, plus public holidays, sick leave and family leave entitlement. A WIETA audit revealed that pickers are paid double the rates of ordinary farm workers and living standards are better than in many urban areas (personal communication, 2009). Links with the local Greenfutures horticultural college have created opportunities for local young people from disadvantaged backgrounds to gain scholarships and training, and subsequently to secure employment in the local area (Ashwell *et al.*, 2006). As a result, knowledge about environmental issues has been disseminated throughout the community.

Plate 10.1 Packing sustainably harvested flowers at Flower Valley, Western Cape Province, South Africa (Tony Binns).

Inputs from Flower Valley's corporate and multilateral supporters have been critical in enabling the project to flourish. Shell has provided support through both its foundation and its separate Biodiversity Division, while M&S's CSR Department has acted alongside the firm's procurement team. The retailer has assisted in developing business plans and providing training related to the implementation of ethical, environmental and health-and-safety codes. This transmission of valuable 'business DNA' (Ward, 2005) has enabled Flower Valley to function more effectively as an enterprise. For instance, it is now successfully managing its part of the supply chain. As Fynsa's managing director states, 'we have a maximum of four days to have the flowers picked, made into a bouquet, put on an aeroplane, delivered to Heathrow in sleeves, and then delivered to the supermarkets' (personal communication, 2009). Prior to the intervention of the Shell Foundation and M&S, Flower Valley struggled to meet such strict deadlines.

The issue of product diversification is also relevant in the context of developing Africa's manufacturing sector, as Collier (2007) has shown. He reports that relatively few African manufacturing firms export their products, while those that do have enjoyed significant increases in their productivity. He suggests that 'domestic markets are too small to support much competition, and so learning from exporting is differentially powerful . . . The problem is how to get firms over that initial hump of competitiveness and enable them to get on the escalator' (Collier, 2007: 167). Targeted government support could therefore be useful in raising confidence within the manufacturing sector.

A further important aspect of the economic dimension in future development scenarios will undoubtedly be the increasing role in Africa of new players and funders, such as China and India (see Box 10.5). It is important to reflect briefly on the nature of this engagement and whether African countries are likely to achieve greater benefits than they have from their longstanding colonial and post-colonial links with countries and transnational companies based in Western Europe and North America.

Box 10.5

A new era for Africa? Engaging with BRICs: the case of China

The future development trajectories of African countries must increasingly be examined in light of the role of the so-called 'BRIC countries', which, it might be argued, are engaged in 'neo-colonial plunder' of the continent to source raw materials for their rapidly growing economies.

In 2001, the global investment banking and management firm Goldman Sachs was probably the first to use the term 'BRICs', when referring to the increasingly important role in world investment trends of four rapidly developing countries: Brazil, Russia, India and China (Wilson and Purushothaman, 2003). Goldman Sachs suggested that the BRICs' economies would become larger than those of the world's six most developed countries within the next forty years, possibly leading to a significant shift in the global balance of power. By 2050, it is likely that the list of the world's largest economies (which comprised the USA, China, Japan, India, Germany and the United Kingdom in 2010) will not include either Germany or the UK, but will include Russia and Brazil (Wilson and Purushothaman, 2003). It is predicted that India's economy could grow most rapidly over the next thirty to fifty years, with an estimated average annual growth rate of about 5 per cent. This would mean that India will overtake Japan by 2032. Meanwhile, economic growth in China, which has averaged an annual 10 per cent for the past thirty years, is predicted to decline to 3.5 per cent per year by the mid-2040s, broadly similar to Brazil, which will probably average 3.6 per cent per annum over the fifty-year period. Future economic growth in Russia could well be affected by its shrinking population, but it is still likely to have the highest per capita GDP in the BRICs group by 2050. Wilson and Purushothaman (2003) suggest that Russia's economy will overtake Italy's in 2018, the UK's in 2027 and Germany's in 2028.

Steadily increasing incomes in the BRICs are leading to significant reductions in poverty, the growth of the middle classes and a corresponding increase in demand for consumer goods. In recent years, all four BRICs have become involved, to a greater or lesser extent, in a wide range of development initiatives in African countries. Probably the most active of the four is China, whose engagement with Africa goes back many decades, most notably involving the construction of the Tazara (or TanZam) railway between 1970 and 1975, which, at a cost of $500 million, was the largest foreign-aid project ever undertaken by China at the time (see Plate 10.2). The motivation for initiating such a large project was to reduce the dependence of landlocked Zambia on trading through the white minority states of

Plate 10.2 Tazara railway terminus in Zambia (Etienne Nel).

Rhodesia (now Zimbabwe) and South Africa. The railway stretches 1870 km from Dar es Salaam on Tanzania's eastern coastline to Kapiri Moshi in Zambia, where it links into the Zambian rail network and the important mining towns of the Copperbelt, and then on to the southern African network.

In the last four decades, China has steadily increased its activity in Africa, and a number of authors have charted the development of Sino-African relations (see, e.g., Alden, 2007; Huse and Muyakwa, 2008; Kornegay, 2008; Mawdsley, 2007). After the long-term involvement of Western European countries in Africa, much media commentary has focused on the West's attitude to China's increasing interest in Africa. In many African countries, complex patterns of influence and activity in political and economic spheres have developed from the colonial past, and there is now some suspicion, in both Africa and Europe, about China's increasing involvement in Africa in terms of both motivation and outcomes. As Alden (2007: 102) comments, 'Much to their consternation, traditional Western actors are finding their once undisputed influence and dominance of Africa are being challenged by aggressive Chinese multinational corporations in collusion with the Chinese state.'

Many Africans are also trying to work out whether they will benefit from the flurry of Chinese initiatives. As Kornegay (2008: 4) comments, 'African actors need to identify where the continent's interests converge and diverge with China's priorities. In short, the African diplomatic strategy toward China for optimally interacting with Beijing's diplomacy toward the continent has to be identified and formulated.' A BBC report (BBC, 2010b) examined the recent growth of Chinese activity in the West African country of Benin, and quoted a local businessman:

> The Chinese are investing, helping African countries, building roads: so many things going on commercially . . . They have paid for many things. We have a big stadium here and many road projects . . . France was here as a colonial power: they were here to rule. China has been here to help us get liberated from colonial powers.

But the increasing Chinese activity has been met with hostility in certain African countries. For example, during the 2006 Zambian presidential elections, the opposition leader, Michael Sata, was critical of worker exploitation in Chinese-owned mines, commenting, 'They ill-treat our people [and] that is unacceptable. We are not going to condone exploiting investors because this country belongs to Zambians' (Foreign Policy, 2006: 1). Although President Levy Mwanawasa was re-elected with 43 per cent of the vote, Sata polled three times as many votes as Mwanawasa in the capital, Lusaka, and won support in many areas where there was significant Chinese investment. On 2 October 2006, following the declaration of the election results, there was rioting on the streets of Lusaka, and Chinese shopkeepers barricaded their properties against gangs of looters. In a report from Lusaka at the time, a UK journalist commented, 'Chinese businessmen are accused of underpaying their workers, ignoring safety rules and driving local companies out of business with cheap and shoddy goods' (Blair, 2006: 1; see Plate 10.3).

In 2006, trade between China and African nations increased by 39 per cent, to \$50.5 billion, compared with African–US trade amounting to \$71.1 billion (Alden, 2007). By 2008, trade between China and Africa had reached a record high of \$106.84 billion (*Guardian*, 2009). China has scrapped tariffs on 190 kinds of

Plate 10.3 Chinese investment in Zambia (Tony Binns).

imported goods from twenty-eight of the least developed African countries. Since 2000, Chinese investment in Africa has accelerated, notably in construction projects and infrastructure development, such as roads and railways, which were badly neglected in the colonial and post-colonial periods. In the first ten months of 2005 alone, Chinese companies invested a total of $175 million in African countries, and Africa has become significantly more important for China as a source of raw materials to supply its growing manufacturing sector (Alden, 2007).

Chinese companies have become involved in oil exploration in countries with controversial political regimes, such as Sudan, in order to meet China's rapidly growing petroleum needs. Between 1996 and 2006, China invested over $15 billion in Sudan's 'pariah regime', mainly in the oil industry and a range of infrastructure projects. Correspondingly, bilateral trade between China and Sudan increased from $890 million in 2000 to $3.9 billion in 2005 (Alden, 2007: 60). These links have frequently been condemned around the world as helping to support internal conflicts, such as in Darfur (see Box 7.2, above). As Alden (2007: 62) observes,

> Increasing oil revenues from the sale of oil to China have allowed Khartoum to purchase sophisticated weaponry as well as develop (with Chinese assistance) its own arms manufacturing capacity, based at the MCM Military Manufacturing Complex and two other sites, which reportedly produce light arms, rocket launchers and anti-tank weapons.

In 2000, the Forum for China Africa Cooperation (FOCAC) was established in Beijing. Some forty-eight African countries were represented in a high-level ministerial meeting in Beijing in 2006, when Chinese President Hu Jintao celebrated fifty years of diplomatic relations with Africa that were initiated when Egypt (under President Nasser) forged links with China in 1956. At this meeting, the 2007–2009 Beijing Action Plan was adopted. Among a wide range of stated objectives, this plan pledged that Chinese aid to Africa would be doubled; preferential loans would be made available; a number of trade and economic cooperation zones would be set up in Africa; 15,000 African professionals would be trained; a conference centre would be built for the African Union; one hundred rural schools and thirty hospitals would be constructed; and RMB300 million would be donated to prevent and treat malaria (Huse and Muyakwa, 2008). Another outcome of the 2006 Beijing meeting was the establishment in June 2007 of the China–Africa Development Fund, with first-phase funding of $1 billion, which will eventually increase to $5 billion. The CAD Fund aims to strengthen the China–Africa strategic partnership and guide and support Chinese enterprises that wish to make direct investment in Africa.

Increasing Chinese investment in Africa has been accompanied by a wave of Chinese migrants who have settled in African countries, often in distinctive communities. In South Africa, for example, there were an estimated 4000 Chinese in 1946; by 2006, this number had increased to between 300,000 and 400,000. In the same year, there were an estimated 100,000 Chinese in Nigeria, 20,000 in Tanzania and possibly 80,000 in Zambia (Alden, 2007).

There are many examples of China's recent activity in Africa, and China's insatiable demand for oil has been a significant factor in many recent initiatives. In addition to oil exploration in Sudan, in 2006 China promised over $9 billion to Angola in credits, loans and infrastructure programmes, and in February of that year Angola overtook Saudi Arabia as China's top source of oil. Whereas China was once self-sufficient in oil, in 2007 it was having to import almost half of its daily consumption as a result of the massive growth of domestic manufacturing and increasing vehicle ownership.

More recently, in February 2010, the Hong Kong-listed China National Offshore Oil Corporation agreed to buy a stake in the Ugandan oil assets of Tullow Oil for $2.5 billion, and said it would entirely fund the development of a 1200-km pipeline to carry the oil to the Kenyan port of Mombasa for export (*Wall Street Journal*, 2010). Meanwhile, in the West African state of Niger, China was accused of propping up the autocratic regime of President Mamadou Tandja, who was overthrown by a military coup on 18 February 2010. In 2008, China's state oil company paid $300 million to the government of Niger as part of a secretive deal (BBC, 2008d). However, since the coup, the new regime has shown enthusiasm for maintaining strong diplomatic relations with China in light of the latter's considerable investment in Niger's oil, uranium and hydro-electric industries.

In terms of the quest for other mineral resources, in 2008 China signed a long-term infrastructure development agreement worth $9 billion with the Democratic Republic of the Congo. At the same time, the DRC's national mining company, Gecamines, eased a major mining company, Katanga Mining, out of two major copper deposits, paying some $825 million in compensation, and then granted the concessions to a Chinese company. In Sierra Leone, in July 2010, the Shandong Iron and Steel Group signed a memorandum of understanding to acquire a 25 per cent share of London-listed African Minerals' Tonkolili iron-ore mine at a cost of $1.5 billion. This initiative could provide employment for some 2000 Sierra Leoneans (*China Daily*, 2010).

In Zambia, the Chinese own three copper-mining operations, including a 150,000-tonne-a-year smelter. During the recession in 2009, when Zambia's earnings from copper sales fell to $2.9 billion from the $3.6 billion recorded in 2008, China invested over $400 million into the country's mining industry. As the *Lusaka Times* reported in January 2010, 'China's intervention has rescued several Zambian copper mines out of bankruptcy and helped create some 2000 jobs, most of which were lost when copper prices slid over the biting effects of the global financial crisis' (*Lusaka Times*, 2010: 1).

China has been active in many African countries in developing infrastructure, including roads, railways, schools, hospitals and public buildings. In the DRC, China will be involved in some $3 billion of infrastructure projects, including a 3400-km highway, a 3200-km rail link with Matadi on the Atlantic Ocean, 31 hospitals, 145 health clinics and two universities (Butts and Bankus, 2009). Meanwhile, in Angola, the China International Fund, based in Hong Kong, is undertaking the rehabilitation of the dilapidated Benguela railway from the coast to the DRC at a cost of $300 million (BBC, 2010a). This link suffered during the Angolan Civil War, but it once provided a valuable conduit for the export of manganese from the mineral-rich DRC to the Atlantic coast.

While most attention on China's investment in Africa has concentrated on large-scale infrastructure development and mineral exploitation, both the Chinese government and its private sector have shown strong interest in developing manufacturing enterprises. The president of the World Bank, speaking in December 2009, envisaged the creation of industrial parks, possibly funded jointly by the World Bank and China. At least 2000 Chinese companies are now operating in Africa, many of them small or medium enterprises (Gu, 2009). In the case of Ghana, for example, of the 340 Chinese investment projects in the country, more than 100 are in manufacturing (*Guardian*, 2009). With a strong work ethic, these businesses typically have a medium- or long-term perspective and are prepared to take risks and accept low profit margins in the early stages. The majority come from a small number of Chinese provinces, notably Zhejiang, Guangdong, Fujian, Jiangsu and Shandong. As Gu (2009: 1) comments, 'the biggest motivations for Chinese entrepreneurs to invest in Africa are accessing local markets, transferring excess capacity from the highly competitive Chinese markets and taking advantage of conditions in Africa – which resemble the Chinese market of the 1980s and 1990s'. Personal and business networks are important, and many firms start as trading outlets before moving into manufacturing and then larger industrial parks. As Gu (2009: 2) says, 'while Chinese central government actively promotes overseas investment by both state-owned enterprises and private enterprises as part of a policy to strengthen industrial restructuring, in practice most private firms have followed their own paths overseas, and commercial rather than policy imperatives'. However, in addition to the anti-Chinese sentiment in Zambia, concern has been expressed elsewhere in Africa about the relatively low levels of local employment generated by Chinese investment. Furthermore, there has been criticism of Chinese entrepreneurs' attitudes to corporate social responsibility, including health and safety at work.

There is no doubt that China is making a tremendous impact on the African continent, and Chinese-sponsored improvements in infrastructure and health and education facilities are likely to have positive effects for many people. However, it is important that African countries should not sell their resources cheaply, and that the rewards from Chinese investment contribute to national development. Despite centuries of European contact and colonialism, Africa in the twenty-first century remains poor and marginalized. It must be asked whether China will make a meaningful contribution to strengthening African economies and improving the quality of life for the millions of Africans living in poverty. Furthermore, will China's current interest in Africa raise the profile of the continent on the world stage and reduce its marginalization?

African countries potentially have much to gain from the new investors, but the terms of engagement will need to be carefully evaluated, for example in the exploitation of natural resources. These new funders are already providing valuable assistance in upgrading education, health, transport, water, sanitation, power and telecommunications, which have been inadequately developed and maintained for many years. For instance, in October 2006, Nigeria signed an $8-billion contract with a Chinese company, CCECC (China Civil Engineering Construction Corporation), to lay new rail track along the main south–north, Lagos–Kano route, with a second phase between the southern oil city of Port Harcourt and Jos in central Nigeria (BBC, 2006). Meanwhile, in post-conflict states such as Sierra Leone, and with assistance from such sources as the European Union and the UK government, there is a massive agenda of upgrading neglected roads and power systems, and rebuilding hospitals and schools, many of which were rebel targets during the civil war (see Chapter 7).

10.5 Political

In many ways, the political dimension will be absolutely crucial to Africa's future development. The acquisition and distribution of key resources for development should be a key responsibility for all levels of government. But transparency in conducting government business and long periods of assured stability are essential for meaningful development impacts to be achieved. Ongoing conflict situations, such as in Sudan, the DRC and northern Uganda, make it difficult to achieve any increase in livelihoods, while in dysfunctional states, such as Somalia, there is a desperate need for both direction and accountability in policy-making (see Chapter 7).

Even in countries that are supposedly 'democratic', with regular free and fair elections, levels of corruption and transparency give cause for concern and militate against effective development planning and implementation. Transparency International's Corruption Perceptions Index reveals that ten of the world's twenty most corrupt states are in Africa, with Equatorial Guinea ranked 168th, Sudan 176th, and Somalia 180th out of 180 states surveyed (Transparency International, 2009; see Plate 10.4). In some cases, bad governments can quickly destroy an economy and worsen the lives of the people, as happened in Zimbabwe under President Robert Mugabe from the mid-1990s. When Mugabe came to power in 1980, Zimbabwe had one of the strongest economies and social welfare systems in the continent, but thirty years later it was a demoralized and bankrupt country with rampant annual inflation (516 quintillion per cent – 516 followed by eighteen zeros – in late 2008; *Wealth Daily*, 2008). Collier (2007) suggests that the impetus for change must come from within society itself, but evidence shows that sustained turnarounds are difficult to achieve, and he reaches the rather gloomy conclusion that it takes an average of fifty-nine years to get out of being a failing state. Therefore, he advocates an alliance between

Plate 10.4 Anti-corruption poster in Sierra Leone (Tony Binns).

the OECD and large NGOs to establish international charters that would 'empower the reformers within their societies, and also enable those countries at the early stages of turnaround to lock in change' (Collier, 2007: 185).

Good development strategies should empower all people, and particularly the most marginalized, such that they feel they can play a role in charting a way forward and reaping the benefits from development initiatives. Civil society is often under-represented in decision-making processes, but in some African countries civil society organizations are developing productive relationships with government and the public and private sectors. Such objectives of empowering citizens were uppermost in Nelson Mandela's ambitious Reconstruction and Development Programme (RDP), which was launched before South Africa's first democratic elections in 1994. As the RDP recognized,

> Our history has been a bitter one dominated by colonialism, racism, apartheid, sexism and repressive labour policies. The result is that poverty and degradation exist side by side with modern cities and a developed mining, industrial and commercial infrastructure. Our income distribution is racially distorted and ranks as one of the most unequal in the world – lavish wealth and abject poverty characterise our society.
>
> (ANC, 1994: 2)

Faced with millions of disadvantaged people who were clamouring for a rapid improvement in their livelihoods, the RDP became the cornerstone of government policy in 1994 and 1995, and aimed to empower South Africa's disadvantaged people, specifically the poor, women, youth and the disabled. Despite persistently high rates of unemployment, the first few years of the ANC government recorded some significant achievements, such that by 1999 some 3 million people had been re-housed – a quarter of those in need. Furthermore, between 1994 and 1998, the number of houses with electricity connections increased from 31 per cent to 63 per cent, and the number of telephone lines increased from 150,000 to 386,426. By 1998, 1.7 million additional households had been supplied with piped water, many new schools and clinics had been built, and primary healthcare and free medical attention had been extended to a large proportion of the population (Lester *et al.*, 2000).

There were also significant changes in governance in South Africa, in an attempt to achieve greater involvement of grassroots communities in decision-making. Most notably, a process of decentralization and strengthening of local government was initiated, as indeed has happened in a number of other African countries in recent years. The post-apartheid South African government recognized the vital role of local government, stating: 'Local government is of critical importance to the RDP. It is the level of representative democracy closest to the people. Local government will often be involved in the allocation of resources directly affecting communities. [It should] ensure maximum participation of civil society and communities in decision-making and developmental initiatives' (ANC, 1994: section 5.12). On 5 December 2000, all South Africans had their first opportunity to vote for post-apartheid municipal councils, elections that 'marked the final stage in the legislative transformation of South African democracy' (Binns and Robinson, 2002: 29).

While bringing government closer to the people seems a laudable objective, particularly in African states characterized by endemic corruption and a lack of free and fair elections, there has been some concern about the effects and achievements of decentralization policies, as Binns *et al.* (2005) have identified. In South Africa, the ANC government devolved responsibility for stimulating development initiatives to local government in what was termed 'developmental local government' (see Chapter 9). Unfortunately, the financial and technical resources for this initiative were not similarly devolved, so many local authorities, faced with limited finance and expertise (due to the legacies of apartheid), experienced much difficulty in devising and supporting local-level development interventions (Nel and Binns, 2003b). A comparative study of Ghana and South Africa revealed that 'the (limited) benefits of decentralisation are mostly being mopped up by urban-based local elites' (Binns *et al.*, 2005: 30).

The results of decentralization policies need to be carefully monitored to ensure that greater local democracy and a sense of empowerment actually occur. In post-conflict Sierra Leone, for example, where local government was effectively abolished by President Siaka Stevens in the 1970s in favour of a highly centralized and increasingly corrupt one-party state, the reinstatement of district councils is seen as an important step towards promoting local democracy and, in particular, allowing young people to have a voice and thereby reducing the disaffection that was a hallmark of the country in the run-up to the civil war (Maconachie and Binns, 2007a).

Finally, at a time when Africa seems as marginalized as ever in terms of political and economic events on the world stage, we might consider whether greater cooperation between the African states might help to project a stronger voice for the continent. Kwame Nkrumah's call in 1957 that 'Africa must unite' might be as relevant today as it was then. In its lifespan of almost four decades, the Organization of African Unity (OAU) proved rather ineffective in dealing with the continent's economic and political problems, and in projecting a unified and more positive image of Africa. It remains to be seen whether its successor, the African Union, will achieve greater success, and whether Muammar Gaddafi's call for a 'United States of Africa' will ever materialize; if, indeed, such a goal is even seen as desirable (see Boxes 10.2 and 10.3).

10.6 Conclusion

We started this book by considering how Africa has been perceived by the rest of the world, both in the past and in more recent times. Unfortunately, the overwhelmingly negative gloom and doom scenarios of a continent seemingly beset with a multitude of problems remain uppermost in many people's minds.

But those of us who have lived and worked in Africa for several decades are well aware that many positive images and achievements have failed to reach the media headlines. Africa is a continent with considerable potential in many respects. We cannot ignore the fact that it has a long way to go in catching up with the rest of the world in terms of a range of development indicators. However, progress *is* being made, as a recent report from the Overseas Development Institute concludes. The report argues that the average proportion of the African population living in poverty declined from 52 per cent to 40 per cent between 1990 and 2008. Furthermore, it points out,

> Ten African countries have already halved their poverty rate, including relatively populous countries such as Ethiopia and Egypt, and post-conflict countries such as Angola. Half of the African countries for which data exist have been reducing the poverty rate by at least two percentage points per year, which puts them on track to meet the MDG target of halving poverty.
>
> (ODI, 2010: 6)

Africa and Africans are undeniably rich in resources and justifiably proud of their rich cultural traditions, and they demonstrate good levels of entrepreneurship, often in the face of very difficult circumstances. Without intending to be trite or patronizing, it must be said that a lot of positive things are happening in Africa, and these should be making the headlines. However, in order to achieve more productive and sustainable livelihoods throughout Africa, it is imperative that the wealth of knowledge and understanding that Africans possess is more widely appreciated and more effectively incorporated into development strategies. Only then will tangible improvements be made to the lives of all Africans, in marked contrast to so many initiatives of the past.

Summary

1. The media have frequently portrayed Africa as a continent beset with problems, with regular reports about 'crises in Africa'.
2. There is an urgent need to improve basic healthcare and education, including placing more emphasis on education for girls.
3. Discourses about land degradation and desertification have frequently blamed these processes on African farmers and pastoralists. However, evidence suggests that, given the prevalence of poverty, low-level technology and environmental conditions, many African communities have developed a range of land-management practices that are remarkably sustainable.
4. Some African countries have experienced rapid economic growth, although few of the benefits have 'trickled down' to the bulk of the population – what might be called 'growth without development'.
5. Africa's agricultural sector has considerable potential, in terms of ensuring local food security and generating foreign-exchange earnings from exports, but there is an urgent need for more investment and diversification in the sector.
6. New investors, such as China and India, have great potential to develop the industrial and agricultural sectors of African countries, but these activities need to be carefully monitored in terms of the nature and scale of benefits that accrue to the host countries and their people.
7. Bad governance militates against sustainable development. There is a need for more grassroots democratic governance, engaging with civil society and empowering marginalized individuals and groups.

Discussion questions

1. Why should the agricultural sector receive particular attention in African development strategies?
2. To what extent might political instability be blamed for low levels of development in some African countries?
3. Will multi-party democratic rule necessarily lead to more rapid development in Africa?
4. To what extent might new investors, such as China and India, assist in developing Africa's potential and contribute to an improvement in standards of living?
5. How might increased international cooperation among African countries play a significant role in promoting future development?

Further reading

Alden, C. (2007) *China in Africa*, London: Zed Books.

Collier, P. (2007) *The Bottom Billion*, Oxford: Oxford University Press.

Murithi, T. (2005) *The African Union, Pan-Africanism, Peacebuilding and Development*, Aldershot: Ashgate.

Potter, R.B, Binns, T., Elliott, J. and Smith, D. (2008) *Geographies of Development*, Harlow: Pearson.

Useful websites

African Union: http://www.africa-union.org/. The official website of the African Union has information on history, news and events, and publications.

AGRA (Alliance for a Green Revolution in Africa): http://www.rockfound.org/initiatives/agra/agra.shtml. Provides an overview of the AGRA strategy, its key aims and target outcomes.

Commission for Africa: http://www.commissionforafrica.info/. Examines the origins of the commission (set up in 2004 by British Prime Minster Tony Blair) and has up-to-date items on such issues as progress towards achieving the Millennium Development Goals.

NEPAD (New Partnership for Africa's Development): http://www.nepad.org/. Includes information on the origins, structure, priorities and key events in NEPAD's history.

Aberra, E. (2006) Alternative strategies in alternative spaces: livelihoods of pastoralists in the peri-urban interface of Yabello, Southern Ethiopia. In McGregor, D., Simon, D. and Thompson, D. (eds) *The Peri-urban Interface: Approaches to Sustainable and Natural and Human Resource Use*. London/Stirling: Earthscan, pp. 116–133.

Abraham, C. (2007) Don't blame climate change for Africa's conflicts. *New Scientist* 196 (2626): 24.

Achvarina, V., Nordas, R., Ostby, G. and Rustad, S.A. (2009) Poverty and child soldier recruitment: a disaggregated study of African regions. *Politische Vierteljahresschrift* 50: 386–413.

Adams, W.M. (1988) Rural protest, land policy and the planning process on the Bakalori Project, Nigeria. *Africa* 58(3): 315–336.

ADB (African Development Bank) (2008) *Selected Statistics on African Countries, 2008*. Abidjan: African Development Bank.

ADB (African Development Bank) (2009) *African Development Report 2008/2009: Conflict Resolution, Peace and Reconstruction in Africa*. Oxford: Oxford University Press.

ADB (African Development Bank) (2010) *Agriculture Sector Strategy 2010–2014*. Abidjan: African Development Bank.

ADB (African Development Bank) and OECD (Organization for Economic Cooperation and Development) (2008) *African Economic Outlook*. Abidjan: African Development Bank.

Addison, T. and Tarp, F. (2010) online: *The Triple Crisis: Finance, Food and Climate Change*. www.wider.unu.edu/publications/newsletter/articles-2010/e (accessed 21 October 2010).

Adedeji, A. (ed.) (1996) *South Africa and Africa: Within or Apart?* London: Zed Books.

Adeyemo, O.K. (2003) Consequences of pollution and degradation of Nigerian aquatic environment on fisheries resources. *The Environmentalist* 23(4): 297–306.

Adger, W.N., Huq, S., Brown, K., Conway, D. and Hulme, M. (2003) Adaptation to climate change in the developing world. *Progress in Development Studies* 3(3): 179–195.

Adriansen, H.K. (2008) Understanding pastoral mobility: the case of Senegalese Fulani. *Geographical Journal* 174(3): 207–222.

Africa Institute of South Africa (1995) *Africa at a Glance, 1995/6*. Pretoria: Africa Institute of South Africa.

Africa News (2009) online: *Africa's Population Now 1 Billion*. http://www.africanews.com/site/Africas_population_now_1_billion/l (accessed 15 August 2010)

Africa Progress Panel (2008) *Africa's Development: Promises and Prospects*. Washington, DC: Africa Progress Panel.

African Economic Outlook (2011) online: *Africa is Recovering but There Are Risks*. http://www.africaneconomicoutlook.org/en/outlook/ (accessed 7 August 2011).

African Union (2007) *Africa Health Strategy 2007–2015*. Addis Ababa: African Union.

African Union (n.d.) online: *African Union Website*. http://www.africa-union.org/ (accessed 22 June 2009)

African Union Commission (2004) *2004–2007 Strategic Framework of the African Union Commission*. Addis Ababa: African Union Commission.

Agnew, J. and Grant, R. (1997) Falling out of the world economy? Theorizing Africa in world trade. In Lee, R. and Wills, J. (eds) *Geographies of Economies*. London: Arnold, pp. 219–330.

AGRA (Alliance for a Green Revolution in Africa) (2009a) online: *Strengthening Food Security: Alliance for a Green Revolution in Africa (AGRA).* http://www.rockfound.org/initiatives/agra/agra.shtml (accessed 16 November 2009).

AGRA (Alliance for a Green Revolution in Africa) (2009b) online: *AGRA Launches Policy Initiative to Empower Africa to Shape Home-Grown Agricultural Policies.* http://www.agra-alliance.org/content/news/detail/1028 (accessed 15 October 2009).

Agrawal, A. (2001) Common property institutions and sustainable governance of resources. *World Development* 29(10): 1649–1672.

Ahern, M.J., Kovats, R.S., Wilkinson, P., Few, R. and Matthies, S. (2005) Global health impacts of floods: epidemiological evidence. *Epidemiology Review* 27(1): 36–45.

Ahmed, I. and Green, R. (1999) The heritage of war and state collapse in Somalia and Somaliland: local level effects, external interventions and reconstruction. *Third World Quarterly* 20(1): 113–128.

Ahwireng-Obeng, F. and McGowan, P.J. (1998) Partner or hegemon? South Africa in Africa. *Journal of Contemporary African Studies* 16(1): 5–38, and 16(2): 165–195.

Aina, T.A. (1988) The construction of housing for the urban poor of Lagos. *Habitat International* 12(1): 31–48.

Akokpari, J. (2008) Conclusion: building a unified Africa. In Akokpari, J., Ndinga-Muvumba, A. and Murithi, T. (eds) *The African Union and its Institutions.* Cape Town: Fanele, pp. 371–381.

Akokpari, J., Ndinga-Muvumba, A. and Murithi, T. (eds) (2008) *The African Union and its Institutions.* Cape Town: Fanele.

Akosah-Sarpong, K. (1998) A continent in transition. *West Africa* 4195: 665–666.

Akukwe, C. (2007) online: *Africa's Development in 2007.* http://www.worldpress.org/print_article.cfin?/article_id=2773&doc (accessed 21 October 2010).

Albertyn R., Bickler, S.W., van As, A.B., Millar, A.J.W. and Rode, H. (2003) The effects of war on children in Africa. *Pediatric Surgery International* 19(4): 227–232.

Alden, C. (2007) *China in Africa.* London: Zed Books.

Alem, A., Jacobsson, L. and Hanlon, C. (2008) Community-based mental health care in Africa: mental health workers' views. *World Psychiatry* 7(1): 54–57.

Al-Hebshi, N.N. and Skaug, N. (2005) Khat (Catha edulis) – an updated review. *Addiction Biology* 10(4): 299–307.

allAfrica.com (2008a) online: *Angola: Government Considers Housing Policy Important to Development.* http://allafrica.com/stories/200811280022.html (accessed 30 December 2008).

allAfrica.com (2008b) online: *Water Running on Empty.* http://allafrica.com/stories (accessed 21 August 2008).

allAfrica.com (2009) online: *Museveni Opposes Gaddafi on African Political Union.* http://allafrica.com/stories/ (accessed 4 July 2009).

allAfrica.com (2010) online: *Africa: Population Explosion – Africa Sitting on a Time Bomb.* http://allafrica.com/stories/printable/201004010036.htm (accessed 15 August 2010).

Allan, W. (1949) How much land does a man require? *Rhodes–Livingstone Papers* 15: 1–23.

Allen, J. (2010) Opening remarks. CEO address, New Zealand Institute of International Affairs Seminar, Wellington, 20 October.

Allen, T. and Heald, S. (2004) HIV/AIDS policy in Africa: what has worked in Uganda and what has failed in Botswana. *Journal of International Development* 16(8): 1141–1154.

Allison, E.H. and Horemans, B. (2006) Putting the principles of the sustainable livelihoods approach into fisheries development policy and practice. *Marine Policy* 30(6): 757–766.

Amis, P. and Lloyd, P. (eds) (1990) *Housing Africa's Urban Poor.* Manchester: Manchester University Press.

ANC (African National Congress) (1994) *The Reconstruction and Development Programme.* Johannesburg: Umanyano Publications.

Andreasson, S. (2005) Orientalism and African development studies. *Third World Quarterly* 26(6): 971–986.

Annan, K. (2000) *Africa: Maintaining the Momentum.* London: The Commonwealth Foundation.

APIC (Africa Policy Information Centre) (1996) online: *Africa on the Internet: Starting Points for Policy Information.* http://www.igc.apc.org/apic/bp/inetall.html (accessed 10 January 2009).

Aryee, G. (1981) The informal manufacturing sector in Kumasi. In Sethuraman, S.V. (ed.) *The Urban Informal Sector in Developing Countries: Employment, Poverty and Environment*. Geneva: International Labour Office.

Ashford, L.S. (2007) *Africa's Youthful Population: Risk or Opportunity?* Washington, DC: Population Reference Bureau.

Ashley, C. and Mitchell, J. (2007) *Assessing How Tourism Revenues Reach the Poor*. London: Overseas Development Institute.

Ashraf, H. (2005) Countries need better information to receive development aid. *Bulletin of the World Health Organization* 83(8): 565–566.

Ashwell, A., Sandwith, T., Barnett, M., Parker, A. and Wisani, F. (2006) *Fynbos Fynmense: People Making Biodiversity Work*. Pretoria: South African Biodiversity Institute.

ASP (Africa Stockpiles Programme) (2009) online: *About the Programme*. http://www.africastock piles.net/about (accessed 25 August 2010).

Atampugre, N. (1993) *Behind the Lines of Stone: The Social Impact of a Soil and Water Conservation Project in the Sahel*. Oxford: Oxfam Publications.

Atta-Mills, J., Alder, J. and Sumaila, U.R. (2004) The decline of a regional fishing nation: the case of Ghana and West Africa. *Natural Resources Forum* 28(1): 13–21.

AVERT (2010a) online: *HIV and AIDS in South Africa*. http://www.avert.org/aidssouthafrica.htm (accessed 16 August 2010).

AVERT (2010b) online: *HIV and AIDS in Uganda*. http://www.avert.org/aids-uganda.htm (accessed 16 August 2010).

Awases, M., Gbary, A., Nyoni, J. and Chatora, R. (2004) *Migration of Health Professionals in Six Countries: A Synthesis Report*. Brazzaville: WHO-AFRO.

Ayeni, B. (1997) Regional development planning and management in Africa. In HABITAT (United Nations Centre for Human Settlements), *Regional Development Planning and Management of Urbanization*. Nairobi: Habitat, pp. 7–62.

Babulo, B., Muys, B., Nega, F., Tollens, E., Nyssen, J., Deckers, J. and Mathijs, E. (2008) Household livelihood strategies and forest dependence in the highlands of Tigray, northern Ethiopia. *Agricultural Systems* 98(2): 147–155.

Baden, S. and Wach, H. (1998) *Gender, HIV/AIDS Transmission and Impacts: A Review of Issues and Evidence*. Brighton: Institute of Development Studies.

Badiane, O. (2008) *Sustaining and Accelerating Africa's Agricultural Growth Recovery in the Context of Changing Global Food Prices*. Washington, DC: IFPRI.

Baharoglu, D., Peltier, N. and Buckley, R. (2005) online: *The Macroeconomic and Sectoral Performance of Housing Supply Policies in Selected MENA Countries: A Comparative Analysis*. http://go.worldbank.org/JIVVDU2490 (accessed 30 December 2008).

Baingana, F.K., Alem, A. and Jenkins, R. (2006) Mental health and the abuse of alcohol and controlled substances. In Jamison, D.T., Feachem, R.G., Makgoba, M.W., Bos, E.R., Baingana, F.K., Hofman, K.J. and Rogo, K.O. (eds) *Disease and Mortality in Sub-Saharan Africa*. Washington, DC: World Bank, pp. 329–350.

Baker, J. (ed.) (1990) *Small Town Africa*. Uppsala: The Scandinavian Institute of African Studies.

Baker, J. and Pedersen, P.O. (eds) (1992) *The Rural Urban Interface in Africa*. Uppsala: The Scandinavian Institute of African Studies.

Barratt Brown, M. (1995) *Africa's Choices: After Thirty Years of the World Bank*. Boulder, CO: Westview Press.

Barrett, C.B., Carter, M.R. and Little, P.D. (2008) *Understanding and Reducing Persistent Poverty in Africa*. Abingdon: Routledge.

Barrett, H.R. Ilbery, B., Browne, A.W. and Binns, T. (1999) Globalisation and the changing networks of food supply: the importation of fresh horticultural produce from Kenya into the UK. *Transactions of the Institute of British Geographers* 24(2): 159–174.

Barwa, S.D. (1995) *Structural Adjustment Programmes and the Urban Informal Sector in Ghana, Issues in Development*. Geneva: Development and Technical Cooperation Department, International Labour Office.

Batterbury, S. (2005) Development, planning, and agricultural knowledge on the Central Plateau of Burkina Faso. In Cline-Cole, R. and Robson, E. (eds) *West African Worlds*. Harlow: Pearson, pp. 259–279.

Batterbury, S. and Warren, A. (2001) The African Sahel 25 years after the great drought: assessing progress and moving towards new agendas and approaches. *Global Environmental Change* 11(1): 1–8.

BBC (British Broadcasting Corporation) (2006) online: *China to Build Nigerian Railway*. http://news.bbc.co.uk/2/hi/africa/6101736.stm (accessed 24 November 2009).

BBC (British Broadcasting Corporation) (2007) online: *Ghana hopes for EU Timber Deal*. http://news.bbc.co.uk/2/hi/business/6983895.stm (accessed 2 October 2009).

BBC (British Broadcasting Corporation) (2008a) online: *French Isles Top EU Jobless Table*. http://news.bbc.co.uk/2/hi/europe/7892773.stm (accessed 17 February 2009).

BBC (British Broadcasting Corporation) (2008b) online: *Africa Free Trade Zone is Agreed*. http://news.bbc.co.uk/2/hi/7684903.stm (accessed 27 November 2009).

BBC (British Broadcasting Corporation) (2008c) online: *Around Kenya: After the Violence*. http://newsvote.bbc.co.uk/mpapps/pagetools/print/news.bbc.co.uk/ (accessed 2 January 2009).

BBC (British Broadcasting Corporation) (2008d) online: *Outcry over China–Niger Oil Deal*. http://news.bbc.co.uk/2/hi/7534315.stm (accessed 7 August 2011).

BBC (British Broadcasting Corporation) (2009) online: *Gaddafi Vows to Push African Unity*. http://newsvote.bbc.co.uk (accessed 4 July 2009).

BBC (British Broadcasting Corporation) (2010) online: *Angolan Railway's Chinese Makeover*. http://news.bbc.co.uk/2/hi/programmes/from_our_own_correspondent/9023642.stm (accessed 7 August 2011).

BBC (British Broadcasting Corporation) (2010b) online: *Benin: Democracy, the Chinese way – or Voodoo?* http://www.bbc.co.uk/news/business-10520699 (accessed 23 July 2010).

Beall, J., Crankshaw, O. and Parnell, S. (2000a) The causes of unemployment in post-apartheid Johannesburg and the livelihood strategies of the poor. *Tijdschrift voor Economische en Sociale Geografie* 91(4): 379–396.

Beall, J., Crankshaw, O. and Parnell, S. (2000b) Local government, poverty reduction and inequality in Johannesburg. *Environment and Urbanization* 12(1): 107–122.

Beauchemin, C. and Bocquier, P. (2004) Migration and urbanisation in francophone West Africa: a review of the recent empirical evidence. *Urban Studies* 41(11): 2245–2272.

Bell, M. (1987) *Contemporary Africa: Development, Culture and the State*. Harlow: Longman.

Best, A.C.C. and de Blij, H.J. (1977) *African Survey*. New York: John Wiley.

Bethony, J., Brooker, S., Albonico, M., Geiger, S.M., Loukas, A., Diemert, D. and Hotez, P.J. (2006) Soil-transmitted helminth infections: ascariasis, trichuriasis, and hookworm. *The Lancet* 367(9521): 1521–1532.

Bigsten, A. and Söderbom, M. (2006) What have we learned from a decade of manufacturing enterprise surveys in Africa? *World Bank Research Observer* 21(2): 241–265.

Bigsten, A. and Shimeles, A. (2004) *Prospects of 'Pro-poor' Growth in Africa*. Research Paper No. 2004/42. Helsinki: United Nations University.

Binns, T. (1982a) Agricultural change in Sierra Leone. *Geography* 67(2): 113–125.

Binns, T. (1982b) *The Changing Impact of Diamond Mining in Sierra Leone*. University of Sussex Research Papers in Geography 9. Brighton: University of Sussex.

Binns, T. (1984) People of the six seasons. *Geographical Magazine* December: 640–644.

Binns, T. (ed) (1988) Geographical perspectives on the crisis in Africa. *Geography* 73(1): 47–73.

Binns, T. (1990) Is desertification a myth? *Geography* 75(2): 106–113.

Binns, T. (1991) The Gambia: tourism versus rural development? *Geography Review* 5(2): 26–29.

Binns, T. (1992) Traditional agriculture, pastoralism and fishing. In Gleave, M.B. (ed.) *Tropical African Development*. London: Longman, 153–191.

Binns, T. (1994) *Tropical Africa*. London: Routledge.

Binns, T. (ed.) (1995) *People and Environment in Africa*. Chichester: John Wiley.

Binns, T. (1998) Geography and development in the 'new' South Africa. *Geography* 83(1): 3–14.

Binns, T. (2008) Dualistic and unilinear concepts of development. In Desai, V. and Potter, R.B. (eds) *The Companion to Development Studies* (2nd edn). London: Hodder Education.

Binns, T. and Funnell, D.C. (1983) Geography and integrated rural development. *Geografiska Annaler* 65B(1): 57–63.

Binns, T., Illgner, P.M. and Nel, E.L. (2001) Water shortage, deforestation and development: South Africa's Working for Water Programme. *Land Degradation and Development* 12(4): 341–355.

Binns, T. and Lynch, K. (1998) Feeding Africa's growing cities into the 21st century: the potential of urban agriculture. *Journal of International Development* 10(7): 777–793.

Binns, T. and Maconachie, R. (2005) Going home in post-conflict Sierra Leone: diamonds, agriculture and re-building rural livelihoods in the Eastern Province. *Geography* 90(1): 67–78.

Binns, T. and Maconachie, R. (2006) Re-evaluating people–environment relationships at the rural–urban interface: How sustainable is the peri-urban zone in Kano, northern Nigeria? In McGregor, D., Simon, D. and Thompson, D. (eds) *The Peri-urban Interface: Approaches to Sustainable Natural and Human Resource Use.* London: Earthscan, pp. 211–228.

Binns, T., Maconachie, R.A. and Tanko, A.I. (2003) Water, land and health in urban and peri-urban food production: the case of Kano, Nigeria. *Land Degradation and Development* 14(5): 431–444.

Binns, T. and Mortimore, M.J. (1989) Ecology, time and development in Kano State, Nigeria. In Swindell, K., Baba, J.M. and Mortimore, M.J. (eds) *Inequality and Development: Case Studies from the Third World.* London: Macmillan, pp. 359–380

Binns, T. and Nel, E. (2001) Gold loses its shine: decline and response in the South African goldfields. *Geography* 86(3): 255–260.

Binns, T. and Nel, E. (2002) Devolving development: integrated development planning and developmental local government in post-apartheid South Africa. *Regional Studies* 36(8): 921–932.

Binns, T., Porter, G., Nel, E. and Kyei, P. (2005) Decentralising poverty? Reflections on the experience of decentralisation and the capacity to achieve local development in Ghana and South Africa. *Africa Insight* 35(4): 21–31.

Binns, T. and Robinson, R. (2002) Sustaining democracy in the 'new' South Africa. *Geography* 87(1): 25–37.

Birchall, J.P. (n.d) online: *Of Grandiose Schemes and Catastrophes – The Groundnut Affair.* http://www.themeister.co.uk/economics/groundnut_scheme.htm (accessed 26 October 2010).

BirdLife International (2008) online: *Working Together for Birds and People.* http://www.birdlife.org/index.html (accessed 29 November 2008).

Black, R. (1998) *Refugees, Environment and Development.* Harlow: Longman.

Blair, D. (2006) online: *Rioters Attack Chinese after Zambian Poll.* http://www.telegraph.co.uk/news/worldnews/1530464/ (accessed 23 July 2010).

Blakely, E.J. (1989) *Planning Local Economic Development.* Newbury Park: Sage.

Bloom, G., Lucas, H., Edun, A., Lenneiye, M. and Milimo, J. (2000) *Health and Poverty in Sub-Saharan Africa.* IDS Working Paper No. 103. Brighton: Institute of Development Studies.

Blowfield, M. and Frynas, J. (2005) Editorial: setting new agendas – critical perspectives on corporate social responsibility in the developing world. *International Affairs* 81(3): 499–513.

Boadi, K., Kuitunen, M., Raheem, K. and Hanninen, K. (2005) Urbanisation without development: environmental and health implications in African cities. *Environment, Development and Sustainability* 7(4): 465–500.

Boaduo, N. (2008) Africa's political, industrial and economic development dilemma in the contemporary era of the African Union. *Journal of Pan African Studies* 2(4): 93–102.

Boko, M., Niang, I., Nyong, A., Vogel, C., Githeko, A. and Medany, M. (2007) Africa. In Parry, M.L., Canziani, O.F., Palutikof, J.P., Linden, P.J. and van der Hanson, C.E. (eds) *Climate Change 2007: Impacts, Adaptation, Vulnerability: Contribution of Working Group II to the Fourth Assessment Report of the Intergovernmental Panel on Climate Change.* Cambridge: Cambridge University Press, pp. 333–367.

Bond, P. (2007) *Looting Africa: The Economics of Exploitation.* London: Zed Books.

Boschi-Pinto, C., Lanata, C.F., Mendoza, W. and Habte, D. (2006) Diarrheal diseases. In Jamison, D.T., Feachem, R.G., Makgoba, M.W., Bos, E.R., Baingana, F.K., Hofman, K.J. and Rogo, K.O. (eds) *Disease and Mortality in Sub-Saharan Africa.* Washington, DC: World Bank, pp. 107–123.

Boserup, E. (1965) *The Conditions of Agricultural Growth.* London: Allen and Unwin (republished London: Earthscan, 1993).

Botha, R.F. (1990) South Africa and Africa. *Africa Insight* 20(4): 215–218.

Bowman, B., Seedat, M., Duncan, N. and Kobusingye, O. (2006) Violence and injuries. In Jamison, D.T., Feachem, R.G., Makgoba, M.W., Bos, E.R., Baingana, F.K., Hofman, K.J. and Rogo, K.O. (eds) *Disease and Mortality in Sub-Saharan Africa.* Washington, DC: World Bank, pp. 361–373.

BP (British Petroleum) (2010) *BP Statistical Review of World Energy.* London: British Petroleum.

Bray, J. (1969) The economics of traditional cloth production in Iseyin, Nigeria. *Economic Development and Cultural Change* 17(4): 540–551.

Bremner, L. (2000) Re-inventing the Johannesburg inner city. *Cities* 17(3): 184–193.

Brockington, D. (1999) Conservation, displacement, and livelihoods: the consequences of eviction for pastoralists moved from the Mkomazi Game Reserve, Tanzania. *Nomadic Peoples* 3(2): 74–96.

Bromley, R. and Gerry, C. (eds) (1979) *Casual Work and Poverty in Third World Cities*. New York: Wiley.

Brooker, S., Hotez, P.J. and Bundy, D.A.P. (2008) Hookworm-related anaemia among pregnant women: a systematic review. *Neglected Tropical Diseases* 2(9): 1–9.

Brooker, S., Peshu, N., Warn, P.A., Mosobo, M., Guyatt, H.L., Marsh, K. and Snow, R.W. (1999) The epidemiology of hookworm infection and its contribution to anaemia among pre-school children on the Kenyan coast. *Transactions of the Royal Society of Tropical Medicine and Hygiene* 93(3): 240–246.

Brown, N.E. (1999) *ECOWAS and the Liberia Experience: Peacekeeping and Self Preservation*. Washington, DC: United States Department of State.

Brown, O., Hammill, A. and McLeman, R. (2007) Climate change as the 'new' security threat: implications for Africa. *International Affairs* 83(6): 1141–1154.

Bryceson, D. (2000) *Rural Africa at the Crossroads: Livelihood Practices and Policies*. ODI Natural Resources Perspectives No. 52. London: Overseas Development Institute.

Brzoska, M. (2007) Appendix 2c: collective violence beyond the standard definition of armed conflict. In *SIPRI Yearbook 2007*. Oxford: Oxford University Press, pp. 94–106.

Bunwaree, S. (2008) NEPAD and is discontents. In Akokpari, J., Nidinga-Muvumba, A. and Murithi, T. (eds) *The African Union and its Institutions*. Cape Town: Fanele, pp. 227–240.

Burkey, S. (1993) *People First: A Guide to Self-Reliant, Participatory Rural Development*. London: Zed Books.

Buch, A. and Dixon, A.B. (2009) South Africa's Working for Water Programme: a win–win situation for environment and development? *Sustainable Development* 17(3): 129–141.

Butts, K.H. and Bankus, B. (2009) *China's Pursuit of Africa's Natural Resources*. Carlisle, PA: Center for Strategic Leadership, US Army War College.

Buvé, A., Bishikwabo-Nsarhaza, K. and Mutangadura, G. (2002) The spread and effect of HIV-1 infection in sub-Saharan Africa. *The Lancet* 359(9322): 2011–2017.

CAADP (Comprehensive Africa Agriculture Development Programme) (2009) online: *About CAADP*. http://www.nepad-caadp.net/about-caadp.php (accessed 16 November 2009).

Cambria Press (2008) online: *The African Union and New Strategies for Development in Africa*. http://www.cambriapress.com/cambriapress.cfm?/template=4&bid=271 (accessed 12 November 2009).

Carney, D. (1998) *Sustainable Rural Livelihoods: What Contribution Can We Make?* London: Department for International Development.

Carney, D. (2002) *Sustainable Livelihoods Approaches: Progress and Possibilities for Change*. London: Department for International Development.

Cashin, P.C., Liang, H. and McDermott, C.J. (1999) Do commodity price shocks last too long for stabilization schemes to work? *Finance and Development* 36(3): 40–43.

CBFF (Congo Basin Forest Fund) (2008) online: *A Global Response to a Global Issue*. http://www.cbf-fund.org/index.php (accessed 28 November 2008).

Chambers, R. (1983) *Rural Development: Putting the Last First*. Harlow: Longman.

Chambers, R. (1997) *Whose Reality Counts? Putting the First Last*. London: ITDG.

Chambers, R. and Conway, G. (1992) *Sustainable Rural Livelihoods: Practical Concepts for the 21st Century*. IDS Discussion Paper 296. Brighton: Institute of Development Studies.

Chen, L. and Hanvoravongchai, P. (2005) HIV/AIDS and human resources. *Bulletin of the World Health Organization* 83(4): 243–245.

Chimbwete, C., Watkins, S.C. and Msiyapazizulu, E. (2005) The evolution of population policies in Kenya and Malawi. *Population Research and Policy Review* 24(1): 85–106.

China Daily (2010) online: *Sierra Leone Welcomes Chinese Steel Mill's Investment*. http://www.chinadaily.com.cn/business/2010-07/17/content_10120051.htm (accessed 24 July 2010).

Christian Aid (2006) online: *The Climate of Poverty: Facts, Fears and Hope.* http://www.christian-aid.org.uk/indepth/605caweek/index.htm (accessed 15 February 2009).

Cissé, O., Gueye, N.F.D. and Sy, M. (2005) Institutional and legal aspects of urban agriculture in French-speaking West Africa: from marginalization to legitimization. *Environment and Urbanization* 17(1): 143–154.

Clancy, J.S. (2008) Urban ecological footprints in Africa. *African Journal of Ecology* 46(4): 463–470.

Clemens, M.A., Kenny, C.J. and Moss, T.J. (2007) The trouble with the MDGs: confronting expectations of aid and development success. *World Development* 35(5): 735–751.

Cline-Cole, R.A. (1989) Inequality and domestic energy in Kano, Nigeria and Freetown, Sierra Leone. In Swindell, K., Baba, J.M. and Mortimore, M.J. (eds) *Inequality and Development: Case Studies from the Third World.* London: Macmillan, pp. 243–268.

Cline-Cole, R.A. (1995) Livelihood, sustainable development and indigenous forestry in dryland Nigeria. In Binns, T. (ed.) *People and Environment in Africa.* Chichester: John Wiley, pp. 171–185.

Cline-Cole, R.A., Falola, J.A., Main, H.A.C., Mortimore, M.J., Nichol, J.E. and O'Reilly, F.D. (1990) *Wood Fuel in Kano.* Tokyo: United Nations University Press.

CMC (Cape Metropolitan Council) (1999) online: *Going Global, Working Local: Cato Manor Development Project.* http://www.cmda.org.za/ (accessed 20 December 2008).

CMDA (Cato Manor Development Association) (2008) online: *Cato Manor Development Project, Durban.* http://www.cmda.org.za/ (accessed 20 December 2008).

Coghlan, B., Ngoy, P., Mulumba, F., Hardy, C., Bemo, N.K., Stewart, T., Lewis, J. and Brennan, R. (2008) *Mortality in the Democratic Republic of Congo: An Ongoing Crisis.* New York: International Rescue Committee.

Cohen, S.A. (2003) Beyond slogans: lessons from Uganda's experience with ABC and HIV/AIDS. *Guttmacher Report on Public Policy* 6(5): 1–3.

Cointreau, S. (2006) *Occupational and Environmental Health Issues of Solid Waste Management Special Emphasis on Middle- and Lower-Income Countries.* Urban Papers No. 2. Washington, DC: World Bank.

Collier, P. (1988) Oil shocks and food security in Nigeria. *International Labour Review* 127(6): 761–782.

Collier, P. (1995) The marginalisation of Africa. *International Labour Review* 134(4/5): 541–557.

Collier, P. (1998) *Living Down the Past: How Europe Can Help Africa Grow.* Studies in Trade and Development No. 2. London: Institute of Economic Affairs.

Collier, P. (2007) *The Bottom Billion: Why the Poorest Countries are Failing and What Can be Done about it*? Oxford: Oxford University Press.

Collier, P. and Hoeffler, A. (2001) *Greed and Grievance in Civil War.* Policy Research Working Paper No. 2355. Washington, DC: World Bank.

Collier, P. and Hoeffler, A. (2004) Greed and grievance in civil war. *Oxford Economic Papers* 56(4): 563–595.

Colwell, R.R. (1996) Global climate and infectious disease: the cholera paradigm. *Science* 274(5295): 2025–2031.

Corbett, E.L., Marston, B., Churchyard, G.J. and De Cock, K.M. (2006) Tuberculosis in sub-Saharan Africa: opportunities, challenges, and change in the era of antiretroviral treatment. *The Lancet* 367(9514): 926–937.

Coulson, A. (1982) *Tanzania: A Political Economy.* Oxford: Clarendon Press.

Counsell, S., Long, C. and Wilson, S. (eds) (2007) *The Environmental, Social and Economic Impacts of Industrial Logging Concessions in Africa's Rainforests.* London: Rainforests Foundation and Forests Monitor.

Cramer, C. (2003) Does inequality cause conflict? *Journal of International Development* 15(4): 397–412.

Critchley, W. (1991) *Looking after our Land: Soil and Water Conservation in Dryland Africa.* Oxford: Oxfam.

Cropley, E. (2009) online: *China Shoves US in Scramble for Africa.* http://www.mg.co.za/article/2009-08-06-china-shoves-us-in-scramble-for-africa (accessed 27 May 2010).

Cropley, E. (2010) online: *Rising Africa Puts SA on the Spot.* http://www.mg.co.za/article/2010-05-26-rising-africa-puts-sa-on-the-spot (accessed 27 May 2010).

CSO (Central Statistics Office, Zambia) (2003) online: *The Monthly.* http://www.zamstats.gov.zm (accessed 27 February 2009).

CSO (Central Statistics Office, Zambia) (2007) *Company Closures and Job Loses in Zambia.* Lusaka: CSO.

Daboah, F., Fatoma, S. and Kuch, M. (2010) Disarmament, demobilisation, rehabilitation and reintegration (DDRR): a case study of Liberia, Sierra Leone, and South Sudan. *New York Science Journal* 3(6): 6–19.

Davidson, B. (1994) *Modern Africa: A Social and Political History.* London: Longman.

Davies, R. (1996) South Africa's economic relations with Africa: current patterns and future perspectives. In Adedeji, A. (ed.) *South Africa and Africa: Within or Apart?* London: Zed Books, pp. 167–192.

DBSA (Development Bank of Southern Africa) (2000) *South Africa: Inter-provincial Comparative Report.* Johannesburg: Halfway House, DBSA.

de Janvry, A. and Sadoulet, E. (2010) online: *The Triple Crisis: What Development Prospects for Africa.* http://www.wider.unu.edu/publications/newsletter/articles-2010/e (accessed 21 October 2010).

de Waal, A. (1989) *Famine that Kills: Darfur, Sudan.* Oxford: Clarendon Press.

de Waal, A. (2005) Briefing: Darfur, Sudan: Prospects for Peace. *African Affairs* 104(414): 127–135

de Wulf, L. (2004) *TradeNet in Ghana: Best Practice of the Use of Information Technology.* Washington, DC: World Bank.

Dean, W.R.J., Hoffman, M.T., Meadows, M.E. and S.J. Milton (1995) Desertification in the semi-arid Karoo, South Africa: review and reassessment. *Journal of Arid Environments* 30(3): 247–264.

Deiemar, G. (2004) online: *Office Du Niger – Reforms, International Network on Participatory Irrigation Management.* http://www.inpim.org/leftlinks/FAQ/Newsletters/N12/n12a5.htm (accessed 15 August 2010).

Denis, E. (1997) Urban planning and growth in Cairo. *Middle East Report* 27(1): 7–12.

Desanker, P.V. and Magadza, C. (2001) Africa. In McCarthy, J.J., Canziani, O.F., Leary, N.A., Dokken, D.J. and White, K.S. (eds) *Climate Change 2001: Impacts, Adaptation and Vulnerability: Contribution of Working Group II to the Third Assessment Report of the Intergovernmental Panel on Climate Change.* Cambridge: Cambridge University Press, pp. 489–531.

Devarajan, S. (2010) Why the surge of renewed interest in Africa. Chief Economist of the Africa Region, World Bank, address to the New Zealand Institute of International Affairs Seminar, Wellington, 20 October.

Dey, J. (1981) Gambian women: unequal partners in rice development projects? *Journal of Development Studies* 17(3): 109–122.

DfID (Department for International Development) (1999) *Sustainable Livelihoods Guidance Sheets.* London: Department for International Development.

DfID (Department for International Development) (2005) online: *Girls' Education: Towards a Better Future for All.* http://www.dfid.gov.uk/pubs/education/girls-education (accessed 21 June 2009).

DfID (Department for International Development) (2007) *Girl's Education.* DfID Practice Paper. London: Department for International Development.

DfID (Department for International Development) (2008a) *DfID Annual Report 2008 – Development: Making it Happen.* London: Department for International Development.

DfID (Department for International Development) (2008b) online: *Press Release on the Launch of the Congo Basin Forest Fund.* http://www.dfid.gov.uk/news/files/pressreleases/congo-basin.asp (accessed 28 November 2008).

DfID (Department for International Development) (2008c) *Africa Conflict Prevention Programme Annual Report 2007/2008.* London: Department for International Development.

Diarra, S.T. (2009) *Mali: Rush for Land Along the Niger.* http://allafrica.com/stories/2010 04230001.html (accessed 15 August 2010).

Divine Chocolate (2011) online: *About Divine.* http://www.divinechocolate.com/about/story.aspx (accessed 9 March 2011).

Dixon, A.B. (2003) *Indigenous Management of Wetlands: Experiences in Ethiopia.* Aldershot: Ashgate.
Dixon, A.B. (2005) Wetland sustainability and the evolution of indigenous knowledge in Ethiopia. *Geographical Journal* 171(4): 306–323.
Dixon, A.B. (2008) The resilience of local wetland management institutions in Ethiopia. *Singapore Journal of Tropical Geography* 29(3): 341–357.
Dixon, C. (ed.) (1987) *Rural–Urban Interaction in the Third World.* London: Developing Areas Research Group.
Dixon, J., Gulliver, A. and Gibbon, D. (2001) *Farming Systems and Poverty: Improving Farmers' Livelihoods in a Changing World.* Rome and Washington, DC: FAO and World Bank.
Draper, P. (2010) Why the surge of renewed interest in Africa. Senior Research Fellow, South African Institute of International Affairs, address to the New Zealand Institute of International Affairs Seminar, Wellington, 20 October.
DSF (Digital Solidarity Fund) (2007) online: *Hewlett-Packard, MPA and the Global Digital Solidarity Fund (DSF) Join Forces to Improve e-Waste Management in Africa.* www.dsf-fsn.org/cms/documents/en/pdf/EWasteGB51.pdf (accessed 1 January 2009).
Durban Metro (2000) online: *Durban Economic Review.* http://www.durban.gov.za/durban/government/planning/about_us (accessed 8 August 2011).
Durban Unicity (2001) *Towards a Long Term, Development Framework for Durban Unicity.* Durban: Durban Unicity.
Durning, A.B (1990) *Apartheid's Environmental Toll.* Washington, DC: Worldwatch Institute.
DWAF (Department of Water Affairs and Forestry, Republic of South Africa) (2008) online: *The Working for Water Project.* http://www.dwaf.gov.za/wfw/ (accessed 1 December 2008).
Dye, C., Harries, A.D., Maher, D., Hosseini, S.M., Nkhoma, W. and Salaniponi, F.M. (2006) Tuberculosis. In Jamison, D.T., Feachem, R.G., Makgoba, M.W., Bos, E.R., Baingana, F.K., Hofman, K.J. and Rogo, K.O. (eds) *Disease and Mortality in Sub-Saharan Africa.* Washington, DC: World Bank, pp. 179–193.
Easterly, W. (2007) Was development assistance a mistake? *American Economic Review* 97(2): 328–332.
Easterly, W. (2009) How the Millennium Development Goals are unfair to Africa. *World Development* 37(1): 26–35.
Eaton, D. and Sarch, M. (1997) *The Economic Importance of Wild Resources in the Hadejia–Nguru Wetlands, Nigeria.* CREED Working Paper No. 12. London: IIED.
ECA (Economic Commission for Africa) (2005a) *Land Tenure Systems and their Impacts on Food Security and Sustainable Development in Africa.* Addis Ababa: Economic Commission for Africa.
ECA (Economic Commission for Africa) (2005b) *The Millennium Development Goals in Africa: Progress and Challenges.* Addis Ababa: Economic Commission for Africa.
ECA (Economic Commission for Africa) (2008) *Economic Report on Africa 2008.* Addis Ababa: Economic Commission for Africa.
Economic Development Unit (n.d.) online: *Durban Economic Development Unit (EDU).* http://www.durban.gov.za/durban/invest/economic-development (accessed 26 October 2010).
Economist, The (2009) The baby bonanza. *The Economist*, 392(8646): 21.
Editors Inc. (2006) *SA 2006–7, South Africa at a Glance.* Greenside: Editors Inc.
Egziabher, T.G. and Helmsing, A.H.J. (2005) *Local Economic Development in Africa.* Maastricht: Shaker.
Ekaya, W.N. (2005) The shift from mobile pastoralism to sedentary crop–livestock farming in the drylands of eastern Africa: some issues and challenges for research. *African Crop Science Conference Proceedings* 7: 1513–1519.
El Araby, M. (2002) Urban growth and environmental degradation: the case of Cairo, Egypt. *Cities* 19(6): 389–400.
El Jack, A. (2003) *Gender and Armed Conflict.* BRIDGE report. Brighton: Institute of Development Studies.
Elbadawi, I. and Sambanis, N. (2002) How much war will we see? Estimating the prevalence of civil war in 161 countries, 1960–1999. *Journal of Conflict Resolution* 46(2): 307–334.

Ellis, F. (1999) *Rural Livelihood Diversity in Developing Countries: Evidence and Policy Implications*. ODI Natural Resources Perspectives No. 40. London: ODI.

Ellis, F. (2000) *Rural Livelihoods and Diversity in Developing Countries*. Oxford: Oxford University Press.

Ellis, F. (2001) online: *Rural Livelihoods, Diversity and Poverty Reduction Policies: Uganda, Tanzania, Malawi and Kenya*. http://www.uea.ac.uk/polopoly_fs/1.1004!wp1.pdf (accessed 9 February 2011).

Ellis, F. (2004) *Occupational Diversification in Developing Countries and Implications for Agricultural Policy*. http://www.uea.ac.uk/polopoly_fs/1.53422!2005%20oc cupational%20div.pdf (accessed 9 February 2011).

Ellis, F. and Sumberg, J. (1998) Food production, urban areas and policy responses. *World Development* 26(2): 213–225.

EMVI (2008) online: *European Malaria Vaccine Initiative*. http://www.emvi.org/ (accessed 6 December 2008).

Enterprise Surveys (2008) online: *Economic Research, Statistics and Reference Data*. http://www.enterprisesurveys.org/ (accessed 27 December 2008).

European Parliament (2006) online: *Report on Fair-Trade and Development (2005/2245(INI))*. http://www.europarl.europa.eu/sides/getDoc.do?pubRef=-//EP//NONSGML+REPORT+A6-2006-0207+0+DOC+PDF+V0//EN (accessed 9 March 2011).

Ewald, J., Nilsson, A. and Narman, A. (2004) *A Strategic Conflict Analysis for the Great Lakes Region*. Stockholm: SIDA.

Ezealor, A.U. and Giles, R.H. (1997) Vertebrate pests of a Sahelian wetland agro-ecosystem: perceptions and attitudes of the indigenes and potential management strategies. *International Journal of Pest Management* 43(2): 97–104.

Facheux, C., Franzel, S. and Tabuna, H. (2007) online: *Tree Crop Development Potentials in Africa – Towards a More Enabling Environment*. http://www.worldagroforestry.org/downloads/publications/PDFs/pp07266.doc (accessed 28 February 2011).

Fage, J.D. (2002) *A History of Africa* (4th edn). London: Routledge.

Fahmi, W. and Sutton, K. (2008) Greater Cairo's housing crisis: contested spaces from inner city areas to new communities. *Cities* 25(5): 277–297.

Fairhead, J. and Leach, M. (1995) Local agro-ecological management and forest–savanna transitions: the case of Kissidougou, Guinea. In Binns, T. (ed.) *People and Environment in Africa*. Chichester: John Wiley, pp.163–170.

Fairhead, J. and Leach, M. (1996) Rethinking the forest–savanna mosaic. In Leach, M. and Mearns, R. (eds) *The Lie of the Land: Challenging Received Wisdom on the African Environment*. Oxford: James Currey, pp. 105–121.

Fairtrade Foundation (2011) online: *Kuapa Kokoo Union*. http://www.fairtrade.org.uk/producers/cacao/kuapa_kokoo_union.aspx (accessed 9 March 2011).

FAO (Food and Agriculture Organization) (1995) *Women, Agriculture and Rural Development: A Synthesis Report of the Africa Region*. Rome: FAO.

FAO (Food and Agriculture Organization) (1996) *World Food Summit Plan of Action*. Rome: FAO.

FAO (Food and Agriculture Organization) (1999a) *Prevention and Disposal of Obsolete and Banned Pesticide Stocks in Africa and the Near East*. Rome: FAO.

FAO (Food and Agriculture Organization) (1999b) *Extensive Pastoral Livestock Systems: Issues and Options for the Future*. Rome: FAO.

FAO (Food and Agriculture Organization) (2004) *Small Scale Fisheries: Assessing their Contribution to Rural Livelihoods in Developing Countries*. FAO Fisheries Circular No. 1008. Rome: FAO.

FAO (Food and Agriculture Organization) (2006) *Food Security and Agricultural Development in Sub-Saharan Africa: Building a Case for More Public Support*. Policy Assistance Series No. 2. Rome FAO.

FAO (Food and Agriculture Organization) (2007) *State of the World's Forests 2007*. Rome: FAO.

FAO (Food and Agriculture Organization) (2008) online: *Africa: Loss of Mangroves Alarming*. http://www.waterconserve.org/shared/reader/welcome.aspx?linkid=92259&keybold (accessed 4 August 2008).

FAO (Food and Agriculture Organization) (2009a) *State of the World's Forests 2009*. Rome: FAO.

FAO (Food and Agriculture Organization) (2009b) online: *The State of Food Insecurity in the World: Economic Crises – Impacts and Lessons Learned*. http://www.fao.org/docrep/012/i0876e/i0876e00.htm (accessed 23 June 2010).

FAO (Food and Agricultural Organization) (2009c) *FAO Statistical Yearbook*. Rome: FAO.

FAO (Food and Agriculture Organization) (2011) online: *FAO Statistics Division*. http://faostat.fao.org/ (accessed 18 April 2011).

Fearon, J.D. and Laitin, D. (2003) Ethnicity, insurgency and civil war. *American Political Science* Review 97(1): 75–90.

Fellmann, J.D., Getis, A. and Getis, J. (2007) *Human Geography*. Boston, MA: McGraw Hill.

Feyissa, T.H. and Aune, J.B. (2003) Khat expansion in the Ethiopian Highlands. *Mountain Research and Development* 23(2): 185–189.

FLO (Fair Trade Labelling Organization) (2010) *Growing Stronger Together: Annual Report 2009–2010*. Bonn: FLO.

Fobil, J.M. and Attuquayefio, D.K. (2003) online: *Remediation of the Environmental Impacts of the Akosombo and Kpong Dams in Ghana, Horizon Solutions*. http://www.solutions-site.org/artman/publish/article_53.shtml (accessed 26 October 2010).

Foreign Policy (2006) online: *China Decides Election in Zambia*. http://blog.foreignpolicy.com/posts/2006/09/06 (accessed 23 July 2010).

Foresight (2011) *The Future of Food and Farming: Challenges and Choices for Global Sustainability*. London: Government Office for Science.

Francis, D. (ed.) (2008) *Peace and Conflict in Africa*. London: Zed Books.

Franke, R.W. and Chasin, B. (1980) *Seeds of Famine: Ecological Destruction and the Development Dilemma in the West African Sahel*. Montclair, NJ: Allanheld, Osmun.

Franzel, S., Phiri, D. and Kwesiga, F. (2002) Assessing the adoption potential of improved fallows in eastern Zambia. In Franzel, S. and Scherr, S.J. (eds) *Trees on the Farm: Assessing the Adoption Potential of Agroforestry Practices in Africa*. Wallingford: CABI, pp. 37–64.

FT.com (2009) online: *Nigeria Dispute Fuels Petrol Shortages*. http://www.ft.com/cms/s/0/0ead29f8-3e40-11de-9a6c-00144feabdc0.html#axzz1TUJIub5A (accessed 17 June 2009).

FVCT (Flower Valley Conservation Trust) (2009) online: *Flower Valley Conservation Trust*. http://www.flowervalley.org.za/cgi-bin/giga.cgi?c=1866 (accessed 8 November 2009).

Gambia, The. (2009) online: *Visit the Gambia*. http://visitthegambia.gm/ (accessed 10 February 2009).

Gandy, M. (2006) Planning, anti-planning and the infrastructure crisis facing metropolitan Lagos. *Urban Studies* 43(2): 371–396.

Garten, J. (1996) The big emerging markets. *Columbia Journal of World Business* Summer: 6–31.

Gebrewold, B. (2008) Democracy and democratisation in Africa. In Francis, D.J. (ed.) *Peace and Conflict in Africa*. London: Zed Books, pp. 148–170.

Geheb, K. and Binns, T. (1997) 'Fishing farmers' or 'farming fishermen'? The quest for nutritional security on the Kenyan shores of Lake Victoria. *African Affairs* 96(382): 73–93.

Geheb, K., Kalloch, S., Medard, M., Nyapendi, A., Lwenya, C. and Kyangwa, M. (2008) Nile perch and the hungry of Lake Victoria: gender, status and food in an East African fishery. *Food Policy* 33(1): 85–98.

Geist, H. (2005) *The Causes and Progression of Desertification*. Aldershot: Ashgate.

Ghana Web (n.d.) online: *History of Aksombo Dam*. http://www.ghanaweb.com/GhanaHomePage/history/aksombo_dam.php (accessed 26 October 2010).

Giarelli, E. and Jacobs, L. (2003) Traditional healing and HIV-AIDS in KwaZulu-Natal, South Africa: to curb the epidemic, South African nurses, physicians, and traditional healers are learning to collaborate. *American Journal of Nursing* 103(10): 36–46.

Gilbert, R., Stevenson, D., Girardet, H. and Stren, R. (1996) *Making Cities Work: The Role of Local Authorities in the Urban Environment*. London: Earthscan.

Ginty, R.M. and Williams, A. (2009) *Conflict and Development*. Abingdon: Routledge.

Gislesen, K. (2006) *A Childhood Lost? The Challenges of Successful Disarmament, Demobilisation and Reintegration of Child Soldiers: The Case of West Africa*. Paper No. 712. Oslo: Norwegian Institute of International Affairs.

Gleditsch, N.P., Wallensteen, P., Eriksson, M., Sollenberg, M. and Strand, H. (2002) Armed conflict 1946–2001: a new dataset. *Journal of Peace Research* 39(5): 615–637.

Gleditsch, K.S. (2007) Transnational dimensions of civil war. *Journal of Peace Research* 44(3): 293–309.

GOK (Government of Kenya) (2007a) *Update on Tourism Statistics*. Nairobi: Ministry of Tourism.

GOK (Government of Kenya) (2007b) *Kenya Vision 2030*. Nairobi: National Economic and Social Council of Kenya.

GOK (Government of Kenya) (2008) *Key Facts and Figures 2008*. Nairobi: Kenya National Bureau of Statistics.

Goma Epidemiology Group (1995) Public health impact of Rwandan refugee crisis: what happened in Goma, Zaire, in July, 1994? *The Lancet* 345(8946): 339–344.

Gooneratne, W. and Mbilinyi, M. (eds) (1992) *Reviving Local Self-Reliance in Africa*. Nagoya: United Nations Centre for Regional Development.

Gooneratne, W. and Obudho, R.A. (1997) *Contemporary Issues in Regional Development Policy*. Aldershot: Avebury.

Gould, W.T.S. (2009) HIV/AIDS in developing countries. In Kitchin, R. and Thrift, N. (eds) *International Encyclopaedia of Human Geography*. Oxford: Elsevier, 173–179.

Government of Uganda (2010) *UNGASS Country Progress Report January 2008–December 2009*. Kampala: Government of Uganda.

GPEI (Global Polio Eradication Initiative) (2009) online: *Monthly Situation Report*. http://www.polioeradication.org/content/general/PolioSitrepJune2009ENG.pdf (accessed 15 August 2009).

Grant, R. and Yankson, P. (2003) City profile: Accra. *Cities* 20(1): 65–74.

Green, R.H. (2001) Planning for post-conflict reconstruction. In Belshaw, D. and Livingstone, I. (eds) *Renewing Development in Sub-Saharan Africa*. London: Routledge, pp. 81–98.

Griffiths, I. (1995) *The African Inheritance*. London: Routledge.

Griffiths, I.L. (1994) *An Atlas of African Affairs* (2nd edn). London: Routledge.

Grown, C., Gupta, G.R. and Pande, R. (2005) Taking action to improve women's health through gender equality and women's empowerment. *The Lancet* 365(9458): 541–543.

GRZ (Government of the Republic of Zambia) (1995) *The Laws of Zambia*. Lusaka: GRZ.

GSS (Ghana Statistical Services) (2000) *Ghana Living Standards Survey: Report of the Fourth Round*. Accra: Ghana Statistical Services.

Gu, J. (2009) *Where Western Business Sees 'Risk', Chinese Entrepreneurs See Opportunity. Brighton:* Institute of Development Studies.

Guardian (2009) online: *China 'Wants to Set up Factories in Africa'*. http://www.guardian.co.uk/world/2009/dec/04/china-manufacturing-factories-africa/print (accessed 24 July 2010).

Guardian (2010a) online: *Africa's Untold Story is of a Booming Continent and a Growing Middle Class*. http://www.guradian.co.uk/business/2010/jul/11/africa-recovery-glo (accessed 8 August 2010).

Guardian (2010b) Tullow Oil given licence to flare Ugandan gas. *Guardian* 16 February.

Guardian (2010c) Iceland volcano: Kenya's farmers losing $1.3m a day in flights chaos. *Guardian* 18 April.

Guardian Weekly (2010): Is Africa under-populated? *Guardian Weekly* 2 April: 45.

Gurjar, B.R., Butler, T.M., Lawrence, M.G. and Lelieveld, J. (2008) Evaluation of emissions and air quality in megacities. *Atmospheric Environment* 42(7): 1593–1606.

Gyau-Boakye, P. (2001) Environmental impacts of the Akosombo Dam and effects of climate change on the lake levels. *Environment, Development and Sustainability* 3(1): 17–19.

HABITAT (UN Centre for Human Settlements) (1996) *An Urbanizing World*. Oxford: Oxford University Press.

Hampwaye, G. (2008) Local Economic Development in the City of Lusaka, Zambia. *Urban Forum* 19(2): 187–204.

Hampwaye, G., Nel, E. and Rogerson, C.M. (2007) Urban agriculture as local initiative in Lusaka, Zambia. *Environment and Planning C: Government and Policy* 25(4): 553–572.

Hanjra, M.A., Ferede, T. and Gutta, D.G. (2009) Reducing poverty in sub-Saharan Africa through investments in water and other priorities. *Agricultural Water Management* 96(7): 1062–1070.

Hardoy, J.E., Mitlin, D. and Satterthwaite, D. (1992) *Environmental Problems in Third World Cities*. London: Earthscan.

Hardoy, J.E. and Satterthwaite, D. (1989) *Squatter Citizen*. London: Earthscan.
Hargreaves, J.D. (1996) *Decolonization in Africa*. London: Longman.
Hart, K. (1973) Informal income opportunities and urban employment in Ghana. *Journal of Modern African Studies* 11(1): 61–89.
Haylamicheal, H.D. and Dalvie, M.A. (2009) Disposal of obsolete pesticides, the case of Ethiopia. *Environment International* 35(3): 667–673.
HDI (Human Development Index) (2009) online: *Data from United Nations Human Development Index*. http://hdr.undp.org/en/statistics/ (accessed 15 August 2009).
Hellum, A. and Derman, B. (2004) Land reform and human rights in contemporary Zimbabwe: balancing individual and social justice through an integrated human rights framework. *World Development* 32(10): 1785–1805.
Hendrix, C.S. and Glaser, S.M. (2007) Trends and triggers: climate, climate change and civil conflict in Sub-Saharan Africa. *Political Geography* 26(6): 695–715.
Hesse, C. and Cotula, L. (2006) *Climate Change and Pastoralists: Investing in People to Respond to Adversity*. London: International Institute for Environment and Development.
Hettne, B. (1995) *Development Theory and the Three Worlds: Towards an International Political Economy of Development*. Harlow: Longman.
Hill, R., Taylor, G. and Temin, J. (2008) *Would You Fight Again? Understanding Liberian Ex-combatant Reintegration*. Washington, DC: United Sates Institute of Peace.
Hillocks, R.J. (2002) Cassava in Africa. In Hillocks, R.J., Thresh, J.M. and Bellotti, A. (eds) *Cassava: Biology, Production and Utilization*. Wallingford: CABI, pp. 41–54.
Homewood, K. (1995) Development, demarcation and ecological outcomes in Maasailand. *Africa* 65(3): 331–350.
Hope, K.R. (2008) *Poverty, Livelihoods and Governance in Africa*. New York: Palgrave.
Hopkins, A.G. (1973) *An Economic History of West Africa*. London: Longman.
Houghton, P. (2004) Khat: a growing concern in the UK. *Pharmaceutical Journal* 272: 162.
HRW (Human Rights Watch) (2002) online: *Fast Track Land Reform in Zimbabwe*. http://www.unhcr.org/refworld/docid/3c8c82df4.html (accessed 27 February 2011).
HRW (Human Rights Watch) (2009) online: *Well Oiled: Oil and Human Rights in Equatorial Guinea*. http://www.hrw.org/sites/default/files/reports/bhr0709web_0.pdf (accessed 12 July 2009).
Huq, S. and Ayers, J. (2007) *Critical List: The 100 Nations Most Vulnerable to Climate Change*. London: International Institute for Environment and Development.
Hunt, P. and De Mesquita, J.B. (2008) online: *Reducing Maternal Mortality: The Contribution of the Right to the Highest Attainable Standard of Health*. http://www.unfpa.org/public/publications/pid/4968 (accessed 15 January 2009).
Huse, M.D. and Muyakwa, S.L. (2008) *China in Africa: Lending, Policy Space and Governance*. Oslo: Norwegian Council for Africa.
Hussein, K. (2002) *Livelihood Approaches Compared: A Multi-agency Review of Current Practice*. London: DfID.
Hussein, K. and Nelson, J. (1998) *Sustainable Livelihoods and Diversification*. IDS Working Paper No. 69. Brighton: Institute of Development Studies.
Hyden, G. (1980) *Beyond Ujamaa in Tanzania: Underdevelopment and an Uncaptured Poverty*. London: Heinemann.
IAC (Inter-Academy Council) (2004) *Realizing the Promise and Potential of African Agriculture*. Amsterdam: Inter-Academy Council.
Ibhawoh, B. and Dibua, J.I. (2003) Deconstructing ujamaa: the legacy of Julius Nyerere in the quest for social and economic development. *African Journal of Political Science* 8(1): 59–83.
IDRC (International Development Research Centre) (2007) *Climate Change Adaptation in Africa (CCAA): Programme Strategy Overview*. Ottawa: IDRC.
IFC (International Finance Corporation) (2007) *The Business of Health in Africa: Partnering with the Private Sector to Improve People's Lives*. Washington, DC: International Finance Corporation/World Bank Group.
IFPRI (International Food Policy Research Institute) (2008) *IFPRI Forum*. Washington, DC: IFPRI.
Iliffe, J. (1995) *Africans: The History of a Continent*. Cambridge: Cambridge University Press.

Illgner, P., Nel, E. and Robertson, M. (1998) Beekeeping and local self-reliance in rural southern Africa. *Geographical Review* 88(3): 349–362.

ILO (International Labour Organization) (1972) *Employment, Incomes and Equality: A Strategy for Increasing Productive Employment in Kenya*. Geneva: ILO.

ILO (International Labour Organization) (1995) *Structural Adjustment Programmes and the Urban Informal Sector in Ghana*. Geneva: ILO.

ILO (International Labour Organization) (2008) *Global Employment Trends: January 2008*. Geneva: ILO.

IMF (International Monetary Fund) (2003) *Ghana: Poverty Reduction Strategy Paper*. Washington, DC: IMF.

Independent (2009) online: *Changing River Course Alters Uganda–DR Congo Border*. http://www.independent.co.uk/environment/changing-river-course-alters-ugandadr-congo-border-1818532.html (accessed 16 November 2009).

Inselman, A.D. (2003) Environmental degradation and conflict in Karamoja, Uganda: the decline of a pastoral society. *International Journal of Global Environmental Issues* 3(2): 168–187.

International Housing Coalition (2007) *Housing Challenges and Opportunities in Sub-Saharan Africa*. Washington, DC: IHC.

IPCC (Intergovernmental Panel on Climate Change) (2007) *Climate Change 2007: Impacts, Adaptation and Vulnerability: Contribution of Working Group II to the Fourth Assessment Report of the Intergovernmental Panel on Climate Change*. Cambridge: Cambridge University Press.

IPPG (Improving Institutions for Pro-Poor Growth) (2009) online: *Institutional Architecture and Pro-Poor Growth in Mali: Research Programme Consortium for Improving Institutions for Pro-Poor Growth.* http://www.ippg.org.uk/iacollective.html (accessed 15 August 2010).

Iyiani, C., Shannon, P. and Binns, T. (2010) Talking past each other: towards HIV/AIDS prevention in Nigeria. *International Social Work* 54(2): 258–271.

Iyiani, C., Shannon, P. and Binns, T. (in press) HIV/AIDS prevention: building on community strengths in Ajegunle, Lagos. *Development in Practice* 21(6).

Jagger, P. and Pender, J. (2003) The role of trees for sustainable management of less-favoured lands: the case of eucalyptus in Ethiopia. *Forest Policy and Economics* 5(1): 83–95.

Jamal, V. and Weeks, J. (1994) *Africa Misunderstood: Or Whatever Happened to the Rural–Urban Gap?* Basingstoke: Macmillan.

Janneh, A. (2006) *Development in Africa and ECA*. New York: United Nations.

Janowitz, B., Chege, J., Thompson, A., Rutenberg, N. and Homan, R. (2000) Community-based distribution in Tanzania: costs and impacts of alternative strategies to improve worker performance. *International Family Planning Perspectives* 26(4): 158–195.

Jenkinsa, P. and Wilkinson, P. (2002) Assessing the growing impact of the global economy on urban development in southern African cities: case studies in Maputo and Cape Town. *Cities* 19(1): 33–47.

Johnston, H.H. (1899) *A history of the colonisation of Africa by alien races*. Cambridge: Cambridge University Press.

Joireman, S.F. (2008) The mystery of capital formation in Sub-Saharan Africa: women, property rights and customary law. *World Development* 36(7): 1233–1246.

Jones, B. (1938) Desiccation and the West African colonies. *Geographical Journal* 91(5): 401–423.

Kahama, G.C., Maliyamkona, T.L. and Wells, S. (1986) *The Challenge of Tanzania's Economy.* London: James Currey.

Kalaba, F.K., Chirw, P.W. and Prozesky, H. (2009) The contribution of indigenous fruit trees in sustaining rural livelihoods and conservation of natural resources. *Journal of Horticulture and Forestry* 1(1): 1–6.

Kampala (City of Kampala, Uganda) (2008) online: *The Kampala Urban Sanitation Project (KUSP).* http://www.kcc.go.ug/city_council_of_kampala_projects_completed_the_kampala_urban_sanitation_project.asp (accessed 31 December 2008).

Kanuma, S. (2003) Local justice. *Developments Magazine* 24: 14–15.

Kar, K. and Chambers, R. (2008) *Handbook on Community-Led Total Sanitation*. London: Plan International.

Kasekende, L., Brixova, Z. and Ndikumana, L. (2010) Africa: Africa's counter-cyclical policy response to the crisis. *Journal of Globalization and Development* 1(1): 1–20.

Kennedy, E. and Bouis, H.E. (1993) *Linkages between Agriculture and Nutrition: Implications for Policy and Research*. Washington, DC: International Food Policy Research Institute.

Kenya Population Council (2009) online: *Overview*. http://www.popcouncil.org/countries/kenya.asp (accessed 16 August 2010).

Kevane, M. and Gray, L. (2008) Darfur: rainfall and conflict. *Environmental Research Letters* 3(3): 1–10.

Khan, M.R., Patnaik, P., Brown, L., Nagot, N. and Salouka, S. (2007). Mobility and HIV-related sexual behaviour in Burkina Faso. *AIDS and Behaviour* 12(2): 202–212.

Kingma, K. (1997) Demobilization of combatants after civil wars in Africa and their reintegration into civilian life. *Policy Sciences* 30(3): 151–165.

Kirigia, J.M. and Barry, S.P. (2008) Health challenges in Africa and the way forward. *International Archives of Medicine* 1(27): 1–3.

Kjekshus, H. (1977) The Tanzanian villagization policy: implementational lessons and ecological dimensions. *Journal of African Studies* 11(2): 269–282.

Kofoworola, O.F. (2007) Recovery and recycling practices in municipal solid waste management in Lagos, Nigeria. *Waste Management* 27(9): 1139–1143.

Konadu-Agyemang, K. (2001) A survey of housing conditions and characteristics in Accra, an African city. *Habitat International* 25(1): 15–34.

Konadu-Agyemang, K. and Panford, K. (2006) *Africa's Development in the Twenty-first Century*. Aldershot: Ashgate.

Konteh, F.H. (2009) Urban sanitation and health in the developing world: reminiscing the nineteenth century industrial nations. *Health and Place* 15(1): 69–78.

Kornegay, F.A. (2008) Africa's strategic diplomatic engagement with China. In Edinger, H., Herman, H. and Jansson, J. (eds) *New Impulses from the South: China's Engagement of Africa*. Stellenbosch: Centre for Chinese Studies, University of Stellenbosch, pp. 3–12.

Kpodo, K. (2008) online: *Update 3: Ghana Assembly Clears VALCO Sale*. http://www.reuters.com/assets/print?aid=USL765668520081107 (accessed 11 April 2010).

Kristjanson, P., Radeny, M., Baltenweck, I., Ogutu, J. and Notenbaert, A. (2005) Livelihood mapping and poverty correlates at a meso-level in Kenya. *Food Policy* 30(5/6): 568–583.

Kyambalesa, H. and Hougnikpo, M.C. (2006) *Economic Integration and Development in Africa*. Aldershot: Ashgate.

Lacina, B. and Gleditsch, N.P. (2005) Monitoring trends in global combat: a new dataset of battle deaths. *European Journal of Population* 21(2/3): 145–166.

Landsberg, C. (2008) The birth and evolution of NEPAD. In Akokpari, J., Ndinga-Muvumba, A. and Murithi, T. (eds) *The African Union and its Institutions*. Cape Town: Fanele, pp. 207–226.

Lawrence, P. (ed.) (1986) *World Recession and the Food Crisis in Africa*. London: James Currey.

Leach, M. and Mearns, R. (eds) (1996) *The Lie of the Land: Challenging Received Wisdom on the African Environment*. Oxford: James Currey.

Lesotho Bureau of Statistics (2008) online: *Summary of the Labour Statistics Report 2001–2005*. http://www.bos.gov.ls/Demography/ (accessed 21 August 2008).

Lester, A., Nel, E. and Binns, T. (2000) *South Africa, Past, Present and Future*. Harlow: Pearson.

Lewis, L.A. and Berry, L. (1988) *African Environments and Resources*. Boston, MA: Unwin Hyman.

Leys, C. (2005) The rise and fall of development theory. In Edelman, M. and Haugerud, A. (eds) *The Anthropology of Development and Globalization: Classical Political Economy to Contemporary Neoliberalism*. Malden: Blackwell, pp. 109–125.

LHWP (Lesotho Highlands Water Project) (2008) online: *Overview of the LHWP*. http://www.lhwp.org.ls/overview (accessed 20 August 2008).

Longhurst, R. (1988) Cash crops and food security. *IDS Bulletin* 19(2): 28–36.

Longman, T. (2009) An assessment of Rwanda's *gacaca* courts. *Peace Review* 21: 304–312.

Luckham, R., Ahmed, I., Muggah, R. and White, S. (2001) *Conflict and Poverty in Sub-Saharan Africa: An Assessment of the Issues and Evidence*. IDS Working Paper No. 128. Brighton: Institute of Development Studies.

Lun, J. (2006) *The African Great Lakes Region: An End to Conflict?* Research Paper No. 06/51. London: House of Commons Library.

Lusaka Times (2010) online: *China Rescues Bankrupt Zambian Copper Mines*. http://www.lusakatimes.com/?p=23037 (accessed 24 July 2010).

Lynch, K. (2005) *Rural–Urban Interaction in the Developing World*. London: Routledge.

Lynch, K. (2008) Rural–urban interaction. In Desai, V. and Potter, R.B. (eds) *The Companion to Development Studies* (2nd edn). London: Hodder, pp. 268–272.

Lynch, K., Binns, T. and Olofin, E. (2001) Urban agriculture under threat: the land security question in Kano, Nigeria. *Cities* 18(3): 159–171.

Lyon, F. (2000) Trust, networks and norms: the creation of social capital in agricultural economies of Ghana. *World Development* 28(4): 663–681.

Maathai, W. (2009) *The Challenge for Africa*. New York: Pantheon.

Maconachie, R. (2007) *Urban Growth and Land Degradation in Developing Cities*. Aldershot: Ashgate.

Maconachie, R. and Binns, T. (2006) Sustainability under threat? The dynamics of environmental change and food production in peri-urban Kano, northern Nigeria. *Land Degradation and Development* 17(2): 159–171.

Maconachie, R. and Binns, T. (2007a) Beyond greed and grievance in Sierra Leone: can diamonds play a role in post-conflict reconstruction? *Zeitschrift für Wirtschaftsgeographie* 51: 163–175.

Maconachie, R. and Binns, T. (2007b) Beyond the resource curse? Diamond mining, development and post-conflict reconstruction in Sierra Leone. *Resources Policy* 32(3): 104–115.

Maconachie, R. and Binns, T. (2007c) 'Farming miners' or 'mining farmers'? Diamond mining and rural development in post-conflict Sierra Leone. *Journal of Rural Studies* 23(3): 367–380.

Magnarella, P.J. (2005) The background and causes of the genocide in Rwanda. *Journal of International Criminal Justice* 3(4): 801–822.

Maharaj, B., and Ramballi, K. (1998) Local economic development strategies in an emerging democracy: the case of Durban in South Africa. *Urban Studies* 35(1): 131–148.

Maier, K. (2001) *This House Has Fallen: Nigeria in Crisis*. Harmondsworth: Penguin.

Mail and Guardian (2009) online: *SA Life Expectancy Decreases*. http://www.mg.co.za/printformat/single/2009-11-19-sa-life-expectancy (accessed 1 December 2009).

Make Poverty History (2010) online: *Make Poverty History*. http://www.makepovertyhistory.org/takeactio/index.shtml (accessed 24 October 2010).

Makinda, S.M. and Okumu, F.W. (2008) *The African Union: Challenges of Globalization, Security and Governance*. Abingdon: Routledge.

Makoni, N. and Mohamed-Katerere, J. (2006) Genetically modified crops. In UNEP, *Africa Environmental Outlook 2: Our Environment, Our Wealth*. Nairobi: UNEP, pp. 300–330.

Malan, J. (2008) Understanding transitional justice in Africa. In Francis, D.J. (ed.) *Peace and Conflict in Africa*. London: Zed Books, pp. 133–147.

Mapolu, H. (n.d.) online: *Tanzania: Imperialism, the State and the Peasantry*. http://www.unu.edu/unupress/unupbooks/uu28ae/uu28ae0h.htm (accessed 26 October 2010).

Mara, D.D. (2003) Water, sanitation and hygiene for the health of developing nations. *Public Health* 117(6): 452–456.

Marcus, R. (1993) *Gender and HIV/AIDS in Sub-Saharan Africa: The Cases of Uganda and Malawi*. BRIDGE Report No. 13. Brighton: Institute of Development Studies.

Marks, R. and Bezzoli, M. (2001) Palaces of desire: Century City, Cape Town and the ambiguities of development. *Urban Forum* 12(1): 27–47.

Mason, P.R. (2009) Zimbabwe experiences the worst epidemic of cholera in Africa. *Journal of Infection in Developing Countries* 3(2): 148–151.

Mathers, C.D., Sadana, R., Salomon, J.A., Murray, C.J.L. and Lopez, A.D. (2001) Healthy life expectancy in 191 countries, 1999. *The Lancet* 357(9269): 1685–1691.

Mathews, K. (2008) Renaissance of pan-Africanism: the AU and the new pan-Africanists. In Akokpari, J., Ndinga-Muvumba, A. and Murithi, T. (eds) *The African Union and its Institutions*. Cape Town: Fanele, pp. 25–39.

Maumbe, B.M. and Swinton, S.M. (2003) Hidden health costs of pesticide use in Zimbabwe's smallholder cotton growers. *Social Science and Medicine* 57(9): 1559–1571.

Mawdsley, E. (2007) China and Africa: emerging challenges to the geographies of power. *Geography Compass* 1(3): 405–421.

May, A. (2006) *Dealing with the Past: Experiences of Transitional Justice, Truth and Reconciliation Processes after Periods of Violent Conflict in Africa*. London: Conciliation Resources.

Mayers, J. (2007) *Trees, Poverty and Targets*. IIED Briefing Paper. London: IIED.

Mazzucato, V. and Niemeijer, D. (2002) Population growth and environment in Africa: local informal institutions, the missing link. *Economic Geography* 78(2): 171–193.

Mbendi (2009) online: *Africa: Oil and Gas*. http://www.mbendi.com/indy/oilg/af (accessed 15 April 2009).

Mbewu, A. and Mbanya, J.C. (2006) Cardiovascular disease. In Jamison, D.T., Feachem, R.G., Makgoba, M.W., Bos, E.R., Baingana, F.K., Hofman, K.J. and Rogo, K.O. (eds) *Disease and Mortality in Sub-Saharan Africa*. Washington, DC: World Bank, pp. 305–327.

McGranahan, G. and Satterthwaite, D. (2006) *Governance and Getting the Private Sector to Provide Better Water and Sanitation Services to the Urban Poor*. IIED Human Settlements Discussion Paper Series. London: IIED.

MDP (Municipal Development Partnership) (2006) *Strategic Plan for the Municipal Development Partnership for Eastern and Southern Africa, 2006–2016*. Harare: MDP.

Mearns, R. (1995) Institutions and natural resource management: access to and control over woodfuel in East Africa. In Binns, T. (ed.) *People and Environment in Africa*. Chichester: John Wiley, pp. 103–114.

Mendelsohn, J. and Dawson, T. (2007) Climate and cholera in KwaZulu-Natal, South Africa: the role of environmental factors and implications for epidemic preparedness. *International Journal of Hygeine and Environmental Health* 211(1/2): 156–162.

Mensah A. (2006) People and their waste in an emergency context: the case of Monrovia, Liberia. *Habitat International* 30(4): 754–768.

Mensah, G.A. (2008) Ischaemic heart disease in Africa. *Heart* 94(7): 836–843.

Meredith, M. (2005) *The State of Africa: A History of Fifty Years of Independence*. London: The Free Press.

Metamorphosis Nigeria (2008) online: *About Metamorphosis Nigeria Limited*. http://metamorphosis-nigeria.org/about.php (accessed 23 December 2008).

Michailof, S., Kostner, M. and Devictor, X. (2002) *Post-Conflict Recovery in Africa: An Agenda for the African Region*. Africa Region Working Paper Series No. 30. Washington, DC: World Bank.

Milligan, R.S. (2002) Searching for symbiosis: pastoralist–farmer relations in north-east Nigeria. Unpublished PhD thesis. Brighton: University of Sussex.

Milligan, R.S. and Binns, T. (2007) Crisis in policy, policy in crisis: understanding environmental discourse and resource-use conflict in northern Nigeria. *Geographical Journal* 173(2): 143–156.

Millington, A.C. (1984) Indigenous soil conservation studies in Sierra Leone. In International Association of Hydrological Sciences, *Challenges in African Hydrology and Water Resources* [proceedings of the Harare Symposium, July 1984]. Harare: IAHS.

Mistry, P.S. (2005) Reasons for sub-Saharan Africa's development deficit that the Commission for Africa did not consider. *African Affairs* 104(417): 665–678.

Mitchell, J. and Ashley, C. (2007) *Can Tourism Offer Pro-poor Pathways to Prosperity?* Briefing Paper No. 22. London: Overseas Development Institute.

Mitchell, J. and Faal, J. (2006) *The Gambian Tourist Value Chain and Prospects for Pro-poor Tourism*. London: Overseas Development Institute.

Mohamoud, A.A. (2007) *Shaping a New Africa*. Amsterdam: KIT.

Molden, D. (2007) *Water for Food, Water for Life: A Comprehensive Assessment of Water Management in Agriculture*. London: IWMI/Earthscan.

Moller, B. (2006) *Religion and Conflict in Africa with a Special Focus on East Africa*. DIIS Report No. 2006:6. Copenhagen: Danish Institute for International Studies.

Monk, C., Sandefur, J. and Teal, F. (2007) Skills and earnings in formal and informal urban employment in Ghana. Unpublished paper. Oxford: Centre for the Study of African Economies, Department of Economics, University of Oxford.

Morna, C.L. (1990) SADCC's first decade. *Africa Report*: 49–52.

Mortimore, M.J. (1989) *Adapting to Drought: Farmers, Famines and Desertification in West Africa*. Cambridge: Cambridge University Press.

Mortimore, M.J. (2003) Changing systems and changing landscapes: measuring and interpreting land use transformation in African drylands. In Mertz, O., Wadley, R. and Christensen, A.E. (eds) *Local Land Use Strategies in a Globalizing World: Shaping Sustainable Social and Natural Environments*. Copenhagen: Institute of Geography, University of Copenhagen, pp. 209–241.

Mortimore, M.J. and Adams, W.M. (1999) *Working the Sahel: Environment and Society in Northern Nigeria*. London: Routledge.

Mortimore, M.J. and Tiffen, M. (1995) Population and environment in time perspective: the Machakos story. In Binns, T. (ed.) *People and Environment in Africa*. Chichester: John Wiley, pp. 69–90.

Moser, C. and Holland, J. (1997) *Household Responses to Poverty and Vulnerability*, Volume 4: *Confronting Crisis in Chawama, Lusaka, Zambia*. UNDP/UN-HABITAT Urban Management Programme. Washington, DC: World Bank.

Moss, T.J. (2007) *African Development: Making Sense of the Issues and Actors*. Boulder, CO: Lynne Rienner.

Mougeot, L.J.A. (ed.) (2005) *Agropolis: The Social, Political and Environmental Dimensions of Urban Agriculture*. London: Earthscan.

Mowforth, M. and Munt, I. (2009) *Tourism and Sustainability: Development, Globalization and New Tourism in the Third World* (3rd edn). Abingdon: Routledge.

Moyo, S. (2000) *The Land Question and Land Reform in Southern Africa*. Zimbabwe: Southern African Regional Institute for Policy Studies.

Moyo, S. (2006) Social movements, civil society and land reform. Paper presented at ISS Conference on Market-oriented Change in Land Policies in Developing and Transition Countries: Understanding the Varied Views and Reactions from Below, The Hague, 12–14 January.

Mullan, Z. (2008) Uganda: losing its grip on HIV/AIDS? *The Lancet Infectious Diseases* 8(8): 471.

Muraya, P.W.K. (2006) Urban planning and small-scale enterprises in Nairobi, Kenya. *Habitat International* 30(1): 127–143.

Murithi, T. (2005) *The African Union, Pan-Africanism, Peacebuilding and Development*. Aldershot: Ashgate.

Murithi, T. (2006) The AU/NEPAD post-conflict reconstruction policy: an analysis. *Conflict Trends* 2006(1): 16–21.

Mutula, S.M. (2008) Digital divide and economic development: a case study of sub-Saharan Africa. *Electronic Library* 26(4): 468–489.

MVI (Malaria Vaccine Initiative) (2008) online: *Malaria Vaccine Initiative PATH*. http://www.malariavaccine.org/about-overview.php (accessed 6 December 2008).

NBI (Nile Basin Initiative) (2008) online: *Nile Basin Initiative: Home*. http://www.nilebasin.org/ (accessed 3 December 2008).

NCCK (National Council of the Churches of Kenya) (1991) *Nairobi Demolitions: What Next?* Nairobi: Church House.

Ndulu, B.J., Cahraborti, L., Lijane, L., Ramachandran, V. and Wolgin, J. (2007) *Challenges of African Growth: Opportunities, Constraints and Strategic Directions*. Washington, DC: World Bank.

Nel, E. (1999) *Regional and Local Economic Development in South Africa*. Aldershot: Ashgate.

Nel, E. (2001) Local economic development: a review of its current status in South Africa. *Urban Studies* 38(7): 1003–1024.

Nel, E. and Binns, T. (2001) Initiating 'developmental local government' in South Africa: evolving local economic development policy. *Regional Studies* 35(4): 355–362.

Nel, E. and Binns, T. (2002) Decline and response in South Africa's Free State goldfields: local economic development in Matjhabeng. *International Development Planning Review* 24(3): 249–269.

Nel, E. and Binns, T. (2003a) Putting 'developmental local government' into practice: the experience of South Africa's towns and cities. *Urban Forum* 14(2/3): 165–184.

Nel, E. and Binns, T. (2003b) Decentralising development in South Africa. *Geography* 88(2): 108–116.

Nel, E., Binns, T. and Bek, D. (2009) Misplaced expectations? The experience of applied local economic development in post-apartheid South Africa. *Local Economy* 24(3): 224–237.

Nel, E., Hill, T. and Maharaj, B. (2005) Durban's pursuit of economic development in the post-apartheid era. In Nel, E. and Rogerson, C.M. (eds) *Local Economic Development in the Developing World*. New Brunswick, NJ: Transaction, pp. 211–230.

Nel, E. and Illgner, P. (2001) Tapping Lesotho's 'white gold': inter-basin water transfer in southern Africa. *Geography* 86(2): 163–167.

Nel, E., Illgner, P., Wilkins, K. and Robertson, M. (2000) Rural self-reliance in Bondolfi, Zimbabwe: the role of bee-keeping. *Geographical Journal* 166(1): 26–34.

Nel, E. and Rogerson, C.M. (eds) (2005) *Local Economic Development in the Developing World.* New Brunswick, NJ: Transaction.

Nelson, N. (1997) How women and men got by and still get by (only not so well): the gender division of labour in a Nairobi shanty town. In Gugler, J. (ed.) *Cities in the Developing World: Issues, Theory and Policy*. Oxford: Oxford University Press, pp. 156–170.

Nelson, R. (1988) *Dryland Management: The 'Desertification' Problem.* Environment Department Working Paper No. 8. Washington, DC: World Bank.

NEPAD (New Partnership for Africa's Development) (2001) online: *New Partnership for Africa's Development.* http://www.nepad.org/ (accessed 22 June 2009).

NEPAD (New Partnership for Africa's Development) (2003) online: *Action Plan of the Environment Initiative of the New Partnership for Africa's Development.* http://www.nepad.org/2005/files/documents/113.pdf (accessed 9 August 2009).

NEPAD (New Partnership for Africa's Development) (2005) *African Post-conflict Reconstruction Policy Framework*. Johannesburg: NEPAD Secretariat.

NEPAD–OECD (New Partnership for Africa's Development–Organization for Economic Cooperation and Development) (2010) *FDI in Africa*. Policy Brief No. 4. New York: UN.

Netting, R.M. (1968) *Hill Farmers of Nigeria: Cultural Ecology of the Kofyar of the Jos Plateau.* Seattle: University of Washington Press.

New African (2008) At last, something is happening in Lagos. *New African* 479: 50–52.

News24 (2009) online: *Zimbabwe's HIV Rate Falls.* http://www.news24.com/printArtcile.aspx?ifram&aid=eb70576c (accessed 25 November 2009).

NFCTA (Nigerian Federal Capital Territory Administration) (2008) online: *Official Website.* http://fct.gov.ng/fcta/ (accessed 17 December 2008).

Ngaruko, F. and Nkurunziza, J.D. (2005) Civil war and its duration in Burundi. In Collier, P. and Sambanis, N. (eds) *Understanding Civil War: Evidence and Analysis*, Volume 1: *Africa.* Washington, DC: World Bank.

Niang, A. (2006) *Towards a Viable and Credible Development in Africa.* Raleigh, NC: Ivy House.

Niemeijer, D. (1998) Soil nutrient harvesting in indigenous *teras* water harvesting in eastern Sudan. *Land Degradation and Development* 9(4): 323–330.

Nigerian Government (1975) *Third National Development Plan, 1975–80.* Lagos: Nigerian Government.

Nigerian Government (2006) online: *Census Results, 2006.* http://www.nigerianstat.gov.ng/Connections/Pop2006.pdf (accessed 15 December 2008).

Njoh, A.J. (2006) African cities and regional trade in historical perspective: implications for contemporary globalization trends. *Cities* 23(1): 18–29.

NRI (Natural Resources Institutes) (2004) *Prospects for Sustainable Tree Crop Development in Sub-Saharan Africa*. Policy Series No 17. Greenwich: Natural Resources Institutes, University of Greenwich.

NTFP–PFM (Non-Timber Forest Products–Participatory Forest Management) (2011) online: *Non-Timber Forest Products and Participatory Forest Management Research and Development Project.* http://forests.hud.ac.uk (accessed 16 March 2011).

Nyerere, J. (1973) *Freedom and Development*. Dar es Salaam: Government Printer.

Nzo, A. (1994) online: *Speech by the South African Minister of Foreign Affairs to the 48th Session, 95th Meeting of the United Nations General Assembly, 23 June.* http://www.gov.za/search97cgi/s97_cgi?action=View&Collection=empty&Collection=Speech95&QueryZip=Nelson+Mandela%2C+inaugural&SortSpec=Score+Desc&SortOrder=desc&SortField=TDEDate&DocOffset=42&AdminScriptName=&ServerKey=&AdminImagePath=%2Fsearch97admimg%2F (accessed 7 August 2011).

Obermeyer, Z., Abbott-Klafter, J. and Murray, C.J. (2008) Has the DOTS strategy improved case finding or treatment success? An empirical assessment. *PLoS ONE* 3(3): e1721.

O'Connor, A.M. (1983) *The African City*. London: Hutchinson.

O'Connor, A.M. (1991) *Poverty in Africa*. London: Belhaven.

ODI (Overseas Development Institute) (2009) *The Global Financial Crisis and Developing Countries*. Working Paper No. 306. London: ODI.

ODI (Overseas Development Institute) (2010) online: *Millennium Development Goals Report Card: Learning from Progress*. http://www.odi.org.uk/resources/download/4908.pdf (accessed 23 June 2010).

OECD (Organization for Economic Cooperation and Development) (2007) *Promoting and Supporting Change in Transhumant Pastoralism in the Sahel and West Africa*. Paris: OECD.

OECD Observer (2008) Africa emerges. *OECD Observer* 267: 31–32.

OECD Observer (2010) Africa's outlook: before the global recession, most of Africa was booming. *OECD Observer* 279: 64–65.

Oliver, R. and Fage, J.D. (1988) *A Short History of Africa*. London: Penguin.

Olonisakin, F. (2008) *Lessons Learned from an Assessment of Peacekeeping and Peace Support Operations in West Africa*. Accra: Kofi Annan International Peacekeeping Training Centre.

Omeje, K.C. (2008) Understanding conflict resolution in Africa. In Francis, D.J. (ed.) *Peace and Conflict in Africa*. London: Zed Books, pp. 68–91.

Omotola, J.S. and Saliu, H. (2009) Foreign aid, debt relief and Africa's development: problems and prospects. *South African Journal of International Affairs* 16(1): 87–102.

Omotoye, R.O., Ekanayake, E.M. and Bawuah, K. (2007) Africa and strategies for future development: more aid or more trade? *International Journal of Business Strategy* 7(3): 127–140.

Omran, A. (1971) The epidemiologic transition: a theory of the epidemiology of population change. *Milbank Memorial Fund Quarterly* 49(4): 509–538.

Onibokun, A.G. and Kumuyi, A.J. (1999) *Managing the Monster: Urban Waste and Governance in Africa*. Ottawa: IDRC.

Onimode, B. (1992) *A Future for Africa*. London: Earthscan.

Orr, A. and Mwale, B. (2001) Adapting to adjustment: smallholder livelihood strategies in southern Malawi. *World Development* 29(8): 1325–1343.

Ostby, G. (2008) Polarization, horizontal inequalities and violent civil conflict. *Journal of Peace* Research 45(2): 143–162.

Ostrom, E. (1990) *Governing the Commons: The Evolution of Institutions for Collective Action*. Cambridge: Cambridge University Press.

Otiso, K.M. (2002) Forced evictions in Kenyan cities. *Singapore Journal of Tropical Geography* 23(3): 252–267.

Otiso, K.M. (2003) State, voluntary and private sector partnerships for slum upgrading and basic service delivery in Nairobi City, Kenya. *Cities* 20(4): 221–229.

Ouedraogo, M. and Kaboré, V. (1996) The *zai*: a traditional technique for the rehabilitation of degraded land in the Yatenga, Burkina Faso. In Reij, C., Scoones, I. and Toulmin, C. (eds) *Sustaining the Soil: Indigenous Soil and Water Conservation in Africa*. London: Earthscan.

Oxfam (1993) *Africa – Make or Break: Action for Recovery*. Oxford: Oxfam.

Oxfam (2007) *Africa's Missing Billions: International Arms Flows and the Cost of Conflict*. Oxford: Oxfam.

Oyuke, J. (2008) online: *Images of Chaos Damage Kenya's Tourism Sector*. http://www.bushdrums.com/news/index.php?shownews=1362 (accessed 1 July 2010).

Painter, D. (2006) *Scaling up Slum Improvement: Engaging Slum Dwellers and the Private Sector to Finance a Better Future*. Washington, DC: TCG International.

Parkhurst, J.O. (2002) The Ugandan success story? Evidence and claims of HIV-1 prevention. *The Lancet* 360(9326): 78–80.

Paarlberg, R. (2010) GMO foods and crops: Africa's choice. *New Biotechnology* 27(5): 609–613.

Pacione, M. (2001) *Urban Geography: A Global Perspective*. London: Routledge.

Parrot, L., Sotamenou, J. and Dia, B.K. (2009) Municipal solid waste management in Africa: strategies and livelihoods in Yaounde, Cameroon. *Waste Management* 29(2): 986–995.

Parulkar, A. (2011) African land, up for grabs. *World Policy Journal* 28(1): 103–110.

Paterson, C., Mara, D. and Curtis, T. (2006) Pro-poor sanitation technologies. *Geoforum* 38(5): 901–907.

Pearce, D., Barbier, E. and Markandya, A. (1990) *Sustainable Development*. London: Earthscan.

Pearn, J. (2003) Children and war. *Journal of Paediatrics and Child Health* 39(3): 166–172.

Peil, M. and Sada, P.O. (1984) *African Urban Society*. Chichester: John Wiley.

Perkins, A. (2010) Is Africa underpopulated? *Guardian Weekly* 2 April.

Pettifor, A.E., Measham, D.M., Rees, H.V. and Padian, N.S. (2004) Sexual power and HIV risk, South Africa. *Emerging Infectious Diseases* 10(11): 1996–2004.

Pickard-Cambridge, C. (1998) Regional rescue plan breaks new ground. *Business Day* 22 April.

Pieterse, J.N. (1998) My paradigm or yours? Alternative development, post-development, reflexive development. *Development and Change* 29(2): 343–373.

Pieterse, J.N. (2001) *Development Theory: Deconstructions/Reconstructions*. London: Sage.

Pike, A., Rodriguez-Pose, A. and Tomaney, J. (2006) *Local and Regional Development*. Abingdon: Routledge.

Polack, S., Brooker, S., Kuper, H., Mariotti, S., Mabey, D. and Foster, A. (2005) Mapping the global distribution of trachoma. *Bulletin of the World Health Organization* 83(12): 913–919.

Porto, G., Chauvin, N.D. and Olarreaga, M. (2011) *Supply Chains in Export Agriculture, Competition, and Poverty in Sub-Saharan Africa*. Washington, DC: World Bank and Centre for Economic Policy Research.

Potter, R., Binns, T., Elliot, J.A. and Smith, D. (1999) *Geographies of Development* (1st edn). Harlow: Longman.

Potter, R.B., Binns, T., Elliott, J. and Smith, D. (2008) *Geographies of Development* (3rd edn). Harlow: Pearson.

Potter, R.B. and Salau, A.T. (eds) (1990) *Cities and Development in the Third World*. London: Mansell.

Pottier, J. (2005) Customary land tenure in sub-Saharan Africa today: meanings and contexts. In Huggins, C. and Clover, J. (eds) *From the Ground up: Land Rights, Conflict and Peace in Sub-Saharan Africa*. Pretoria: Institute for Security Studies, pp. 55–75.

Potts, D. (1995) Shall we go home? Increasing urban poverty in African cities and migration processes. *Geographical Journal* 161(3): 245–264.

Potts, D. (2005) Counter-urbanisation on the Zambian Copperbelt? Interpretations and implications. *Urban Studies* 42(4): 583–609.

Potts, D. (2008) The urban informal sector in sub-Saharan Africa: from bad to good (and back again?). *Development Southern Africa* 25(2): 151–167.

Poulton, R. and Harris, M. (eds) (1988) *Putting People First: Voluntary Organisations and Third World Organisations*. London: Macmillan.

Power, M. (2003) *Rethinking Development Geographies*. London: Routledge.

Pratt, C. (1999) Julius Nyerere: reflections on the legacy of his socialism. *Canadian Journal of African Studies* 33(1): 137–152.

PRB (Population Reference Bureau) (2009) online: *2009 World Population Data Sheet*. http://www.prb.org/Publications/Datasheets/2009/2009wpds.aspx (accessed 20 November 2009).

Pretty, J. and Ward, H. (2001) Social capital and the environment. *World Development* 29(2): 209–227.

Prüss, A., Kay, D., Fewtrell, L. and Bartram, J. (2008) Estimating the burden of disease from water, sanitation, and hygiene at a global level. *Environmental Health Perspectives* 110(5): 537–542.

Qamar, M.K. (2001) The HIV/AIDS epidemic: an unusual challenge to agricultural extension services in sub-Saharan Africa. *Journal of Agricultural Education and Extension* 8(1): 1–11.

Ramadhani, T., Otsyina, R. and Franzel, S. (2002) Improving household incomes and reducing deforestation: the example of rotational woodlots in Tabora District, Tanzania. *Agriculture, Ecosystem and the Environment* 89(3): 227–237.

Ramin, B.M. and McMichael, A.J. (2009) Climate change and health in sub-Saharan Africa: a case based perspective. *EcoHealth* 6(1): 52–57.

Ravallion, M. (2004) *Pro-poor Growth: A Primer*. Washington, DC: Development Research Group, World Bank.

RBMP (Roll Back Malaria Partnership) (2008) online: *African Ministers of Health Discuss Key Malaria Issues at RBM Board Meeting*. http://www.rbm.who.int/ (accessed 6 December 2008).

RDP (Reconstruction and Development Programme) (1996) *Working for Water!* Waterkloof: Western Cape Province Water Conservation Programme.

Reid, H. and Satterthwaite, D. (2007) *Climate Change and Cities: Why Urban Agendas are Central to Adaptation and Mitigation*. London: International Institute for Environment and Development.

Reid, H., MacGregor, J., Sahlen, L. and Stage, J. (2007) *Counting the Cost of Climate Change in Namibia*. London: International Institute for Environment and Development.
Reij, C. (2007) Unrecognised success stories in Africa's drylands: a spectacular case of regeneration in Niger. *Haramata* 52: 14–15.
Reij, C., Scoones, I. and Toulmin, C. (eds) (1996) *Sustaining the Soil: Indigenous Soil and Water Conservation in Africa*. London: Earthscan.
Reij, C. and Waters-Bayer, A. (2001) *Farmer Innovation in Africa*. London: Earthscan.
Richards, P. (1985) *Indigenous Agricultural Revolution: Ecology and Food Production in West Africa*. London: Hutchinson.
Richards, P. (1986) *Coping with Hunger: Hazard and Experiment in an African Rice-farming System*. London: Allen and Unwin.
Richards, P. (2003) *The Political Economy of Internal Conflict in Sierra Leone*. Working Paper No. 21. The Hague: Netherlands Institute of International Relations.
Riddell, J.B. (1974) Periodic markets in Sierra Leone. *Annals of the Association of American Geographers* 64(4): 541–548.
Riddell, J.C. and Campbell, D.J. (1986) Agricultural intensification and rural development: the Mandara Mountains of north Cameroon. *African Studies Review* 29(3): 89–106.
Riddell, R.C. (1990) *Manufacturing Africa: Performance and Prospects of Seven Countries in Sub-Saharan Africa*. London: James Currey.
Riley, S. (1988) Structural adjustment and the new urban poor: the case of Freetown. Paper presented at Workshop on the New Urban Poor in Africa, School of Oriental and African Studies, University of London, May.
Rimmer, D. (ed.) (1991) *Africa 30 Years on*. London: Royal African Society/James Currey.
Rizzo, M. (2006) What was left of the Groundnut Scheme? Development disaster and labour market in southern Tanganyika 1946–1952. *Journal of Agrarian Change* 6(2): 205–238.
Robinson, J. (2008) Developing ordinary cities: city visioning processes in Durban and Johannesburg. *Environment and Planning A* 40(1): 74–87.
Rodriguez-Pose, A. and Tijmstra, S. (2005) *Local Economic Development as an Alternative Approach to Economic Development in Sub-Saharan Africa*. Washington, DC: Municipal Development Programme/World Bank.
Rogerson, C.M. (1997) Local economic development and post-apartheid reconstruction in South Africa. *Singapore Journal of Tropical Geography* 18(2): 175–195.
Rogerson, C.M. (2001) In search of the African miracle: debates on successful small enterprise development in Africa. *Habitat International* 25(1): 115–142.
Rogo, K.O., Oucho, J. and Mwalali, P. (2006) Maternal mortality. In Jamison D.T., Feachem, R.G., Makgoba, M.W., Bos, E.R., Baingana, F.K., Hofman, K.J. and Rogo, K.O. (eds) *Disease and Mortality in Sub-Saharan Africa*. Washington, DC: World Bank, pp. 223–236.
Rondinelli, D.A. (1990) Housing the poor in developing countries. *American Journal of Economics and Sociology* 49(2): 153–166.
Ross, M. (1999) The political economy of the resource curse. *World Politics* 51(2): 297–322.
Rosser, A. (2006) *The Political Economy of the Resource Curse: A Literature Survey*. IDS Working Paper No. 268. Brighton: Institute of Development Studies.
Rostow, W.W. (1960) *The Stages of Economic Growth: A Non-Communist Manifesto*. Cambridge: Cambridge University Press.
RSA (Republic of South Africa) (1996) *The Constitution of the Republic of South Africa*. Act No. 108 of 1996. Pretoria: Republic of South Africa.
RSA (Republic of South Africa) (2008) online: *Department of Minerals and Energy: Mining Statistics*. http://www.dme.gov.za/minerals/mineral_stats.stm (accessed 29 December 2008).
Sachs, J. (2005) *The End of Poverty: How We Can Make it Happen in our Lifetime*. London: Penguin.
Salih, M.A.M. (2008) Poverty and human security in Africa: the liberal peace debate. In Francis, D.J. (ed.) *Peace and Conflict in Africa*. London: Zed Books, pp. 171–184.
Samli, A.C. (2008) Entrepreneurship economic development and quality of life in third-world countries. *Applied Research in Quality of Life* 3(3): 203–213.
Sandford, S. (1983) *Management of Pastoral Development in the Third World*. Chichester: John Wiley.
SARPN (Southern African Regional Poverty Network) (2004) online: *Fact Sheet No.1: Poverty in*

South Africa. http://www.sarpn.org.za/documents/d0000990/P1096-Fact_Sheet_No_1_Poverty.pdf (accessed 12 December 2008).

Sarch, M.T. and Allison, E.H. (2000) online: *Fluctuating Fisheries in Africa's Inland Waters: Well-adapted Livelihoods, Maladapted Management*. http://osu.orst.edu/dept/IIFET/2000/papers/sarch.pdf (accessed 2 March 2011).

Sarfo-Mensah, P. (2005) *Exportation of Timber in Ghana: The Menace of Illegal Logging Operations*. Kumasi: Fondazione Eni Enrico Mattei.

Sauve, N., Dzokoto, A., Opare, B., Kaitoo, E., Khonde, N., Mondor, M., Bekoe, V. and Pepin, J. (2002) The price of development: HIV infection in a semi-urban community of Ghana. *Journal of Acquired Immune Deficiency Syndromes* 29(4): 402–408.

Schuppan, T. (2009) E-government in developing countries: experiences from sub-Saharan Africa. *Government Information Quarterly* 26(1): 118–127.

Schuurman, F.J. (1993) *Beyond the Development Impasse: New Directions in Development*. London: Zed Books.

Scoones, I. (1995a) *Living with Uncertainty: New Direction in Pastoral Development in Africa*. London: ITDG.

Scoones, I. (1995b) Policies for pastoralists: new directions for pastoral development in Africa. In Binns, T. (ed.) *People and Environment in Africa*. Chichester: John Wiley, pp. 23–30.

Scoones, I. (1998) *Sustainable Rural Livelihoods: A Framework for Analysis*. IDS Working Paper No. 72. Brighton: Institute of Development Studies.

Scoones, I. (2006) Can GM crops prevent famine in Africa? In Devereux, S. (ed.) *New Famines*. Abingdon: Routledge, pp. 312–335.

Scoones, I. and Chibudu, C. (1996) *Hazard and Opportunities: Farming Livelihoods in Dryland Africa*. London: Zed Books.

Seck, D. and Busari, D.T. (2009) *Growth and Development in Africa*. Trenton: Africa World Press.

Seers, D. (1969) The meaning of development. *International Development Review* 11(4): 2–6.

Seers, D. (1979) The new meaning of development. In Lehmann, D. (ed.) *Development Theory: Four Critical Studies*. London: Frank Cass, pp. 25–30.

Sen, A. (1999) *Development as Freedom*. Oxford: Oxford University Press.

Setel, P.W., Sankoh, O., Rao, C., Velkoff, V.A., Mathers, C., Gonghuan, Y., Hemed, Y., Jha, P. and Lopez, A.D. (2005) Sample registration of vital events with verbal autopsy: a renewed commitment to measuring and monitoring vital statistics. *Bulletin of the World Health Organization* 83(8): 611–617.

Shell Foundation (2009) online: *Trading up: M&S Partnership – Flowers from South Africa*. http://www.shellfoundation.org/pages/core_lines.php?p=corelines_inside_content&page=trading&newsID=108 (accessed 8 November 2009).

Sidaway, J.D. (1998) The (geo)politics of regional integration: the example of the Southern African Development Community. *Environment and Planning D: Society and Space* 16: 549–576.

Simatele, D.M. and Binns, T. (2008) Motivation and marginalization in African urban agriculture: the case of Lusaka, Zambia. *Urban Forum* 19(1): 1–21.

Simon, D., McGregor, D. and Thompson, D. (2006) Contemporary perspectives on the peri-urban zones of cities in developing areas. In McGregor, D., Simon, D. and Thompson, D. (ed.) *The Peri-urban Interface: Approaches to Sustainable Natural and Human Resource Use*. London: Earthscan, pp. 3–17.

Singerman, D. (1995) *Avenues of Participation: Family Politics and Networks in Urban Quarters of Cairo*. Princeton, NJ: Princeton University Press.

SIPRI (Stockholm International Peace Research Institute) (2010) online: *SIPRI Military Expenditure Database*. http://www.sipri.org/databases (accessed 9 June 2010).

Sjöstedt, M. (2007) Land policies and property rights in sub-Saharan Africa: a comparative study of credible commitments. Paper presented at the 48th ISA Convention, Chicago, IL, February–March.

SJSU (San Jose State University) (n.d) online: *The Tanganyikan Groundnut Scheme*. http://www.sjsu.edu/faculty/watkins/groundnt,htm (accessed 26 October 2010).

Smith, O.B. (2001) *Overview of Urban Agriculture in Western African Cities*. Ottawa: IDRC.

Söderbom, M. and Teal, F. (2003) *How Can Policy towards Manufacturing in Africa Reduce*

Poverty? A Review of the Current Evidence from Cross-country Firm Studies. Oxford: Centre for the Study of African Economies, Department of Economics, University of Oxford.

Sol Plaatje Municipality (2001) *Local Economic Development Strategic Plan*. Kimberley: Sol Plaatje Municipality.

Solomon, C. (2008) *Disarmament, Demobilisation and Reintegration in West Africa: An Overview of Sierra Leone and Liberia*. Bradford: Centre for International Cooperation and Security, University of Bradford.

SouthAfrica.info (2008) online: *Gateway to the Nation*. http://www.southafrica.info/ (accessed 14 April 2008).

Stamp, L.D. (1940) The southern margin of the Sahara: comments on some recent studies on the question of desiccation in West Africa. *Geographical Review* 30(2): 297–300.

Stanback, J., Mbonye, A. and Bekiita, M. (2007) Contraceptive injections by community health workers in Uganda: a nonrandomized community trial. *Bulletin of the World Health Organization* 85(10): 768–773.

Statistics South Africa (1975–1999) *Quarterly Bulletin of Statistics*. Pretoria: Statistics South Africa.

Statistics South Africa (1999) *Census in Brief*. Pretoria: Statistics South Africa.

Statistics South Africa (2008) online: *Mining Statistics*. http://www.dme.gov.za/minerals/mineral_stats.stm (accessed 29 December 2008).

Stebbing, E.P. (1935) The encroaching Sahara: the threat to the West African colonies. *Geographical Journal* 85(6): 506–519.

Stebbing, E.P. (1938) The man-made desert in Africa: erosion and drought. *African Affairs* 37(146): 144–181.

Stern, N. (2006) *The Economics of Climate Change: The Stern Review*. Cambridge, Cambridge University Press.

Stock, R. (2004) *Africa South of the Sahara* (2nd edn). New York: Guilford Press.

Stockmayer, A. (1999) Decentralization: global fad or recipe for sustainable local development? *Agriculture and Rural Development* 6(1): 3–6.

Stokes, S. (2003) online: *Measuring Impacts of HIV/AIDS on Rural Livelihoods and Food Security*. http://www.fao.org/sd/2003/PE0102a_en.htm (accessed 19 April 2009).

Stone, G.D. (1996) *Settlement Ecology: The Social and Spatial Organization of Kofyar Agriculture*. Tucson: University of Arizona Press.

Stone, G.D. (1998) Keeping the home fires burning: the changed nature of householding in the Kofyar homeland. *Human Ecology* 26(2): 239–265.

Storey, A. (1986) online: *Food vs. Cash Crops in Africa*. http://www.trocaire.org/resources/tdr-article/food-vs-cash-crops-africa (accessed 1 April 2011).

Street, J.M. (1969) An evaluation of the concept of carrying capacity. *Professional Geographer* 21(2): 104–107.

Sullivan, R. (2006) The politics of population policy adoption in sub-Saharan Africa. Paper presented to the African Sociological Association Conference, Montreal, 10 August.

Sutton, K. and Fahmi, W. (2001) Cairo's urban growth and strategic master plans in the light of Egypt's 1996 population census results. *Cities* 18(3): 135–149.

Swift, J. (1996) Desertification: narratives, winners and losers. In Leach, M. and Mearns, R. (eds) *The Lie of the Land: Challenging Received Wisdom on the African Environment*. Oxford: James Currey, pp. 73–90.

Swindell, K. (1981) *The Strange Farmers of The Gambia: A Study in the Redistribution of African Population*. Norwich: Geo Books.

Swindell, K. (1988) Agrarian change and peri-urban fringes in tropical Africa. In Rimmer, D. (ed.) *Rural Transformation in Tropical Africa*. London: Belhaven, pp. 98–115.

Tanner, T. and Mitchell, T. (2007) *Pro-poor Climate Change Adaptation: A Research Agenda*. Brighton: Institute of Development Studies.

Taylor, D.R.F. and Mackenzie, F. (eds) (1992) *Development from within: Survival in Rural Africa*. London: Routledge.

Thomas, D.S.G., Twyman, C., Osbahr, H. and Hewitson, B. (2007) Adaptation to climate change and variability: farmer responses to intra-seasonal precipitation trends in South Africa. *Climatic Change* 83(3): 301–322.

Thornton, A. (2008) Beyond the metropolis: small town case studies of urban and peri-urban agriculture in South Africa. *Urban Forum* 19(3): 243–262.

Tiffen, M. and Mortimore, M.J. (1990) *Theory and Practice in Plantation Agriculture: An Economic Review*. London: Overseas Development Institute.

Tiffen, M., Mortimore, M.J. and Gichuki, F. (1994) *More People, Less Erosion: Environmental Recovery in Kenya*. Chichester: Wiley.

Timberlake, L. (1985) *Africa in Crisis*. London: Earthscan.

Time (2009) online: *Cutting off a Continent?* http://www.time.com/time/specials/packages (accessed 11 July 2009).

Time (1975) online: *Tanzania: Ujamaa's Bitter Harvest.* www.time.com/time/printout/0,8816, 912757,00.html (accessed 26 October 2010).

Times of Zambia (2007) online: *Zambia's Mining Sector Marks Tremendous Growth.* http://craigeisele.wordpress.com/2007/06/29 (accessed 19 June 2009).

Timise, T. (2009) online: *SA Faces Major Crisis over AIDS Orphans.* http://www.mg.co.za/print format/sing;e/2009-11-30-sa-faces-major (accessed 1 December 2009).

Tradearabia (2010) online: *Coffee Helps Boost Ethiopia Exports to $2bn.* http://www.trade arabia.com/news/food_184199.html (accessed 7 August 2011).

Transparency International (2009) online: *Corruption Perceptions Index, 2009.* http://www. transparency.org/ (accessed 24 November 2009).

Trapnell, C.G. and Clothier, J.N. (1937) *The Soils, Vegetation and Agriculture Systems of North-western Rhodesia*. Lusaka: Government of Northern Rhodesia.

Tuck, C. (2000) 'Every car or movable object gone': the ECOMOG intervention in Liberia. *African Studies Quarterly* 4(1): 1–16.

Turner, T. (2007) *The Congo Wars: Conflict, Myth and Reality*. London: Zed Books.

Turok, B., Kekana, D., Turok, M., Maganya, E., Noe, J., Onimode, B., Chikora, J. and Suliman, M. (eds) (1993) *Development and Reconstruction in South Africa*. Johannesburg: Institute for African Alternatives.

Turok, I. and Watson, V. (2001) Divergent development in South African cities: strategic challenges facing Cape Town. *Urban Forum* 12(2): 119–138.

Turpie, J., Smith, B., Emerton, L. and Barnes, B. (1999) *Economic Value of the Zambezi Basin Wetlands*. Cape Town: IUCN.

Ukaga, O. and Afoaku, O.G. (2005) *Sustainable Development in Africa*. Trenton, NJ: Africa World Press.

UN (United Nations) (2001) *Road Map Towards the Implementation of the United Nations Millennium Declaration*. Washington, DC: United Nations.

UN (United Nations) (2006) online: *Political Declaration on HIV/AIDS: Resolution Adopted by the General Assembly.* http://data.unaids.org/pub/Report/2006/20060615_HLM_PoliticalDeclaration_ARES60262_en.pdf (accessed 23 March 2010).

UN (United Nations) (2007) online: *Africa and the Millennium Development Goals: 2007 Update.* http://unstats.un.org/unsd/mdg/resources/static/Products/Africa/Africa-MDGs07.pdf (accessed 15 October 2010).

UN (United Nations) (2008a) online: *Millennium Development Goals Progress Chart 2008.* http://www.un.org/millenniumgoals/2008highlevel/pdf/newsroom/MDG_Report_2008_Progress_Chart_en_r8.pdf (accessed 28 April 2011).

UN (United Nations) (2008b) online: *The Millennium Development Goals Report, 2008.* http://www.undp.org/mdg/ (accessed 6 December 2008).

UN (United Nations) (2010a) *World Population Prospects: The 2009 Revision*. New York: Population Division, Department of Economic and Social Affairs, United Nations.

UN (United Nations) (2010b) online: *Secretary-General SG/SM/13014, SC/9985, AFR/2013.* http://www.un.org/News/Press/docs/2010/sgsm13014.doc.htm (accessed 7 April 2011).

UN (United Nations) (2010c) *The Millennium Development Goals Report*. New York: United Nations.

UN (United Nations) (2010d) online: *Extra Push Needed on Aid, Trade and Debt to Meet Global Anti-poverty Goals.* http://www.un.org/millenniumgoals/pdf/MGR_PR-En.pdf?NewsID=35832&Cr=UNICEF&Cr1 (accessed 23 October 2010).

UNAIDS (2007) online: *2007 AIDS Epidemic Update*. http://www.unaids.org/en/KnowledgeCentre/HIVData/EpiUpdate (accessed 21 November 2010).

UNAIDS (2008) *2008 Report on the Global AIDS Epidemic*. Geneva: UNAIDS.

UNAIDS (2009) *Global Facts and Figures 09*. Geneva: UNAIDS.

UNCCD (United Nations Convention to Combat Desertification) (n.d.) online: *Website*. http://www.unccd.int/main.php (accessed 8 August 2011).

UNCOD (United Nations Conference on Desertification) (1978) online: *Round-up, Plan of Action and Resolutions*. http://www.ciesin.org/docs/002-478/002-478.html (accessed 31 August 2008).

UNCTAD (United Nations Conference on Trade and Development) (2009) *Economic Development in Africa 2009*. Press release, PR/2009/022 25/06/09. Geneva: UNCTAD.

UNDP (United Nations Development Programme) (1994) *Human Development Report, 1994*. Oxford: Oxford University Press.

UNDP (United Nations Development Programme) (2005) *Human Development Report 2005: International Cooperation at a Crossroads: Aid, Trade and Security in an Unequal World*. New York: UNDP.

UNDP (United Nations Development Programme) (2007) *Human Development Report, 2007/2008*. Basingstoke: Palgrave Macmillan.

UNDP (United Nations Development Programme) (2009) *Human Development Report, 2009*. Basingstoke: Palgrave Macmillan.

UNDP (2010) *Human Development Report 2010: The Real Wealth of Nations: Pathways to Human Development*. Geneva: UNDP.

UNECA (United Nations Economic Commission for Africa) (2001) online: *Population and Development*. http://www.uneca.org/csd.population.pdf (accessed 15 August 2010).

UNECA (United Nations Economic Commission for Africa) (2006) *International Migration and Development: Implications for Africa*. Addis Ababa: UNECA.

UNEP (United Nations Environment Programme) (1984) *General Assessment of Progress in the Implementation of the Plan of Action to Combat Desertification 1978–1984*. Report of the Executive Director, Governing Council, 12th Session, UNEP/GC.12/9. Nairobi: UNEP.

UNEP (United Nations Environment Programme) (2002) online: *Africa Environment Outlook: Past, Present and Future Perspectives*. http://www.grida.no/aeo/ (accessed 21 August 2008).

UNEP (United Nations Environment Programme) (2005) The impact of refugees and internally displaced persons on local environmental resources. *Environmental Emergencies News* (newsletter) 5 (December).

UNEP (United Nations Environment Programme) (2007) *Sudan: Post-conflict Environmental Assessment*. Nairobi: UNEP.

UNEP (United Nations Environment Programme) (2008) *Africa: Atlas of Our Changing Environment*. Nairobi: UNEP.

UNEP (United Nations Environment Programme) (2009) *From Conflict to Peace Building: The Role of Natural Resources and the Environment*. Nairobi: UNEP.

UNESC/ECA (United Nations Economic and Social Council/Economic Commission for Africa) (2008) online: *Assessing Progress in Africa towards the Millennium Development Goals: Report 2008*. http://www.uneca.org/cfm/2008/docs/AssessingProgressinAfricaMDGs.pdf (accessed 30 April 2010).

UNFPA (United Nations Population Fund) (2007) online: *State of World Population 2007*. http://www.unfpa.org/swp/2007/english/introduction.html (accessed 4 September 2008).

UN-HABITAT (United Nations Human Settlements Programme) (1987) *Case Study of Sites and Services Schemes in Kenya: Lessons from Dandora and Thika*. Nairobi: UN-HABITAT.

UN-HABITAT (United Nations Human Settlements Programme) (2003) *The Challenge of Slums: Global Report on Human Settlements*. London: Earthscan.

UN-HABITAT (United Nations Human Settlements Programme) (2008) *The State of African Cities, 2008*. Nairobi: UN-HABITAT.

UN-HABITAT (United Nations Human Settlements Programme) (2011) online: *Nigeria*. http://www.unchs.org/categories.asp?catid=223 (accessed 8 August 2011).

UNHCR (United Nations High Commissioner for Refugees) (2010) online: *UNHCR Statistical Online Population Database*. http://apps.who.int/globalatlas/ (accessed 12 February 2011).

UNICEF (United Nations Children's Fund) (2006) *Meeting the MDG Drinking Water and Sanitation Target: The Urban and Rural Challenge of the Decade*. Geneva: UNICEF/WHO.

UNICEF (United Nations Children's Fund) (2008) *A Snapshot of Drinking Water and Sanitation in Africa*. Geneva: UNICEF/WHO.

UNICEF (United Nations Children's Fund) (2009) *Quarterly WASH Report, July–September 2009*. Freetown: UNICEF.

UN News Service (2009) online: *Stronger Agriculture Sector Key to Brighter Future for Africa, Migiro Tells Leaders*. http://www.un.org/apps/news/ (accessed 4 July 2009).

UNPD (United Nations Population Division) (2002) *World Urbanization Prospects: The 2001 Revision*. New York: UNPD.

UNTAP (UN Technical Assistance Programme) (1964) *Metropolitan Lagos*. Lagos: Ministry of Lagos Affairs.

Unwin, T. (2002) War and development. In Desai, V. and Potter, R.B. (eds) *The Companion to Development Studies*. London: Arnold, pp. 440–444.

US Census Bureau (2003) online: *Nations of the World*. http://nationmaster.com/country (accessed 9 August 2010).

US Census Bureau (2009) online: *Census Data for Countries and Areas of Africa: 1945 to 2014*. http://www.census.gov/ipc/www/cendates/cenafric.html (accessed 16 August 2010).

USGS (US Geological Survey) (2009) *Minerals Yearbook, 2006*. Washington, DC: US Government Printing Office.

USGS (US Geological Survey) (n.d.) online: *What is a Desert?* http://pubs.usgs.gov/gip/deserts/what/ (accessed 6 August 2011).

van Wesenbeeck, C., Keyzer, M. and Nube, M. (2009) Estimations of under-nutrition and mean calorie intake in Africa: methodology, findings and implications. *International Journal of Health Geographics* 8(37): 8–37.

Vautravers, A.J. (2009) Why child soldiers are such a complex issue. *Refugee Survey Quarterly* 27(4): 96–107.

Von Freyhold, M. (1979) *Ujamaa Villages in Tanzania: Analysis of a Social Experiment*. London: Heinemann.

Wade, R. (2004) US and European relations with developing countries: aid, trade and investment. In Wells, S. and Kunhardt, L. (eds) *The Crisis in Transatlantic Relations*. Bonn: Centre for European Integration Studies, pp. 43–56.

Wall Street Journal (2010) online: *Cnooc to Buy Uganda Oil Stake*. http://online.wsj.com/article/SB10001424052748704533204575046481584946258.html (accessed 24 July 2010).

Wallensteen, P. and Sollenberg, M. (2001) Armed conflict 1989–2000. *Journal of Peace Research* 38(5): 629–644.

Ward, H. (2005) online: *Corporate Social Responsibility: A Step towards Stronger Involvement of Business in MEA Implementation?* http://www.oecd.org/dataoecd/63/46/35173055.pdf (accessed 8 November 2009).

Warren, A. and Batterbury, S.P.J. (2004) Desertification. In Forsyth, T.J. (ed.) *The Routledge Encyclopedia of International Development*. London: Routledge, pp. 148–150.

Warren, D.M., Slikkerveer, L.J. and Brokensha, D. (1995) *The Cultural Dimension of Development: Indigenous Knowledge Systems*. London: ITDG.

Watkins, K. (1995) *The Oxfam Poverty Report*. Oxford: Oxfam.

Watts, S. (2007) Water and health: the case of Guinea worm disease (Dracunculiaisis). *Geography* 92(2): 149–153.

Wealth Daily (2008) online: *Zimbabwe Inflation Rate*. http://www.wealthdaily.com/articles/zimbabwe-inflation-rate/2029 (accessed 24 November 2009).

WEF (World Economic Forum) (2008) *Strengthening the Management of Health Systems in Sub-Saharan Africa: Working with Governments*. Geneva: World Economic Forum.

Wetlands International (2008) online: *New Life for the Conservation of the Hadejia– Nguru Wetlands*. http://wow.wetlands.org/HANDSon/Nigeria/tabid/131/language/en-US/Default.aspx (accessed 30 November 2008).

WFWP (Working for Water Programme) (2000) *Annual Report 1999/2000*. Cape Town: WFWP.

WGCCD (Working Group on Climate Change and Development) (2006) online: *Africa: Up in Smoke 2*. http://www.upinsmokecoalition.org (accessed 26 August 2008).

WHO (World Health Organization) (2002) *The Health Effects of Indoor Air Pollution Exposure in Developing Countries*. Geneva: WHO.

WHO (World Health Organization) (2004a) *The World Health Report 2004: Changing History*. Geneva: WHO.

WHO (World Health Organization) (2004b) *Inheriting the World: The Atlas of Children's Health and the Environment*. Geneva: WHO.

WHO (World Health Organization) (2004c) *Malaria and HIV/AIDS Interactions and Complications: Conclusions of a Technical Consultation Convened by WHO, 23–25 June, 2004*. Geneva: WHO.

WHO (World Health Organization) (2005a) *The World Health Report 2005: Making Every Mother and Child Count*. Geneva: WHO.

WHO (World Health Organization) (2005b) online: *Promoting Evidence-based Sexual and Reproductive Health Care: Progress in Reproductive Health 71*. http://www.who.int/hrp/publications/progress71.pdf (accessed 24 March 2009).

WHO (World Health Organization) (2005c) online: *WHO Declares TB an Emergency in Africa*. http://www.who.int/mediacentre/news/releases/2005/africa_emergency/en/ (accessed 24 April 2010).

WHO (World Health Organization) (2006a) *The African Regional Health Report*. Brazzaville: WHO-EARO.

WHO (World Health Organization) (2006b) *The Stop TB Strategy: Building on and Enhancing DOTS to Meet the TB Related Millennium Development Goals*. Geneva: WHO.

WHO (World Health Organization) (2006c) online: *Africa Health Workforce Observatory: Country Factsheets*. http://www.afro.who.int/hrh-observatory/country_information/fact_sheets/index.html (accessed 5 October 2010).

WHO (World Health Organization) (2006d) *Sierra Leone: Mortality Country Fact Sheet 2006*. Geneva: WHO.

WHO (World Health Organization) (2007a) *Women, Ageing and Health: A Framework for Action*. Geneva: WHO.

WHO (World Health Organization) (2007b) *Maternal Mortality in 2005*. Geneva: WHO.

WHO (World Health Organization) (2008a) *The Global Burden of Disease: 2004 Update*. Geneva: WHO.

WHO (World Health Organization) (2008b) Cholera 2007. *WHO Weekly Epidemiological Record* 83: 269–284.

WHO (World Health Organization) (2008c) *Epidemiological Fact Sheet on HIV and AIDS: Swaziland: 2008 Update*. Geneva: WHO.

WHO (World Health Organization) (2008d) *Epidemiological Fact Sheet on HIV and AIDS: Botswana: 2008 Update*. Geneva: WHO.

WHO (World Health Organization) (2008e) Progress in global measles control and mortality reduction, 2000–2007. *Weekly Epidemiological Record* 83: 441–448.

WHO (World Health Organization) (2008f) online: *Ougagadougou Declaration on Primary Health Care and Health Systems in Africa: Achieving Better Health for Africa in the New Millennium*. http://www.afro.who.int/phc_hs_2008/documents/En/Ouagadougou%20declaration%20version%20Eng.pdf (accessed 3 November 2009).

WHO (World Health Organization) (2008g) *Ouagadougou Declaration: Proposed Framework for Implementation*. Geneva: WHO.

WHO (World Health Organization) (2009a) *World Health Statistics 2009*. Geneva: WHO.

WHO (World Health Organization) (2009b) *Global Tuberculosis Control 2009: Epidemiology, Strategy, Financing*. Geneva: WHO.

WHO (World Health Organization) (2009c) *Global Status Report on Road Injuries*. Geneva: WHO.

WHO (World Health Organization) (2011) *Causes of Death 2008 Summary Tables*. Geneva: WHO.

WIETA (Wine and Agricultural Ethical Trade Association) (2009) online: *WIETA Home*. http://www.wieta.org.za/ (accessed 8 November 2009).

Willis, K. (2005) *Theories and Practices of Development* (1st edn). London: Routledge.

Willis, K. (2011) *Theories and Practices of Development* (2nd edn). London: Routledge.

Wilson, D. and Purushothaman, R. (2003) *Dreaming with BRICs: The Path to 2050*. Global Economics Paper No. 99. New York: Goldman Sachs.

Wilson, D.C., Velis, C. and Cheeseman, C. (2006) Role of informal sector recycling in waste management in developing countries. *Habitat International* 30(4): 797–808.

Wilson, H.S. (1994) *African Decolonization*. London: Edward Arnold.

World Bank (1998) *World Development Report 1998/99*. Oxford: Oxford University Press.

World Bank (2002a) *Reinvesting in African Smallholder Agriculture: The Role of Tree Crops in Sustainable Farming Systems*. Washington, DC: World Bank.

World Bank (2002b) online: *IMF/ World Bank PRSP Comprehensive Review*. http://web.world bank.org/WBSITE/EXTERNAL/TOPICS/EXTPOVERTY/EXTPRS/0.contemtMDK (accessed 24 October 2010).

World Bank (2007a) online: *Africa Action Plan*. http://siteresources.worldbank.org/DEVCOMMINT/Documentation/21289631/DC2007-0008(E)-AfricaActionPlan.pdf (accessed 4 December 2008).

World Bank (2007b) *Africa Development Indicators, 2007*. Washington, DC: World Bank.

World Bank (2007c) *World Development Indicators, 2007*. Washington, DC: World Bank.

World Bank (2008a) *Africa Development Indicators 2007*. Washington, DC: World Bank.

World Bank (2008b) online: *Nile Basin Initiative*. http://web.worldbank.org/WBSITE/EXTERNAL/COUNTRIES/AFRICAEXT/EXTREGINI/EXTAFRNILEBASINI/0,,contentMDK:21066483~pagePK:64168427~piPK:64168435~theSitePK:2959951,00.html (accessed 3 December 2008).

World Bank (2008c) *The Little Data Book on Africa, 2007*. Washington, DC: World Bank.

World Bank (2008d) *The World Bank's Commitment to HIV/AIDS in Africa: Our Agenda for Action*. Washington, DC: World Bank.

World Bank (2008e) *World Development Report 2008: Agriculture for Development*. Washington, DC: World Bank.

World Bank (2009a) online: *Africa Development Indicators, 2009*. http://data.worldbank.org/indicator/NV.AGR.TOTL.ZS (accessed 7 August 2011).

World Bank (2009b) *The State and Peace-building Fund: Addressing the Unique Challenges of Fragility and Conflict*. Washington, DC: World Bank.

World Bank (2009c) *World Development Report, 2009*. Washington, DC: World Bank.

World Bank (2009d) online: *World Development Indicators 2009*. http://data.worldbank.org/indicator/MS.MIL.XPND.GD.ZS (accessed 12 September 2010).

World Bank (2009e) online: *Zoellick Calls for 'Vulnerability Fund' ahead of Davos Forum*. http://web.worldbank.org/WBSITE/EXTERNAL/NEWS/0 (accessed 22 June 2009).

World Bank (2009f) online: *Food Crisis: What the World Bank is Doing*. http://web.worldbank.org/html/extdr/foodprices/bankinitiatives.htm (accessed 22 June 2009).

World Bank (2009g) online: *World Bank Approves Fast Track Assistance Programs*. http://web.worldbank.org/WBSITE/EXTERNAL/NEWS/0 (accessed 22 June 2009).

World Christian Encyclopedia (2001) online: *Global Statistics for All Religions*. http://www.bible.ca/global-religion-statistcs (accessed 15 August 2010).

World Tourism Organization (2006) online: *World Tourism Highlights: 2006 Edition*. http://www.unwto.org/facts (accessed 12 February 2010).

World Tourism Organization (2009) online: *Facts*. http://unwto.org/facts (accessed 1 January 2009).

Worldwatch Institute (2007) *2007 State of the World: Our Urban Future*. Washington, DC: Worldwatch Institute

WRI (World Resources Institute) (1998) *World Resources 1998–1999: Environmental Change and Human Health*. Washington, DC: World Resources Institute.

Yemane, B. (2003) *Food Security Situation in the Pastoral Areas of Ethiopia*. Oxford: Oxfam.

Ziring, L., Plano, J.C. and Olton, R. (1995) *International Relations: A Political Dictionary*. London: ABC-Clio.

Index